Youssef Belmabkhout, Kyle E. Cordova (Eds.)

Reticular Chemistry and Applications

Also of Interest

Rare Earth Chemistry
Rainer Pöttgen, Thomas Jüstel and Cristian A. Strassert (Eds.), 2020
ISBN 978-3-11-065360-1, e-ISBN 978-3-11-065492-9

Organoselenium Chemistry
Brindaban C. Ranu and Bubun Banerjee, 2020
ISBN 978-3-11-062224-9, e-ISBN 978-3-11-062511-0

Transition Metals and Sulfur – A Strong Relationship for Life
Martha E. Sosa Torres and Peter M. H. Kroneck, 2020
ISBN 978-3-11-058889-7, e-ISBN 978-3-11-058975-7

Chemistry of the Non-Metals
Syntheses - Structures - Bonding - Applications
Ralf Steudel, 2020
ISBN 978-3-11-057805-8, e-ISBN 978-3-11-057806-5

Reticular Chemistry and Applications

Metal-Organic Frameworks

Edited by
Youssef Belmabkhout, Kyle E. Cordova

DE GRUYTER

Editors
Dr. Youssef Belmabkhout
Mohammed VI Polytechnic University
Lot 660
43150 Ben Guerir
Morocco
Youssef.belmabkhout@um6p.ma

Kyle E. Cordova
Royal Scientific Society
70 Ahmad Al-Tarawneh Street
Jubeiha, 11941 Amman
Jordan
Kyle.cordova@rss.jo

ISBN 978-1-5015-2470-7
e-ISBN (PDF) 978-1-5015-2472-1
e-ISBN (EPUB) 978-1-5015-1630-6

Library of Congress Control Number: 2022947144

Bibliographic information published by the Deutsche Nationalbibliothek
The Deutsche Nationalbibliothek lists this publication in the Deutsche Nationalbibliografie; detailed bibliographic data are available on the Internet at http://dnb.dnb.de.

Cover image: Kyle Cordova and Ashley M. Osborn
Typesetting: VTeX UAB, Lithuania
Printing and binding: CPI books GmbH, Leck

www.degruyter.com

Foreword

The beauty of science lies in its reimagining and realization of the possible. As someone who has dedicated my life to the pursuit of knowledge yet lacks the practical skills to create novel molecules in a lab, I wonder about this creative potential and work tirelessly to ensure that knowledge may transform lives and foment positive change. I feel honored to work with scientists at the forefront of advanced research for social transformation and I am proud to call myself a science enabler who has embraced and championed our shared enterprise.

My small nation of Jordan has been at the center of timeless human conversations on innovation and the application of knowledge for thousands of years and the very same is true of our brother kingdom of Morocco. Indeed, our countries are central and historic parts of a region where so many communities and civilizations have had to innovate to survive, where for thousands of generations, creative adaptation and cultural exchange have gone hand in hand. Together, our ancestors have reimagined and realized so much.

With a deep knowledge of past achievements and much hope for a better future, I have always believed that the greatest resource of any nation is its people, and its future prosperity lies in unlocking their innovative capacity to explore and create. This book, I believe, will greatly assist that quest to realize innovation through education and research. This is a vital technical guide and an inspiration, with expert contributions from around the world. It is coedited by scientists representing Jordan and Morocco in a novel and extraordinary field of science (reticular chemistry) that was invented by a Jordanian-Palestinian.

Over the past decade, I have come to appreciate reticular chemistry for a number of reasons. The beauty of the crystalline frameworks, which are its outputs, remind me of the mosaic patterns of Islamic art—simply put, I am in awe of their simple beauty, their periodicity and their symmetry. But more than that, the utility of reticular materials to solve real-world problems (from water scarcity, security and independence to realizing new forms of renewable energy and addressing the causes of climate change, among others) has inspired me. Finally, the fact that reticular chemistry is studied, practiced and researched in thousands of laboratories around the world today shows that this field is an exemplar of cooperative science that is illuminating the vast potential of creative innovation to solve problems, to make life better for all of us and to teach us more about ourselves as a human family.

What we create and what we leave behind will define our legacy. We must always remember that with every action, we earn a place in our descendants' halls of history.

https://doi.org/10.1515/9781501524721-201

I hope that this book will help to show future generations that we were worthy of our shared human gift of innovation. This is my affirmed truth and for this reason I am supremely proud of this publication. I hope you will find it to be as rich and stimulating as I have.

Amman, July 2022 Her Royal Highness Princess Sumaya bint El Hassan

Contents

List of Contributing Authors

Mohd Basyaruddin Abdul Rahman
Integrated Chemical BioPhysics Research
Faculty of Science
Universiti Putra Malaysia
43400 UPM Serdang
Selangor
Malaysia
Institute of Nanoscience and Nanotechnology
Universiti Putra Malaysia
43400 UPM Serdang
Selangor
Malaysia
E-mail: basya@upm.edu.my

Ala'a Al-Ghourani
Materials Discovery Research Unit
Advanced Research Centre
Royal Scientific Society (RSS)
Amman 11941
Jordan
E-mail: ala.ghourani@rss.jo

Bassem Al-Maythalony
Materials Discovery Research Unit
Advanced Research Centre
Royal Scientific Society (RSS)
Amman 11941
Jordan
E-mail: bassem.maythalony@rss.jo

Rawan Al Natour
Department of Chemistry
Faculty of Arts and Sciences
American University of Beirut
P. O. Box 11-0236 Riad El Solh
Beirut 1107-2020
Lebanon
E-mail: rawan.natour@kaust.edu.sa

Mohamed H. Alkordi
Center for Materials Science
Zewail City of Science and Technology
Giza 12578
Egypt
E-mail: malkordi@zewailcity.edu.eg

Ayalew H. Assen
Technology Development Cell (TechCell)
Applied Chemistry and Engineering Research Centre of Excellence (ACER CoE)
Mohammed VI Polytechnic University (UM6P)
Lot 660 – Hay Moulay Rachid
43150 Ben Guerir
Morocco
Department of Chemistry
College of Natural Science
Wollo University
Dessie
Ethiopia
E-mail: ayalew.assen@um6p.ma

Youssef Belmabkhout
Applied Chemistry and Engineering Research Centre of Excellence (ACER CoE)
Mohammed VI Polytechnic University (UM6P)
Lot 660 – Hay Moulay Rachid
43150 Ben Guerir
Morocco
Technology Development Cell (TechCell)
Mohammed VI Polytechnic University (UM6P)
Lot 660 – Hay Moulay Rachid
43150 Ben Guerir
Morocco
E-mail: Youssef.belmabkhout@um6p.ma

Christopher Copeman
Department of Chemistry and Biochemistry,
and Centre for NanoScience Research
Concordia University
7141 Sherbrooke St W.
Montréal, QC
Canada

Kyle E. Cordova
Materials Discovery Research Unit
Advanced Research Centre (ARC)
Royal Scientific Society
11941 Amman
Jordan
E-mail: kyle.cordova@rss.jo

Patrick Damacet
Department of Chemistry
Faculty of Arts and Sciences
American University of Beirut
P. O. Box 11-0236 Riad El Solh
Beirut 1107-2020
Lebanon
E-mail: pld01@mail.aub.edu

P. Rafael Donnarumma
Department of Chemistry and Biochemistry,
and Centre for NanoScience Research
Concordia University
7141 Sherbrooke St W.
Montréal, QC
Canada

Worood A. El-Mehalmey
Center for Materials Science
Zewail City of Science and Technology
Giza 12578
Egypt
E-mail: wadel@zewailcity.edu.eg

El-Sayed M. El-Sayed
State Key Laboratory of Structural Chemistry
Fujian Institute of Research on the Structure of Matter
Chinese Academy of Sciences
Fuzhou
P. R. China
University of the Chinese Academy of Sciences
Beijing
P. R. China
Chemical Refining Laboratory
Refining Department
Egyptian Petroleum Research Institute
Nasr City, Cairo
Egypt
E-mail: elsayedm.elsayed@fjirsm.ac.cn

G. S. Fanourgakis
Department of Chemistry
University of Crete
Voutes Campus
GR-70013 Heraklion
Crete
Greece
E-mail: fanourg@uoc.gr

G. E. Froudakis
Department of Chemistry
University of Crete
Voutes Campus
GR-70013 Heraklion
Crete
Greece
E-mail: frudakis@uoc.gr

Rana R. Haikal
Center for Materials Science
Zewail City of Science and Technology
Giza 12578
Egypt
E-mail: rraouf@zewailcity.edu.eg

Karen Hannouche
Department of Chemistry
Faculty of Arts and Sciences
American University of Beirut
P. O. Box 11-0236 Riad El Solh
Beirut 1107-2020
Lebanon
E-mail: kgh09@mail.aub.edu

Mohamad Hmadeh
Department of Chemistry
Faculty of Arts and Sciences
American University of Beirut
P. O. Box 11-0236 Riad El Solh
Beirut 1107-2020
Lebanon
E-mail: mohamad.hmadeh@aub.edu.lb

Ashlee J. Howarth
Department of Chemistry and Biochemistry,
and Centre for NanoScience Research
Concordia University
7141 Sherbrooke St W.
Montréal, QC
Canada
E-mail: ashlee.howarth@concordia.ca

Zhehao Huang
Department of Materials and Environmental Chemistry
Stockholm University
Stockholm 10691
Sweden
E-mail: zhehao.huang@mmk.su.se

Hicham Idriss
Institute of Functional Interfaces (IFG)
Karlsruhe Institute of Technology (KIT)
Hermann-von-Helmholtz-Platz
176344 Eggenstein-Leopoldshafen
Germany
E-mail: hicham.idriss@kit.edu

Mohamed Infas
Department of Chemistry
Khalifa University
Building 2, SAN Campus Abu Dhabi 127788
United Arab Emirates
E-mail: mohamed.mohideen@ku.ac.ae

James Kegere
Department of Chemistry
United Arab Emirates University
Al-Ain 15551
United Arab Emirates
E-mail: 202090111@uaeu.ac.ae

Ying Siew Khoo
School of Chemical and Energy Engineering
Faculty of Engineering
Universiti Teknologi Malaysia
81310 Skudai
Johor
Malaysia
E-mail: yingsiew520@gmail.com

Woei Jye Lau
School of Chemical and Energy Engineering
Faculty of Engineering
Universiti Teknologi Malaysia
81310 Skudai
Johor
Malaysia
E-mail: lwoeijye@utm.my

Muhammad Alif Mohammad Latif
Integrated Chemical BioPhysics Research
Faculty of Science
Universiti Putra Malaysia
43400 UPM Serdang
Selangor
Malaysia
Centre of Foundation Studies for Agricultural Science
Universiti Putra Malaysia
43400 UPM Serdang
Selangor
Malaysia
Institute of Nanoscience and Nanotechnology
Universiti Putra Malaysia
43400 UPM Serdang
Selangor
Malaysia
E-mail: aliflatif@upm.edu.my

Long H. Ngo
NTT Hi-Tech Institute
Nguyen Tat Thanh University
Ho Chi Minh City
Vietnam
E-mail: longnh@ntt.edu.vn

Ha L. Nguyen
Department of Chemistry and Berkeley Global Science Institute
University of California Berkeley Berkeley
CA 94720
United States
Joint UAEU–UC Berkeley Laboratories for Materials Innovations
United Arab Emirates University
Al-Ain 15551
United Arab Emirates
E-mail: nguyen.lh@berkeley.edu

Tung T. Nguyen
NTT Hi-Tech Institute
Nguyen Tat Thanh University
Ho Chi Minh City
Vietnam
E-mail: tungnt@ntt.edu.vn

Victor Quezada-Novoa
Department of Chemistry and Biochemistry,
and Centre for NanoScience Research
Concordia University
7141 Sherbrooke St W.
Montréal, QC
Canada

A. P. Sarikas
Department of Chemistry
University of Crete
Voutes Campus
GR-70013 Heraklion
Crete
Greece
E-mail: chemp1160@edu.chemistry.uoc.gr

Thach N. Tu
Nguyen Tat Thanh University
Ho Chi Minh City 755414
Vietnam
Faculty of Chemistry and Pharmacy
University of Regensburg
93053 Regensburg
Germany
E-mail: tnthach@ntt.edu.vn

Mostafa Yousefzadeh Borzehandani
Integrated Chemical BioPhysics Research
Faculty of Science
Universiti Putra Malaysia
43400 UPM Serdang
Selangor
Malaysia
E-mail: mostafa.yousefzadeh@gmail.com

Daqiang Yuan
State Key Laboratory of Structural Chemistry
Fujian Institute of Research on the Structure of Matter
Chinese Academy of Sciences
Fuzhou
P. R. China
University of the Chinese Academy of Sciences
Beijing
P. R. China
E-mail: ydq@fjirsm.ac.cn

Kyle E. Cordova

1 Introduction to reticular chemistry and its applications

1.1 History

If one dives deep into the nuances of the chemical literature, much can be argued about the origins of metal-organic frameworks (MOFs). Many research groups from around the world have laid claim to the earliest discovery of MOFs (ideation or practical conception) and those seminal contributions from Robson [1], Yaghi [2, 3], O'Keeffe [4], Fujita [5], Zaworotko [6] and Kitagawa [7] among others, are worthy of praise. Regardless, the history of the chemical science that underlies MOFs, termed reticular chemistry—the study of linking discrete molecular building blocks (inorganic and/or organic) through strong chemical bonds into extended, crystalline architectures—dates back to the invention of coordination chemistry by Alfred Werner who was awarded the Nobel Prize in Chemistry in 1913 [8]. The result of Werner's work on understanding the spatial arrangement of atoms in discrete, 0-dimensional (0D) coordination complexes led to conceptual transfer from designing 0D structures to 2D and 3D extended structures in the form of Hofmann clathrates [9, 10]. Advancements were then made in the targeted synthesis of such 2D and 3D coordination compounds in the form of the modulation of distance between layers in the 2D compounds and in the connecting of layers to achieve 3D structures. However, it was at this point that several issues began to spring forth: (i) all compounds reported were purely inorganic with limited chemical functionalization being possible and (ii) although the concept of tuning the metrics of a given Hofmann clathrate was proven possible, precise control over these metrics was yet to be fully realized. Given that the structures were crystalline meant that they were well understood and this provided an opportunity to expand beyond purely inorganic building units to employing organic units that could be chemically functionalized and their metrics precisely controlled. In the 1950s, Saito's group exploited the use of organic linkers to construct extended, solid-state materials, termed coordination polymers. This group demonstrated the concept of topological design by simply varying the length of the organic linker used in producing a series of bis(alkylnitrilo)-linked Cu^I coordination polymers of varying dimensionality [11–13]. Specifically, they demonstrated that they could produce a 1D, 2D or 3D coordination polymer simply by moving from succinonitrile to glutaronitrile and adiponitrile as organic linkers, respectively. In the 1990s, Robson followed this groundbreaking work by designing more complex organic linkers and targeting

Kyle E. Cordova, Materials Discovery Research Unit, Advanced Research Centre (ARC), Royal Scientific Society, 11941 Amman, Jordan, e-mail: kyle.cordova@rss.jo

https://doi.org/10.1515/9781501524721-001

binodal topologies (i. e., having two vertices in the topology), by which the principle of "building blocks" was proven to truly work [1]. Indeed, Robson demonstrated the ability to systematically change the metrics of the building blocks through changing their geometry, size and points of extension (i. e., coordination type and number) to realize materials of diverse structure-type. The groups of Fujita, Yaghi and Zaworotko followed this work in the 1990s by switching from cyano-based coordinating groups to employing organic linkers with the neutrally-charged pyridine-based coordinating group [5, 6, 14]. The solvothermal conditions used with these organic linkers led to the discovery of many coordination polymers and proved useful as the synthetic basis for realizing MOFs to this day.

The work reported up to this point had focused on establishing the geometric design principles for transforming 0D discrete coordination complexes to 1D, 2D and 3D coordination polymers through the proper selection and use of inorganic and organic building blocks. Although conceptually important, all synthesized materials were architecturally, thermally and chemically unstable—properties that prevented their use for practical purposes. Addressing the stability challenge rested solely on the "crystallization challenge," whereby coordination polymers needed to be constructed from strong, chelating, coordinate covalent bonds. However, when pursued, such compounds precipitated as amorphous materials because microscopic reversibility (i. e., "error-correcting") could not be established during the crystallization process due to a higher bond energy for M-charged linker bonds (typically up to 400 kJ mol^{-1}) when compared to M-donor bonds (typically up to 200 kJ mol^{-1}). In 1995, Yaghi demonstrated that extended structures could, in fact, be crystallized when charged, chelating linkers were used [2]. The critical discovery here being that the reaction kinetics needed to be slowed down via the slow release of base (in this case pyridine), which slowly deprotonated the carboxylate-based organic linker to establish reversibility. With the synthesis of $Co(BTC)(Py)_2$ (BTC = benzene-tricarboxylate; Py = pyridine), stronger, charged bonds with bridging and chelating coordination modes were shown to afford more stable extended network materials, now termed MOFs.

To further enhance the stability of MOFs, Yaghi pursued the replacement of single metal ion building blocks that had been predominant in all extended structures reported thus far with polynuclear clusters, termed secondary building units (SBUs). The motivation was that SBUs realized through charged linkers could provide MOF structures with more architectural rigidity and stability. Indeed, in 1998, Yaghi published the first MOF, termed MOF-2 [$Zn(BDC)(H_2O)$], whose internal free space (i. e., pores) could be completely evacuated without the overall structure collapsing [3]. Establishing permanent porosity for such structures demonstrated the almost endless potential of designing 2D and 3D MOFs by combining different SBUs with different charged organic linkers. Indeed, the discovery of the fundamental geometric design principles for realizing extended structures in combination with the use of charged linkers and SBUs ushered in a new era in synthetic inorganic chemistry.

1.2 Reticular chemistry as a field of study

Expanding the scope of the charged organic linker and SBU approach in synthesizing more and more MOF materials, such as the iconic MOF-5, provided further proof that the stability challenge had been overcome [4]. This led to a new field of study being invented —a field called reticular chemistry. By definition, the word "reticular" means to resemble a net in appearance or construction or to be net-like, and when combined with "chemistry," the field is defined as the study of logically constructed chemical structures that are held together by strong bonds [15, 16]. This field of study has not only led to the flourishing of MOFs, but has also brought into existence many other class of materials including, zeolitic imidazolate frameworks (ZIFs—metal-organic analogues of zeolites), covalent organic frameworks (COFs—extended structures made entirely of light atoms), metal triazolates (METs), metal catecholate frameworks (CATs), metal phosphonate and sulfonate frameworks, hybrid extended structures, among others [16–21].

Although the number of original research articles, reviews and textbooks (technical and educational) has exponentially grown since the first MOF report and from when the term reticular chemistry came into our scientific lexicon, it is our belief that this edited book is comprised of the most up-to-date contributions from a diverse cast of renown scholars from all over the world on the topic of reticular chemistry. The scope of this monograph picks up from that first report of MOF-2 and extends through the maturing science underlying the design, synthesis, characterization and post-synthetic modification of MOFs to their applications in gas capture, storage and separation, renewable energy and catalysis, and water, among others. Accordingly, based on almost 30 years of research having been reported on the chemistry and applications of MOFs, it is our intention for this monograph to detail the most important knowledge and know-how available to help any reader embarking on a project based on MOFs such that this work will serve as a reference for those old and new reticular chemists alike. The book is divided practically in multiple parts, which are ordered in a manner that the reader experiences a natural progression in the knowledge and understanding being presented.

1.3 Understanding the fundamentals of reticular chemistry and its applications

The first part (Chapters 1–4) showcases the hallmarks of MOFs and the reason they have attracted so much interest in research laboratories across the world. The aesthetic beauty of MOF structures is derived from their order, periodicity and symmetry. These elements are, in turn, derived from the geometry principles that underly their construction. In reticular chemistry, one can either rationally design (*a priori*) a targeted structure or reverse-engineer a structure that was serendipitously discovered to understand its constituent parts. In their chapter, Yuan and El-Sayed touch on these ideas with prominent examples being provided. To realize one's design of a new MOF in the laboratory requires a fundamental appreciation of the different synthetic conditions and techniques that can be used to practically and tangibly achieve a designed architecture. Yuan and El-Sayed deliver a master course on all of the synthetic approaches, variables and conditions that have been reported—from conventional solvothermal synthesis that includes many different liquid- and solid-phase methods to the state-of-the-art sustainable (green) synthesis techniques. Along the way, the authors provide all necessary information for a new reticular chemist to leave with a fundamental understanding of the different reaction conditions (and their effect) on MOF synthesis including, the effect of solvent, pH, temperature, molar ratio of reactants and other parameters.

In silico techniques are finding a prominent place in reticular chemistry due to the era of big data and the treasure trove of structures and properties of MOFs reported. In their chapter, Sarikas, Fanourgakis and Froudakis present the expediency of computational screening and how machine learning can significantly accelerate the discovery process even with a relatively small amount of data. The authors begin their contribution with a tutorial of the machine learning landscape, touching on the fundamental differences of supervised versus unsupervised and reinforcement learning. Furthermore, they demonstrate a general overview for how to build a machine learning pipeline by defining the critical aspects that must be considered during data collection and algorithm selection (e. g., decision trees, neural networks). In machine learning, the combination of multiple algorithms creates a stronger one and for this, examples are given in random forest and gradient boosted trees. Finally, Froudakis et al. focus the latter half of their chapter to applying machine learning to MOF research. Detailed examples are provided for using such approaches in screening and predicting MOFs for gas storage and separation. As the authors state so succinctly in their conclusion, "...this chapter serves as an introduction for material science researchers to the world of machine learning. The basic terminology, techniques and workflow were presented, allowing new practitioners to familiarize with the field of machine learning in an easy and comprehensive manner."

MOFs are crystalline solids—a property that works toward a reticular chemist's advantage; with crystalline solids as a product means that one of the most powerful structural characterization techniques can be used to elucidate the atomic structures, X-ray diffraction. In their chapter, Infas and Huang delineate the crystal growth and crystallization process of MOFs highlighting key synthetic elements to pay attention to while describing the key characteristics of a single crystal suitable for X-ray diffraction analysis. They then administer a brief course on the fundamentals of X-ray diffraction with context emphasis being placed on how such fundamentals are applied to the structure determination, refinement and solution of MOF crystals. Wisely, the authors describe, in detail, this process from both a single-crystal and powder X-ray diffraction perspective. To demonstrate how far X-ray diffraction, as a technique, has been applied to the study of MOFs, the chapter details advanced studies on the crystallization mechanisms of MOFs through time-resolved *in situ* measurements. Infas and Huang end their substantive chapter on what has become widely seen as the future of structure solution in reticular chemistry: electron diffraction. The rate-limiting step in reticular chemistry has always been obtaining MOF crystals of suitable size. With evolving electron diffraction techniques, single crystals that are too small for X-ray diffraction are now capable of providing structural information at the atomic level. In a rather insightful final portion of the chapter, the authors present the principles and protocols of electron diffraction and lay out the structure determination process from data processing and structure solution to structure refinement and further applications in the study of MOFs.

The second part (Chapters 5–9) highlights the potential usefulness of MOFs for solving some of the world's most vexing challenges pertaining to gases. In a chapter whose depth matches its breadth of coverage, Assen details MOFs' applicability to the challenging air separation, light hydrocarbons separation and noble gas purification processes by correlating structural properties to the physical properties and demands of those specific gas separations. Representative examples are presented throughout, which provide benchmarks for newcomers to the field to assess their own compound's performance. Maythalony and Lau expand upon Assen's chapter by demonstrating the effectiveness of processing MOFs into pure membranes or mixing them with polymers to form mixed-matrix membranes. The fabrication strategies presented provide a real tutorial for those pursuing this topic leaving the reader with a solid footing in starting or expanding upon their own membrane research program. Understanding how to fabricate MOF-based membranes is important, but not without the context for which they are applied. Maythalony and Lau then take a deep dive into the versatility by which membranes can be utilized for both gas separations and water purification. Similar to Assen before them, they provide benchmark data for the membranes and their separation performances.

At this point, it almost goes without saying, that the world is facing a serious challenge in addressing climate change. A recognized critical step is to prevent further emissions of harmful greenhouse gases, like carbon dioxide, from entering the atmosphere. Carbon capture is where MOFs, and related materials, stand the best chance to make an impact. Nguyen and Kegere describe in great deal the differences that exist between conventional materials used in carbon capture and MOFs. From their discussion, it was made abundantly clear the advantages that MOFs have in terms of structure and performance over zeolites, activated carbons and aqueous alkanolamine absorbents. The authors then take the reader on a tour-de-force through the approaches taken for applying MOFs toward carbon dioxide capture—from structural properties and performance and the processes of carbon dioxide capture to the key design elements that must be considered for future MOF construction. The question that is always posed once carbon dioxide is captured is, what should be done with it? Nguyen and Kegere dedicate a large portion of their chapter to enumerating MOFs for conversion, whether that be via photocatalysis, electrocatalysis or photo-electrochemical conversion.

In parallel with carbon dioxide emission reduction, it is widely recognized that society has to wean itself off the use of petroleum-based fuels. Natural gas is a transition fuel source that is cleaner than petroleum and hydrogen is the cleanest fuel as upon combustion it produces no greenhouse gases. MOFs are once again well positioned to support this transition through their high-porosity and ultrahigh uptake capacities. Tu et al. tackle this topic by starting their chapter with the fundamentals of adsorption before detailing MOFs for onboard natural gas storage. Similar to the previous chapters, Tu et al. present the key metrics for the highest performing MOFs and detail strategies for improving upon these achievements with a particular focus on structural properties such as coordinatively unsaturated metal sites, porosity, pore volume, functionalization and pore metrics. Critical to realizing MOFs for onboard natural gas storage is to understand and then design structures that balance volumetric and gravimetric working capacities. Indeed, Tu et al. do a masterful job at delineating the important points for this. Attention is then turned in the second half of their chapter to MOFs for onboard hydrogen storage. The authors once again cover the important fundamentals, present the benchmark materials with their performance metrics and illustrate the structural features that are most important to consider when pursuing MOFs for this application.

From cost, environmental and energy viewpoints, the production of specialty chemicals is exigent. In the conventional processes, several metal-based reagents have been developed that can selectively and efficiently drive the necessary transformations. However, the requirement of stoichiometric amounts of the metal catalysts and accumulation of a considerable amount of waste remain inevitable. One solution to this challenge is to employ heterogeneous catalysts. As a result of their inherent structural flexibility, MOFs bear similar catalytic features as zeolites, including porosity and diversity in pore sizes and shapes, yet have the distinct advantage of containing functionalizable organic linkers. Functionalization is possible either by pre-MOF synthesis linker design or through post-synthetic modification on the framework. Furthermore, the large surface areas, structured internal pore environment and high density of accessible active sites facilitate both reactivity and selectivity. One theory of higher efficiency in catalysis is the formation of a "near attack conformation," which means the substrate molecules are forced to come into close proximity with catalytic sites (similar to what is proposed to occur in MOF pores), therefore, increasing the rate by many folds. In their chapter, Hmadeh et al. take the reader on a journey to show them the advantages of using the metal nodes as site-selective catalytic sites, functionalized linkers as site-selective active sites and how post-synthetic modification can introduce new sites as well as exploiting synergistic effects between the metal nodes and organic linkers. The authors end their riveting chapter with presenting strategies for creating composites where the "whole is greater than the sum of its parts."

The final part of the monograph (Chapters 10–13) focuses itself on emerging applications of MOFs—applications that are less mature than the previous section but represent exciting hope and potential. Alkordi and colleagues set the scene of their chapter on MOFs for energy conversion and storage via water electrolysis by presenting the current status and challenges in moving from fossil fuel-based energy to renewables. The authors then provide fundamental concepts of the water splitting reactions such that reader has the context from which to understand the applicability of MOFs. They then detail, at great length, the different roles that MOFs play in the hydrogen evolution reaction (e. g., pristine MOF electrocatalysts, porous scaffolds/supports and sacrificial precursors) and follow this with a similar approach for MOFs in the oxygen-evolution reaction. Alkordi et al. conclude their chapter with a clear statement that this is a burgeoning application—one in which the rich chemistry of MOFs is well positioned to advance for years to come.

More than 50 % of the world's population lives in water-stressed areas. Water resources are being depleted faster than can be replenished, drinking water is quickly becoming scarce, and climate change is only exacerbating the problem. MOFs have emerged as viable candidates for addressing a wide-ranging assortment of water issues. In their chapter, Howarth and colleagues demonstrate absolute knowledge of the diverse challenges facing the water security and finesse a brilliant study on the various MOFs used in the adsorption and removal of contaminants or pollutants (e. g., organic, inorganic, and dual removal of organic and inorganic compounds) in water systems. Where adsorption is not feasible, the authors detail how catalytic MOFs can be used to degrade the pollutants. As an emerging field with few examples reported, Howarth wisely presents desalination as a potential application showcasing how much work remains to be done in order to progress this science. Finally, the authors turn their attention to water adsorption in MOFs showcasing the application of atmospheric water harvesting that has received enormous attention as of late. Atmospheric water harvesting represents the promise of MOFs for achieving absolute water security and independence for anyone, anywhere, at any time [22].

The final chapter of the monograph situates itself on addressing and optimizing the structural properties of MOFs for use in biomedical applications. Latif et al. tackle this topic through a deep understanding and illustrating of guest molecules' interactions with the frameworks of MOFs—guest molecules that span from biomolecules (e. g., proteins, enzymes, peptides, carbohydrates, DNA, nucleic acids and/or lipids). Once a solid understanding of those interactions is gleaned, the authors dive into how the resulting systems can be exploited for use in targeted delivery and demonstrate how computational techniques can ease and structure the synthetic work that is needed to realize such systems. From a combination of experimental and computational studies, Latif et al. describe, in great detail, the many factors that affect host-guest interactions in MOFs—from the effect of cage and pore size, functionalization of the linkers and coordinatively unsaturated metal sites, to diffusivity and loading capacity. It is a remarkable chapter with much to learn and apply to new research programs.

At a fundamental level, reticular chemistry offers an intellectually stimulating journey through discovery, rational design, structural characterization and technology-driven properties. The breadth of science, techniques and applications experienced through reticular chemistry is unseen in other fields. Through this monograph, readers will come to appreciate the importance of this experience as each chapter showcases a wide variety of topics and highlights the endless impacts that MOFs have been proven to deliver. It is our hope that readers will find this contribution valuable to their pursuit of knowledge in the field of reticular chemistry. With this, we would like to extend our deepest appreciation to all of the expert contributors for their outstanding work and we wish all of the readers nothing but success in their own research pursuits.

Bibliography

[1] Hoskins BF, Robson R. Design and construction of a new class of scaffolding-like materials comprising infinite polymeric frameworks of 3D-linked molecular rods. A reappraisal of the $Zn(CN)_2$ and $Cd(CN)_2$ structures and the synthesis and structure of the diamond-related frameworks $[N(CH_3)_4][Cu^IZn^{II}(CN)_4]$ and Cu^I[4,4',4'',4'''-tetracyanotetraphenylmethane]$BF_4 \cdot xC_6H_5NO_2$. J Am Chem Soc. 1990;112:1546–54.

[2] Yaghi OM, Li G, Li H. Selective binding and removal of guests in a microporous metal-organic framework. Nature. 1995;378:703–6.

[3] Li H, Eddaoudi M, Groy TL, Yaghi OM. Establishing microporosity in open metal-organic frameworks: Gas sorption isotherms for Zn(BDC) (BDC = 1,4-Benzenedicarboxylate). J Am Chem Soc. 1998;120:8571–2.

[4] Li H, Eddaoudi M, O'Keeffe M, Yaghi OM. Design and synthesis of an exceptionally stable and highly porous metal-organic framework. Nature. 1999;402:276–9.

[5] Fujita M, Kwon YJ, Washizu S, Ogura K. Preparation, clathration ability, and catalysis of a two-dimensional square network material composed of cadmium(II) and 4,4'-bipyridine. J Am Chem Soc. 1994;116:1151–2.

[6] Subramanian S, Zaworotko M. Porous solids by design: $[Zn(4,4'\text{-bpy})_2(SiF_6)]_n \cdot x$DMF, a single framework octahedral coordination polymer with large square channels. Angew Chem Int Ed. 1995;34:2127–9.

[7] Kondo M, Yoshitomi T, Matsuzaka H, Kitagawa S, Seki K. Three-dimensional framework with channeling cavities for small molecules: $\{[M_2(4,4'\text{-bpy})_3(NO_3)_4] \cdot xH_2O\}_n$ (M = Co, Ni, Zn). Angew Chem Int Ed. 1997;36:1725–7.

[8] Werner A. 1,2-Dichloro-tetrammin-kbaltisalze. (Ammoniak-violeosalze). Eur J Inorg Chem. 1907;40:4817–25.

[9] Hofmann K, Kuspert F. Verbindungen von kohlenwasserstoffen mit metallsalzen. Zeit Anorg Chem. 1897;15:204–7.

[10] Powell HM. The structure of molecular compounds. Part IV. Clathrate compounds. J Chem Soc. 1948:61–73.

[11] Kinoshita Y, Matsubara I, Saito Y. The crystal structure of bis(succinonitrilo)copper(I) nitrate. Bull Chem Soc Jpn. 1959;32:741–7.
[12] Kinoshita Y, Matsubara I, Saito Y. The crystal structure of bis(glutaronitrilo)copper(I) nitrate. Bull Chem Soc Jpn. 1959;32:1216–21.
[13] Kinoshita Y, Matsubara I, Higuchi T, Saito Y. The crystal structure of bis(adiponitrilo)copper(I) nitrate. Bull Chem Soc. 1959;32:1221–6.
[14] Yaghi OM, Li H. Hydrothermal synthesis of a metal-organic framework containing large rectangular channels. J Am Chem Soc. 1995;117:10401–2.
[15] Yaghi OM, O'Keeffe M, Ockwig NW, Chae HK, Eddaoudi M, Kim J. Reticular synthesis and the design of new materials. Nature. 2003;423:705–14.
[16] Furukawa H, Cordova KE, O'Keeffe M, Yaghi OM. The chemistry and applications of metal-organic frameworks. Science. 2013;341:1230444.
[17] Zha X, Li X, Al-Omari AA, Liang CC, Al-Ghourani A, Abdellatief M, Yang J, Nguyen HL, Al-Maythalony B, Shi Z, Cordova KE, Zhang YB. Zeolite NPO-type azolate frameworks. Angew Chem Int Ed. 2022;e202207467.
[18] Nguyen NTT, Furukawa H, Gandara F, Nguyen HT, Cordova KE, Yaghi OM. Selective capture of carbon dioxide under humid conditions by hydrophobic chabazite-type zeolitic imidazolate frameworks. Angew Chem Int Ed. 2014;40:10821–4.
[19] Nguyen NTT, Lo TNH, Nguyen HTD, Le TB, Cordova KE, Furukawa H. Mixed-metal zeolitic imidazolate frameworks and their selective capture of wet carbon dioxide over methane. Inorg Chem. 2016;55:6201–7.
[20] Nguyen NTT, Furukawa H, Gandara F, Trickett CA, Jeong HM, Cordova KE, Yaghi OM. Three-dimensional metal-catecholate frameworks and their ultrahigh proton conductivity. J Am Chem Soc. 2015;137:15394–7.
[21] Nguyen HL, Gandara F, Furukawa H, Doan TLH, Cordova KE, Yaghi OM. A titanium-organic framework as an exemplar of combining the chemistry of metal- and covalent-organic frameworks. J Am Chem Soc. 2016;138:4330–3.
[22] Almassad HA, Abaza RI, Siwwan L, Al-Maythalony B, Cordova KE. Environmentally adaptive MOF-based device enables continuous self-optimizing atmospheric water harvesting. Nat Commun. 2022;13:4873.

El-Sayed M. El-Sayed and Daqiang Yuan

2 Reticular design and synthesis strategies of metal-organic frameworks

2.1 Introduction

As society's needs in terms of energy increase due to the global upsurge of population, there is a growing trend among the scientific community to develop advanced innovative materials to address energy and environment-related challenges. Such materials represent the heart of processes whose structural properties can be genuinely linked to unprecedented performance in energy-related applications [1]. Meanwhile, there is a clear worldwide political awareness to limit the global temperature rise to 1.5 degrees by controlling greenhouse gas emissions. Consequently, a growing number of undergoing initiatives aim to construct a plethora of environmentally friendly, cost-effective and simple-to-prepare materials to capture or convert greenhouse gas emissions, mainly carbon dioxide (CO_2), into value-added products. Among these materials, metal-organic frameworks (MOFs) have garnered considerable attention for more than two decades. Such materials are auspicious in various applications such as catalysis, gas separation, adsorption, sensing, piezo/ferroelectric, among others [2]. Such expanded utility is attributed to numerous characteristics that MOFs possess, including crystallinity, porosity, large, adjustable pore sizes/surface areas, shape modularity and tunability.

MOFs are constructed from the coordinative assembly of secondary building units (SBUs) derived from metal ions and organic linkers based on linkages arising from carboxylates, pyridines, azolates and catecholates, among others. MOFs, by design, mostly exist in two- and three-dimensional networks developed by articulating building blocks with different molecular geometries, which offer a plethora of targeted nets [3, 4]. The organic linkers used are mostly rigid, which produce robust architectures; however, when flexible linkers are employed, MOFs with unpredictable crystal structures and properties may be produced. In recent years, the chemistry of MOFs has been enriched with the feasibility of obtaining MOF compounds with characteristics toward specific applications by applying crystal engineering principles. This was achieved by adjusting MOF properties in terms of their pore size and shape, network topology and

El-Sayed M. El-Sayed, State Key Laboratory of Structural Chemistry, Fujian Institute of Research on the Structure of Matter, Chinese Academy of Sciences, Fuzhou, P. R. China; and University of the Chinese Academy of Sciences, Beijing, P. R. China; and Chemical Refining Laboratory, Refining Department, Egyptian Petroleum Research Institute, Nasr City, Cairo, Egypt
Daqiang Yuan, State Key Laboratory of Structural Chemistry, Fujian Institute of Research on the Structure of Matter, Chinese Academy of Sciences, Fuzhou, P. R. China; and University of the Chinese Academy of Sciences, Beijing, P. R. China, e-mail: ydq@fjirsm.ac.cn

https://doi.org/10.1515/9781501524721-002

surface area. Such a targeted design of a MOF from the physicochemical point of view requires a deep understanding of the molecular as well as the intermolecular interactions in the resulting net. The former interactions come from coordination between organic linkers and metal clusters, while the latter is derived from weak bonding within the networks due to hydrogen bonds, π–π interactions and/or van der Waal interactions.

MOF synthesis requires screening for several parameters, including temperature, pH, molar ratio, concentration, solvent type, reaction vessel, metal ion source and metal-to-linker ratio [5]. Optimizing such factors is essential to obtain large single crystals with high stability amenable to subsequent characterization. Modulated synthesis is a beneficial approach to improving the crystallinity of resulting MOFs by adding monocarboxylic acids during the synthesis, such as acetic acid, formic acid, benzoic acid, trifluoroacetic acid, hydrochloric acid and/or nitric acid. These acids compete with the main organic linkers for the coordination assembly with metal clusters to slow down nucleation and crystal growth. The solvent choice is essential to afford a targeted MOF structure with high crystallinity and porosity rather than microcrystalline powder or gel-like products. Single or mixed solvent systems are used in MOF synthesis, most notably *N*,*N*'-dimethylformamide (DMF), *N*,*N*'-dimethylacetamide (DMA), *N*,*N*'-diethylformamide (DEF) and *N*-methyl-2-pyrrolidone (NMP). When used in MOF synthesis, high boiling point solvents are replaced by low boiling point solvents and low surface tensions such as methanol, ethanol and dichloromethane, which an activation process can remove. The efficiency of the exchange process, followed by activation, can optimize the surface area of the prepared MOF. Once proper, large, stable single crystals are formed, they undergo structural examination using single-crystal X-ray diffraction (SCXRD) analysis to resolve the underlying crystal structure. The resultant crystal structure can be modified once external stimuli are applied, such as temperature, light or guest molecules [6].

Deployment of MOFs in industrial settings requires tremendous technical and financial efforts with respect to scaling up the synthetic processes. The scaling up step that is one of many steps in a technical/economical maturation process requires discovering pathways to reduce the production cost of MOFs while reducing negative environmental impact. Different implemented approaches are employed, such as green synthetic routes, inexpensive precursors, mild reaction settings with quantitative MOF yields and ensuring stability. In this regard, the replacement of MOF precursors with green, sustainable and cost-effective ones has been a target of academic research since 2013 [7]. The green approach can be achieved by using sustainable sources such as biomass or recycling discarded products containing MOF ingredients such as waste polymers. From a different perspective, a focus on room temperature synthesis is feasible for MOF commercialization; however, in laboratory settings, this yields amorphous or nonporous products requiring a compromise between mild reaction temperature and targeted properties for MOF products. The same issues can be found when waste materials are used as MOF precursors, which produce MOFs with reduced crystallinity and surface areas in addition to disordered morphologies.

In the following sections, we present the design principles of MOFs using different metal clusters and organic linkers with varying coordination modes. We then detail the synthetic methods typically employed in MOF synthesis and the emerging green approaches that have the ability to facilitate deployment. By the end of this chapter, many early-career researchers will acquire fundamental knowledge of the design, synthesis and commercialization of MOFs. Figure 2.1 depicts the workflow for the activities conducted by reticular chemists in the synthesis, activation and analysis of MOFs.

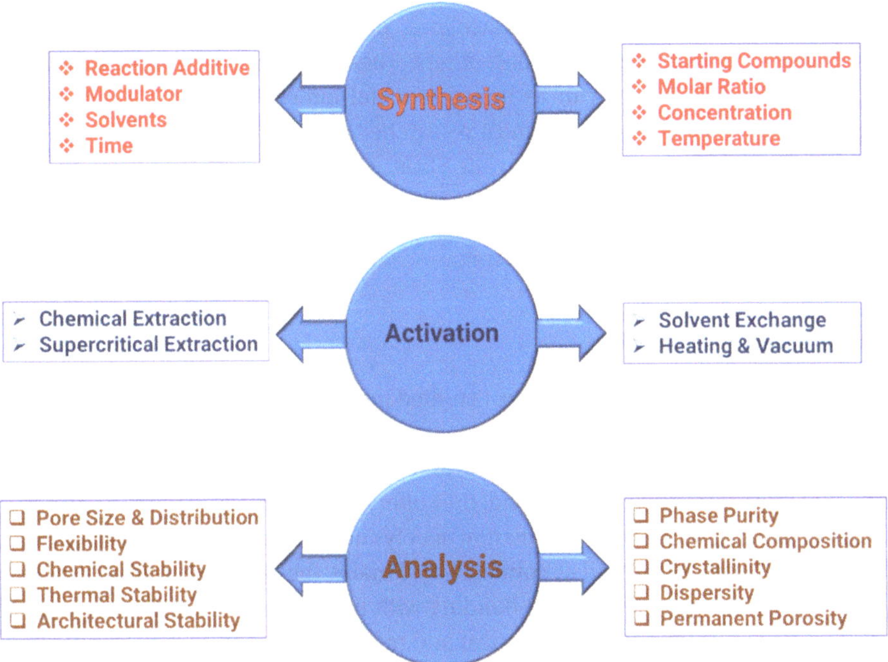

Figure 2.1: Standard practices of reticular chemistry from synthesis to analysis. Reproduced from [3].

2.2 Reticular design of metal-organic frameworks

Based on geometry-guided presynthesis design, MOFs are constructed following the principles of reticular chemistry, in which well-defined molecular building units are rationally selected and linked by strong bonds to produce a vast array of crystalline extended networks [8]. MOF structures can be tuned in terms of pore size, shape and functionality, making them tailorable for different applications. The reticular design of periodic MOFs primarily depends on building blocks, targeted nets and isoreticular chemistry. Understanding the structural characteristics of these building blocks can rationally direct MOF formation. The topology principle is employed to simplify and systematize MOFs based on their nets, which are built from the linkage between vertices (metal clusters/metal ion centers) and edges (organic linkers) (Figure 2.2) [9]. If the underlying topology of a given MOF must be determined, the MOF structure should be deconstructed into its individual building units prior to synthesis. The main ingredients of MOFs are inorganic components, also known as secondary building units (SBUs), in addition to organic parts or linkers. *A priori* design of MOF structures with targeted topologies is accomplished due to the synthetic outcome based on the exact geometry of individual building units. The organic linkers bear different binding groups, including carboxylates, imidazolates, phosphates, sulfonates, pyrazolates, catecholates and tetrazolates, among others. These binding groups are considered points of extension, which can be linked to metal ion centers to provide extended frameworks. Carboxylates are the most researched among these groups as they tend to produce polynuclear metal carboxylate SBUs, given their chelating nature. In addition to directionality and robust bonding, SBUs, as a building unit, endow MOFs with chemical, thermal, architectural and mechanical stabilities.

2.2.1 Organic linkers

Carboxylate-based linkers are the most studied linkers in MOF chemistry due to their charged nature leading to various advantages over neutral linkers [10, 11]. First, they produce neutral frameworks as they neutralize the positive charges of metal centers without requiring counter ions. Second, they offer relative structural rigidity and directionality. Third, they lead to the formation of polynuclear clusters or SBUs that possess fixed overall geometry as well as connectivity. Finally, the charged nature of carboxylates produces strong coordination bonding with metal centers of SBUs, resulting in MOFs possessing high thermal, chemical and mechanical stability. The organic linkers used in MOF synthesis have different topicity (i. e., coordination modes or points of extension), ranging from ditopic (two points of extension) to dodecatopic (twelve points of extension) linkers. By applying the principle of isoreticular expansion, these linkers can be elongated or shortened to provide isoreticular structures (MOFs with different pore metrics but the same underlying topology) with larger or smaller pore sizes, respectively, and varying surface areas, while maintaining the same underlying topology. Figures 2.3 and 2.4 present the linkers with different connectivities used with various metal clusters or SBUs for MOF formation leading to structures with different geometrical shapes or nets.

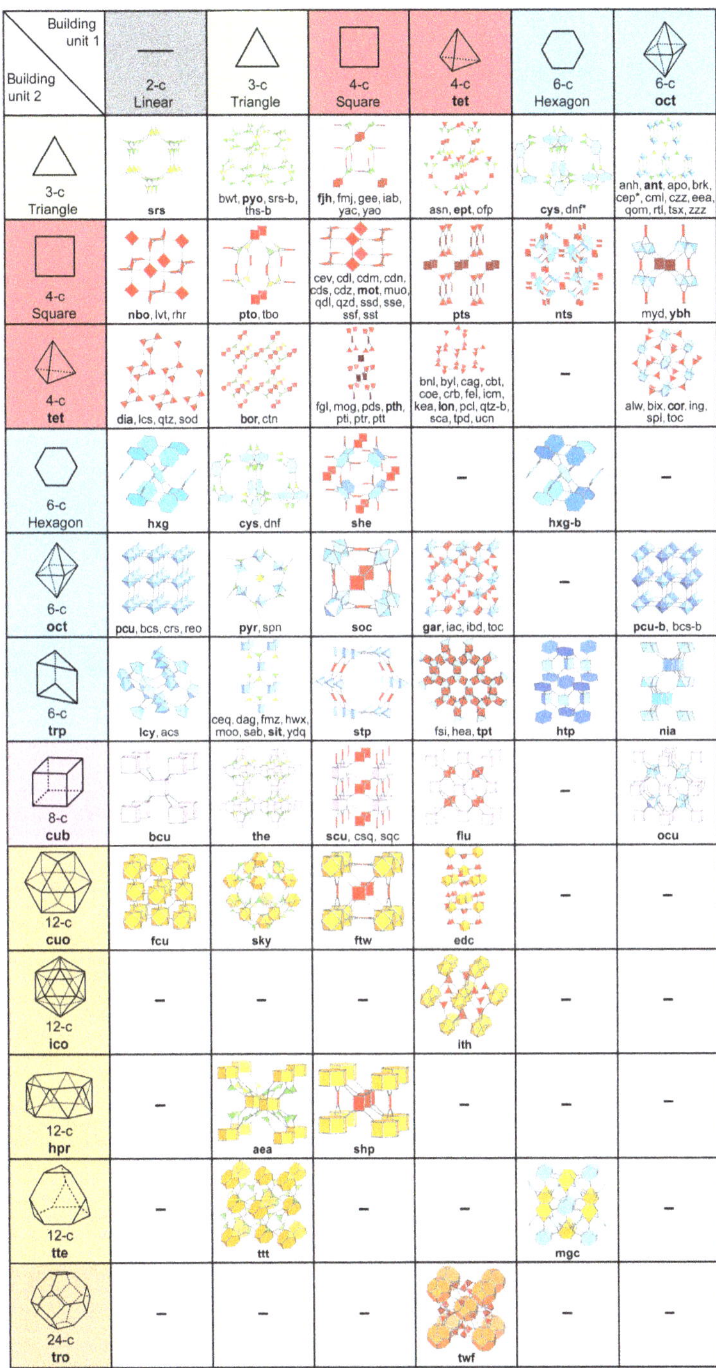

Figure 2.2: The reticular chemistry periodic table displaying predicted nets assembled from different building units. Reproduced from [9].

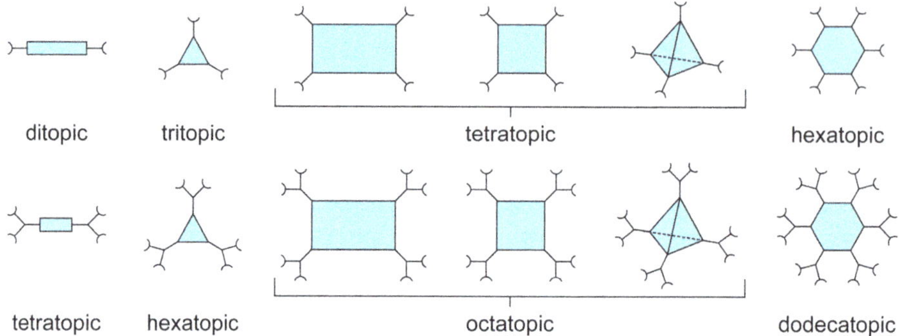

Figure 2.3: Different topicities held by the organic linkers used for MOF synthesis. Reproduced with permission from [12].

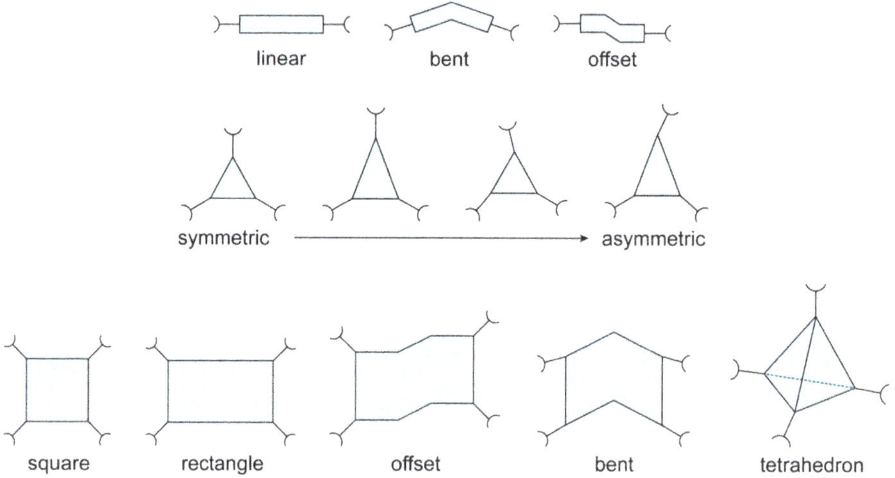

Figure 2.4: Ditopic, tritopic and tetratopic linker geometries for MOF synthesis. Reproduced with permission from [12].

2.2.2 Secondary building units (SBUs)

SBUs refer to the inorganic part of a MOF structure. These can be synthesized *in situ* and are assembled with polydentate binding groups of organic linkers to produce the final MOF architecture [9]. Inorganic SBUs endow MOFs with structural richness due to their versatile geometries and high connectivity (e. g., 3–12), which significantly outperform single metal nodes. The rigid nature of SBUs and well-designed organic linkers ultimately lead to robust structures with predictable net topologies. The connectivity of SBUs (termed "x-c," where x = the coordination number of the overall SBU) typically ranges from 3-c to 12-c SBUs, but sometimes forms infinite rods (Figure 2.5).

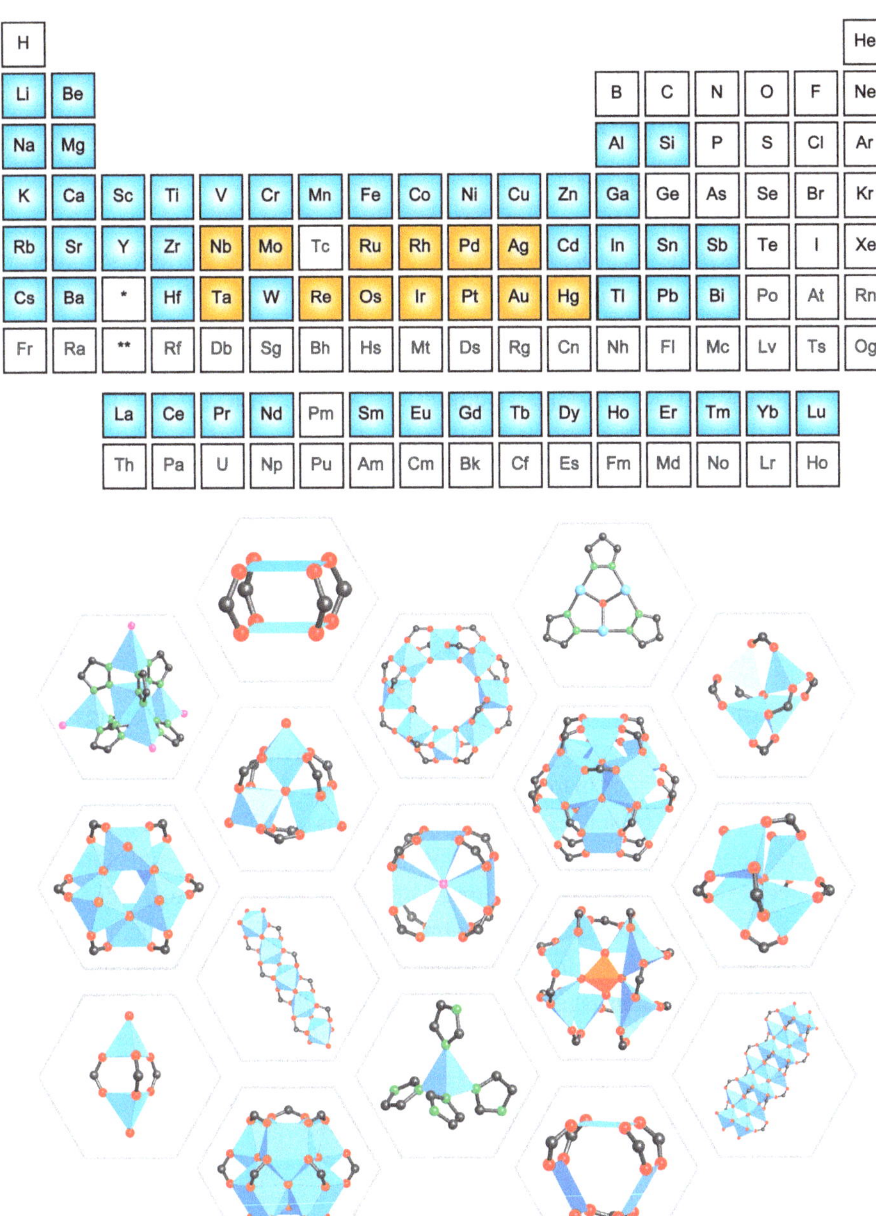

Figure 2.5: The periodic table demonstrating the incorporated metals reported to have been used for MOF construction. Reproduced with permission from [12].

2.3 MOF synthesis

2.3.1 Conventional synthesis

2.3.1.1 Liquid-phase synthesis

In liquid-phase synthesis, MOFs are produced via two scenarios. The first is mixing an organic ligand, metal ion source and solvent in a suitable reactor. The second scenario depends on the separate dissolution of metal salts and organic linkers in a selected solvent, with subsequent comixing in a third reactor. Numerous parameters can affect the liquid-phase synthesis, including solubility, reactivity and stability of MOF precursors, as well as polarity, boiling point and dielectric constants of employed solvents. In liquid-phase synthesis, the molar ratio of MOF precursors and their concentrations, reaction temperature, reaction time, solution pH and metal ion sources can govern the morphology, crystal size, topology and phase purity of the resulting MOF.

2.3.1.1.1 Slow vapor diffusion

Historically, slow vapor diffusion was the most prevalent method to synthesize MOFs. Although this method is applied at room temperature without intensive energy input, its MOF products require long reaction times, which may reach weeks. In this method, after mixing MOF precursors in low boiling solvents, the solution is left to slowly evaporate, resulting in MOF crystals. Alternatively, solutions of MOF precursors can be layered over one another or separated by a solvent layer resulting in crystals forming at the interface of the different layers (Figure 2.6). These approaches are practiced when precursors are insoluble or sensitive to thermal methods. Although reducing reaction times can be accomplished by increasing the concentration of MOF constituents, the quality of produced crystals is reduced, which explains the low number of MOFs produced by this approach.

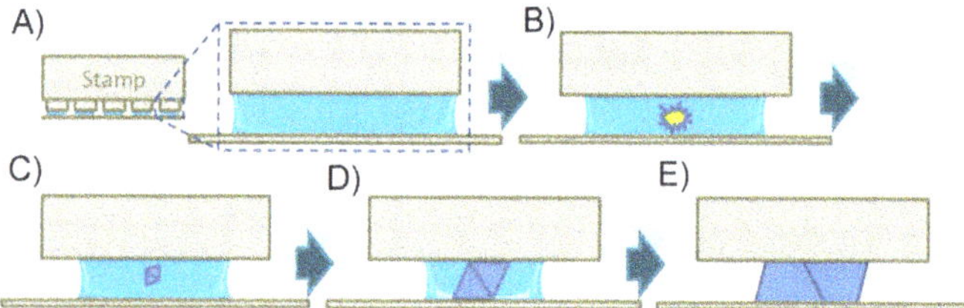

Figure 2.6: Solvent evaporation synthesis of HKUST-1 MOF crystals. Reproduced with permission from [13].

2.3.1.1.2 Solvothermal method

The synthesis of MOFs primarily relies on the solvothermal method, in which MOF precursors are mixed and heated under autogenous pressure in sealed autoclaves or vials of different sizes (Figure 2.7). This synthetic method is preferable for obtaining single crystals with large sizes yet is unfavorable for sensitive MOF precursors due to the prescribed high temperatures (50–260 °C) and long reaction times (hours to days). The synthesis method is referred to as solvothermal when organic solvents (e. g., DMF, DMA, DEF, NMP, methanol, ethanol, acetone or their binary or ternary mixtures) are used and is called hydrothermal when water is utilized as the solvent. When ionic liquids are employed as solvents, the synthesis method is referred to as ionothermal synthesis. The synthesis of MOFs containing inert metal ions, such as Cr, Fe and Al, typically requires high temperatures, which may reach 400 °C, and Teflon-lined autoclaves to conduct such reactions. The autogenous pressure above the boiling point of the solvents employed in these closed reactors can assist in dissolving the slightly soluble precursors, especially high molecular weight linkers. This increases the crystallinity of the produced MOFs, affording regular morphological shapes. However, it is noted that high temperatures may cause structural transformation or decomposition of the final MOF products. Once large single crystals are observed via an optical microscope, the reaction mixture is left for natural or programmed cooling at a specified speed rate. If collected or left in the mother liquor for stability concerns, the crystals are subjected to subsequent characterization using SCXRD.

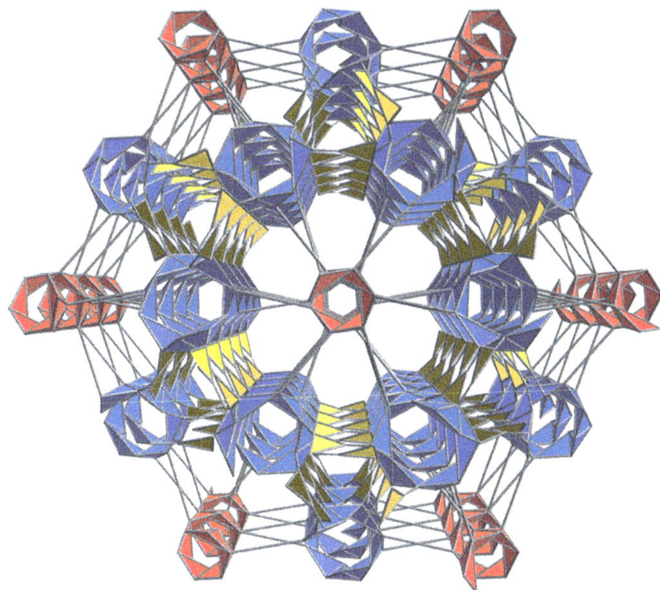

Figure 2.7: A topological view of JOU-25-R structure along *c*-axis synthesized by solvothermal synthesis. Reproduced with permission from [14].

2.3.1.1.3 Electrochemical method

This method is advantageous for rapid MOF powder synthesis at mild reaction temperatures. In fact, electrochemical synthesis could be key for further scaling-up MOFs. In the process, anodic dissolution is responsible for providing metal ions into the reaction mixture that contains organic linkers and electrolytes, and a pH value is increased when hydroxyl anions increase in the solution (Figure 2.8). MOF crystals commence nucleation near the electrode surface, producing a thin MOF layer. Numerous parameters govern the quality of MOF products, such as applied voltage, specific current, spacing between electrodes, synthesis time, solvent type, linker chemistry, water content and electrolyte concentration. Since no pressure is built up during the reaction, electrochemical synthesis allows for greater control of reactant concentrations.

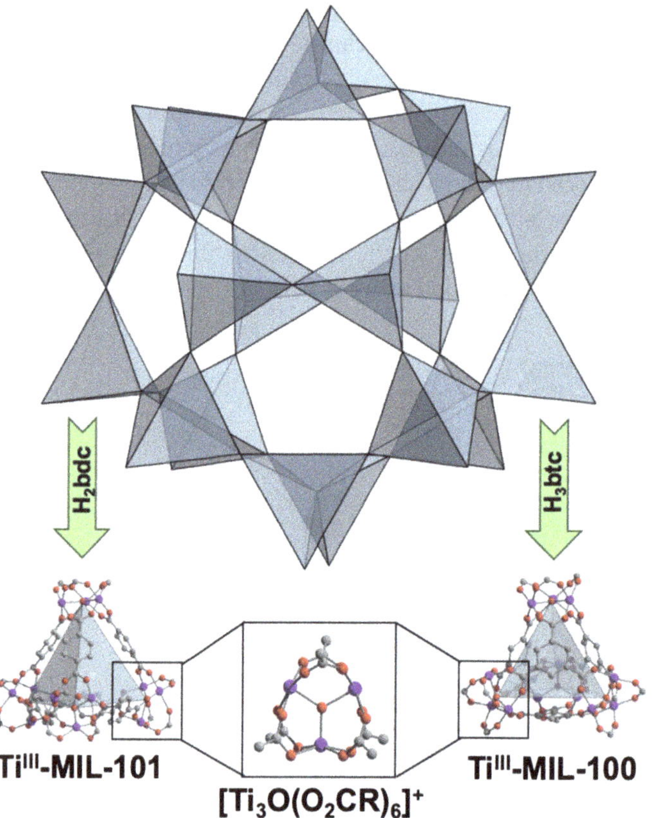

Figure 2.8: Electrochemically synthesized Ti^{III}-MIL-100. Reproduced with permission from [15].

Despite the advantages held by electrochemical synthesis, it does have limitations. For instance, due to the difficulty in controlling the production of thin MOF films, their thickness, particle sizes and shapes are difficult to manage. In addition, nonnegligible amounts of impure MOFs can be produced due to the deposition of reduced metal ions on the electrode surface.

2.3.1.1.4 Sonochemical method

In this method, high-quality MOFs are produced using ultrasound energy (20 kHz–10 MHz), which speeds up the reaction between organic linkers and metal ions in each solvent. The reaction mixture is accelerated toward crystal growth due to its transformation into local hot spots encompassing high temperature and pressure that can accelerate the crystal growth process. The sonochemical method is viable for producing MOF products in a short duration without requiring high temperature and pressure, which is typical in most liquid-phase methods (Figure 2.9).

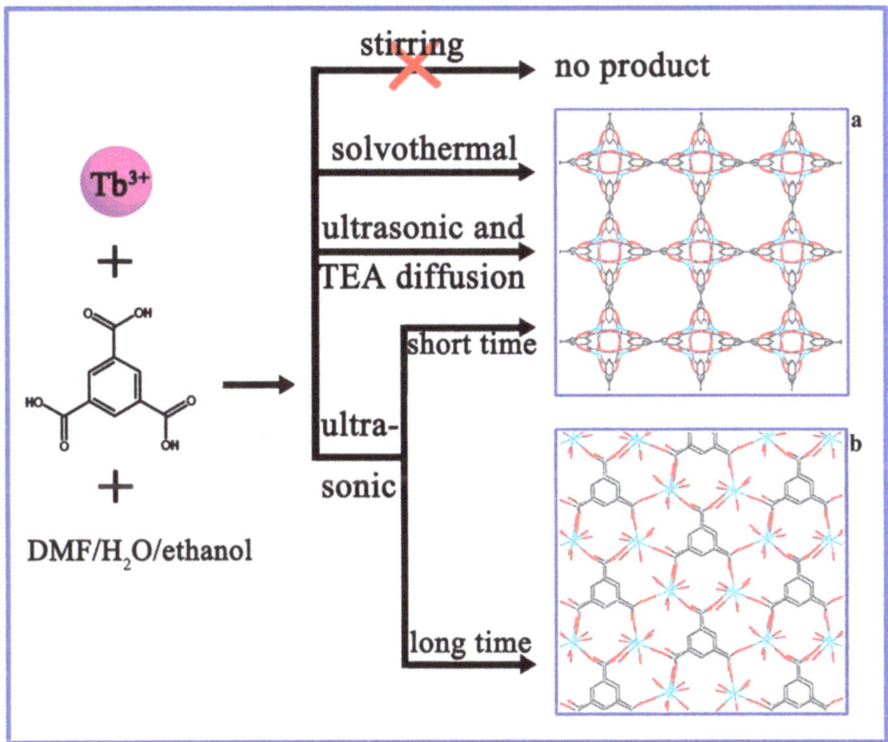

Figure 2.9: Ultrasonically synthesized Tb-BTC MOF. Reproduced with permission from [16].

2.3.1.1.5 Microwave method

Microwave-assisted synthesis is a rapid method to produce small-sized MOF products with high productivity, phase selectivity and controlled morphology (Figure 2.10). In this method, the mobile electric charges in MOF systems, such as polar solvents,

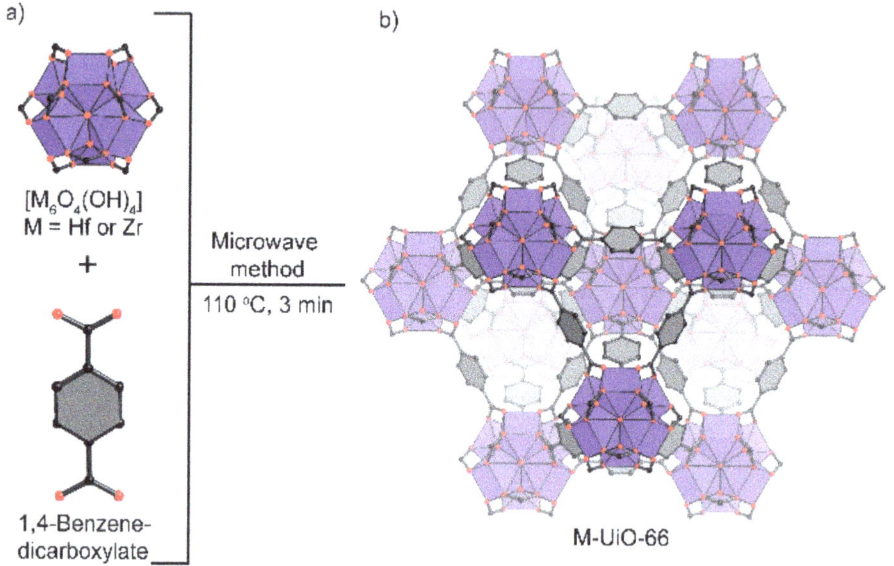

Figure 2.10: Hf/Zr-UiO-66 MOF synthesized by the microwave method. Reproduced with permission from [18].

metal ions and deprotonated organic linkers, undergo interactions with electromagnetic waves. This produces MOFs with high crystallinity and low yield when a short time is applied or MOFs with low crystallinity and high yield when a prolonged reaction is executed. Metal-oxide formation is expected when the reaction mixture contains concentrated precursors [17]. This highlights the significant impact of reaction time and precursor concentration on the final properties of the produced MOFs. Although microwave-assisted synthesis is a rapid, clean and cost-effective approach toward MOF synthesis, several disadvantages remain clear. For instance, using closed systems with high pressures can cause explosions unless advanced equipment is employed. Second, rapid microwave heating can produce concerns about safety or health when nonsophisticated equipment was employed. Third, only low boiling solvents can be employed in MOF synthesis under a microwave-assisted approach, which affects the final properties of MOFs.

2.3.1.1.6 Ionothermal method

As a subcategory of solvothermal synthesis, the ionothermal method can provide several benefits over conventional synthesis. Ionothermal synthesis makes use of ionic liquids (ILs) as solvents in MOF syntheses, with many advantages, such as high thermal stability, high solubility for organic compounds, nonflammability and low vapor pressure. Conventional synthetic methods have several disadvantages due to the need for severe reaction settings, large amounts of organic solvents and energy-intensive processes. Due to these limitations, ionothermal synthesis can be a viable method for producing many MOFs and other porous materials. For example, due to their low volatility, ILs save energy in the synthetic process due to reduced vapor pressures and boiling points. In the ionothermal process, ILs are employed as solvents or templates to direct the synthetic products by providing anions/cations as counterions. Although ionothermal synthesis possesses some merits over conventional synthetic methods, several disadvantages make this synthetic route realistic for a limited number of systems. For instance, ILs possess high dynamic viscosity, precluding the solute diffusion in solution, which slows down stirring or filtration. In addition, ILs can be crystallized out of the reaction mixture at undefined melting points. The ionothermal synthesis is also considered a costly process as it includes IL synthesis followed by MOF synthesis. Finally, resolving the final MOF structures is challenging due to the difficulty of completely removing ILs from the final products. This ensures that selecting a proper synthetic method for MOFs depends on the nature of precursors and expected final products, allowing the ionothermal to be successful only in some cases (Figure 2.11).

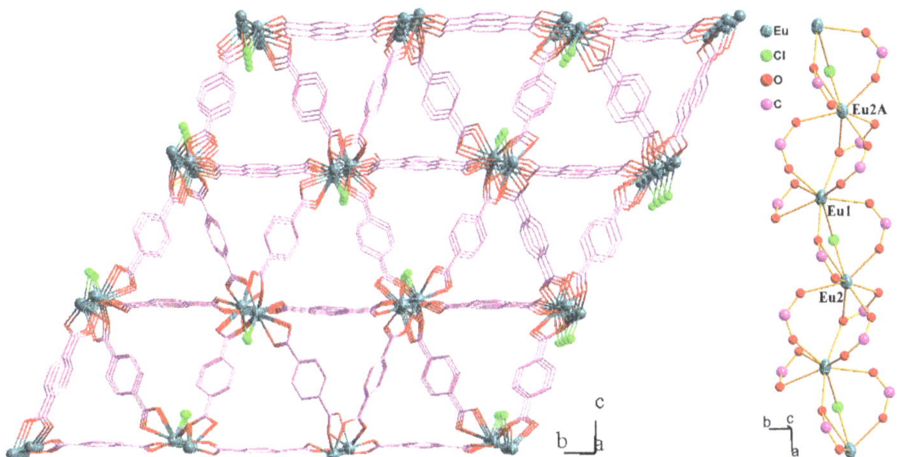

Figure 2.11: Eu-MOF synthesized by the ionothermal synthesis. Reproduced with permission from [19].

2.3.1.1.7 Flow chemistry method

Flow chemistry is employed to synthesize MOF products with tunable properties by providing ultrafast reaction and continuous process via temperature, residence time or sludge volume. The system underlying this method includes four main parts: (i) the syringe system for injecting solvents, metal ion solutions and organic ligand solutions, (ii) microtubular reactor, (iii) temperature-controlled oil bath and (iv) glass vial in the outlet of the reactor for collecting the final products. By flowing the reaction mixture in a reactor, crystalline MOFs are precipitated and collected in the glass vial.

2.3.1.2 Solid-phase synthesis

To synthesize MOFs from low soluble reactants, mechanochemical synthesis by manual grinding or automatic ball mills is employed at room temperature using solvent-free conditions or liquid-assisted grinding. The mechanochemical synthesis is environmentally friendly and a green alternative for liquid-phase intensive synthesis producing MOFs with high purity and efficiency in short durations under normal temperatures and pressures. This method facilitates mass transfer, reduced particle size, molten reagents and accelerated reaction time. The mechanochemical synthesis can be practically viable in MOF commercialization, producing a high yield under normal conditions. However, like the other methods, it has some disadvantages, such as producing amorphous MOFs unsuitable for SCXRD to determine the final crystal structure.

Although using conventional synthetic methods toward MOFs requires large amounts of solvents, energy-intensive conditions, sealed autoclaves to produce autogenous pressure and prolonged reaction times, these methods are always capable of producing single crystals that enable their structural determination. This implies that when MOFs are synthesized for the first time, solvothermal/hydrothermal methods should be the first option to pursue to thoroughly investigate their chemistry for further synthetic approaches under normal conditions. Other synthetic methods, when used first, mainly produce amorphous MOFs or MOFs with very small crystals unsuitable to SCXRD for structural determination. This represents an obstacle in elaborating structure properties-application relationships, hiding essential information about the chemistry of a given produced MOFs. However, these new synthetic methods are reliable in industrialization at normal pressures and temperatures without special equipment like autoclaves (Table 2.1).

Table 2.1: Selected examples of MOFs synthesized by different approaches.

MOFs	Metal source	Linker	Synthetic methods	Ref.
HKUST-1	$Cu(NO_3)_2 \cdot 3H_2O$	H_3BTC	Slow evaporation	[13]
JOU-25	$In(NO_3)_3 \cdot 4.5H_2O$	H_3TCA	Solvothermal synthesis	[14]
Ti-MIL-100	$TiCl_4$	H_2BDC	Electrochemical synthesis	[15]
Tb-MOF	$Tb(NO_3)_3 \cdot 6H_2O$	H_3BTC	Sonochemical synthesis	[16]
Hf-UiO-66	$HfCl_4$	H_2BDC	Microwave synthesis	[18]
Eu-MOF	$EuCl_3.6H_2O$	H_2BDC	Ionothermal synthesis	[19]
Zr-UiO-66	$ZrCl_4$	H_2BDC	Flow chemistry synthesis	[20]
Fe-MOF	$Fe(NO_3)_3 \cdot 9H_2O$	H_3BTC	Mechanochemical synthesis	[21]

H_3BTC = 1,3,5-benzenetricarboxylic acid or trimesic acid.
H_3TCA = 4,4′,4″-triphenylamine tricarboxylic acid.
H_2BDC = 1,4-benzenedicarboxylic acid.

2.3.2 Sustainable (green) synthesis

Conventional MOF synthesis, as described above, often uses different toxic solvents and metal sources that can cause safety issues, corrosion or environmental issues upon disposal. As a consequence, identifying new MOF precursors that are nontoxic, sustainable and green is of immense importance toward commercialization [7]. In this regard, several attempts have been reported to use green solvents other than DMF or metal ion sources instead of nitrates or chlorides. Moreover, post-consumer products have been employed as linker sources, such as waste plastic materials derived from polyethylene terephthalate (PET), polylactic acid (PLA) and polybutylene terephthalate (PBT). Other waste metal sources such as electronic and oil refinery waste can provide different metals for further use with commercial linkers or waste-based linkers.

2.3.2.1 MOFs based on waste polymers

2.3.2.1.1 PET-derived MOFs

2.3.2.1.1.1 MOFs based on a one-pot synthesis

PET is the major source of terephthalic acid (H_2BDC), from which many MOFs have been synthesized. PET is thermally depolymerized into H_2BDC molecules, which can be used with different metal ions to construct various MOFs for different applications. Stepwise synthesis by converting PET first into H_2BDC followed by MOF synthesis has been practiced, but several steps for isolating or purifying the products in addition to salt waste formation are detrimental to the H_2BDC yield, which in turn negatively impact the MOF yield. In this regard, several groups have executed the one-pot scenario for MOF synthesis from waste PET to produce divalent, trivalent and tetravalent MOFs (Table 2.2). The main idea for achieving such a one-pot reaction is to select certain metal ions that form metal clusters at the same hydrolysis temperature of PET

Table 2.2: Examples of waste-derived MOFs synthesized by a one-pot method from waste PET.

MOFs	Metal sources	Linker sources	Conditions	Ref.
MIL-53(Al)	$AlCl_3{\cdot}(H_2O)_6$	Waste PET	200 °C	[22]
MIL-47(V)	VCl_3	Waste PET	200 °C	[22]
MIL-47	VCl_3	Waste PET	200 °C for 96 h	[23]
MIL-53(Cr)	$CrCl_3{\cdot}6H_2O$	Waste PET	160 °C for 72 h	[23]
MIL-53(Al)	$Al(NO_3)_3{\cdot}9H_2O$	Waste PET	160 °C for 72 h	[23]
MIL-53(Ga)	$Ga(NO_3)_3{\cdot}9H_2O$	Waste PET	210 °C for 5 h	[23]
MIL-101(Cr)	$CrCl_3{\cdot}6H_2O$	Waste PET	220 °C for 8 h	[23]
Cr-MOF	$CrCl_3{\cdot}6H_2O$	Waste PET	210 °C for 8 h	[27]
hcp UiO-66(Zr)	$ZrCl_4$	Waste PET	160 °C for 24 h	[24]

at 200 °C. PET depolymerization into H_2BDC can then be conducted simultaneously with MOF synthesis. This approach was accomplished to synthesize some MIL series, including Al-, Cr- and V-MOFs. For instance, MIL-53(Al) and MIL-47(V) were obtained through a one-pot synthesis by depolymerizing PET into H_2BDC, followed by assembly into MOFs in the absence of toxic DMF [22]. Another series of terephthalate MOFs was also prepared, including MIL-53(Cr, Al, Ga), MIL-101(Cr) and MIL-47(V), using depolymerized PET in a one-pot reaction [23]. To accelerate the reactions, hydrofluoric acid (HF) or hydrochloric acid (HCl) was employed, having the dual functions of being an acid catalyst to rapidly hydrolyze PET into H_2BDC and as a modulator for MOF synthesis. In a recent report, Serre et al. have employed PET as a direct linker source without isolating the depolymerization step to synthesize **hcp** UiO-66 [24]. The reaction was conducted using zirconium tetrachloride ($ZrCl_4$) and the less toxic acetone instead of toxic DMF in the presence of formic acid as a modulator (Figure 2.12). The **hcp** UiO-66 was then employed in cyclohexane/benzene separation, exhibiting good performance. For a full green process, a waste-derived MOF was prepared and employed for insecticide removal [25]. This report applied *in situ* MOF growth for UiO-66(Zr) on a PET surface using optimized hydrolysis instead of depolymerization to produce highly stable PET@UiO-66 as a hybrid material for imidacloprid removal from aqueous system, which demonstrated high adsorption capacity and rate (Figure 2.13).

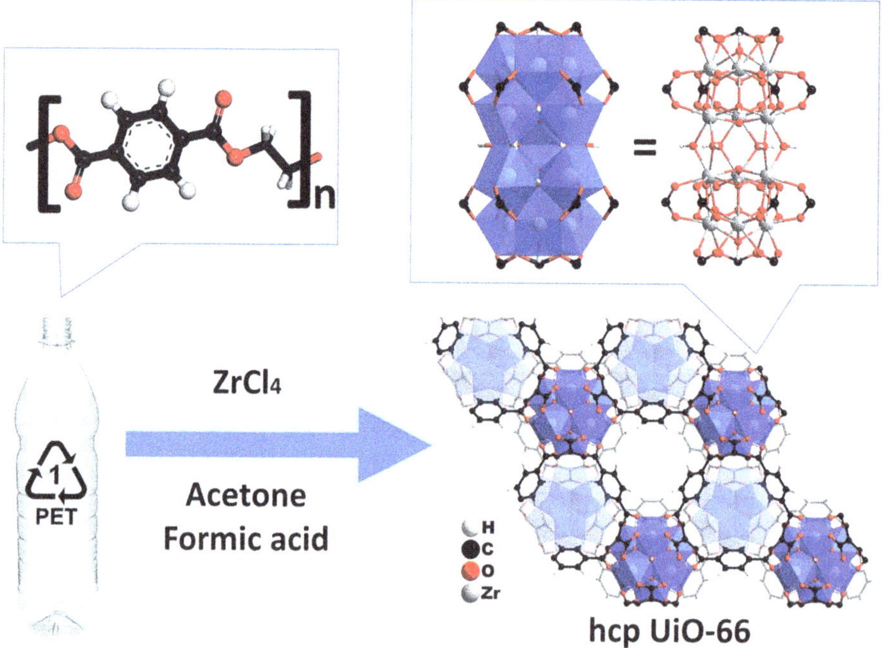

Figure 2.12: Direct synthesis of **hcp** UiO-66 from waste PET. Reproduced with permission from [24].

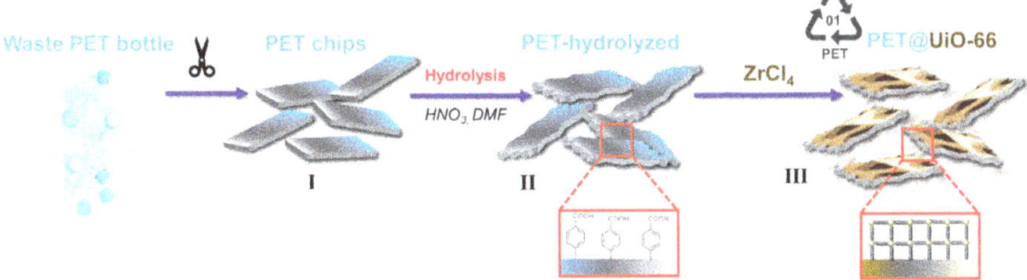

Figure 2.13: Synthesis of PET@UiO-66 derived from waste PET. Reproduced with permission from [25].

A recent report detailed the synthesis of new waste-derived nickel catalysts for dry methane reforming (DRM) [26]. The catalysts were synthesized by introducing nickel using incipient wetness impregnation into MIL-53(Al) derived from PET and commercial H_2BDC. After calcination and reduction, PET-derived $Ni^0–Al_2O_3$ and BDC-derived $Ni^0–Al_2O_3$ were obtained. Both catalysts displayed identical physicochemical properties, and PET-derived $Ni^0–Al_2O_3$ presented similar CH_4 and CO_2 conversions, selectivity and stability as BDC-derived $Ni^0–Al_2O_3$. This example also contributes to the feasibility of recycling discarded PET waste for integration into a circular economy approach toward different environmental and energy applications.

2.3.2.1.1.2 MOFs based on a two-step synthesis

In stepwise synthesis, depolymerized PET bottles were isolated and purified before reacting with commercial metal ion sources to produce different MOFs (Table 2.3). The purified H_2BDC reacted with zinc acetate dihydrate and copper nitrate trihydrate to synthesize MOF-5 and Cu terephthalate [28]. In another report, Cu-BDC was synthesized and employed as an adsorbent to remove methylene blue (MB) from an aqueous system [29]. Other reports have synthesized a series of divalent (Cu, Zn, Fe, Co, Ni and Sn), trivalent (Tb) and tetravalent (Zr) MOFs for various applications [30–36] (Figure 2.14). Notably, waste-derived MOF-5, synthesized from waste PET and zinc nitrate hexahydrate, was carbonized into highly porous N-doped mesoporous carbon (N-MC) [37]. In addition to the high surface area, N-MC showed high specific capacitance, thus ensuring the auspicious use of waste PET in future energy materials.

2.3.2.1.2 PLA-derived MOFs

To synthesize cost-effective LA-derived MOFs, the employment of recycled PLA as a linker source is a viable approach. In this regard, three homochiral PLA-derived MOFs were successfully prepared, including ZnBLD, MOF 1201 and MOF 1203 [38] (Figure 2.15). The reactions were carried out by *in situ* depolymerization of polylactic acid with simultaneous MOF synthesis in a one-pot reaction.

Table 2.3: Examples of waste-derived MOFs synthesized by a two-step approach from PET-derived BDC.

MOFs	Metal sources	Linker sources	Conditions	Ref.
MOF-5	$Zn(CH_3COO)_2·2H_2O$	PET-derived BDC	RT for 2.5 h	[28]
Cu-BDC	$Cu(NO_3)_2·3H_2O$	PET-derived BDC	110 °C for 36 h	[28]
Cu-BDC	$Cu(NO_3)_2·3H_2O$	PET-derived BDC	100 °C for 24 h	[29]
Cu-BDC	$Cu(NO_3)_2$	PET-derived BDC	RT for 48 h	[30]
$[Fe_2(BDC)_2(DABCO)]$	$FeCl_3$	PET-derived BDC/DABCO	100 °C for 48 h	[31]
$[Co_2(BDC)_2(DABCO)]$	$Co(NO_3)_2·6H_2O$	PET-derived BDC/DABCO	120 °C for 48 h	[31]
$[Ni_2(BDC)_2(DABCO)]$	$Ni(NO_3)_2·6H_2O$	PET-derived BDC/DABCO	110 °C for 24 h	[31]
$[Cu_2(BDC)_2(DABCO)]$	$Cu(NO_3)_2·3H_2O$	PET-derived BDC/DABCO	120 °C for 48 h	[31]
$[Zn_2(BDC)_2(DABCO)]$	$Zn(NO_3)_2·6H_2O$	PET-derived BDC/DABCO	120 °C for 48 h	[31]
Tb-BDC	$Tb(NO_3)_3·6H_2O$	PET-derived BDC	N/A	[32]
UiO-66(Zr)	$ZrCl_4$	PET-derived BDC	120 °C for 4 h	[33]
UiO-66(Zr)	$ZrCl_4$	PET-derived BDC	120 °C for 4 h	[35]
UiO-66(Zr)	$ZrO(NO_3)_2$	PET-derived BDC	150 °C for 4 h	[36]
Sn(II)-MOF	$SnCl_2·2H_2O$	PET-derived BDC	170 °C for 24 h	[34]
MOF-5	$Zn(NO_3)_2·6H_2O$	PET-derived BDC	100 °C for 24	[37]

BDC = benzene-1,4-dicarboxylate.
DABCO = 1,4-diazabicyclo[2.2.2]octane.
PET = polyethylene terephthalate.

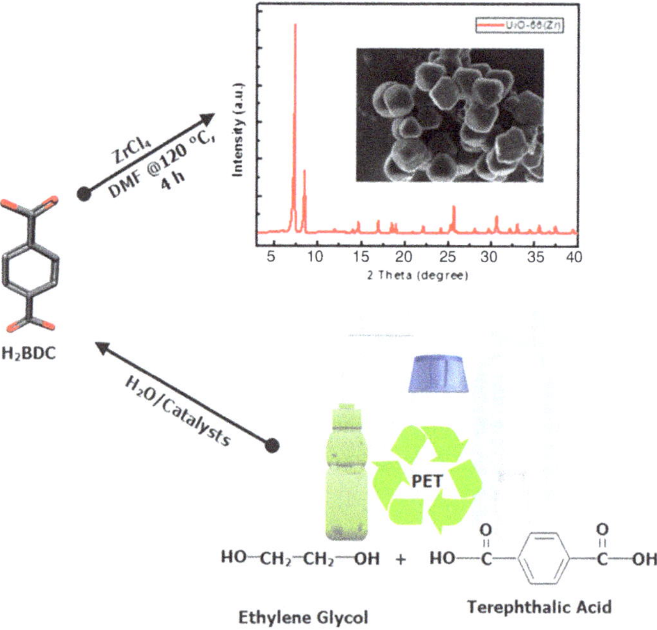

Figure 2.14: Synthesis of UiO-66 from colored waste PET. Reproduced with permission from [35].

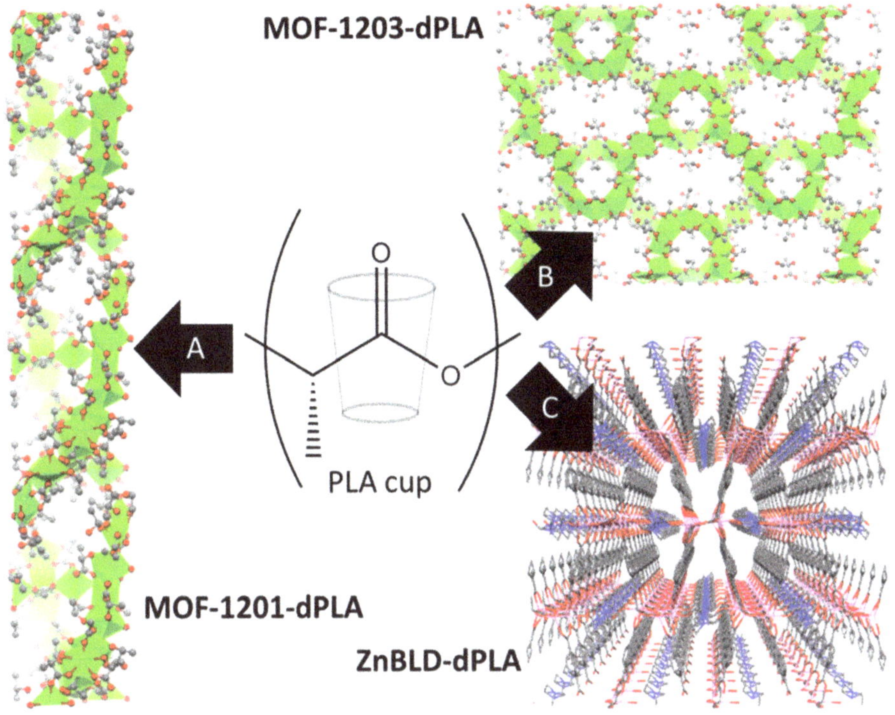

Figure 2.15: Synthesis of waste homochiral PLA-derived MOFs. Reproduced from [38].

2.3.2.2 MOFs based on waste metal sources

As described above, the main organic building unit of MOFs is the organic linkers, which can be derived from the thermal depolymerization of waste polymers such as PET or PLA. The second main building unit of MOF synthesis is metal precursors, which can also be derived from waste sources. In this regard, regenerating valuable metals from spent containments is a feasible method for metal recovery (Table 2.4).

Table 2.4: Examples of waste-derived MOFs based on waste metal sources.

MOFs	Metal sources	Linker sources	Conditions	Ref.
MOF-5	Alkaline battery waste	H_2BDC	RT for 2 h	[39]
V-BDC and V-NDC	Oil refinery waste (carbon black waste)	H_2BDC and H_2NDC	200 °C for 15 h.	[40]
MIL-53(Al)	Aluminum foil + Coca-Cola cans	H_2BDC	220 °C for 24 h	[41]
MIL-96(Al)	Aluminum foil + Coca-Cola cans	Trimesic acid	220 °C for 24 h	[41]
MIL-53(Al)	High-alumina fly ash	H_2BDC	220 °C for 72 h	[42]

H_2BDC = 1,4-benzenedicarboxylic acid.
H_2NDC = 1,4-naphthalenedicarboxylic acid.

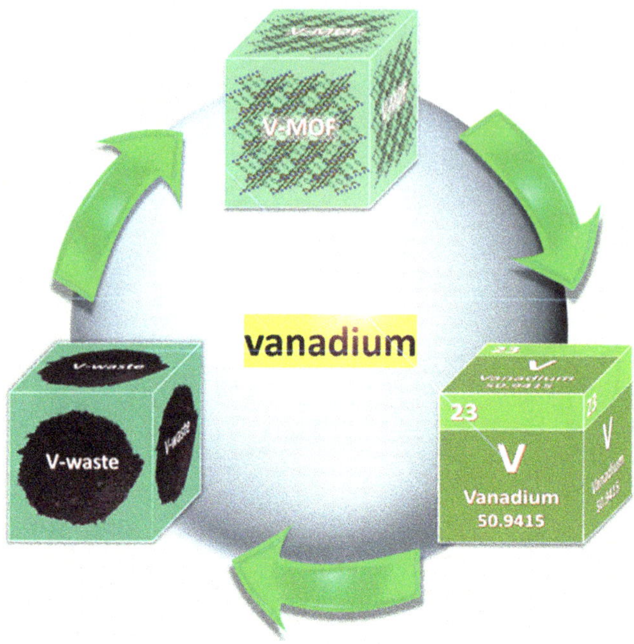

Figure 2.16: Synthesis of V-MOFs from waste V-waste. Reproduced with permission from [40].

For instance, alkaline battery waste was utilized as a source for zinc ions, which were subsequently reacted with H_2BDC into waste-derived MOF-5 [39]. In another report, it was found that carbon black waste (oil refinery waste) containing a high concentration of vanadium can be chemically leached and mixed with H_2BDC and 1,4-naphthalenedicarboxylic acid (H_2NDC) to produce waste-derived vanadium MOFs (Figure 2.16) [40]. Additionally, a highly pure aluminum foil was employed as a waste metal source for synthesizing supported and nonsupported MIL-53(Al). Moreover, terephthalic and trimesic acids were combined with waste Coca-Cola cans to produce nonsupported MIL-53(Al) and MIL-96(Al) MOFs, respectively [41]. Finally, as a sustainable aluminum source, high-alumina fly ash (HAFA) was assembled with commercial terephthalic acid to produce eco-MIL-53(Al), which was employed for methylene blue removal [42].

2.3.2.3 MOFs based on waste organic linkers and waste metal sources

The "waste-to-MOF" approach depends on using only waste polymers for providing organic linkers to be reacted with commercial metal precursors or vice versa. However, a more feasible and economic process can be realized when organic linkers and metal precursors are provided from waste sources, allowing for large-scale produc-

Table 2.5: Examples of waste-derived MOFs by combining waste organic linkers and waste metal sources.

MOFs	Metal sources	Ligand sources	Conditions	Ref.
MIL-53(Al)	Waste lithium-ion-battery (LiB)	Waste PET	70 or 90 °C for 12 h	[43]
MIL-53(Al)	Waste aluminum foil/can	Waste PET	220 °C for 72 h	[44]
Ca-MOF	Waste chicken eggshells	Waste PET	210 °C for 15 h	[45]
Ni-BDC	Electroplating sludge (EPS)	Waste PET	140 °C for 24 h	[46]

tion (Table 2.5). For example, PET waste plastic bottles were depolymerized to produce H_2BDC, which was assembled with aluminum ions derived from a lithium-ion battery (LiB) waste to produce waste-derived MIL-53(Al) [43]. This MOF was also prepared using PET water/beverage bottles and household aluminum foil, requiring one-third of the cost of the same MOF when prepared using commercial precursors from Sigma Aldrich [44]. This methodology demonstrates the feasibility of linking particular waste streams to the sustainable synthesis of inexpensive, high-efficiency materials to resolve socioeconomic problems. In a different report, waste-derived Cu-BDC was successfully prepared using two waste streams for Cu ions and H_2BDC linkers. The copper ions were provided by printed circuit board (PCB) wastewater and the H_2BDC linkers were supplied by alkali reduction wastewater (AR). By comparing waste-derived Cu-BDC to the commercially synthesized Cu-BDC, the waste-derived MOF exhibited higher CO_2 uptake values due to its interlamellar voids contributing to the adsorption properties. Similarly, waste chicken eggshells were employed as a benign and green precursor for calcium carbonate, which was then coupled with PET plastic bottles to afford a waste-derived calcium-based MOF using hydrothermal or mechanochemical methods [45]. The waste chicken eggshells were also used to prepare squarate- and fumarate-based MOFs (Figure 2.17). Another report provides waste-derived MOF based on waste PET and waste electroplating sludge (EPS) as a nickel source [46].

2.3.2.4 MOFs based on green solvents

To fully synthesize MOFs from green sources, a proper selection for the solvent is essential due to its role in establishing the final frameworks. As is widely known, DMF is the most predominant solvent used in MOF synthesis; however, DMF is toxic, and consuming large amounts can threaten the environment and preclude its use in MOF commercialization. In this regard, the replacement of DMF with lower boiling solvents, methanol, ethanol, acetone, acetic acid, among others, is ideal. The most feasible approach for making this transition is to use safer, sustainable and cost-effective solvents derived from readily available resources. For instance, a bio-derived solvent called Cyrene (dihydrolevoglucosenone) derived from waste cellulose was deployed as a green solvent to synthesize a series of MOFs [47]. Although DMF-based MOFs

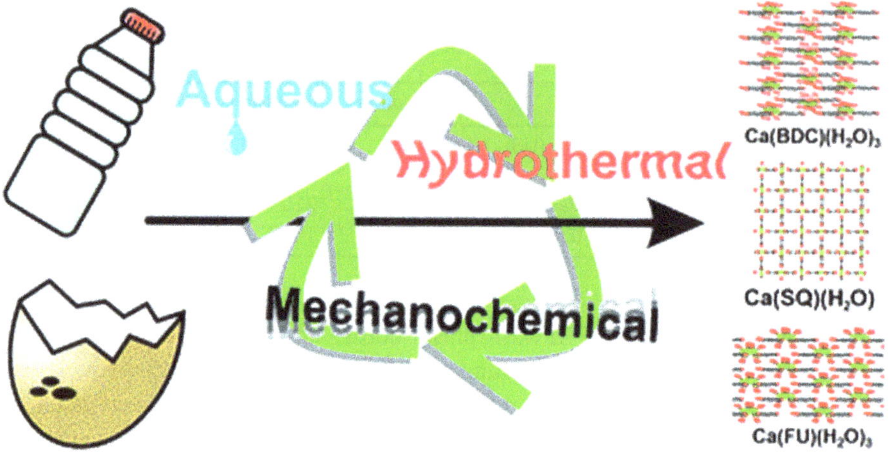

Figure 2.17: Synthesis of waste-derived Ca-based MOFs. Reproduced with permission from [45].

exhibited higher surface area than Cyrene-based MOFs, Cyrene represents a promising easy-to-access solvent to make MOF synthesis sustainable. Although investigating MOF synthesis from waste solid materials has garnered much interest in organic linkers or metal precursors, green solvents in MOF synthesis are rarely reported. A recent report investigated a bio-derived solvent called STEPOSOL MET-10U (N,N-dimethyl-9-decenamide) to synthesize diverse MOFs [48]. This solvent is produced from olefin metathesis using renewable feedstocks (plant oils).

2.3.3 Effect of reaction conditions on MOF synthesis

2.3.3.1 Effect of solvent

As an influential factor that affects MOF synthesis, the solvent type, quality and basicity are determinants for the structure, properties and functions of the final MOF. This is because solvents, as direct or indirect crystal growth media, impact the coordination behaviors of metals and linkers, ensuring their role as structure-directing agents. This influence appears when different solvent systems may coordinate to metal centers or behave like guest molecules in the final framework. The solvent's boiling point is important as it dictates when basic molecules (amines) are liberated in the reaction system to deprotonate the carboxylate linkers for initiating the construction of the final architecture. This ensures that the extent of deprotonation of carboxylate linkers is governed by two factors: (i) proper choice of solvents, such as DMF, DMA and DEF and (ii) the basicity of solvent. Indeed, differing degrees of organic linker deprotonations in the presence of DMF and DEF have led to structural changes in the final frame-

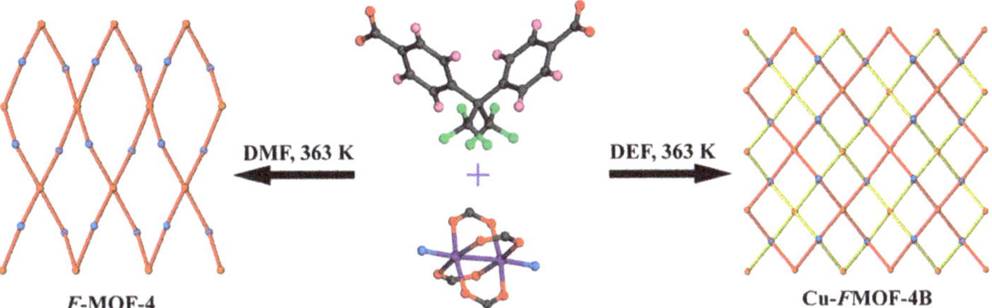

Figure 2.18: Solvent-induced structural changes in F-MOF-4 and Cu-F-MOF-4B frameworks. Reproduced with permission from [49].

works of F-MOF-4 and Cu-F-MOF-4B, which impacts their gas adsorption properties (Figure 2.18) [49].

Furthermore, the coordination sphere in the synthesized MOFs undergoes changes when different solvents (DMF, DEF, and DMA) are applied to the reaction between biphenyl tricarboxylic acid (H_3BPT) and $Cd(NO_3)_2{\cdot}4H_2O$ (Figure 2.19) [50]. The obtained coordination architectures demonstrate that DMF and DMA are incorporated in the structure lattice, where they behave as coordinative capping ligands.

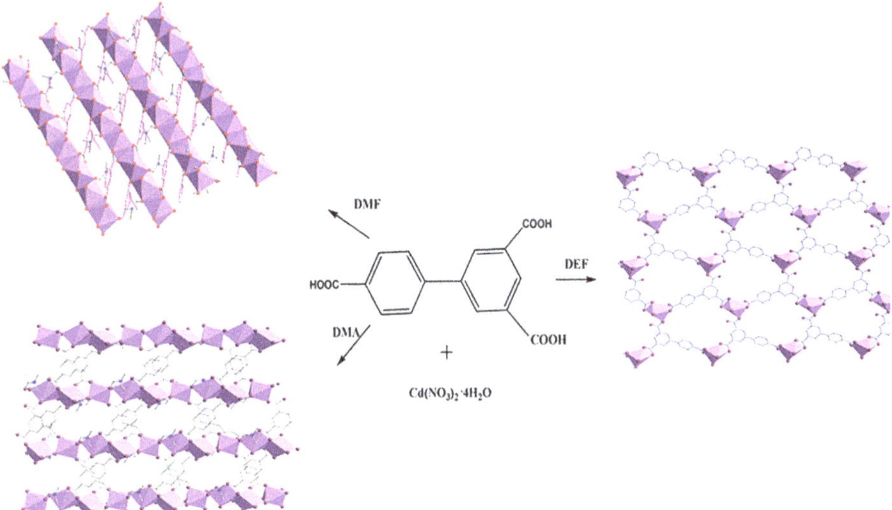

Figure 2.19: Solvent-induced structural changes in Cd-based MOFs. Reproduced with permission from [50].

When selecting solvents, various factors significantly impact the coordination environment and structure of MOFs. For example, using a mixed solvent strategy affects the coordination modes of the employed organic linkers. Using DMF/EtOH led to a 3D pillar-layered framework, and a coordination polymer with a 1D zig-zag chain structure was obtained when anhydrous methanol alone was employed [51]. This example highlights the role of solvent polarity in changing a 1D chain structure to a 3D framework due to induced changes in the linker coordination modes. The dimensionality of the prepared networks can also be attributed to the relative coordinative ability of solvents to metal centers, like the high affinity of water (unlike DMF) to coordinate with Mg that have coordinatively unsaturated sites [52]. From another perspective, different MOF structures can be formed when different solvent volumetric ratios are employed [53]. Such structural diversity is obtained as the solvent system governs dissimilar coordination environments around the central metal ion leading to different MOF products. For example, the structural dimensionality of a series of MOFs was tuned when different H_2O/EtOH ratios were utilized [54]. Indeed, a 0D structure was transformed into a 3D architecture by increasing the EtOH ratio from 0 to 100%. Moreover, the steric hindrance held by solvents due to their molecular sizes can result in different MOF structures [55]. The larger the steric hindrance of a solvent, the higher the difficulty in coordination with metals resulting in different dimensionalities in MOF structures and impairing the crystal growth process. Since the accessible pore volumes of solvents vary due to solvent size, the prepared MOFs demonstrate different pore volumes, implying a size-controlled strategy [56]. In addition to the template impact of solvents behaving as structure-directing agents [57], solvation and desolvation processes can cause crystal-to-crystal transformation resulting in new MOFs unattainable by conventional methods [58]. Lastly, solvent quality/purity is essential when carboxylate linkers are used in MOF synthesis [59]. This is because solvents, in some cases, hydrolyze in air, resulting in counterions that adopt a template role in MOF formation. For example, DMF and DEF readily decompose to form diethyl ammonium cation (NH_2Et^{2+}) and dimethylammonium cation (NH_2Me^{2+}), respectively. This was confirmed when reactions were charged with (NH_2Et_2)Cl and (NH_2Me_2)Cl in addition to fresh DMF and DEF solvents, and the same results were obtained.

2.3.3.2 Effect of pH

As demonstrated, the pH (acidity or basicity) of the reaction governs the crystal growth of inorganic-organic hybrids. Based on the acid-base approach, MOF synthesis is influenced by pH values that favor the deprotonation of carboxylate linkers to metal centers or produce hydroxy linkers. Accordingly, tuning the pH value in each MOF reaction is critical for controlling the degree of deprotonation of the used linkers, which determines their coordination modes/preferences or conformations. Indeed, this can result in MOFs with different compositions and dimensionalities. In a report, higher pH values led to higher connectivity of organic linkers that led to interpenetrating or complicated networks, whereas lower pH values resulted in a simple un-interpenetrated 3D framework [60]. Moreover, the degree of deprotonation significantly impacts the color of synthesized MOFs. For this, by adjusting the pH value, partially (low pH) or fully (high pH) deprotonated organic linkers demonstrate different connectivity or coordination modes, which gives rise to different colors of the prepared crystals [61]. Another interesting report revealed that base-induced heteronucleation of a given MOF led to different phases being obtained [62]. By using different amounts of NaOH and keeping the other reaction conditions constant, different MOF phases were produced. This is because NaOH has a dual function; it works as a basic medium to deprotonate the organic linkers and as a coordinating agent. The coordination ability of NaOH during the crystallization process was demonstrated due to the presence of OH ions in the generated phases. Finally, tuning MOF synthesis using pH control in the presence of the same building units but with other reaction conditions can lead to supramolecular isomerism and give rise to dissimilar structures [63].

2.3.3.3 Effect of temperature

As demonstrated in MOF synthesis since this class of materials' inception, the solvothermal/hydrothermal method is the best approach to realizing large single crystals with high quality. MOF precursors undergo high solubility when subjected to a temperature over 100 °C and autogenous pressure over 1 atm in sealed autoclaves or vials. This method is ideal for sparingly soluble precursors in the solvent system at room temperature but unsuitable for sensitive or less thermally stable reactants. The advantages of the solvothermal/hydrothermal method over other synthetic methods for MOF preparation make it the first choice when new MOFs are attempted to be synthesized for the first time. Due to the harsh reaction conditions in the solvothermal/hydrothermal synthesis, the obtained MOFs are likely to exhibit high thermal stability over those obtained from room-temperature synthesis and possess open frameworks with empty channels, but this is not always the case [64]. This is because high temperature mainly produces thermodynamically favored products, while low temperature mostly produces kinetically favored products. Typically, high temperatures are favored to obtain high dimensional frameworks with high density and extended networks. The impact of controlling temperature was investigated on succinate-based MOFs, in which two supramolecular isomers (monoclinic and triclinic) were obtained at different temperature settings [65]. This example implies that tuning the supramolecular isomerism in coordination polymers can be accomplished by changing the reaction temperature. The temperature also impacts the size, shape and morphology of these MOFs, yielding small primary nanosheets at 120 °C and layered microrods at 160 °C [66]. As previously stated, when the temperature is considered the only independent variable, high temperature leads to easy deprotonation of the organic linkers, making them multi-coordinated to metal centers. However, temperature-dependent high-dimensional MOFs are not always formed at high temperatures, as demonstrated in examples of Ni- or Cd-based MOFs [67, 68].

2.3.3.4 Effect of molar ratio of reactants

From a topological point of view, the crystal structure of a given MOF is dependent on the molar ratio of reactants. In practice, the molar ratio is screened between organic linkers and metal precursors, starting with identical ratios followed by increasing metal precursors over organic linkers or vice versa. Metal/linker combinations are screened using different ratios under otherwise identical settings, and the final products are confirmed via thorough characterization. When the best metal/linker ratio is discovered, MOF products can be scaled-up to prepare enough amount for characterizations and applications. By using one metal/linker ratio and screening other reaction parameters, several MOF structures can be produced. This ensures that the metal/linker ratio significantly impacts the final structures. In pillared MOFs, when using a colinker in the reaction system, the co-inker/metal ratio can also be screened [69].

2.3.3.5 Effect of other parameters

MOF synthesis is governed by a list of parameters that should be screened to optimally synthesize structures suitable to be identified by SCXRD. In addition to the above-discussed parameters, time is vital to achieving stable single crystals. Kinetically-favored MOF products can be obtained in a shorter time than thermodynamically-favored MOFs produced at longer times. Reaction time can lead to structural variation by transformation from one structure to another during the course of a reaction. When reaction time exceeds the optimal duration for producing MOF crystals, irregular or unstable frameworks may be obtained. This implies that the attempt to conduct a given MOF reaction can use similar reaction times as were discovered during the first trials; otherwise, different reaction times can be applied until obtaining MOF crystals. The concentration of reactants should be screened by increasing/decreasing the solvent amount or increasing/decreasing reactants with leaving enough space in the reaction reactor over the reaction mixture. Also, modulators are important to produce single crystals rather than gel-like or powder materials. It was demonstrated that modulators should also be screened while conducting MOF reactions to identify which modulator can obtain single crystals. The modulators include acetic acid, hydrochloric acid, benzoic acid, nitric acid, trifluoroacetic acid, hydrofluoric acid and oxalic acid, among others, and they function by competing with the main organic linkers to coordinate with metal ions and to slow down the crystallization process, thus allowing for defect correction (Figure 2.20). Moreover, when the volume/amount of modulators exceeds the proper limit to obtain single crystals, the dissolution of these crystals can happen, which ensures the importance of screening them properly. When it comes to using metal precursors, the most commonly used ones are nitrate

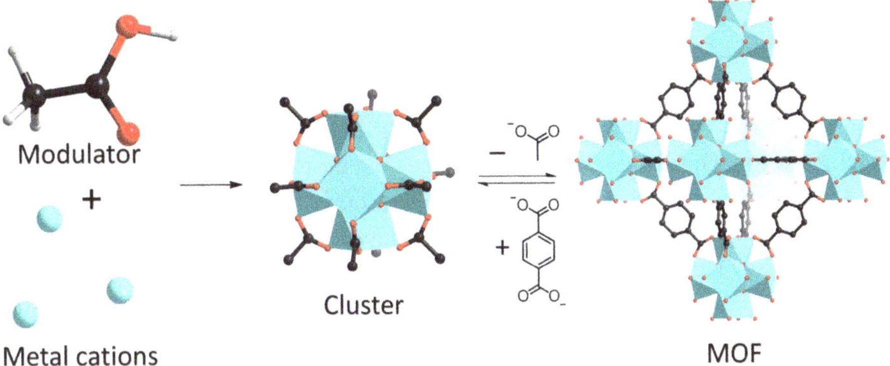

Figure 2.20: Modulating the synthesis of MOFs. Reproduced with permission from [71].

salts, which are highly soluble in organic solvents and facilitate the reaction assembly in short times. However, the explosive nature of nitrates precludes their use on an industrial scale requiring replacing them with benign precursors like metal acetate salts. In another regard, instead of using protonated organic carboxylate linkers, the employment of organic carboxylate salts as anion linkers has been demonstrated as a reliable, inexpensive, rapid and environmentally friendly process toward the MOF synthesis [70]. These reactions can be performed in water and at room temperature, which are critical parameters required in industrialization. This approach facilitates the solubility of organic linkers, their deprotonation and subsequent crystallization. Although the availability of organic linkers salts limits this approach, this can be overcome using protonated organic linkers in the presence of a base.

2.4 Conclusion

The MOF field has garnered incredible attention in the past two decades due to the versatile utility of the produced framework structures. According to Yaghi et al., MOFs are designed and discovered based on three approaches. The first one relies on designing and synthesizing new organic linkers and screening their ability to produce different MOFs by assembling them with different metals. In this approach, several trial-and-error reactions should be conducted until achieving the optimized condition. Yaghi stated that at least 50 reactions with different conditions should be attempted until reaching the best conditions at which MOF products can be crystallized. Such conditions can be inspired by already reported MOF procedures. In the second approach, several researchers adopt established MOF systems for illustrating chemical principles or conducting reactions inside them to study their correlated properties. This approach is faster than the first one as the crystallization conditions of employed MOFs are figured out by others, making MOFs reproducible. The third one, or the engineering approach, relies on the application-guided design of MOFs. For instance, when researchers address specific applications, they can work backward by establishing MOFs with requirements for such applications, such as using specific functional groups. This approach is likely to produce new MOFs or even get already made MOFs. The three approaches are productive and appealing based on the researchers' needs.

Following the above-mentioned approaches, new organic linkers should be introduced to the MOF field. For instance, N-oxide-based linkers are rarely reported in MOFs, and they can be obtained by oxidation of many pyridine-based linkers to be explored in different applications. In addition, aldehyde-based linkers are rarely presented and can produce MOFs amenable to several condensation reactions or linking framework units for extended porous networks. Typically, quaternary ammonium-based MOFs are rarely addressed, which can be very beneficial in exchange reactions. Furthermore, soluble UiO-Zr MOFs can be attained by reacting with ammonium hydroxide, which can be employed for efficient homogeneous catalysis. Other linkers based on anthracene, quinoline and carbazole are promising categories for developing versatile MOFs. Building new linkers containing different rings with specific functionality can develop MOFs with multifunctional properties, such as pyridine, triazole, tetrazole, N-oxide pyridine and others in one ligand system. Focusing on amino Zr-MOFs, a vast array of post-synthetic modifications for targeted applications in terms of sensing, separations, catalysis, etc. As recently reported, examining unusual reactions using modified MOFs is mandatory, such as dark catalysis.

For practical applications, room temperature synthesis is the main target of realizing high-quality MOFs with quantitative yields. Among different organic linkers, 2,6-naphthalene dicarboxylic acid (H_2NDC) has been employed to produce several MOF structures. However, waste polyethylene naphthalate (PEN) has never been investigated for sustainable MOF synthesis by depolymerization into H_2NDC. In addition, exploring bio-solvents with similar formamide moieties or using alcohols has the potential to develop sustainable MOFs with similar properties to those based on conventional precursors.

The MOF library is rich with simple structures; however, moving to a higher structural complexity level requires tedious adjustment of reaction conditions and rational design to develop more complex structures, such as in the case of using mixed-metal or mixed-ligand strategies. The case is the same when highly connected networks are addressed, such as those >12. Moreover, surface-mounted metal-organic frameworks (SURMOFs) developed as thin-film layered structures over solid substrates require laborious efforts to convert the conventional MOF synthesis conditions to those suitable for growing multiple layers with high homogeneity, controlled thickness, porosity and crystallinity. In SURMOFs, the developed film on a substrate surface is only the acceptable form to develop specific applications that cannot be accessed by powered MOF form. This encourages researchers to focus on these more complex materials by attempting *de novo* synthesis of SURMOFs without the dependence of their MOF counterparts.

Bibliography

[1] Chuhadiya S et al. Metal organic frameworks as hybrid porous materials for energy storage and conversion devices: A review. Coord Chem Rev. 2021;446:214115.

[2] Zeeshan M, Shahid M. State of the art developments and prospects of metal-organic frameworks for energy applications. Dalton Trans. 2022;51:1675–723.

[3] Gropp C et al. Standard practices of reticular chemistry. ACS Cent Sci. 2020:1255–73.

[4] Lyu H et al. Digital reticular chemistry. Chem. 2020;6:2219–41.

[5] Seetharaj R et al. Dependence of solvents, pH, molar ratio and temperature in tuning metal organic framework architecture. Arab J Chem. 2019;12:295–315.

[6] Nagarkar SS, Desai AV, Ghosh SK. Stimulus-responsive metal-organic frameworks. Chem Asian J. 2014;9:2358–76.

[7] El-Sayed ESM, Yuan D. Waste to MOFs: sustainable linker, metal, and solvent sources for value-added MOF synthesis and applications. Green Chem. 2020;22:4082–104.

[8] Chen Z et al. Reticular chemistry for highly porous metal-organic frameworks: The chemistry and applications. Acc Chem Res. 2022;55:579–91.

[9] Kalmutzki MJ, Hanikel N, Yaghi OM. Secondary building units as the turning point in the development of the reticular chemistry of MOFs. Sci Adv. 2018;4:eaat9180.

[10] Ghasempour H et al. Metal-organic frameworks based on multicarboxylate linkers. Coord Chem Rev. 2021;426:213542.

[11] Ali Akbar Razavi S, Morsali A. Linker functionalized metal-organic frameworks. Coord Chem Rev. 2019;399:213023.

[12] Yaghi OM, Kalmutzki MJ, Diercks CS. Building units of MOFs. In: Introduction to Reticular Chemistry. Wiley; 2019. p. 57–81.

[13] Ameloot R et al. Direct patterning of oriented metal-organic framework crystals via control over crystallization kinetics in clear precursor solutions. Adv Mater. 2010;22:2685–8.

[14] Zhang H et al. Assembly of a rod indium–organic framework with fluorescence properties for selective sensing of Cu^{2+}, Fe^{3+} and nitroaromatics in water. CrystEngComm. 2022;24:667–73.

[15] Antonio AM, Rosenthal J, Bloch ED. Electrochemically mediated syntheses of titanium(III)-based metal-organic frameworks. J Am Chem Soc. 2019;141:11383–7.

[16] Xiao JD et al. Rapid synthesis of nanoscale terbium-based metal-organic frameworks by a combined ultrasound-vapour phase diffusion method for highly selective sensing of picric acid. J Mater Chem A. 2013;1:8745–52.

[17] Isaeva VI, Kustov LM. Microwave activation as an alternative production of metal-organic frameworks. Russ Chem Bull. 2016;65:2103–14.

[18] Thi Dang Y et al. Microwave-assisted synthesis of nano Hf- and Zr-based metal-organic frameworks for enhancement of curcumin adsorption. Microporous Mesoporous Mater. 2020;298:110064.

[19] Cao HY et al. Ionothermal syntheses, crystal structures and luminescence of three three-dimensional lanthanide-1,4-benzenedicarboxylate frameworks. Inorg Chim Acta. 2014;414:226–33.

[20] Waitschat S, Wharmby MT, Stock N. Flow-synthesis of carboxylate and phosphonate based metal-organic frameworks under non-solvothermal reaction conditions. Dalton Trans. 2015;44:11235–40.

[21] Pilloni M et al. Liquid-assisted mechanochemical synthesis of an iron carboxylate metal organic framework and its evaluation in diesel fuel desulfurization. Microporous Mesoporous Mater. 2015;213:14–21.

[22] Deleu WPR et al. Waste PET (bottles) as a resource or substrate for MOF synthesis. J Mater Chem A. 2016;4:9519–25.

[23] Lo SH et al. Waste polyethylene terephthalate (PET) materials as sustainable precursors for the synthesis of nanoporous MOFs, MIL-47, MIL-53(Cr, Al, Ga) and MIL-101(Cr). Dalton Trans. 2016;45:9565–73.
[24] Zhou L et al. Direct synthesis of robust hcp UiO-66(Zr) MOF using poly(ethylene terephthalate) waste as ligand source. Microporous Mesoporous Mater. 2019;290:109674.
[25] Semyonov O et al. Smart recycling of PET to sorbents for insecticides through in situ MOF growth. Appl Mater Today. 2021;22:100910.
[26] Karam L et al. PET waste as organic linker source for the sustainable preparation of MOF-derived methane dry reforming catalysts. Mater Adv. 2021;2:2750–8.
[27] Ren J et al. Green synthesis of chromium-based metal-organic framework (Cr-MOF) from waste polyethylene terephthalate (PET) bottles for hydrogen storage applications. Int J Hydrog Energy. 2016;41:18141–6.
[28] Manju et al. Post consumer PET waste as potential feedstock for metal organic frameworks. Mater Lett. 2013;106:390–2.
[29] Doan VD et al. Utilization of waste plastic pet bottles to prepare copper-1, 4-benzenedicarboxylate metal-organic framework for methylene blue removal. Sep Sci Technol. 2020;55:444–55.
[30] Rahmani A, Rahmani H, Zonouzi A. Cu(BDC) as a catalyst for rapid reduction of methyl orange: room temperature synthesis using recycled terephthalic acid. Chem Pap. 2018;72:449–55.
[31] Senthil Raja D et al. Synthesis of mixed ligand and pillared paddlewheel MOFs using waste polyethylene terephthalate material as sustainable ligand source. Microporous Mesoporous Mater. 2016;231:186–91.
[32] Zhang F et al. Waste PET as a reactant for lanthanide MOF synthesis and application in sensing of picric acid. Polymers. 2015;11:2019.
[33] Dyosiba X et al. Preparation of value-added metal-organic frameworks (MOFs) using waste PET bottles as source of acid linker. Sustain Mater Technol. 2016;10:10–3.
[34] Ghosh A, Das G. Facile synthesis of Sn(II)-MOF using waste PET bottles as an organic precursor and its derivative SnO_2 NPs: Role of surface charge reversal in adsorption of toxic ions. J Environ Chem Eng. 2021;9:105288.
[35] Dyosiba X et al. Feasibility of varied polyethylene terephthalate wastes as a linker source in metal-organic framework UiO-66(Zr) synthesis. Ind Eng Chem Res. 2019;58:17010–6.
[36] Singh S et al. Nanocuboidal-shaped zirconium based metal organic framework for the enhanced adsorptive removal of nonsteroidal anti-inflammatory drug, ketorolac tromethamine, from aqueous phase. New J Chem. 2018;42:1921–30.
[37] Ubaidullah M et al. Fabrication of highly porous N-doped mesoporous carbon using waste polyethylene terephthalate bottle-based MOF-5 for high performance supercapacitor. J Energy Stor. 2021;33:102125.
[38] Slater B et al. Upcycling a plastic cup: one-pot synthesis of lactate containing metal organic frameworks from polylactic acid. Chem Commun. 2019;55:7319–22.
[39] Vellingiri K et al. The utilization of zinc recovered from alkaline battery waste as metal precursor in the synthesis of metal-organic framework. J Clean Prod. 2018;199:995–1006.
[40] Zhan G et al. Effective recovery of vanadium from oil refinery waste into vanadium-based metal-organic frameworks. Environ Sci Technol. 2018;52:3008–15.
[41] Joshi JN et al. Household aluminum products as insoluble precursors for directed growth of metal-organic frameworks. Cryst Growth Des. 2019;19:5097–104.
[42] Yuan N et al. High-alumina fly ash as sustainable aluminum sources for the in situ preparation of Al-based eco-MOFs. Colloids Surf A, Physicochem Eng Asp. 2022;640:128421.
[43] Lagae-Capelle E et al. Combining organic and inorganic wastes to form metal-organic frameworks. Materials. 2020;13:441.

[44] Panda D et al. Lab cooked MOF for CO_2 capture: A sustainable solution to waste management. J Chem Educ. 2020;97:1101–8.
[45] Crickmore TS et al. Toward sustainable syntheses of Ca-based MOFs. Chem Commun. 2021;57:10592–5.
[46] Song K et al. Efficient upcycling electroplating sludge and waste PET into Ni-MOF nanocrystals for the effective photoreduction of CO_2. Environ Sci Nano. 2021;8:390–8.
[47] Sherwood J et al. Dihydrolevoglucosenone (Cyrene) as a bio-based alternative for dipolar aprotic solvents. Chem Commun. 2014;50:9650–2.
[48] Marino P et al. A step toward change: A green alternative for the synthesis of metal-organic frameworks. ACS Sustain Chem Eng. 2021;9:16356–62.
[49] Pachfule P et al. Solvothermal synthesis, structure, and properties of metal organic framework isomers derived from a partially fluorinated link. Cryst Growth Des. 2011;11:1215–22.
[50] Li LN et al. Solvent-dependent formation of Cd(II) coordination polymers based on a C2-symmetric tricarboxylate linker. Cryst Growth Des. 2012;12:4109–15.
[51] Cheng ML et al. Two coordinated-solvent directed zinc(II) coordination polymers with rare gra topological 3D framework and 1D zigzag chain. Inorg Chem Commun. 2011;14:300–3.
[52] Banerjee D et al. Synthesis and structural characterization of magnesium based coordination networks in different solvents. Cryst Growth Des. 2011;11:2572–9.
[53] Cui P et al. Two solvent-dependent zinc(II) supramolecular isomers: Rare kgd and lonsdaleite network topologies based on a tripodal flexible ligand. Cryst Growth Des. 2011;11:5182–7.
[54] Mazaj M et al. Control of the crystallization process and structure dimensionality of Mg–benzene–1,3,5-tricarboxylates by tuning solvent composition. Cryst Growth Des. 2013;13:3825–34.
[55] Dong BX, Gu XJ, Xu Q. Solvent effect on the construction of two microporous yttrium–organic frameworks with high thermostability viain situ ligand hydrolysis. Dalton Trans. 2010;39:5683–7.
[56] Zhou X et al. Solvents influence on sizes of channels in three fry topological Mn(ii)-MOFs based on metal–carboxylate chains: syntheses, structures and magnetic properties. CrystEngComm. 2013;15:8125–32.
[57] Ghosh SK, Kitagawa S. Solvent as structure directing agent for the synthesis of novel coordination frameworks using a tripodal flexible ligand. CrystEngComm. 2008;10:1739–42.
[58] Liu B et al. Two solvent-dependent zinc(II) supramolecular isomers: structure analysis, reversible and nonreversible crystal-to-crystal transformation, highly selective CO_2 gas adsorption, and photoluminescence behaviors. CrystEngComm. 2012;14:6246–51.
[59] Burrows AX et al. Solvent hydrolysis and templating effects in the synthesis of metal-organic frameworks. CrystEngComm. 2005;7:548–50.
[60] Yuan F et al. Effect of pH/metal ion on the structure of metal-organic frameworks based on novel bifunctionalized ligand 4′-carboxy-4,2′:6′,4″-terpyridine. CrystEngComm. 2013;15:1460–7.
[61] Luo L et al. pH-Dependent cobalt(II) frameworks with mixed 3,3′,5,5′-tetra(1H-imidazol-1-yl)-1,1′-biphenyl and 1,3,5-benzenetricarboxylate ligands: synthesis, structure and sorption property. CrystEngComm. 2013;15:9537–43.
[62] Wang HN et al. pH-Induced different crystalline behaviors in extended metal-organic frameworks based on the same reactants. Dalton Trans. 2013;42:6294–7.
[63] Han ML et al. Temperature and pH driven self-assembly of Zn(II) coordination polymers: crystal structures, supramolecular isomerism, and photoluminescence. CrystEngComm. 2014;16:1687–95.
[64] Bernini MC et al. The effect of hydrothermal and non-hydrothermal synthesis on the formation of holmium(III) succinate hydrate frameworks. Eur J Inorg Chem. 2007;2007:684–93.

[65] de Oliveira CAF et al. Effect of temperature on formation of two new lanthanide metal-organic frameworks: Synthesis, characterization and theoretical studies of Tm(III)-succinate. J Solid State Chem. 2013;197:7–13.
[66] Sarawade P, Tan H, Polshettiwar V. Shape- and morphology-controlled sustainable synthesis of Cu, Co, and In metal organic frameworks with high CO_2 capture capacity. ACS Sustain Chem Eng. 2013;1:66–74.
[67] Chen J et al. Polynuclear core-based nickel 1,4-cyclohexanedicarboxylate coordination polymers as temperature-dependent hydrothermal reaction products. Cryst Growth Des. 2006;6:664–8.
[68] Zhang J, Bu X. Temperature dependent charge distribution in three-dimensional homochiral cadmium camphorates. Chem Commun. 2008;2008:444–6.
[69] Wu YP et al. Stoichiometry of N-donor ligand mediated assembly in the Zn(II)-Hfipbb system: From a 2-fold interpenetrating pillared-network to unique (3,4)-connected isomeric nets. Cryst Growth Des. 2011;11:3850–7.
[70] Sánchez-Sánchez M et al. Synthesis of metal-organic frameworks in water at room temperature: salts as linker sources. Green Chem. 2015;17:1500–9.
[71] Yuan S et al. Stable metal-organic frameworks: Design, synthesis, and applications. Adv Mater. 2018;30:1704303.

A. P. Sarikas, G. S. Fanourgakis, and G. E. Froudakis

3 Metal-organic frameworks in the age of machine learning

3.1 Introduction

Inherent properties of MOFs such as their high porosity and surface area, make these materials suitable for applications involving adsorption. MOFs have been considered prominent candidates for gas storage and separation, with their performance being impressive in both processes [1–5].

Structures can be generated by combining different metal clusters and organic linkers, collectively known as building blocks. The chemical and structural diversity of MOFs triggered the experimental synthesis of new materials [6, 7]. This gave researchers the opportunity to explore and assess new candidates, in search of top-performing materials for the aforementioned applications. Notably, the structures of more than 100,000 experimentally synthesized MOFs have been deposited in the Cambridge Structural Database (CSD) [8]. At the same time, the available "search space" of MOFs has been further enlarged via in silico design. Computational design gave birth to a new family of candidates, known as hypothetical MOFs, namely MOFs whose synthesis takes place in silico. One of the first databases of hypothetical MOFs was constructed by [9], where 102 different building blocks were combined to generate 137,953 hypothetical structures. In contrast to this bottom-up synthesis, top-down approaches have also been developed. An example of such an approach is the topologically-based crystal constructor (ToBaCCo) [10], in which topological blueprints are used as templates for the construction of hypothetical MOFs. Recently, more than 100 trillions of hypothetical MOFs were assembled by means of an advanced porous materials constructor, known as pormake [11]. The enormous expansion of the available "search space" is inevitably associated with a new challenge: how the top-performing materials for a given application can be identified from this huge structures space in an *efficient way*?

Clearly, it is impractical to experimentally synthesize and evaluate all of these materials even for one application since a single experimental study requires from days to weeks. Computational screening, where quantum mechanical calculations and molecular simulations are used for the assessment of material's performance, is a more efficient approach. Density functional theory calculations are often employed to accurately describe the interactions between MOFs and guest gases. These results are then used for the parameterization of empirical force-fields, which in turn are used for

A. P. Sarikas, G. S. Fanourgakis, G. E. Froudakis, Department of Chemistry, University of Crete, Voutes Campus, GR-70013 Heraklion, Crete, Greece, e-mails: chemp1160@edu.chemistry.uoc.gr, fanourg@uoc.gr, frudakis@uoc.gr

https://doi.org/10.1515/9781501524721-003

the calculation of the adsorption isotherms by grand canonical Monte Carlo (GCMC) simulations. Improvements in computational power and advanced methods have significantly reduced the time needed to evaluate a single candidate. Screening based on GCMC simulations has been found extremely useful in the identification of top-performing MOFs for gas storage and separation applications [12–16]. Nevertheless, a similar brute computational screening approach still remains impractical given the vast structures space that must be explored. An alternative to multiscale approaches able to bypass the previous challenges is *Machine Learning*. The latter can be employed to identify the optimal materials in a completely data-driven way, requiring only a relatively small amount of data.

3.2 The machine learning landscape

Machine Learning can be defined as "the science (and art) of programming computers so they can *learn from data*" [17]. The aim of ML is the construction of predictive models based on some data or the uncovery of trends and patterns within the data. The tool that extracts information (or learns) from the data is called the *machine learning algorithm*. A machine learning system consists of a data set, an algorithm and the output, which can be a model or a pattern, and is obtained after *training* the algorithm on the data set. Depending on the amount and type of human supervision that machine learning systems get during their training phase, they can be classified to three main categories: supervised learning, unsupervised learning and reinforcement learning.

3.2.1 Supervised learning

The goal of supervised learning is to learn a function that maps a set of input variables $(x_1, x_2, \ldots, x_n)$ to an output variable y. In machine learning parlance, the input variables are called *predictors* (also known as *descriptors* or *features*) and the output variable is called a *label*. The latter can be either categorical or continuous and is known as *class* or *target*, respectively. Based on the label's type, supervised learning tasks are split into two categories: *classification* (predicting a class) and *regression* (predicting a target).

In supervised learning, the data set has the form $\{\mathbf{x}_i, y_i\}_{i=1}^N$, where N is the number of data points. The predictive model is obtained after training the algorithm on a subset of the data set, which is known as the *training set*. To evaluate model's predictive ability, also known as the *generalization error*, the remaining data points form another set, which is referred to as the *test set*, whose target values are compared to model's predictions.

In summary, a labeled data set is split into a training and test set. The algorithm is trained on the training data and the model is evaluated on the test data. Many of

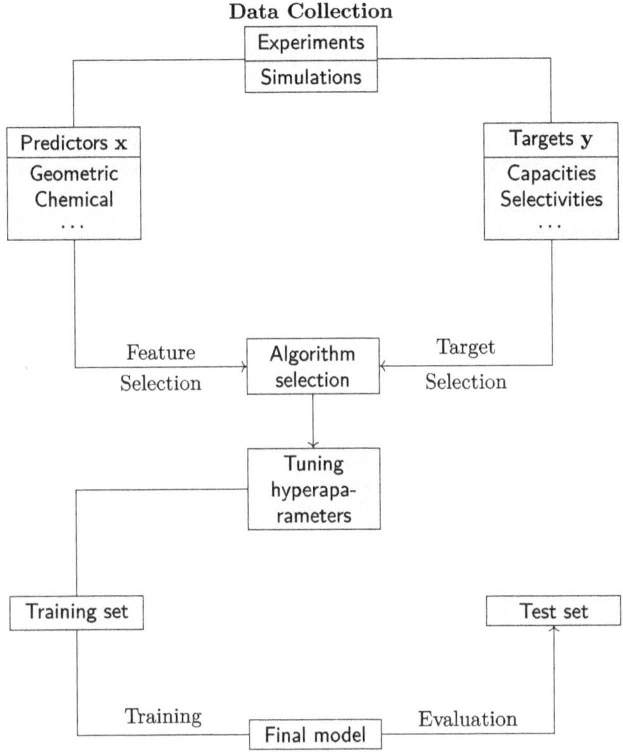

Figure 3.1: Workflow for building a machine learning model.

the machine learning techniques that have been applied to chemistry fall under the umbrella of regression. The general workflow required to build a regression model are examined in the next section and is schematically represented in Figure 3.1.

3.2.2 Unsupervised learning

In unsupervised learning, the data set is unlabeled and has the form $\{\mathbf{x}_i\}_{i=1}^N$. The goal is to identify hidden patterns or groupings within the data. It should be noted that in unsupervised learning there is no need for a test set and the data set serves as the training set. A typical unsupervised task is *clustering*, where the *data points are organized into clusters* according to the similarity of their input variables. The grouping is such that points of the same cluster are similar to each other as much as possible whereas points of different clusters differ as much as possible. Another task of unsupervised learning is *anomaly detection*, where the purpose is to *identify data points that exhibit deviations from the expected behavior*. Anomaly detection techniques can be employed to prevent frauds, detect manufacturing defects or remove outliers from a data set before it used for the training of another algorithm (e. g., a regression algorithm) [17].

In supervised learning tasks like regression, the data set may contain uninformative or correlated predictors. The use of many uninformative inputs variables increases the risk for high generalization error (low performance) whereas the inclusion of correlated features increases the computational cost without providing any useful information. These problems can be tackled by an unsupervised technique called *dimensionality reduction*, an important tool that helps to simplify the data preserving as much information as possible. Dimensionality reduction methods can be divided into *feature selection* and *feature extraction*. The former selects an optimal subset of the original set of predictors and the latter creates new predictors from the original set by merging correlated features.

3.2.3 Reinforcement learning

Reinforcement learning is a *feedback-based* machine learning technique where the learning system, also known as the *agent*, makes observations about the *environment* and then selects and performs actions. Based on these actions, the agent receives *rewards* and the goal is to find the optimal strategy or *policy* in order to maximize the rewards in the long run. Policy guides the agent and determines what action it should choose given a *state* of the environment.

Reinforcement learning finds applications in self driving cars, robotics and gaming. In problems of chemical interest, it has been applied for the generation of candidate drug molecules in a *de novo* way [18, 19] and for the optimization of reaction conditions in organic synthesis [20]. Also, reinforcement learning has been used to tune and control the shape of polymer molecular weight distributions in atom transfer radical polymerization (ATRP) [21]. In this way, the agent plays the role of a trained human that supervises the reaction system and makes actions (e. g., changing the concentrations of some reactants) that lead to the desired distribution (Figure 3.2).

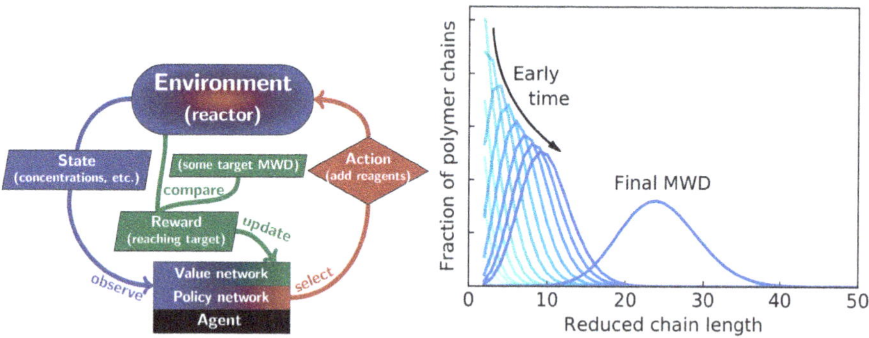

Figure 3.2: An application of reinforcement learning in chemistry. (left) An intelligent agent controls the ATRP reactor [21]. (right) Evolution of polymer molecular weight distribution toward the desired shape [21].

3.3 The machine learning pipeline

3.3.1 Data collection

The first step for building a ML model is the collection of data. This can be achieved by utilizing available data via molecular simulations or through experimental studies. The next, and perhaps the most crucial step, is the selection of the predictors and the target. The selection of target depends on the application at hand. For example, in gas storage common targets include absolute or working capacities (gravimetric or volumetric) in different temperature and pressure conditions. It is important that there is a meaningful relation between the predictors and the target, so that the algorithm is able to extract it. It should be noted that the predictors must be representative of the candidate material. That is, the predictors must uniquely characterize the framework, giving the algorithm the opportunity to discover the structure-property relationship. Also, a very short time should be required for their computation, given that trillions of materials have to be studied. Thus, the set of predictors must be selected in a reasonable manner. For example, if the target is the gas storage capacity of the framework, then a reasonable choice of predictors would be quantities that provide sufficient information about the adsorption of gas molecules in the solid phase, such as void fraction and surface area [22]. For adsorption based applications, the predictors that have been selected in machine learning studies can be grouped into three categories: *geometrical*, *chemical* and *energy-based*.

In the category of geometrical predictors are included features describing the pore environment of the framework. Examples of such predictors are density, void fraction, pore volume, gravimetric surface area and volumetric surface area. These features combined give a macroscopic description of the material. Another set of geometrical predictors, which describes microscopically the pore environment, includes pore limiting diameter, largest cavity diameter, dominant pore diameter and pore size distribution [22]. The aforementioned predictors can be experimentally measured, while computationally they can be calculated relatively fast by using open source computational tools like Zeo++ [23, 24] and PoreBlazer [25] or through GCMC simulations [25].

Owing to the chemical diversity of MOFs, predictors that take into account the chemical environment of the pore have also been introduced. An example of a chemical predictor is the atom or building block count, which stores the number of different atoms or building blocks per unit volume of the material [22]. Other chemical predictors include atomic properties such as electronegativity and atomic number or even atom types [26], taking into consideration the chemical environment diversity found within the pores of different MOFs. Similar to the geometrical features, this type of predictors can be easily calculated, by just identifying and counting the chemical moieties contained within the unit cell of the MOF.

Energy-based predictors try to capture in a direct manner the host-guest interactions of MOFs with the adsorbed gas molecules. This class of predictors provides a better insight of the host-guest energy landscape than the simple chemical predictors. Examples of energy-based predictors contain energy grid-points within the MOF unit cell [27] and hypothetical probes of different sizes or charge distributions [28]. Such hypothetical probes aim to construct the potential energy surface of each framework, giving a significant information about the adsorption behavior of the host material. Notably, by considering probes of different sizes, the geometry of the pore is also indirectly taken into account. Probes can be efficiently calculated together with the void fraction with a negligible additional computational cost. Therefore, the overall time needed for the calculation of predictors still remains significantly lower than the time required for the in silico or experimental evaluation of the material in question.

3.3.2 Algorithm selection

The selection of the predictors $\mathbf{x}$ and target y is based on the idea that there is a relationship between them, which can be expressed as

$$y = f(\mathbf{x}).$$

Our task is to find a function $\hat{f}$ that approximates f. The machine learning algorithm provides the means for determining such a function. The latter is the final model and is obtained after the algorithm is trained on the training set.

The training phase of the algorithm corresponds to a *minimization of a loss function*:

$$\text{Loss} = \mathcal{L}(\hat{y}_i; y_i)$$

where the index i goes over all observations in the training set. The set of approaches (algorithms) for estimating f can be grouped into two categories: *parametric* and *nonparametric methods.*

3.3.2.1 Parametric and nonparametric methods

In parametric methods, we make an assumption about the shape or the functional form of f. For example, we can assume a linear relationship between x and y, i. e.,

$$y = \mathbf{w}^{\top}\mathbf{x} + \beta.$$

The selection of a linear function implies that $\hat{y}$ is only function of $\hat{\mathbf{w}}$ and $\hat{\beta}$ and the same applies for the loss function. Consequently, the final model is obtained after finding the optimum parameters, i. e.,

$$(\hat{\mathbf{w}}, \hat{\beta}) = \underset{\mathbf{w},\beta}{\arg\min}\, \mathcal{L}(\mathbf{w}, \beta)$$

By assuming a parametric form for f, we simplify the estimation of f, as it is a lot easier to find a set of parameters than trying to fit an arbitrary function [29]. Nevertheless, such an approach have the following disadvantage: the functional form we assume for f is usually different than its true form, leading to a poor model.

In nonparametric methods, there is not an explicit assumption about the shape of f. These methods are more *flexible* compared to parametric methods and can be used to fit a wider range of functional forms. Instead of parameters estimation, during the training phase of nonparametric methods the functional form that best fits the data points, without being too rough or wiggly [29] is determined. It should be noted that the ability of nonparametric methods to fit a wide range of functions comes at a cost: to provide an accurate estimate of f, these algorithms require more data compared to parametric ones. Examples of parametric methods include linear regression, neural networks and support vector machines. The last two algorithms, although parametric, can be used to model complex functions. On the other hand, typical examples of nonparametric methods are decision trees, random forests and k-nearest neighbors.

A natural question that arises is why we need so many different learning methods rather than just a *single best method* [29]? The answer is that there is *no free lunch* in machine learning [30]. That is, *there is no machine learning algorithm that dominates all others for all possible data sets*. For a specific data set, one particular learning algorithm may perform best while for a different but similar data set some other algorithm may be the optimal choice [29]. There is no guarantee *a priori* that a given method will work better than the other. In order to know the optimal method, we have to evaluate them all and select the one that performs best. Since this is practically impossible, we can make some assumptions about our data, evaluate few reasonable models and pick the top-performing one.

3.3.2.2 Overfitting and underfitting

Depending on the approach employed during its training, a ML predictive model may perform well on the training data, (the *training error* is *low*), but on the test data, namely on the data unseen by the ML algorithm during its training, the predictions are poor (the *test error* is *high*). This discrepancy in performance implies that the ML predictive model can poorly generalize in unseen cases and is usually attributed to a phenomenon known as *overfitting*. This occurs when the model is very complex with respect to the size and the noise of the training data. In other words, the model *memorizes* the training data instead of learning to generalize from a trend. Some options to address overfitting are:

- Collection of more training data
- Reduction of the noise in the training set (e. g., removing outliers)
- Selection of a simpler model (e. g., a linear model instead of a high-degree polynomial) or restricting the model

The restriction of a model to become simpler in order to reduce the risk of overfitting is called *regularization* [17]. The amount of regularization is controlled by a *set of hyperparameters*. A hyperparameter is a parameter of the learning algorithm and not of the model, therefore, *it is not affected by the algorithm*. It should be set prior to training and remain constant during the training phase [17]. In contrast, *the parameters of a model*, e. g., the weights $\mathbf{w}$ and bias β of a linear model, *are free to vary during training*. The value of the hyperparameter *controls the flatness of the model with respect to the training data*. As such, if its value is too large, then the final model would be flat. That is, the model will not overfit the training data but is less likely that it will find a good estimate of f [17]. *Hyperparameters tuning* is an important step of building a reliable machine learning model.

The inverse of overfitting is called *underfitting* and it happens when the model is too simplistic to actually represent the data. This results in high error in both training and test set. Possible strategies to fix this problem are:

- Selection of a more powerful model (e. g., a polynomial instead of a linear)
- Selection of better (or more) predictors
- Reduction of the model's constraints (e. g., reducing the value of the regularization hyperparameter)

A simple example that illustrates the overfitting and underfitting of a ML model is provided in Figure 3.3 together with the case that the ML model works reasonably well on both training and test data sets.

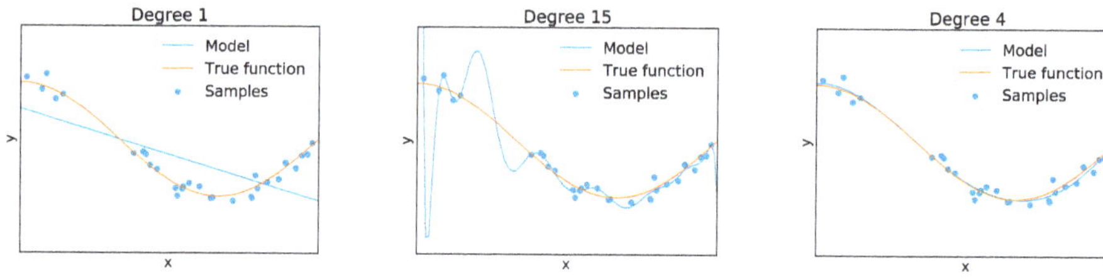

Figure 3.3: A polynomial model of degree 1 is not sufficient to fit the training data (left graph). On the other side, a polynomial model of degree 15 overfits the training data (middle graph). A polynomial of degree 4 provides a good estimate of the true function (right graph).

3.3.2.3 Hyperparameters tuning

Suppose we have to decide between two competing models (e. g., a linear model and a polynomial model). We can use the test set to get an estimate of their generalization error and select the one that performs best. Further, assume that the polynomial model performs better but we need to apply some regularization to avoid overfitting [17]. As mentioned above, a hyperparameter must be set prior to training. *How then should we choose the value of the hyperparameter*?

One way is to select n different values for the hyperparameter, build n different models and measure n test errors. Suppose that the optimal value has been found. Nevertheless, using the model for instances outside the test set leads to poor predictions. This discrepancy in accuracy is due to adapting hyperparameters to produce the best model *for this specific set*. As a result, it is less likely that the model will perform well on new instances [17]. *Holdout validation* and *k-fold cross-validation* are two workarounds to alleviate this issue.

Holdout validation

In holdout validation, a subset of the training data (called *validation set*) is used for evaluating models with different hyperparameter values and selecting the best one. Initially, models with various hyperparameters values are trained on the remaining data of the initial training set (*reduced training set*) and the model that performs best on the validation set is chosen. Finally, the hyperparameters of the best model are used for the whole training set and the final model is obtained. The latter is assessed on the test set to get an estimate of the generalization error. Although this strategy is computationally cheap, it has a major drawback. *Results from holdout validation may heavily depend on how the training-validation set partitioning was made*, i. e., which data points end up in the reduced training set and which on the validation set. In particular, for cases that a relatively small data set is used, a suboptimal hyperparameter value may be selected by this approach.

K-fold cross-validation

Cross-validation is one way to improve over the holdout validation. In this approach, the training set is split into k subsets and the holdout validation procedure is repeated k times (folds). At each iteration, one of the k subsets is selected as the validation set and the remaining $k-1$ subsets are joined to form a training set. At each iteration, during the training of the algorithm its hyperparameters are optimized. The overall performance of the final model is determined by averaging the results across the k iterations. By increasing the number of folds, we can get a more accurate picture of model's performance but this comes at a cost: *the training time is multiplied by the number of the k-folds.*

The aforementioned described pipeline must be implemented on computers, which means that the practitioner must possess some programming skills. Additionally, a new practitioner may not know which algorithm/hyperparameters to select/tune, rendering difficult the implementation of the pipeline. Nevertheless, platforms such as Just Add Data (JAD) [31], automate the ML pipeline with minimal user intervention, democratizing ML to nonexpert practitioners. Material science researchers can enter the world of ML without compromising productivity or falling into common analysis pitfalls [32].

3.4 Machine learning algorithms

3.4.1 Decision trees

Decision trees (DT) are supervised learning algorithms that are widely used for regression and classification. They make predictions (decisions) based on a *hierarchy of* `if-then-else` *questions*. This flowchart-like structure corresponds to *splitting the predictor space into regions R* as shown in Figure 3.4. These regions are known as *terminal nodes* or *leaves* of the tree. An *internal node* in the tree, i. e., the point along the tree where the predictor space splits, represents a question.

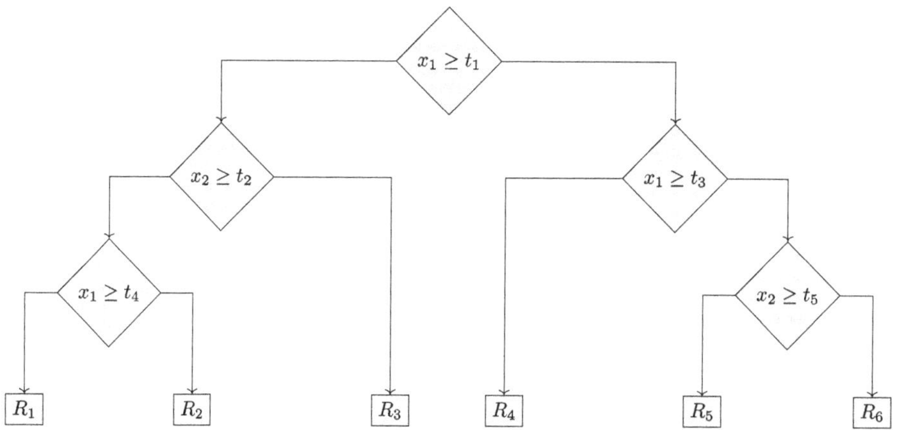

Figure 3.4: Decision tree for two predictors.

In general, the procedure for building a regression or classification tree includes the following steps [29]:

1. The predictor space $\mathcal{X}$, i. e., the set of all possible values for $x_1, x_2, \ldots, x_n$, is divided into J regions, $R_1, R_2, \ldots R_J$.
2. For every observation inside R_j, the prediction $\hat{y}_{R_j}$ is the same and it is given by the mean (in regression tasks) or the mode (in classification tasks) of the target values for the training observations in R_j.

The regions $R_1, R_2, \ldots, R_J$ that arise from step one could have in theory any shape. For sake of simplicity and interpretability of the model, the partitioning is chosen to be such that the predictor space is divided into high-dimensional rectangles [29] or simply *boxes*. The regression tree is obtained after finding the regions $R_1, R_2, \ldots, R_J$ that minimize the residual sum of squares (RSS), which is equal to

$$\sum_{j=1}^{J} \sum_{i \in R_j} (y_i - \hat{y}_{R_j})^2$$

where $\hat{y}_{R_j}$ is the mean target value of the training observations inside the jth box. Notice that instead of the residual sum of squares, other loss functions, such as the absolute error, can be used as well. For classification tasks, appropriate loss functions such as *Gini index* and *entropy* should be employed.

It should be realized at this point that is computationally infeasible to consider all possible divisions of the predictor space. Therefore, a *top-down, greedy* approach is adopted, which is known as *recursive binary splitting* [33].

Top-down The approach *begins from the top (root) of the tree* where all training observations belong to a single box. The predictor space is consecutively split and every split gives birth to new branches [33] as shown in Figure 3.4.

Greedy At each stage, the split is made based on the *locally optimal choice* instead of selecting a split that will lead to a greater reduction in RSS in a future step [33].

The decision tree algorithm for regression is presented in Algorithm 1. A *cutpoint* is denoted as t and the notation $\{\mathbf{x} \mid x_p \geq t\}$ describes the region of the predictor space where x_p takes a value greater than or equal to t. After the regions $R_1, R_2, \ldots, R_J$ have been found, the model predicts the target value of a new observation $\mathbf{x}$, depending on the region R it lies on.

If a leaf contains just one observation ($k = 1$), it is called *pure*. Forcing a decision tree to have pure leafs, that is small values for k, leads to overfitting. To avoid this problem, two general procedures are followed: stopping the building of the tree early (*prepruning*) or removing nodes that provide little information (*post-pruning*). Some hyperparameters that control prepruning are the maximum depth and the maximum number of leaves.

Algorithm 1 Decision tree algorithm (regression).

1. $\forall(p,t)$ define the half-planes:

$$R_m(p,t) = \{\mathbf{x} \mid x_p < t\} \quad \text{and} \quad R_n(p,t) = \{\mathbf{x} \mid x_p \geq t\}$$

   and search the values of (p,t) that minimize:

$$\sum_{\mathbf{x}_i \in R_m} (y_i - \hat{y}_{R_m})^2 + \sum_{\mathbf{x}_i \in R_n} (y_i - \hat{y}_{R_n})^2$$

2. Repeat the first step in the resulting regions.

$$\vdots$$

3. Termination condition (e. g., no leaf contains more than k observations).

3.4.2 Neural networks

Neural networks (NN) are a family of algorithms that mimic the function of a biological brain. They can be used for supervised and unsupervised learning tasks. The basic unit of neural networks is the *neuron*. The latter accept some inputs, process them via a mathematical function and produces an output.

A neural network is composed of *successive layers of neurons* where every neuron is connected with all others via *weights* and *biases*, similar to how neurons in a biological brain communicate with each other via *synapses*. The layers of the network are divided into three types:

Input Layer This layer accepts the values of predictors.

Hidden Layer One or more layers where the inputs undergo processing via some mathematical functions.

Output Layer It contains the prediction, which can be a scalar or a vector, depending on the number of neurons. Usually, output layer contains just one neuron.

A schematic representation of a neuron and of a neural network is illustrated in Figure 3.5. Every neuron executes the following task:

$$\sum_{k=1}^{m}(w_{ik}o_{ik} + b_{ik}) = \text{Input} \longrightarrow \underbrace{\text{Activation}}_{\mathcal{A}(\text{Input})} \longrightarrow \text{Output}$$

where w_{ik}, o_{ik}, b_{ik} represent the weights, outputs and biases, respectively, that connect the neuron with all neurons of the previous layer. If the previous layer corresponds to the input layer, the outputs o_{ik} are replaced by $x_1, x_2, \ldots, x_p$. The training phase of the algorithm results in learning the values of $\mathbf{w}$ and $\mathbf{b}$. A neural network can be repre-

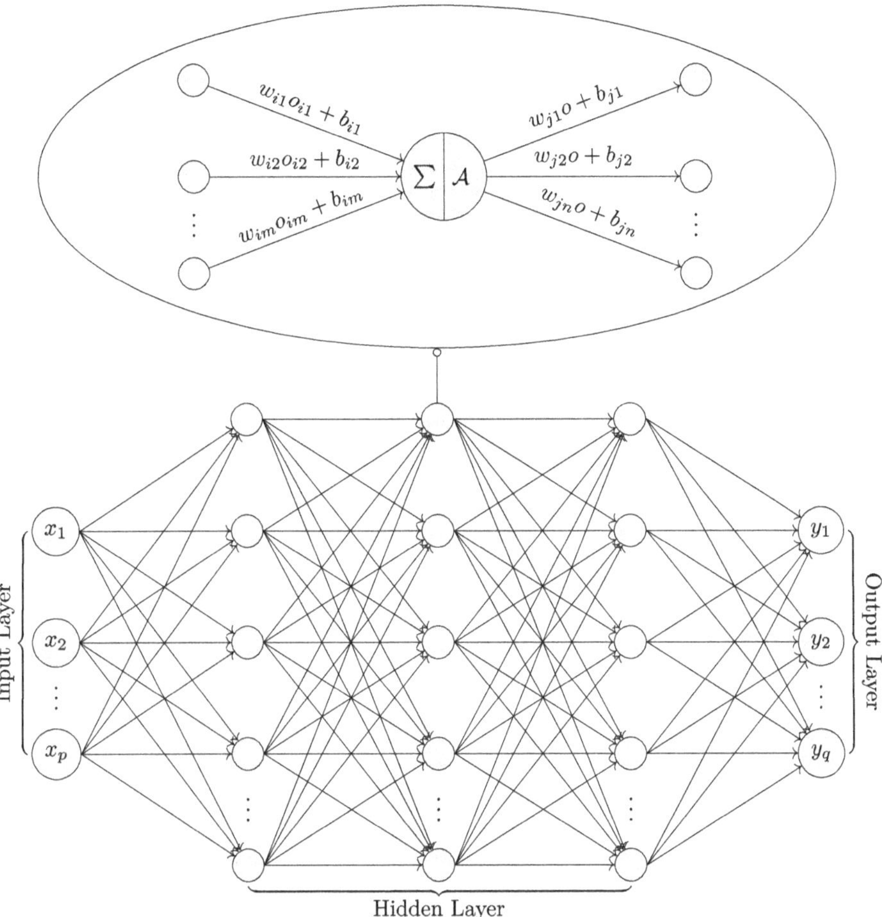

Figure 3.5: Schematic representation of a neural network and its basic unit, the neuron.

sented as a composite function of l functions:

$$(g_l \circ g_{l-1} \circ \cdots \circ g_2 \circ g_1)(\mathbf{x})$$

which entails that if a nonlinear element is not present, it is just a linear function. The ability of neural networks to model complex *nonlinear* functions is attributed to the presence of a nonlinear element called *activation function*, denoted as $\mathcal{A}$. Typical examples of $\mathcal{A}$ include the hyperbolic tangent and the sigmoid functions. It should be noted that number of hidden layers and the number of neurons in each layer, i. e., the *architecture of the network*, are both hyperparameters of the algorithm along with the activation function. The neural network algorithm for regression is presented in Algorithm 2.

Algorithm 2 Neural network algorithm (regression).

For a given network architecture and loss function $\mathcal{L}$:

1. Initialize weights and biases with random values:
$$\mathbf{p} = (\mathbf{w}, \mathbf{b})$$
2. Calculate
$$\nabla\mathcal{L}(\mathbf{w}, \mathbf{b})$$
and update
$$\mathbf{p} \leftarrow \mathbf{p} - \eta \cdot \nabla\mathcal{L}$$
3. Repeat the second step.
$$\vdots$$
4. Termination condition (e. g., number of updates).

In order to efficiently compute the gradient, neural networks utilize an algorithm known as *backpropagation*. The latter takes advantage of the chain rule and computes the gradient one layer at a time, iterating backwards from the last layer to avoid redundant calculations of intermediate terms in chain rule [34]. The *learning rate* denoted by η in Algorithm 2 is also a hyperparameter of the algorithm and determines the size of step taken in the direction of the gradient. A small value of learning rate results in slow convergence to a minimum (local or global) whereas a large value will make the learning jump out over minima (local or global).

3.4.3 Ensemble learning

In machine learning, *the combination of multiple algorithms to create a stronger one* is called *ensemble learning*. Two main categories of ensemble learning are *bagging* and *boosting*. The main difference between them lies on the way they combine the *base learners*, i. e., the members of the ensemble. Bagging algorithms combine base learners in a *parallel* fashion whereas boosting algorithms combine them in a *sequential* manner.

3.4.3.1 Random forest

A typical example of a bagging algorithm is *random forest* (RF), which is an ensemble of decision trees, each of them is slightly different from the others. It can be used for classification and regression tasks, just as decision trees. Random forests are based on the idea that every tree make good predictions but may overfits some parts of the training set. If many decision trees are generated, all of which have good performance and overfit different parts of the training set, then overfitting can be reduced by averaging (mean for regression and mode for classification) their predictions.

In order the trees to be different from each other, *randomness* must be introduced during their building, hence the name of the algorithm. Randomness is introduced by two mechanisms: first, each tree is built using different training data obtained by random sampling with replacement from the training set, a method known as *bootstrap sampling*. Second, at each internal node a *subset of predictors* is randomly selected and the best split is determined considering only this subset. The random forest algorithm for regression is described in Algorithm 3.

Algorithm 3 Random forest algorithm (regression).

1. Select the number of trees to build, N.
2. For $i = 1$ to N:
   (a) Generate a bootstrap sample of the training set.
   (b) Build a decision tree as described in Algorithm 1 but with a subset of predictors at each split.
3. Final model:

$$\hat{y} = \frac{1}{N}\sum_{i=1}^{N}\hat{y}_i$$

The greater the number of trees in the ensemble, the greater the reduction in overfitting. However, after a point the performance gain does not compensate to the increase in training time [35]. Random forests have become very popular due to their out-of-the-box performance. Although they have several hyperparameters, high performance is usually achieved without even tuning them [36].

3.4.3.2 Gradient boosted trees

Gradient boosted trees (GBT) is a boosting algorithm, which combines many *shallow trees* and can be used for both regression and classification. Owing to their low depth, the decision trees have low predictive ability. The idea is to combine them sequentially where *each tree tries to correct the errors of the previous trees* [35]. The gradient boosted trees algorithm for regression is described in Algorithm 4.

Algorithm 4 Gradient boosted trees algorithm (regression).

Given a training set $\{(\mathbf{x}_i, y_i)\}_{i=1}^N$ and loss function $\mathcal{L}(y, F(\mathbf{x}))$:

1. Initialize model with a constant value:

$$F_0(\mathbf{x}) = \arg\min_{\gamma} \sum_{i=1}^{N} \mathcal{L}(y_i, \gamma)$$

2. For $m = 1$ to M:
   (a) Calculate the pseudo-residuals:

$$r_{im} = -\left[\frac{\partial \mathcal{L}(y_i, F(\mathbf{x}_i))}{\partial F(\mathbf{x}_i)}\right]_{F(\mathbf{x})=F_{m-1}(\mathbf{x})}$$

   (b) Train a tree h_m on $\{(\mathbf{x}_i, r_{im})\}_{i=1}^N$.
   (c) Update the model:

$$F_m(\mathbf{x}) = F_{m-1}(\mathbf{x}) + \eta \cdot h_m(\mathbf{x})$$

3. Final model $F_M(\mathbf{x})$.

The pseudo-residuals express the errors of the previous trees. This can be easily verified by considering RSS as a loss function, where

$$r_{im} \propto (\hat{y}_{im-1} - y_i) = F_{m-1}(\mathbf{x}_i) - y_i$$

with F_{m-1} the prediction of the previous model for the ith observation. Utilizing partial derivatives instead of residuals allows the algorithm to adopt a more general character and work with different loss functions. As with neural networks, a hyperparameter of gradient boosted trees (owing to the use of gradient) is the learning rate, denoted with η. This hyperparameter controls the amount of error that each tree is allowed to correct. The number of trees M and learning rate are two hyperparameters that together regulate the complexity of the model. For example, a lower value for learning rate means that more trees are needed to achieve the same model complexity.

It should be emphasized that the update step corresponds to a *gradient descent*. Instead of following the gradient in a parameter space, like in the case of neural networks, the descent happens in a *function space*. Finally, increasing the number of trees M means that more and more errors are corrected, which leads to overfitting, in contrast with random forests where increasing M leads to better performance.

3.5 Applications of ML in MOFs research

3.5.1 Gas storage

MOFs have been regarded as an attractive solution for gas storage problems, especially in the fields of natural and hydrogen gas storage, with the latter focusing on mobile applications. A representative example of this class of porous materials is NU-1501-Al, which was found to exhibit remarkable gas capacities. This material yielded high usable capacities for both methane and hydrogen, 238 v_{STP}/v and 46.2 g L^{-1}, respectively [37]. Although such performance comes close to the target usable capacities set by the Department of Energy (DoE) (263 v_{STP}/v for CH_4 and 50 g L^{-1} for H_2) [38], further enhancements are needed since a significant amount of performance is compromised when MOF crystals are processed into real storage systems [22]. The huge number of available data make the problem of finding materials with the aforementioned criteria well suited for machine learning, which can greatly accelerate the search process.

Machine learning methods for studying methane capacity in MOFs were first utilized by [39]. In this work, a large-scale quantitative structure-property relationship (QSPR) analysis was performed on the hypothetical MOFs database constructed by Wilmer et al. [9]. In silico characterization by GCMC simulations were performed for each hypothetical structure within the database, at 1, 35 and 100 bar at 298 K. In addition to methane capacities, geometrical features such as void fraction, dominant pore diameter, gravimetric surface area, volumetric surface area and framework density were also obtained. Predictive models at different thermodynamic conditions (Figure 3.6) were built by means of three machine learning techniques: multilinear regression (MLR), DT and nonlinear support vector machine (SVM). High pressure models demonstrated remarkable predictive accuracy (e. g., R^2 = 0.94 for SVM at 100 bar)

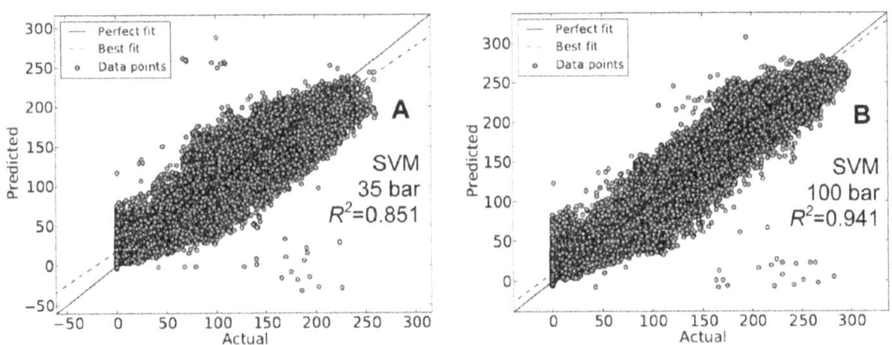

Figure 3.6: Parity plots of the MLR predicted vs. GCMC reference CH_4 capacities at 35 bar (left) and 100 bar (right) using the SVM model. If the prediction is below the diagonal (the line of the equation $y = x$), the model underestimates the capacity whereas if it is above the diagonal, the model overestimates it. Perfect predictions should lie exactly on the diagonal. Reproduced with permission from the work [39] Copyright 2013 American Chemical Society.

compared to the models in the low pressure regime (R^2 = 0.72 for SVM), possibly indicating that geometrical predictors alone are inadequate to describe CH_4 adsorption behavior at lower pressures. In a follow-up work by Fernadez and coworkers [40], the chemical character of MOFs was considered by employing atomic property weighted radial distribution functions as predictors. Improved ML predictions were observed in the low pressure regime (R^2 = 0.83). A remarkable conclusion can be drawn from these two works: methane's adsorption behavior at lower pressures can be accurately predicted by models when both geometrical and chemical predictors are combined, whereas at higher pressures, geometrical predictors alone exhibit satisfactory performance.

In the effort to improve further the predictive performance of ML-based models, geometrical and chemical predictors were combined for the adsorption analysis of CH_4 at 35 bar and 298 K [41]. In this work, several chemical predictors were introduced, taking into account the chemical composition of the framework: type and number of each atom, degree of unsaturation, metal to carbon ratio, halogen to carbon ratio, nitrogen to oxygen ratio and degree of electronegativity. Various machine learning algorithms namely, DT, SVM, Poisson and RF, were trained on 8 % of a data set consisting of 130,400 hypothetical MOFs and tested on the remaining 92 %. This work demonstrated that when geometrical and chemical predictors are jointly used, predictive models can attain impressive accuracy (R^2 = 0.98 for RF), higher than models solely based on one class of predictors. Furthermore, by performing feature importance with their RF model, they identified density, void fraction, surface area, pore diameter, metallic percentage relative to carbon and degree of unsaturation per carbon as the most crucial features for predicting CH_4 capacities.

It is of great importance that the resulting ML model is transferable, i. e., it has the ability to generalize well on different families of materials, as it relieves the necessity of training an ML algorithm each time that a new class of materials is to be explored. Ensuring transferability requires predictors that are able to represent as many materials families as possible. For example, predictors based on simple counts of building blocks (e. g., functional groups) hinder the transferability of the resulting ML model, since it would be inapplicable to a data set containing building blocks unseen during its training phase. Moving in the direction of maximizing transferability, Fanourgakis et al. [26] introduced a new class of chemical predictors on the basis of *atom types*. With this approach, much of the material's chemical identity is retained allowing the construction of accurate ML models while transferability is left uncompromised, since each new material can be broken down to its constituent atom types. The RF algorithm was trained with these novel predictors and used to predict CH_4 and CO_2 uptake capacities at various thermodynamic conditions. It was found that R^2 and mean absolute error (MAE) values were dramatically improved compared to other set of predictors, including chemical ones and much smaller training sets (even by an order of magnitude) were required to reach a specified threshold accuracy (Figure 3.7). Their method

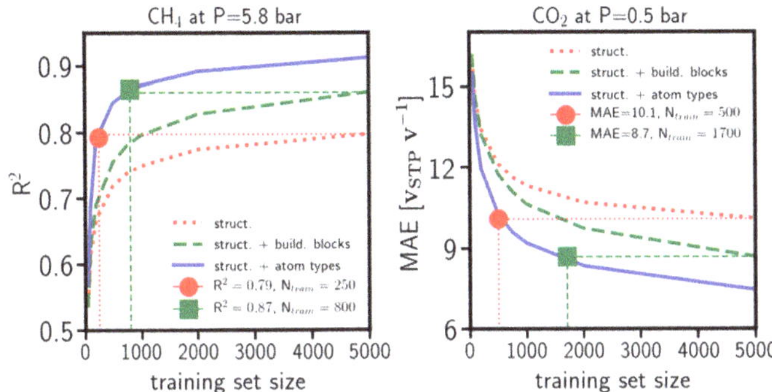

Figure 3.7: R^2 (left) and MAE (right) as a function of the training set size for CH_4 at 5.8 bar and CO_2 at 0.5 bar, respectively. The coordinates of the red circle (structural predictors only) and green square (structural predictors + building blocks) indicate the training set size (*x* coordinate) that ML models built with structural predictors + atom types (blue line), need to reach a specified threshold accuracy (*y* coordinate). This limit corresponds to the accuracy (as measured either by R^2 or MAE) of ML models with structural predictors only (red line) or structural predictors + building blocks (green line) when N_{train} = 5,000. Reproduced with permission from the work [26] Copyright 2020 American Chemical Society.

really shined when a RF model trained on a data set composed only of MOFs was employed to predict CH_4 adsorption capacities for COFs, namely a completely different family of materials. The model's predictions were very accurate (R^2 = 0.88), reflecting the universal character of the atomic type predictors.

In another study [28], a novel set of chemical predictors was proposed in order to enhance the predictive accuracy of ML models, regarding CH_4 adsorption in nanoporous materials. The newly introduced predictors are based on the interaction energies of the framework with hypothetical probes of different sizes (called *Vprobes*), particularly four in number. Although these particles have van der Waals radii that do not necessarily map to real atoms, their physical meaning is analogous to that of real ones, such as the helium atom void fraction. The necessity of various probes instead of one was attributed to the insufficiency of a single probe (e. g., the helium void fraction) to distinguish between frameworks of different topologies and capture detailed structural characteristics, such as pore size distribution. To test the validity of their proposed approach, the RF algorithm was employed while the predictions were evaluated using GCMC results of 4,764 MOFs, selected from the chemically and structurally diverse CoRE database [6]. ML models with different sets of descriptors were constructed, namely structural features only, probes atoms only and a combination of the two, with their performance being assessed by the 10-fold cross validation method. Impressively, the results revealed that incorporating the suggested predictors into the ML models elevates their accuracy compared to their probes-free analogs, particularly at the low pressure regime where structural-only based ML models are known to underperform.

In a follow up work [42], Fanourgakis and his coworkers appended to the aforementioned set of hypothetical probes (Vprobes) two new probe types, referred to as *Qprobe* and *Dprobe*, to take into account the charge distribution of the guest molecule. Qprobe atoms carry a point charge on their center and as such they possess an extra interaction term compared to Vprobe atoms, which participate only in van der Waals dispersion interactions. On the other hand, Dprobe atoms are neutrally charged, carrying an electric dipole, since they contain two charges of opposite sign separated by a certain distance. The various types of probe atoms are illustrated in Figure 3.8 (left). Similar to their previous work, the sets of probe atoms were combined with six structural predictors while various ML models were obtained after training the RF algorithm with a data set of 2,932 CoRE MOFs. Remarkably, their ML models were very accurate, showing R^2 up to 0.92 for CO_2 (at T=300 K and P=10 bar), 0.94 for H_2S (at T=300 K and P=10 bar), 0.95 for CH_4 (at T=298 K and P=65 bar) and 0.97 for H_2 (at T=298 K and P=50 bar). Furthermore, ML models with predictors from only one type of probe atoms (Vprobes, Qprobes or Dprobes) were constructed to sort out the relation between predictive performance and probe type as shown in Figure 3.8 (right). This analysis revealed that when there are no electrostatic interactions between the guest species and the framework, employing Vprobes resulted in the best performance. In converse, when electrostatic interactions between the guest species and the framework are present (guest species possess dipole or quadrupole moment), Dprobes gave the best results. A highlight of this work was the achieved performance regarding CO_2

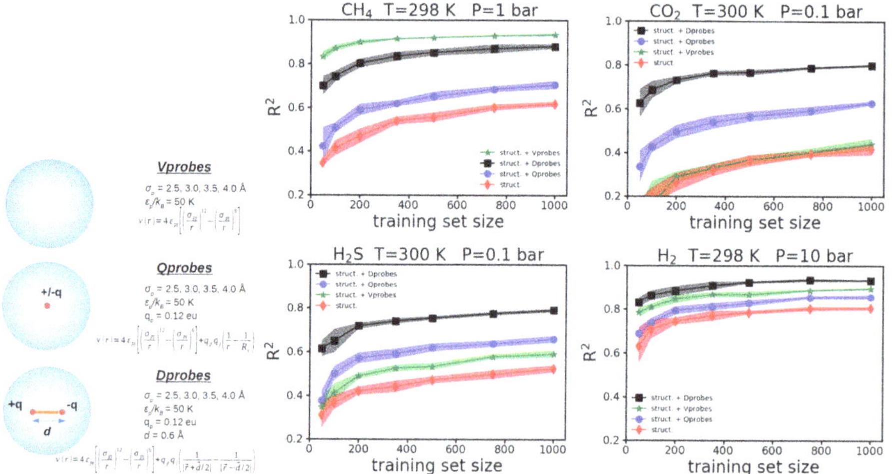

Figure 3.8: (left) Schematic representation of the probe types. (right) R^2 as a function of the training set size for different combinations between structural predictors and probe types. Vprobes perform better when electrostatic interactions between the framework and the guest species are absent (CH_4), whereas Dropbes give better results in the presence of them (CO_2, H_2S, H_2). Reproduced with permission from the work [42] Copyright 2020 American Chemical Society.

uptake at low pressures (R^2 = 0.84), which was significantly improved compared to previous works.

An other type of energy based descriptors was proposed by Bucior et al. [27], together with a data-driven approach for accelerating materials screening and analyzing the H_2 and CH_4 adsorption behavior in MOFs. The new set of predictors was derived from the histogram of the interaction energy of one H_2 or CH_4 molecule with the framework, capturing in this way the energetics of host-guest interactions (Figure 3.9). By overlaying a grid on the unit cell of each MOF and then partitioning each grid point into a bin of the respective energy width, the host-guest potential energy landscape was efficiently sampled. The selected ML model was LASSO, owing to its simplicity, interpretability and robustness to overfitting compared to the commonly used MLR. The aforementioned model was able to predict gas uptakes in several MOF databases to an accuracy within 3 g L^{-1}. As a case study, they applied the developed method to predict the H_2 volumetric working capacity from a subset of CSD and flag the top 1,000 materials. Among these materials, 51 of them exhibited working capacity greater than 45 g L^{-1}, as calculated by GCMC simulations. Finally, one of the top performers, MFU-4l, was experimentally synthesized and showed H_2 working capacity of 47 g L^{-1} for storage at 100 bar, 77 K and delivery at 2 bar, 160 K. Notably, at the core of this method are the host-guest interactions, which are the main factors governing physisorption.

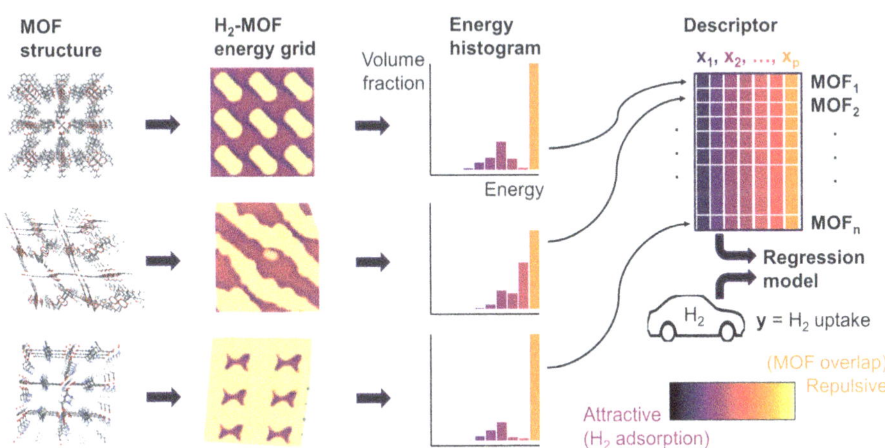

Figure 3.9: Workflow for the construction of the unit cell's potential energy histogram. The latter is employed for the prediction of H_2 and CH_4 uptake capacities [27].

Accuracy of ML-based models is not the only concern when assessing ML-assisted screening of large databases. Reducing computational cost by minimizing the amount of reference data (e. g., GCMC simulations) needed by the ML algorithms, that is increasing efficiency, is of equal importance for the identification of top performing materials. Moving in this direction, a self-consistent ML-based approach was proposed for the fast and accurate screening of large databases, aiming to reduce the required number of molecular simulations as much as possible [43]. In the proposed method, an initial training set with 100 materials is formed by randomly sampling from a database. The top materials, as predicted by the machine learning model, are added to the initial training set and new predictions are made. This iterative procedure continues until the top predicted materials are already in the training set, and the method is considered to be converged. In a case study, two databases consisting of 4,763 CoRE MOFs and 69,840 hypothetical covalent organic frameworks [44] (COFs), respectively, were selected. The property under examination was the CH_4 uptake capacity at 1, 5.8 and 65 bar and 298 K. Notably, the final model identified 70–80 out of the top 100 materials in both databases, utilizing only a small fraction from the available material pool (< 300 MOFs and < 400 COFs).

One of the first attempts to incorporate machine learning techniques for studying hydrogen storage in MOFs (and other nanoporous materials) was made by [45]. The target property was the net deliverable energy for H_2 storage while the database under consideration [46] contained more than 850,000 nanoporous materials. In order to perform an efficient screening, a neural network with five geometrical features as descriptors was trained to predict H_2 uptake, significantly reducing the amount of required GCMC simulations. The ML-assisted screening was performed in an evolutionary manner, where an initial round of GCMC simulation results served to train the network. Predictions from the latter were made for the whole database, and the 1,000 top-performing materials (according to the ML model) were used together with the previous ones for the construction of a new training set. This procedure of GCMC simulations followed by NN training was repeated for two more times after which no new candidates were identified by the model. According to the predictions of this approach, the best performance was achieved by a hypothetical MOF, with a volumetric capacity of $40\,g\,L^{-1}$. Two remarks can be made from this work. First, employing ML-assisted screening can dramatically reduce the computational cost compared to brute force screening. Second, a representative training set of all the materials under study is of great importance, if the ML models is intended to achieve high predictive performance.

Although MOFs are promising adsorbents for hydrogen storage, volumetric capacities of known materials are in general low. Thus, MOFs with high volumetric capacities are desired, but their identification remains a challenge. Searching for such novel materials, a large data set of 918,734 structures, sourced from 19 MOF databases, was explored by means of machine learning [47]. By employing seven geometrical predictors, namely, density, pore volume, void fraction, gravimetric surface area, volumetric surface area, largest cavity diameter and pore limiting diameter, the extremely randomized tree (ERT) algorithm was able to identify 8,282 materials, with the potential to outperform the state-of-the-art known ones (regarding H_2 storage under usable conditions). Most of the identified structures were hypothetical structures, which for pressure swing conditions possessed low density (less than 0.31 g cm^{-3}), together with high surface areas (greater than 5,300 m^2 g^{-1}), void fractions (≈ 0.90) and pore volumes (greater than 3.3 cm^3 g^{-1}). It should be noted that prior to selecting the ERT algorithm, different ML algorithms were trained and assessed while the accuracy of ML models as a of function of training set size was also analyzed. From this analysis, it was concluded that accuracy dependence on training set size can be well approximated by a simple power law. Additionally, accuracy dependence on the number and combination of predictors was investigated, by evaluating all possible combinations of the seven predictors. After a univariate feature importance analysis (see Figure 3.10), it was concluded that the most important features for predicting H_2 gravimetric and volumetric working capacities were the pore volume and the void fraction, respectively.

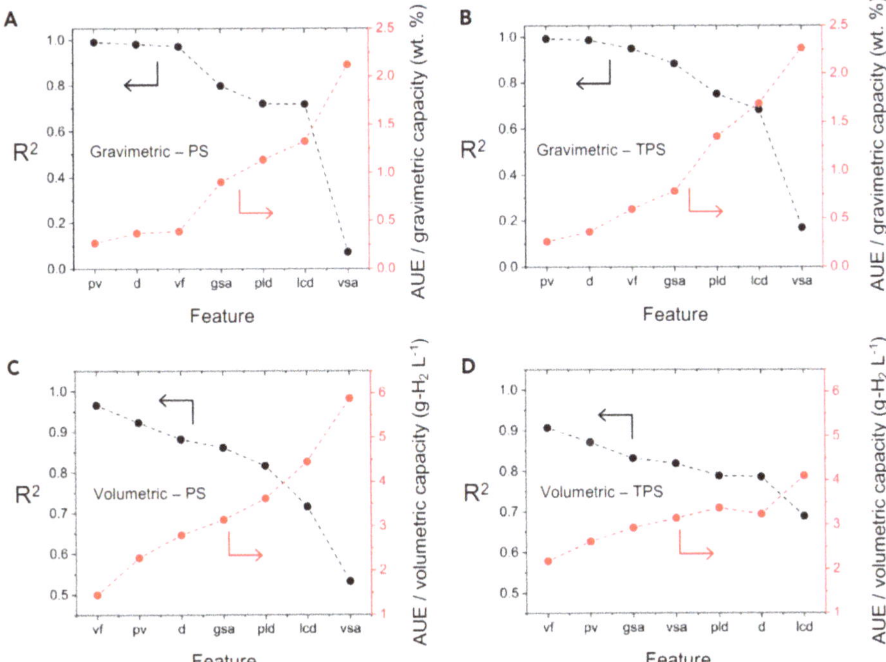

Figure 3.10: Univariate feature importance for UG at PS (top-left), UG at TPS (top-right), UV at PS (bottom-left) and UV at TPS (bottom-right). For each individual geometrical feature, distinct ERT models were built and the performance was assessed by R^2 (left-axis) and absolute unsigned error (right axis, AUE). UG, usable gravimetric; UV, usable volumetric; PS, pressure swing between 100 bar and 5 bar at 77 K; TPS, temperature-pressure swing between 100 bar at 77 K and 5 bar at 160 K; vf, void fraction; pv, pore volume; gsa, gravimetric surface area; vsa, volumetric surface area; pld, pore limiting diameter; d, density; lcd, largest cavity diameter [47].

3.5.2 Gas separation

MOFs have also been considered as attractive candidates for gas separation problems. Their narrow but tunable pore windows, typically in the range of few nanometers or Angstroms, renders them suitable for molecular sieving. For instance, ZIF-8 have demonstrated excellent separation performance for C_2/C_3 and alkane/alkene mixtures [48–50]. Beyond that, by employing MOFs with pores containing binding sites that favors the adsorption of one species over another, advanced separations can be achieved. Fe-MOF-74 is such a material, achieving ethylene/ethane and propylene/propane selectivity that ranges between 13–18 and 13–15, respectively [51]. Until now, mixed-gas GCMC simulations were the main tool for assessing adsorptive separation performance. As in the case of gas storage, employing GCMC simulations for the brute force screening is prohibitive, considering the size of the available material pool. Machine learning enables rapid screening by utilizing a substantially smaller amount of data compared to brute force approaches.

ML-assisted screening for separation performance was first introduced by [52], exploring a database of 670,000 structures from the Nanoporous Materials Genome. The system under study was a 20:80 Xe/Kr mixture at 298 K and 1 atm of total pressure. Simon and his coworkers used seven predictors as input for their RF model, with five of them being among the usual geometrical features. The other two were the surface density, defined as the mass of atoms per accessible surface area and the Voronoi energy, defined as the average energy of Xe at the accessible Voronoi nodes. Mixed gas GCMC simulations were performed for a diverse set of 15,000 materials to obtain their Xe/Kr selectivity and to serve as training data during the ML-based study. Out of bag ML predictions were far more accurate in the lower selectivity region compared to the higher one, with the difference attributed to the data scarcity and/or to the inadequacy of the selected predictors in this performance region. Feature importance analysis revealed the Voronoi energy and the void fraction as the first and second most important predictors, respectively. Furthermore, 4,066 of the remaining 655,000 materials were predicted to posses selectivity greater than 11. Predictions for this promising set of materials were verified by GCMC calculations, showing a RMSE of 2.2. Extra improvement of the ML model were achieved, by including the selectivities of these promising materials in the initial training set. The new model led to the identification of 461 materials exhibiting predicted selectivities higher than 11. An additional repetition of the previous procedure did not result in the discovery of any new promising structures. A deeper analysis of the results revealed that by considering only structural features it was not possible to construct a predictive model, which will indicate a promising adsorbent, regarding Xe/Kr separation. This underscores the importance of incorporating ML algorithms to learn the structure-selectivity relationship in a high-dimensional space.

In another work [53], the role of pore chemistry and topology on the CO_2 capture potential of over 400 hypothetical MOFs was analyzed by feeding various machine learning algorithms namely, gradient boosted trees (GBT), MLR, RF, SVM, NN, DT and RF, with results of GCMC simulations for pure CO_2, CO_2/H_2 and CO_2/N_2 mixtures. The thermodynamic conditions chosen were similar to those found in industry. In order to focus on a specific region of the materials space, both training and test set encompassed structures from all possible combinations of 16 topologies and 13 functionalized molecular building blocks. The obtained models were used for predicting CO_2 capture metrics of 31 parent MOFs and their functionalized derivatives, whereas the DT model was further employed to forecast the enhancement or deterioration of CO_2 capture metrics upon functionalization. Notably, feature importance analysis obtained from the GBT model showed that feature rankings depend on the composition of the gas mixture. Last but not least, analysis of the results unveiled that functionalization of MOFS with thiol, cyano, amino and nitro functional groups often improves CO_2 capture metrics of the parent structures.

Structural defects in MOFs may have major impact on their properties, potentially resulting in enhanced performance compared to their defect-free counterparts. Wu and his coworkers [54] adopted multiple ML methods to model the performance of defective UiO-66 structures with regards to ethane/ethylene separation. Prior to their ML analysis, a library of 425 defective structures was generated in silico with varying missing-linker concentration and spatial distribution. These two defect-related characteristics were properly addressed by introducing two new predictors, the missing-linker ratio and missing-linker short-range order (SRO) accounting for defects concentration and distribution, respectively. Target properties included ethane/ethylene selectivity, mixed-gas working capacities, bulk and shear modulus. The latter two take into account sorbent's mechanical stability, an important parameter for operation under real conditions. All but one target properties were modeled by linear regression (LR), with the exception of ethane/ethylene selectivity, which was modeled by RF. All linear models achieved an outstanding accuracy ($R^2 = 0.99$) whereas the RF model showed a lower performance ($R^2 = 0.92$). Unexpectedly, feature importance analysis highlighted that SRO shows little response while missing-linker ratio contributes to a broad range of target properties. This indicates that focusing on controlling defects concentration instead of their spatial distribution, should be the primary concern when optimizing separation performance of defective MOFs. Afterwards, DT classifiers with different combinations of four geometrical features as input were built to extract the criteria for optimal ethane/ethylene separation in defective UiO-66 MOfs (Figure 3.11). It was found that gravimetric surface area and pore volume are the two most sensitive factors for performance variation. Remarkably, in this work two new predictors capturing crystal defects in MOFs were introduced and mechanical stability was examined along with material's performance.

In all the aforementioned works, ML methods were employed to find a direct structure-property relationship. The latter is subsequently utilized to screen large databases, aiming to discover novel materials. Machine learning can go a step further *by reversing the exploration route*, giving rise to the inverse design of materials. This

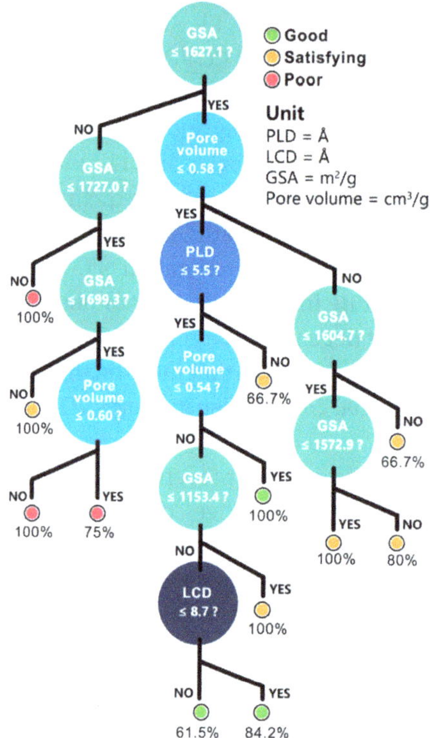

Figure 3.11: Decision tree with the best combination of predictors for the determination of characteristics that lead to optimal ethane/ethylene separation at 0.1 bar. The classes “Good,” “Satisfying” and “Poor” take into account ethane working capacity, ethane selectivity and bulk modulus at once. The terminal values of the tree are the highest probabilities that the structure under consideration will fall into the specific performance class. Reproduced with permission from the work [54] Copyright 2020 American Chemical Society.

approach, which falls under the umbrella of *generative modeling*, was adopted by [55] to automate the design of MOFs in a manner that enhances their performances, regarding CO_2/CH_4 and CO_2/N_2 separation. Random functionalizations of 14,000 CoRE MOFs and further modification of the functionalized structures, resulted in an augmented data set of approximately 2 million MOFs. Next, all MOFs in the data set were decomposed as tuples (edges, vertices, topologies) and for a fraction of the data set (around 45,000 randomly selected structures) mixed-gas uptake properties were calculated via GCMC simulations. This semi-labeled data set served to train a supramolecular variational autoencoder (SmVAE) in a semi-supervised fashion. Notably, the SmVAE showed excellent capability of capturing structural features and reassembling MOFs structures. More importantly, the optimized structures generated by SmVAE exhibited higher or similar performance compared to state-of-the-art materials, regarding mixed-gas working capacity and selectivity.

3.6 Conclusions

The overwhelming number of synthesized MOFs, either experimentally or in silico, has established ML as the main tool for their efficient exploration. ML takes over when data analysis becomes cumbersome or humanly impossible. Nevertheless, human intervention is still necessary and important for one specific reason—the ML algorithms need to be fed with data and these must be provided by the user. The latter is responsible for providing the algorithm with clean and informative data. Otherwise "*garbage in, garbage out*" applies, which reflects the necessity for good predictors. With regards to MOFs research, various predictor types have been introduced, namely geometrical, chemical and energy-based, each one aiming to capture a different aspect of the underlying framework. It is foreseeable that new types of predictor will be introduced, allowing ML models to capture complex structure-property relationships more accurately.

With the advent of ML era, the development and refinement of traditional computational tools, such as molecular simulations, may seems unnecessary. However, this is far from true: ML predictive models capable of providing high quality results that will reduce the necessity of expensive and time consuming experiments and will assist toward the discovery of new materials with extraordinary properties, require high quality simulations results. Since a suitable ML algorithm can produce fruitful results with only a handful of data (compared to brute-force approaches), feature research may focus on the performance of more accurate simulations instead of the performance of a lot of simulations. As a result, several approximations in molecular modeling that were necessary so far in order to reduce the computational cost of simulations may be avoided. Such an obvious approximation is the length of simulations: longer GCMC or molecular dynamics simulations will reduce the statistical uncertainties and will provide more accurate results. A second approximation is related to the description of the physical system, namely the force-field that is employed to account for the intermolecular interactions. The employment of more accurate force fields (e. g., many-body instead of pairwise-additive potentials) was prohibited so far, given their high computational cost and the enormous amount of materials that had to be studied. Also, phenomena such as the charge-transfer and the flexibility of the MOFs frameworks may be considered in the future more often.

The presented studies were solely based on adsorption related problems, such as gas storage and separation, but they exemplify the impact of ML in MOFs research. These works show that ML has drastically transformed the way researchers tackle a problem, providing them with useful insights and paving the way for tailor-made materials. Notably, generative modelling, a type of unsupervised learning, opens the door for design of materials with user-specified properties.

Finally, this chapter serves as a gentle introduction for material science researchers to the world of ML. The basic terminology, techniques and workflow were presented, allowing new practitioners to familiarize with the field of ML in an easy and comprehensive manner. Although this chapter focused mainly on one of the three categories of ML, namely supervised learning, we hope that it will trigger the interest of the reader to explore the remaining ones as well.

Bibliography

[1] Mason JA, Veenstra M, Long JR. Evaluating metal–organic frameworks for natural gas storage. Chem Sci. 2014;5:32–51.

[2] Suh MP, Park HJ, Prasad TK, Lim D-W. Hydrogen storage in metal–organic frameworks. Chem Rev. 2012;112(2):782–835.

[3] Bobbitt NS, Mendonca ML, Howarth AJ, Islamoglu T, Hupp JT, Farha OK, Snurr RQ. Metal–organic frameworks for the removal of toxic industrial chemicals and chemical warfare agents. Chem Soc Rev. 2017;46:3357–85.

[4] Herm ZR, Wiers BM, Mason JA, van Baten JM, Hudson MR, Zajdel P, Brown CM, Masciocchi N, Krishna R, Long JR. Separation of hexane isomers in a metal-organic framework with triangular channels. Science. 2013;340(6135):960–4.

[5] Hanikel N, Prévot MS, Yaghi OM. Mof water harvesters. Nat Nanotechnol. 05 2020;15(5):348–55.

[6] Chung YG, Camp J, Haranczyk M, Sikora BJ, Bury W, Krungleviciute V, Yildirim T, Farha OK, Sholl DS, Snurr RQ. Computation-ready, experimental metal–organic frameworks: A tool to enable high-throughput screening of nanoporous crystals. Chem Mater. 2014;26(21):6185–92.

[7] Chung YG, Haldoupis E, Bucior BJ, Haranczyk M, Lee S, Zhang H, Vogiatzis KD, Milisavljevic M, Ling S, Camp JS, Slater B, Siepmann JI, Sholl DS, Snurr RQ. Advances, updates, and analytics for the computation-ready, experimental metal–organic framework database: Core mof 2019. J Chem Eng Data. 2019;64(12):5985–98.

[8] Moghadam PZ, Li A, Wiggin SB, Tao A, Maloney AGP, Wood PA, Ward SC, Fairen-Jimenez D. Development of a cambridge structural database subset: A collection of metal–organic frameworks for past, present, and future. Chem Mater. 2017;29(7):2618–25.

[9] Wilmer CE, Leaf M, Lee CY, Farha OK, Hauser BG, Hupp JT, Snurr RQ. Large-scale screening of hypothetical metal-organic frameworks. Nat Chem. 02 2012;4(2):83–9.

[10] Colón YJ, Gómez-Gualdrón DA, Snurr RQ. Topologically guided, automated construction of metal–organic frameworks and their evaluation for energy-related applications. Cryst Growth Des. 2017;17(11):5801–10.

[11] Lee S, Kim B, Cho H, Lee H, Lee SY, Cho ES, Kim J. Computational screening of trillions of metal–organic frameworks for high-performance methane storage. ACS Appl Mater Interfaces. 2021;13(20):23647–54.

[12] Simon CM, Kim J, Gomez-Gualdron DA, Camp JS, Chung YG, Martin RL, Mercado R, Deem MW, Gunter D, Haranczyk M, Sholl DS, Snurr RQ, Smit B. The materials genome in action: identifying the performance limits for methane storage. Energy Environ Sci. 2015;8:1190–9.

[13] Gómez-Gualdrón DA, Colón YJ, Zhang X, Wang TC, Chen Y-S, Hupp JT, Yildirim T, Farha OK, Zhang J, Snurr RQ. Evaluating topologically diverse metal–organic frameworks for cryo-adsorbed hydrogen storage. Energy Environ Sci. 2016;9:3279–89.

[14] Moghadam PZ, Islamoglu T, Goswami S, Exley J, Fantham M, Kaminski CF, Snurr RQ, Farha OK, Fairen-Jimenez D. Computer-aided discovery of a metal-organic framework with superior oxygen uptake. Nat Commun. 04 2018;9(1):1378.

[15] Jeong W, Lim D-W, Kim S, Harale A, Yoon M, Suh MP, Kim J. Modeling adsorption properties of structurally deformed metal-organic frameworks using structure–property map. Proc Natl Acad Sci. 2017;114(30):7923–8.

[16] Banerjee D, Simon CM, Plonka AM, Motkuri RK, Liu J, Chen X, Smit B, Parise JB, Haranczyk M, Thallapally PK. Metal-organic framework with optimally selective xenon adsorption and separation. Nat Commun. 06 2016;7(1):ncomms11831.
[17] Géron A. Hands-on Machine Learning with Scikit-Learn and TensorFlow: Concepts, Tools, and Techniques to Build Intelligent Systems. Sebastopol, CA: O'Reilly Media; 2017.
[18] Popova M, Isayev O, Tropsha A. Deep reinforcement learning for de-novo drug design. CoRR. 2017. abs/1711.10907.
[19] Olivecrona M, Blaschke T, Engkvist O, Chen H. Molecular de-novo design through deep reinforcement learning. J Cheminformatics. 9 2017;9(1):48.
[20] Zhou Z, Li X, Zare RN. Optimizing chemical reactions with deep reinforcement learning. ACS Cent Sci. 2017;3(12):1337–44.
[21] Li H, Collins CR, Ribelli TG, Matyjaszewski K, Gordon GJ, Kowalewski T, Yaron DJ. Tuning the molecular weight distribution from atom transfer radical polymerization using deep reinforcement learning. Mol Syst Des Eng. 2018;3:496–508.
[22] Chong S, Lee S, Kim B, Kim J. Applications of machine learning in metal-organic frameworks. Coord Chem Rev. 2020;423:213487.
[23] Willems TF, Rycroft CH, Kazi M, Meza JC, Haranczyk M. Algorithms and tools for high-throughput geometry-based analysis of crystalline porous materials. Microporous Mesoporous Mater. 2012;149(1):134–41.
[24] Ongari D, Boyd PG, Barthel S, Witman M, Haranczyk M, Smit B. Accurate characterization of the pore volume in microporous crystalline materials. Langmuir. 2017;33(51):14529–38.
[25] Sarkisov L, Harrison A. Computational structure characterisation tools in application to ordered and disordered porous materials. Mol Simul. 2011;37(15):1248–57.
[26] Fanourgakis GS, Gkagkas K, Tylianakis E, Froudakis GE. A universal machine learning algorithm for large-scale screening of materials. J Am Chem Soc. 2020;142(8):3814–22.
[27] Bucior BJ, Bobbitt NS, Islamoglu T, Goswami S, Gopalan A, Yildirim T, Farha OK, Bagheri N, Snurr RQ. Energy-based descriptors to rapidly predict hydrogen storage in metal–organic frameworks. Mol Syst Des Eng. 2019;4:162–74.
[28] Fanourgakis GS, Gkagkas K, Tylianakis E, Klontzas E, Froudakis G. A robust machine learning algorithm for the prediction of methane adsorption in nanoporous materials. J Phys Chem A. 2019;123(28):6080–7.
[29] James G, Witten D, Hastie T, Tibshirani R. An Introduction to Statistical Learning: with Applications in R. Springer Texts in Statistics. New York: Springer; 2014.
[30] Wolpert DH. The lack of a priori distinctions between learning algorithms. Neural Comput. 10 1996;8(7):1341–90.
[31] Tsamardinos I, Charonyktakis P, Papoutsoglou G, Borboudakis G, Lakiotaki K, Zenklusen JC, Juhl H, Chatzaki E, Lagani V. Just add data: automated predictive modeling for knowledge discovery and feature selection. npj Precisi Oncol. 2022;6(1):6.
[32] Borboudakis G, Stergiannakos T, Frysali MG, Klontzas E, Tsamardinos I, Froudakis GE. Chemically intuited, large-scale screening of mofs by machine learning techniques. npj Comput Mater. 2017;3:1–7.
[33] Hastie T, Tibshirani R, Friedman JH. The Elements of Statistical Learning: Data Mining, Inference, and Prediction. Springer Series in Statistics. Springer; 2009.
[34] Goodfellow I, Bengio Y, Courville A. Deep Learning. MIT Press; 2016. http://www.deeplearningbook.org.
[35] Müller AC, Guido S. Introduction to Machine Learning with Python: A Guide for Data Scientists. O'Reilly Media; 2016.
[36] Boehmke B, Greenwell B. Hands-On Machine Learning with R. 2019.
[37] Chen Z, Li P, Anderson R, Wang X, Zhang X, Robison L, Redfern LR, Moribe S, Islamoglu T, Gómez-Gualdrón DA, Yildirim T, Stoddart JF, Farha OK. Balancing volumetric and gravimetric uptake in highly porous materials for clean energy. Science. 2020;368(6488):297–303.

[38] Allendorf MD, Hulvey Z, Gennett T, Ahmed A, Autrey T, Camp J, Seon Cho E, Furukawa H, Haranczyk M, Head-Gordon M, Jeong S, Karkamkar A, Liu D-J, Long JR, Meihaus KR, Nayyar IH, Nazarov R, Siegel DJ, Stavila V, Urban JJ, Veccham SP, Wood BC. An assessment of strategies for the development of solid-state adsorbents for vehicular hydrogen storage. Energy Environ Sci. 2018;11:2784–812.
[39] Fernandez M, Woo TK, Wilmer CE, Snurr RQ. Large-scale quantitative structure–property relationship (qspr) analysis of methane storage in metal–organic frameworks. J Phys Chem C. 2013;117(15):7681–9.
[40] Fernandez M, Trefiak NR, Woo TK. Atomic property weighted radial distribution functions descriptors of metal–organic frameworks for the prediction of gas uptake capacity. J Phys Chem C. 2013;117(27):14095–105.
[41] Pardakhti M, Moharreri E, Wanik D, Suib SL, Srivastava R. Machine learning using combined structural and chemical descriptors for prediction of methane adsorption performance of metal organic frameworks (mofs). ACS Comb Sci. 2017;19(10):640–5.
[42] Fanourgakis GS, Gkagkas K, Tylianakis E, Froudakis G. A generic machine learning algorithm for the prediction of gas adsorption in nanoporous materials. J Phys Chem C. 2020;124(13):7117–26.
[43] Fanourgakis GS, Gkagkas K, Tylianakis E, Froudakis G. Fast screening of large databases for top performing nanomaterials using a self-consistent, machine learning based approach. J Phys Chem C. 2020;124(36):19639–48.
[44] Mercado R, Fu R-S, Yakutovich AV, Talirz L, Haranczyk M, Smit B. In silico design of 2d and 3d covalent organic frameworks for methane storage applications. Chem Mater. 6 2018;30(15):5069–86.
[45] Thornton AW, Simon CM, Kim J, Kwon O, Deeg KS, Konstas K, Pas SJ, Hill MR, Winkler DA, Haranczyk M, Smit B. Materials genome in action: Identifying the performance limits of physical hydrogen storage. Chem Mater. 2017;29(7):2844–54.
[46] Nanoporous materials genome center. https://cse.umn.edu/chem.
[47] Ahmed A, Siegel D. Predicting hydrogen storage in mofs via machine learning. Patterns. 2021;2:100291.
[48] Pan Y, Lai Z. Sharp separation of c2/c3 hydrocarbon mixtures by zeolitic imidazolate framework-8 (zif-8) membranes synthesized in aqueous solutions. Chem Commun. 2011;47:10275–7.
[49] Bux H, Chmelik C, Krishna R, Caro J. Ethene/ethane separation by the mof membrane zif-8: Molecular correlation of permeation, adsorption, diffusion. J Membr Sci. 2011;369(1):284–9.
[50] Zhang C, Dai Y, Johnson JR, Karvan O, Koros WJ. High performance zif-8/6fda-dam mixed matrix membrane for propylene/propane separations. J Membr Sci. 2012;389:34–42.
[51] Bloch ED, Queen WL, Krishna R, Zadrozny JM, Brown CM, Long JR. Hydrocarbon separations in a metal-organic framework with open iron(ii) coordination sites. Science. 2012;335(6076):1606–10.
[52] Simon CM, Mercado R, Schnell SK, Smit B, Haranczyk M. What are the best materials to separate a xenon/krypton mixture? Chem Mater. 2015;27(12):4459–75.
[53] Anderson R, Rodgers J, Argueta E, Biong A, Gómez-Gualdrón DA. Role of pore chemistry and topology in the co_2 capture capabilities of mofs: From molecular simulation to machine learning. Chem Mater. 8 2018;30(18):6325–37.
[54] Wu Y, Duan H, Xi H. Machine learning-driven insights into defects of zirconium metal–organic frameworks for enhanced ethane–ethylene separation. Chem Mater. 2020;32(7):2986–97.
[55] Yao Z, Sánchez-Lengeling B, Bobbitt NS, Bucior BJ, Kumar SGH, Collins SP, Burns T, Woo TK, Farha OK, Snurr RQ, Aspuru-Guzik A. Inverse design of nanoporous crystalline reticular materials with deep generative models. Nat Mach Intell. 01 2021;3(1):76–86.

Mohamed Infas and Zhehao Huang

4 Structure elucidation and advanced characterization techniques

4.1 Introduction

Metal-organic frameworks (MOFs) have gained increased attention in the last two decades partly due to the precise visualization of their atomic arrangements in their structures. Understanding the structure-property relationship is critical to MOF chemistry in order to target advanced materials to be deployed in advanced technology. One of the advantages of MOFs is their ordered structures that allow link their chemical modularity to structural properties. Accordingly, benefiting from MOFs' features can be achieved only if their structure is carefully determined. Various advanced characterization techniques have been employed in order to characterize different MOFs with the structure elucidation of MOFs being primarily carried out by different diffraction techniques. In this chapter, we provide an overview of the current progress in structure elucidation and advanced characterization methods of MOFs using X-ray diffraction and electron diffraction techniques. We elaborate on the theoretical background of several techniques and provide case studies where coupling of advanced characterization techniques was implemented to elucidate structural features of MOFs.

4.2 Crystal growth and crystallization

A crystal structure can be defined as a particular repeating arrangement of atoms, molecules or ions of a lattice exhibiting a long-range order and symmetry. Crystallization can be regarded as a two-step process that includes nucleation and crystal growth. Nucleation is the stage where the birth of new crystals occur by phase separation and the crystal growth is where the size of the crystals grows larger. In MOF chemistry, the nucleation stage is where the secondary building unit replaces the anions that are originally attached to the metals [1–4]. It creates the self-assembled metastable phases with the interconnected secondary building units [1–4]. Thereafter, the growth period continues with the development of the crystal structure of the MOF. Finally, nano-crystalline and metastable phases in the synthetic solution will be

Mohamed Infas, Department of Chemistry, Khalifa University, Abu Dhabi 127788, United Arab Emirates, e-mail: mohamed.mohideen@ku.ac.ae
Zhehao Huang, Department of Materials and Environmental Chemistry, Stockholm University, Stockholm 10691, Sweden

https://doi.org/10.1515/9781501524721-004

consumed in order to create a stationary stage where the crystals grow homogenously in size and phase [3].

The success of a diffraction experiment depends on the quality of the specimen that is being studied. Comprehending the fundamental chemistry of MOF formation is of utmost importance to have acceptable control over the structure and properties of MOFs. The synthetic conditions of MOFs play an important role in obtaining suitable quality single crystals as the crystal growth takes place during the synthetic reactions for extended structures [5, 6]. Therefore, optimization of these conditions is critical in order to improve the quality of the crystals. Among different techniques that have been used to address this issue, engagement of modulating agents is one of the methods that has been widely studied. It has been reported that the modulating agents such as monocarboxylates (e. g., formic acid, benzoic acid, acetic acid, among others) play an important role in controlling the size and shape of MOF crystals [7, 8]. The differences in crystal size and shape occur due to the competition between modulating agents and the linkers during the synthesis process [9, 10]. For example, it has been reported that UiO-66 facilitates larger crystals in the presence of modulating agents [7]. The influence of benzoic acid and acetic acid as modulating agents have been studied on UiO-66 (Zr-bdc; bdc = benzene-1,4-dicarboxylate) and UiO-67 (Zr-bpdc; bpdc = biphenyl-4,4-dicarboxylate). The size of the UiO-66 crystals was tuned and the product changed from intergrown to individual crystals by varying the amount of the modulating agents [7]. A similar effect has been observed for UiO-67 as well [7]. However, in MOF synthesis the products are obtained more often as powders. For example, certain synthetic approaches such as fast crystallization, mechanochemical reactions or microwave assisted synthesis result in polycrystalline products rather than products with suitable single crystals [11].

4.3 X-ray diffraction

As crystals have a regular repeating pattern, they can diffract radiation such as X-rays that have wavelengths similar to interatomic distances in crystalline solids. The scattered beams can interfere constructively or destructively. X-ray diffraction technique is a non-destructive tool that facilitates elucidation of the structural information of crystalline samples. It requires high energy X-rays to extract details on the atomic level. X-ray diffraction is a technique that is widely used for the identification and characterization of MOFs.

This segment provides an overview of the theoretical background of single crystal X-ray diffraction (SC-XRD) and powder X-ray diffraction (PXRD) techniques. Single crystal diffraction involves collecting the intensity of diffraction spots from a crystal rotated in the X-ray beam whereas the powder X-ray diffraction is about the analysis of the diffraction of a polycrystalline sample. The aim of this portion is to confer how these techniques are involved in structure determination and characterization of different MOFs.

4.3.1 Structure solution from single crystal X-ray diffraction

SC-XRD provides information about the internal lattice of MOF samples including unit cell dimensions, bond lengths, bond angles and details of site-ordering [12–14]. A suitable crystal should be designated that extinguishes light uniformly under polarized light in order to make the measurements. The crystal should be mounted in the instrument in such a way that all orientations of the lattice planes can be accessed in order to collect the information about reflections from all crystal planes. Moreover, it is necessary to rotate the crystal during the data collection to move all reciprocal lattice points through the Ewald sphere. Initially, data will be collected for a short period to determine the scattering capability of the crystal and to predict whether the crystal is suitable for full data collection. The unit cell of the crystal can be determined by using these initial images, which should then be compared with the information already in databases in order to avoid the recollection of data of a known MOF. Full data collection can be continued if a crystal is suitable and diffracts well. Once diffraction frames have been collected, the data is integrated to correct intensities to each hkl value that is assigned to each spot of the diffractogram [12–14]. The unit cell is recalculated from the full data set and symmetry equivalent reflections are merged. The intensity data must be corrected for a number of systematic errors such as absorption and polarization. The experimental intensity data is related to the structure factor as depicted in equation (4.1):

$$I \propto |F(hkl)|^2 \tag{4.1}$$

What is observed experimentally is the intensities of the diffracted waves and, therefore, the information about the phase of the waves will be lost [12–14]. This is known as the "phase problem." It is vital to solve the phase problem in order to determine the crystal structure. The phase problem is addressed by obtaining a unique set of phases and combining it with structure factors that allow the diffraction data to be converted into a 3D map of electron density. The two most common methods of addressing this issue are the Patterson method and direct methods.

4.3.1.1 Patterson method [12–14]

The Patterson method is based on a simple concept, which is ignorant of the phase problem. A Patterson function is the convolution of the electron density with itself and it displays a set of interatomic vectors. It can be acquired experimentally by Fourier transformation of the observed intensity as indicated in equation (4.2):

$$Puvw = \frac{1}{V}\sum_{hkl} |F_{hkl}|^2 \cos 2\pi(hu + kv + lw) \tag{4.2}$$

where, u, v and w are the coordinates of the vector. Since the electron density is calculated using a set of squares of the numbers of electrons of the scattering atoms, the Patterson method is usually utilized when the unit cell contains heavier atoms.

4.3.1.2 Direct methods [12–14]

Direct methods is the most frequently employed method in determining MOF crystal structures using SC-XRD. It treats the structure factor directly from the observed F_{hkl} using mathematical models. The concept is based on two main properties of electron density function where: (i) electron density is always positive or zero; and (ii) scattering centers in a crystal are composed of discrete atoms. The non-negativity of electron density is extended to three dimensions, which are described in equation (4.3):

$$s(h_1, k_1, l_1)s(h_2, k_2, l_2) \approx s(h_1 + h_2, k_1 + k_2, l_1 + l_2) \tag{4.3}$$

where s is the sign of the phase in question and (h_1, k_1, l_1), (h_2, k_2, l_2), $(h_1{+}h_2, k_1{+}k_2, l_1{+}l_2)$ are referred to strong reflections. In determining the crystallographic structure, direct methods work on promising combinations of phases for the strongest reflections, which, in turn, are used in a Fourier transform to find electron density patterns that resemble the atomic arrangements.

4.3.1.3 Structure refinement [12]

Once a suitable model is constructed, the next step is to refine the model to a best fit by using a least squares refinement method. The least squares refinement method compares the structure factors of the calculated density with experimentally observed data. Experimental parameters, unit cell parameters and atomic positions can be altered in order to find the model that minimizes the M, which shows the minimum differences between the experimental intensities and the theoretical structure factors as shown in equation (4.4):

$$M = \sum w(F_o^2 - F_c^2)^2 \tag{4.4}$$

where F_o is the observed structure function and F_c is the structure function calculated from the theoretical model. The weighting factor, w, will indicate the accuracy of the measurement based on its standard uncertainty.

The quality of the refined model with the experimental data is determined by using residual factors (*R*-factors). The most commonly used *R*-factor is the unweighted *R*-factor based on *F*, given by equation (4.5):

$$R1 = \frac{\sum ||\mathbf{F}_o| - |\mathbf{F}_c||}{\sum |\mathbf{F}_o|} \tag{4.5}$$

On the other hand, the weighted *R*-factor is based on F^2, given by equation (4.6):

$$wR2 = \sqrt{\frac{\sum[w(\mathbf{F}_o^2 - \mathbf{F}_c^2)^2]}{\sum[w(\mathbf{F}_o^2)^2]}} \tag{4.6}$$

where *w* is the weighting of each measured reflection based on uncertainties. The *R*-factor should attain a minimum during the refinement. Once the refinement is complete, the goodness-of-fit, *S* should be close to 1.0 for an accurately weighted refinement. When $S < 1$, it suggests that the model is better than the data, which is an indication of errors in the refinement, or data. Goodness-of-fit *S* is given by equation (4.7):

$$S = \left[\frac{\sum w(F_o^2 - F_c^2)^2}{(N_R - N_p)}\right]^{1/2} \tag{4.7}$$

where, N_R is the number of independent reflections and N_P is number of refined parameters.

A low *R*-value is an indication of a good model. However, there is a possibility to obtain an erroneous model with a low *R*-value. Therefore, it is important to make certain that the structure makes sense chemically. Sensible bond lengths and angles, site occupancy and the coordination number of atoms should be rationalized by the chemistry of the MOF being studied.

4.3.2 Structure elucidation using powder X-ray diffraction

Solving a crystal structure using powder X-ray diffraction data is relatively complicated compared to single crystal analysis. This is because in a SC-XRD pattern, data are distributed in the reciprocal 3D space whereas in a PXRD pattern the information is available in 1D space [15]. In order to solve a structure using PXRD, it is important to produce a high quality pure polycrystalline sample with high crystallinity so that it may produce a diffraction data set with high resolution. Subsequently, this will enable accurate peak positions and intensities in order to determine the unit cell and space group of the material. Once the preliminary structure is obtained by using the PXRD data, then the atomic positions can be refined by means of Rietveld refinement [16, 17]. Hence, it is important to have a reliable preliminary model in order to have an acceptable final structure. In the course of the crystal structure determination, the following steps should be meticulously carried out in order to obtain an acceptable final structure: (i) indexing and space group determination, (ii) structure solution and (iii) structure refinement.

4.3.2.1 Indexing and space group determination [16, 17]

The first step of the PXRD structure solution is the determination of cell parameters. This can be done by indexing the PXRD pattern. Indexing is an accurate analysis of peak positions in the diffractogram. In order to do this, the exact peak positions will be extracted and trial unit cell parameters are screened. When selecting the peaks for indexing, it is important to select peaks from the lower diffraction angles in order to minimize the peak overlap. Indexing step is crucial in determining the unit cell parameters in order to solve the structure. Nevertheless, the following factors can affect the indexing process: (i) the quality and purity of the sample, (ii) number of strong peaks at low angle and (iii) resolution of the instrument.

Once the indexing is completed, the intensity data will be subsequently used to determine the space group and structure solution. In this step, the refinement variables will be used to obtain a good fit with the PXRD pattern. The variables such as peak positions, peak intensities, peak width and shapes will be considered together with the contextual intensity profile. This profile fitting step helps to get accurate lattice parameters. It is important to note that there will not be a structural model in the profile fitting at this stage and it will be merely fitting the parameters against the experimental data to get an optimal fit. Once the intensities are extracted, a space group can be determined by applying the rules of systematic absences and symmetry elements.

4.3.2.2 Structure solution [18, 19]

Structure solution will be carried out by using one of the following methods: (i) The traditional method utilizes the integrated intensities $I(hkl)$ of each reflection of the diffraction pattern similar to the direct methods, Patterson method or charge flipping method in a single crystal structure solution. This is a consistent and rapid method that has superior success rate in determining the accurate structure. Reliability depends on the large number of reflection intensities recorded at the atomic resolution. However, one of the drawbacks of this approach is that there is a possibility of obtaining inaccurate data due to the peak overlap, which in turn, results in the failure of the structure solution. (ii) The direct-space method is based on a global optimization technique. It handles the peak overlap problem. In this method, random trial structures are generated and the simulated powder patterns that are obtained through these structures will be compared with the experimental pattern. However, it is important to use any structural information that is being suspected in the structural motifs in the calculations of trial structures. The weighted R_{wp} and unweighted R_p profile R factors are the parameters that are used to assess the quality of the structures and are described in Equations (4.8) and (4.9):

$$R_{wp} = 100 \times \sqrt{\frac{\sum_i w_i (y_i - y_{ci})^2}{\sum_i w_i y_i^2}} \tag{4.8}$$

$$R_p = 100 \times \frac{\sum_i |y_i - y_{ci}|}{\sum_i |y_i|} \tag{4.9}$$

where y_i is the intensity of the ith point in the experimental PXRD, y_{ci} is the intensity of the jth point in the calculated PXRD and the w_i is the weighting factor for the jth point. The value of R_{wp} decreases for a good agreement between the experimental and calculated PXRD patterns. Thus, the aim is to find the trial structure corresponding to the lowest R-factor.

4.3.2.3 PXRD structure refinement [17]

Least squares methods are used in the Rietveld refinement process to find a good fit between the experimental and calculated powder XRD patterns. Therefore, a good fit between the XRD profile and the variables of the structural model will be evaluated. In the Rietveld refinement, atomic coordinates, atomic thermal parameters and site occupancies will be used as structural parameters. The weighted profile R-factor will be used as a tool to assess this criterion. Figure 4.1 summarizes the key stages of structure determination from PXRD.

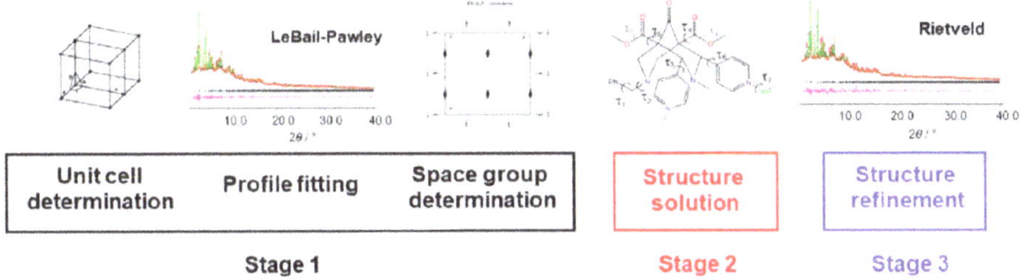

Figure 4.1: Key stages in the structure determination from PXRD. Stage 1 has been divided into three sub stages highlighting the importance of each point. Reproduced from ref. [11], with permission from the Royal Society of Chemistry, copyright 2020.

4.3.3 Structure determination of MOFs

The structures of most MOFs reported so far are solved by using SC-XRD, which is a relatively uncomplicated technique. However, as mentioned in the previous section, obtaining single crystals of suitable quality and size is not always a possibility. In such situations, a structure solution from PXRD data is pursued. There are numerous examples of MOFs that have been reported with precise structural analysis using PXRD data. In some situations, structural information is obtained using diverse advanced characterization techniques and this information is coupled with XRD data in order to obtain the most accurate MOF structures. In this section, different examples of structure determination of MOFs are discussed.

The direct methods have been employed by using PXRD data for the structure solution of various MOFs. One such example is UiO-66, which was synthesized by using $ZrCl_4$ and 1,4-benzenedicarboxylic acid in dimethylformamide [20]. UiO-66 crystallizes as inter-grown cubic crystals, which makes structure determination difficult by using single crystal x-ray diffraction methods due to its minute proportion. Hence, extraction of peak positions, pattern indexing and Rietveld refinement were carried out using PXRD data in order to solve the structure. The asymmetric unit of the structure was determined with a face centered cubic unit cell and a space group of *Fm-3m* by using the direct methods. The resultant structural model was further refined by using Rietveld refinement, which is shown in Figure 4.2a to obtain a satisfactory structural model with $R_{Bragg} = 0.011$ and profile factors $R_P = 0.016$ and $R_{WP} = 0.022$ [20].

The structural analysis of UiO-66 revealed that it consists of an inner $Zr_6O_4(OH)_4$ core where each Zr atom is eight coordinated and the triangular faces of the Zr_6-octahedron are alternatively capped by μ_3-O and μ_3-OH groups. These are further bridged by carboxylates forming $Zr_6O_4(OH)_4(CO_2)_{12}$ clusters resulting in a silhouette of a Maltese star as shown in Figure 4.2b [20].

The same process was carried out with a different method to determine the structure of metal-triazolates (METs) [21]. These structures have been obtained using the charge-flipping method, which is an alternative method to obtain MOF structures instead of direct methods [22]. In the charge-flipping method single crystal diffraction data was employed initially and subsequent modifications were carried out using histograms to match the chemical compositions. The indexing of MET-6 material using PXRD resulted in a cubic unit cell with parameter $a = 17.67$ Å [21]. Pawley refinement was performed on the diffractogram to obtain the integrated intensities. The charge-flipping algorithm was then applied on these intensities and the refined unit cell parameters to generate the electron density maps to solve the structure of MET-6 material. As illustrated in Figure 4.3, MET-6 adapts a diamond-type structure where metal ions are octahedrally coordinated to the nitrogen atoms of triazolate in a way that five metal centers are joined, bridging the triazolates to form super-tetrahedral units that lie at the vertices [21].

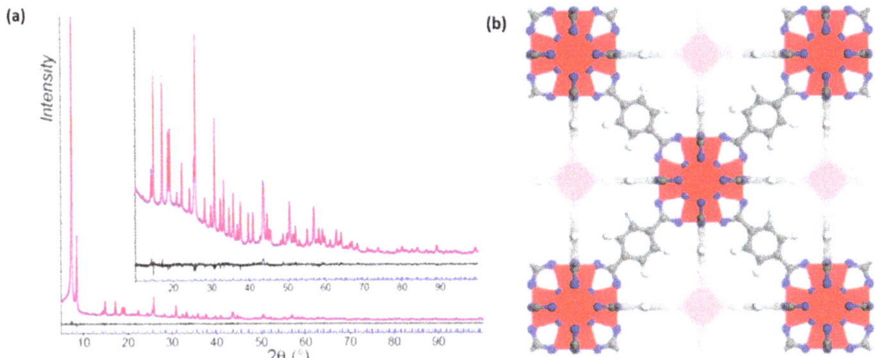

Figure 4.2: (a) Final Rietveld Plot of the desolvated Zr-BDC structure (UiO-66 desolvated) showing observed (blue circles), calculated (red line), and difference (black line) curves. A zoom at high angles is shown as inset. (λ = 1.5405981 Å). (b) Zr-MOF with 1,4-benzene-dicarboxylate (BDC) as linker, UiO-66. Reproduced from ref. [20], with permission from the American Chemical Society, copyright 2008.

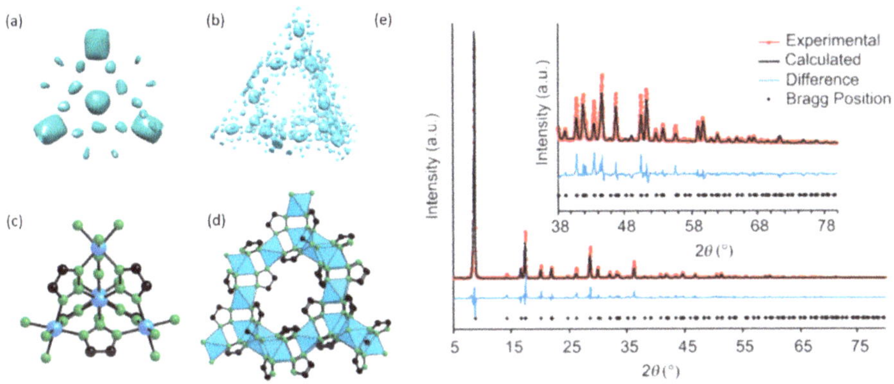

Figure 4.3: Electron density maps were generated by applying the charge flipping method to PXRD data, clearly showing the position of the metal atoms and the triazole ring (a). The full unit cell is shown in (b). The crystal structure was refined accordingly. The tetrahedral SBU is shown in (c). The polyhedral representation of the framework is shown in (d). Metal atoms are represented as blue spheres (c) or polyhedra (d), nitrogen and carbon atoms are green and black spheres, respectively. Hydrogen atoms are omitted for clarity.(e) Rietveld refinement of the MET-6 compound showing the experimental (red), calculated (black) and difference (blue) patterns. Bragg positions are marked as black crosses. Inset: zoom of the high angle area. Reproduced from ref. [21], with permission from the Wiley-VCH, copyright 2012.

Ultrahigh porosity is one of the unique features of MOFs. MOFs with larger pores often end up with large unit cells. Formation of suitable, high-quality single crystals of such MOFs is extremely arduous. One such example is MIL-100, a MOF based on chromium trimesate that has a formula of $Cr_3F(H_2O)_3O[C_6H_3(CO_2)_3]_2{\cdot}nH_2O$ [24]. MIL-

100 was synthesized and fully characterized by an *ab initio* method using a computational algorithm based on the global optimization technique [23]. The assembly of the secondary building unit was monitored computationally, which explored how the organic linker and the inorganic cluster could connect together to direct the periodic structure. Along with the crystallographic features, the simulated PXRD patterns that were generated through virtual structures were compared with the experimental pattern. Highly crystalline green solids of MIL-100 were indexed as a cubic cell with lattice parameter a = 72.906 Å and the space group was found as *Fd-3m* [24]. In the structure solution, it was established that inorganic trimers are linked with benzene-1,3,5-tricarboxylate linker to form a potential hybrid model, which, in turn, build a super-tetrahedron (Figure 4.4b–d). The vertices of the super tetrahedron were occupied by the inorganic cluster while the linker occupies the faces. These super-tetrahedra are connected through the vertices to give rise to a corner sharing 3D network that resembles the MTN-type zeolitic topology. As shown in Figure 4.4a, both experimental and simulated PXRD patterns were in good agreement with a successful Rietveld refinement of R_{wp} = 7.61 % [24]. The structural analysis of MIL-100 reveals that the material is comprised of two cages. The larger cage is formed with about 28 corner-sharing super-tetrahedra with a diameter of 29 Å. The smaller cage is made up of ca. 20 super-tetrahedra with a diameter of 25 Å [24].

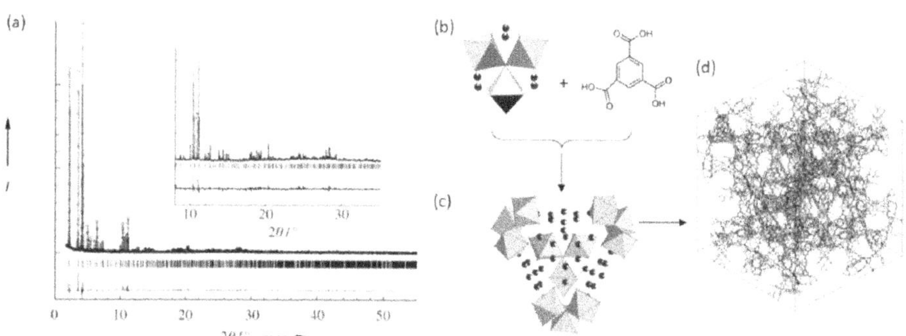

Figure 4.4: (a) Rietveld refinement of MIL-100 (1). (b) Building blocks consisting of the inorganic trimer of metal octahedra chelated by three carboxylic functions and trimesic acid is used to create the supertetrahedron. (c) View of the supertetrahedron with the trimesic acid in the faces of the tetrahedron. (d) View of the unit cell of MIL-100. Reproduced from ref. [24], with permission from the Wiley-VCH, copyright 2004.

Ultrasound irradiation of STAM-1 [25] leads to the formation of a new layered coordination polymer, STAM-2 [26], that forms only as nanocrystals. The quality of the PXRD pattern was significantly reduced with much broader Bragg reflections in the STAM-2 indicative of a loss in crystallinity and reduction of crystallite size making the structure solution by conventional Bragg diffraction methods alone difficult. Pair distribution function (PDF) analysis [27, 28] was used to identify the Cu-based secondary building unit in the material in order to constrain the model for an *ab initio* structure solution and show that the final structure can be successfully refined against the PDF data.

By indexing the observed PXRD pattern, a triclinic unit cell with parameters of a = 11.252 Å, b = 10.614 Å, c = 5.744 Å, α = 86.5°, β = 97.7° and γ = 106.8° were found [26]. PDF analysis was employed to probe the secondary building unit of STAM-2. The paddle wheel dimer in STAM-1 is characterized by a short Cu–Cu of ~2.6 Å and was absent in STAM-2 whereas the intense peaks at 3.0 and 3.4 Å suggested similarity of Cu–Cu distances of ~3.0 and 3.4 Å in Cu-SIP-3 (Figure 4.5a,b) [25, 26, 29]. However, relative intensities of peaks were not comparable indicating that the SBU was similar, but not identical to the one present in Cu-SIP-3 [26, 29]. Further investigation of the peaks in this solution revealed two potential oxygen atoms at interatomic distances of ~2 Å from the proposed copper atoms, which was helpful in determining the coordination around the Cu ions, and to identify the secondary building unit as chain of vertex shared copper tetramers. This structural model was then used as a starting point for Rietveld analysis against the PXRD data and was then refined against the PDF data set (Figure 4.5c). The structure of STAM-2 can be considered as a layered coordination

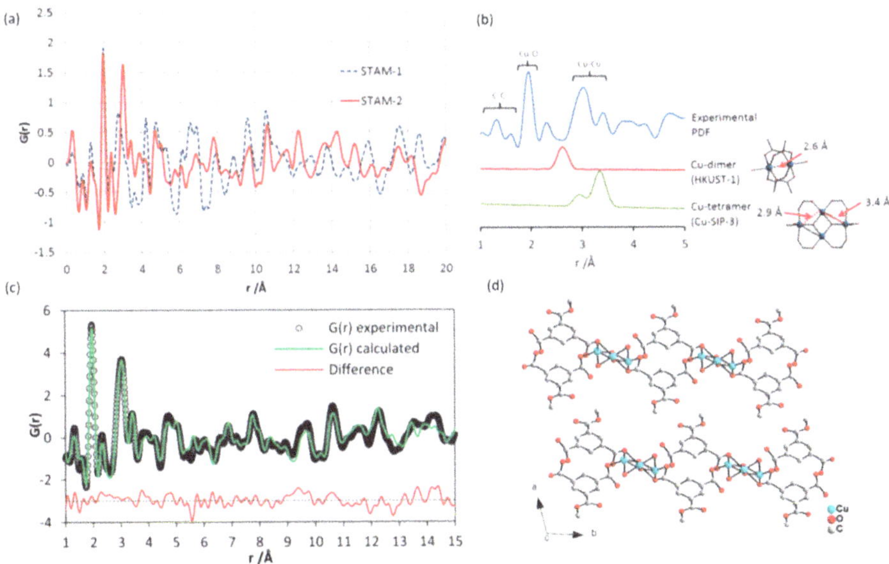

Figure 4.5: (a) PDF for STAM-1 (dotted line) and STAM-2 (red line), (b) Experimental PDF of STAM-2 (top) compared to simulated partial Cu-Cu PDFs for two known Cu-SBUs, the paddle wheel found in STAM-1 and HKUST-1 (middle) and the copper tetramer found in Cu-SIP-3 (bottom), (c) Refinement of the STAM-2 model against the PDF data. The black circles are the experimental data, the green line is the calculated PDF from the model and the red line is the difference between the two, offset by −3 for clarity, (d) The overall structure of STAM-2. Reproduced from ref. [26], with permission from the Royal Society of Chemistry, copyright 2014.

polymer network, comprising of the chain of vertex shared copper tetramers shown in Figure 4.5d. The chains run parallel to the crystallographic *c*-axis and are linked into layers via coordination to the two carboxylate groups of the monomethylbenzenetricarboxylate (mmBTC) linker [26].

4.3.4 Diffraction studies on the formation mechanism of MOFs

A fundamental understanding of the formation mechanism of MOFs is highly important to structural and morphological control in the synthesis in order to tune their properties. Therefore, different diffraction techniques have been employed in order to understand the crystallization process of MOFs.

The structure of ZIF-8 and its crystallization mechanism has been studied by using a combination of advanced characterization techniques such as *ex situ* powder diffraction, selected area electron diffraction (SAED), *in situ* small angle and wide-angle scattering (SAXS/WAXS) and time-resolved static light scattering [30–36]. The formation and gradual disappearance of prenucleation clusters of ZIF-8 in methanol during its early crystal growth stages was identified, which suggests the involvement of nanocrystal formation (Figure 4.6e) [34]. On the contrary, another study suggested a different mechanism as shown in Figure 4.6f that involved a phase transformation from a 2D ZIF-L phase in the presence of other solvents such as water, dimethylformamide and ethanol [35]. While the former studies were limited to methanol, the latter was more common in most of the other solvents.

Similarly, deployment of different metal precursors also has shown different crystallization pathways [31, 32]. For example, when a Zn-nitrate precursor was employed in the presence of an excess linker, ZIF-8 crystallized through a transformation of the metastable semicrystalline to crystalline by following the Avrami's kinetic regime that is illustrated in Figure 4.6 a–d [31]. However, when Zn-carbonate was employed instead, a nanosized ZnO wurtzite phase was identified at the early stage that was coexisting with ZIF-8 nanocrystals [32]. These studies clearly show that the diversity of crystallization mechanisms is strongly dependent on the synthetic parameters. Hence, deployment of complementary diffraction techniques is very useful for having in-depth studies of different MOFs [30].

Time resolved *in situ* energy dispersive X-ray diffraction (EDXRD) is a powerful characterization tool that allows nondestructive penetration of a high intensity X-ray beam through reaction vessels under elevated temperature. EDXRD experiments have been conducted for the characterization of numerous MOF systems including HKUST-1, UiO-66, MIL-100(Mn) and MOF-14 [37–40]. From the EDXRD studies, HKUST-1 crystallizes from a homogenous solution of DMF and ethanol after no detectable induction period [37]. The kinetics of the crystallization were studied at varied temperatures ranging from 85–125 °C in order to understand the crystallization mechanism. As shown in Figure 4.7a, a typical Avrami kinetic model resulted suggesting that the

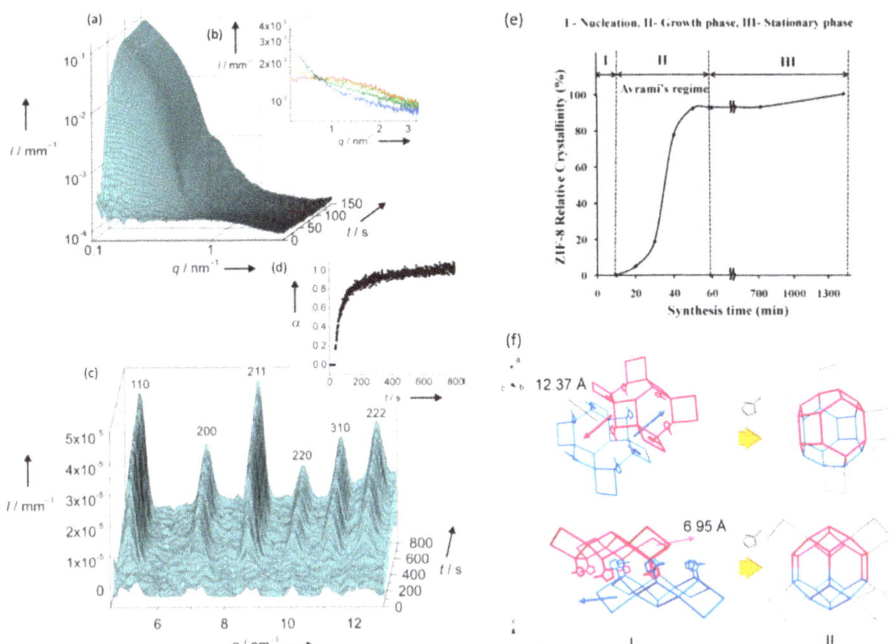

Figure 4.6: Time-resolved scattering patterns during ZIF-8 nanocrystal formation: (a) SAXS patterns for the first 150 s. The time interval between succeeding patterns is 1 s. (b) High-q region of selected SAXS patterns originating from the small particles (clusters). The time at which each pattern was measured is indicated by color: red 10 s, light green 30 s, dark green 50 s, blue 70 s. (c) WAXS patterns between 1 and 800 s. The time interval between succeeding patterns is 1 s. (d) Plot of the extent of crystallization a versus time t as produced from the integrated intensity of the 211 reflections in the WAXS patterns. Reproduced from ref. [34], with permission from the Wiley-VCH, copyright 2011. (e) Kinetics of transformation of ZIF-8 as a function of time. (f) Proposed Phase Transformation Pathways. Zn–MeIm–Zn bonds have been represented by sticks, and H atoms have been omitted for clarity. (upper): Sliding of semi-SOD adjacent crystal layers along the [100] direction. (lower): Sliding of adjacent semi-SOD crystal layers along the [010] direction. I: breakage of H-bonds and displacement of the adjacent layers away from each other. II: Breakage of Zn–MeIm bonds and formation of new Zn–MeIm–Zn bonds. Reproduced from ref. [31] & [35] with permission from the American Chemical Society, copyright 2010 & 2014.

crystallization is largely governed by the formation of nucleation sites rather than crystal growth at the sites or the diffusion of reactive species to the sites [37]. It further demonstrated that the HKUST-1 is a thermodynamically stable structure due to the trend of increased solvothermal stability with time of crystallization.

A similar study was conducted for MIL-53 (iron(II) terephthalate) using EDXRD at 150 °C [37]. Although the ultimate product that was expected is MIL-53, studies revealed that the crystallization was led by the transitory appearance of another crystalline phase, MOF-235 as illustrated by Figure 4.7b [37]. In MIL-53 trans-linked octahedral Fe, units are cross-linked by the dicarboxylates of the linker resulting in the over-

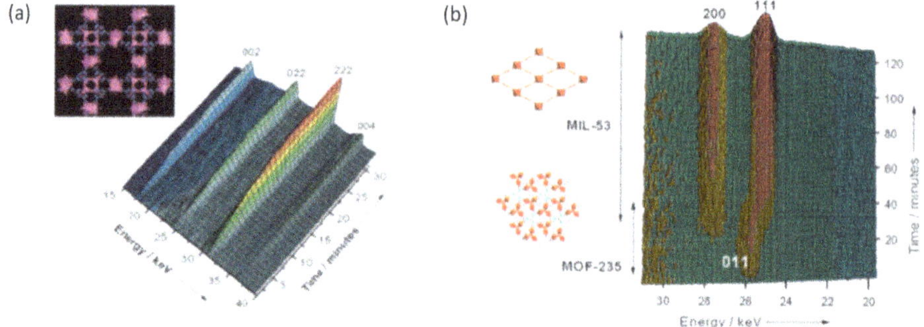

Figure 4.7: (a) Time-resolved in-situ EDXRD data measured during the crystallization of the copper carboxylate HKUST-1 at 125 °C. Inset: view of the structure of HKUST-1 with five-coordinate Cu-based units as pink polyhedral, (b) Time-resolved EDXRD measured during the crystallization of MIL-53 at 150 °C, with a crystalline transient phase seen at short reaction times. Reproduced from ref. [37], with permission from the Wiley-VCH, copyright 2016.

all structure, whereas in MOF-235, the overall framework is constructed from trimers of iron(III) oxy-octahedra linked by terephthalate ligands. While at low temperature, the intermediate has a longer lifetime, MIL-53 was eventually produced a result of prolonged heating [37].

4.4 Electron diffraction

Nanocrystals that were once considered too small for single crystal analysis are now providing structural information at the atomic level due to the development of three-dimensional electron diffraction (3DED) methods [41–44]. Therefore, 3DED has become a valuable and complementary technique to conventional X-ray diffraction techniques nowadays. 3DED studies have usually been conducted by using a transmission electron microscope (TEM). Using condenser lenses, the incoming electron beam passes through the sample nanocrystals. The nanocrystals, therefore, diffract electrons and through the pole piece of the objective lens electron diffraction patterns form on the back focal plane. Using intermediate and projector lenses, the electron diffraction patterns can further be magnified and focused on the detector where data is recorded.

Although the analysis of 3DED data is similar to X-ray data in many ways, there are fundamental differences regarding the formation of electron diffraction. X-ray is an electromagnetic wave, which does not have a resting mass, while the electron shares the wave-particle duality and it has a resting mass of $m = 9.10956 \times 10^{-28}$ g. Electrons interact with the nuclei and electrons in the crystals through electrostatic potential, which includes elastic scattering and inelastic scattering. Typically, electron scattering in crystalline materials can be simplified as a one-body Schrödinger equation for the incident electron as described in equation (4.10):

$$\left[-\frac{\hbar^2}{2m}\nabla^2 + \varphi(\mathbf{r})\right]\psi(\mathbf{r}) = E\psi(\mathbf{r}) \tag{4.10}$$

where h is the Planck constant, m is the resting mass of an electron, $\varphi(\mathbf{r})$ is the electrostatic potential and $\psi(\mathbf{r})$ is the wave function of one incident electron. The electron charge distribution can be calculated from the atomic wave function, which is, in turn, is provided by solving the Dirac equation. Another fundamental difference between X-ray and electron diffraction lies on the radius of their Ewald spheres. For electrons, the curvature is very large with respect to reciprocal-lattice vectors. Therefore, for the diffraction spots in a typical electron diffraction pattern, some of them do not exactly satisfy Bragg's condition. The difference is known as excitation error, s_g.

The electron diffraction intensity can be calculated if we assume an idealized case of single scattering of electrons by the electrostatic potential. The requirement of single scattering is known as the kinematic diffraction approximation or weak phase object (WPOA). It refers to the conditions of recording experimental data from very thin specimens (i. e., nanocrystals ca. <50 to 100 nm in thickness), do not contain heavy atoms (e. g., Pb or Au), and crystal symmetry where not all atoms scatter in phase (e. g., in the symmetry of *Fm*-3*m*) [45]. Similar to X-ray, the kinematical approximation indicates that the intensities are proportional to the square of structure factor F_{hkl}, $I_{hkl} \sim |F_{hkl}|^2$.

However, in practice, very few MOF crystals can fulfil the requirement of a weak phase object. Therefore, for 3DED, the *R*-values of the final refinement after applying the kinematic diffraction approximation are usually higher than those from X-ray data, though little difference can be observed on the framework structure [46]. Dynamical refinement is a method developed to tackle this challenge [47]. In summary, atoms in crystals are weakly scattering objects for X-rays, but they are strongly scattering objects for electrons. The interaction between matter and electron is 3-8 orders of magnitudes larger than that for X-rays. It thus allows us to obtain sufficient intensity from nanocrystals.

4.4.1 Protocols of 3DED

To obtain 3DED data, different protocols have been developed (Table 4.1).

In the 2000s, the first data acquisition protocols to obtain 3DED data sets were developed by stepwise tilting of the sample at fixed angular steps in a TEM, which is known as automated diffraction tomography (ADT) [48, 58] and rotation electron diffraction (RED) [50, 51]. After each tilt step, the goniometer stops to enable collection of an ED pattern (Figure 4.8). Following the goniometer tilt, a series of ED patterns can

Table 4.1: List of 3DED protocols.

Protocol name	Data collection strategy	Data collection speed	EM mode	Beam tilt	Beam procession	Publication year	Ref.
ADT/PEDT	Stepwise	Slow	STEM	No	Yes	2007	[48, 49]
RED	Stepwise	Slow	TEM	Yes	No	2010	[50, 51]
EDT	Stepwise	Slow	TEM	Yes	No	2013	[52]
MicroED	Stepwise	Slow	TEM	No	No	2013	[53]
Rotation electron diffraction	Continuous	Fast	TEM	No	No	2013	[54]
MicroED	Continuous	Fast	TEM	No	No	2014	[55]
Fast-EDT	Continuous	Fast	TEM	No	No	2015	[56]
cRED	Continuous	Fast	TEM	No	No	2018	[57]

be obtained. 3D reciprocal lattice can then be reconstructed from the collected ED sequence, which is the raw diffraction data for single crystal analysis. Different modes of TEM can be used for data collection. For example, ADT works in scanning transmission electron microscopy (STEM) mode using a nanosized beam and RED works in TEM mode using selected area electron diffraction (SAED) patterns. However, due to the accuracy limitation of the goniometers, there is a gap between two steps resulting in missing wedges in the data. This missing information can be filled by using precession electron diffraction (PED) [59] or fine beam tilting [11] (Figure 4.8). In the protocol of precession-assisted electron diffraction tomography (PEDT), the beam is tilted from the central axis of a TEM and scanning on a conical surface with the vertex fixed on the sample (Figure 4.8). In RED, the beam tilt step can be accurately adjusted to <0.1° by using the TEM deflection coils (Figure 4.8). Therefore, a RED data set combines large goniometer tilt steps (i. e., 2.0°) followed by a series of beam tilt steps (i. e., 0.1°) to fill the gap (i. e., 2.0°). Due to the instability of the goniometers, the crystal can move out of the electron beam. Thus, in PEDT and RED, image mode is used to track and recenter the crystal to maximize the reciprocal space coverage.

The most recently developed 3DED methods are based on continuous data collection, such as continuous rotation electron diffraction (cRED) [57, 60], integrated electron diffraction tomography (Fast-EDT) [56] and microcrystal electron diffraction (MicroED) [53, 55]. Contrary to the stepwise methods, the goniometer in the continuous methods rotates at a constant rate and does not stop during the entire data collection (Figure 4.8). As a result, the angular step between frames is determined by the exposure time and the goniometer rotation rate. The development of continuous methods brings a remarkable advantage that a high rotation rate can be used for data acquisition. It is clear that increasing the data collection rate can reduce the total time in which the sample is exposed to the electron beam. Therefore, one major challenge in electron diffraction, beam damage, can be minimized. This is important to allow

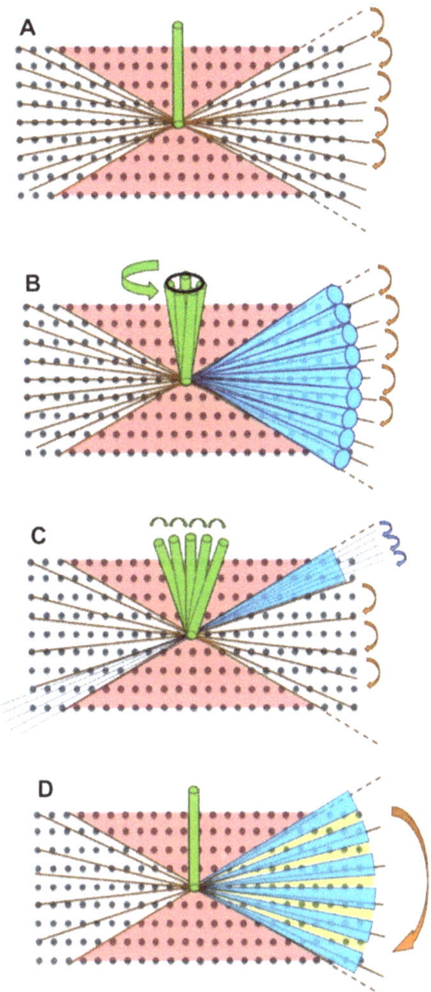

Figure 4.8: Schematic illustrations of 3DED data collection protocols. The stepwise acquisition performed by (a) goniometer tilt only, (b) combination of goniometer tilt and precession electron beam, and (c) combination of goniometer tilt and beam tilt. (d) The continuous rotation method. Brown, goniometer tilt; green, electron beam; blue, data range; yellow, detector readout time [44]. Reproduced from ref. [44], with permission from the American Chemical Society, copyright 2019.

3DED to study soft materials, such as MOFs and COFs. Furthermore, because the missing wedges are directly sampled, the continuous methods provide a more accurate extract of integrated intensities.

4.4.2 Structure determination

4.4.2.1 Data processing

The 3DED protocols share the same concept that the 3D reciprocal space is sampled by a sequence of ED patterns. Several software, including *REDp* [51], *ADT3D* [61], *PETS* [62] and *EDT-PROCESS* [52], have been developed to process 3DED data to reconstruct and visualize 3D reciprocal space, determine unit cells and space groups, index reflections and integrate reflection intensities (Figure 4.9). In addition, the recent development of continuous methods enables utilization of software developed for X-ray crystallography, such as XDS [63, 64] and DIALS [65, 66], for processing and analyzing 3DED data sets. Notably, due to the limitation of the tilting range of goniometer, which can lead to a low completeness, merging data from crystals with different orientations can effectively improve data completeness.

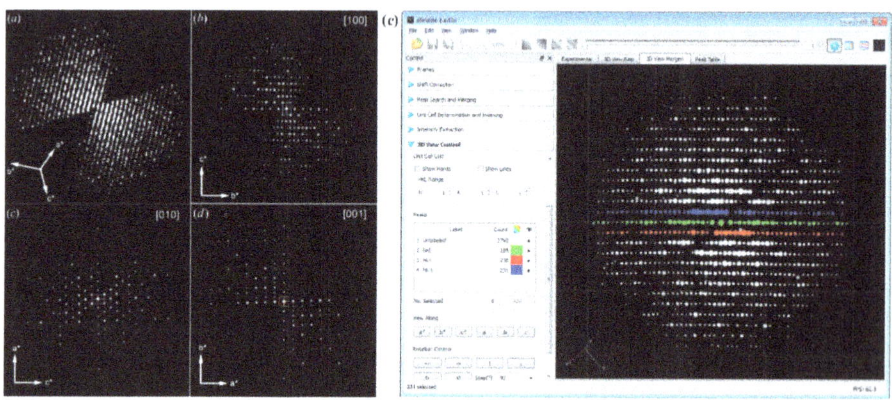

Figure 4.9: (a) The reconstructed 3D reciprocal lattice of zeolite silicalite-1 from the RED data. (b)–(d) Slice cuts from the reconstructed 3D reciprocal lattice showing the (b) 0kl, (c) h0l and (c) hk0 planes. (e) The graphical user interface of the RED data processing program [51]. Reproduced from ref. [51], with permission from the International Union of Crystallography, copyright 2013.

4.4.2.2 Structure solution

As the described 3DED protocols share a common concept that data are collected through a range of arbitrary orientations, it leads to fewer geometrically related reflections and dynamical effects can be drastically reduced. Thus, the intensities extracted from 3DED data can be used as kinematical approximation for *ab initio* structure determination.

For a perfect crystal, Fourier transforming real-space crystal lattice results in a reciprocal lattice, in which the positions of diffracted beams are determined. The amplitude and phase of the diffracted beams are determined by structure factor $\boldsymbol{F}_{hkl}$, which is the vector sum of waves from overall atoms within the unit cell. The structure factor can be expressed by equation (4.11):

$$\mathbf{F}_{hkl} = F_{hkl}e^{i\alpha_{hkl}} = \sum_{j}^{N} f_j e^{[-2\pi i(hx_j+ky_j+lz_j)]} \tag{4.11}$$

where α_{hkl} is the phase of the diffracted beam, f_j is the atomic scattering factor and x_j, y_j, z_j are the atomic coordinates. While the intensity of a diffracted beam is directly related to the amplitude of the structure factor, the phase information is absent by using diffraction techniques. To obtain the phase information and solve the crystal structure of inorganic materials, the direct methods are the most used method [67]. Based on modified electron potential maps, it is thus possible to extract enough of the atom positions to solve the structure.

Real-space methods have also been used for structure solution, especially for MOFs and COFs, because the organic molecules within the structure can normally be characterized by other techniques. Simulated annealing is one of the most effective real-space methods to solve structures using electron data. It is based on a simulation process that a crystal solidifies from a melt. Generally, a slow process can lead to a crystal while a fast quench process leads to a glass form. During the process of simulated annealing, parameters are randomly assigned to a possible structural model, from which structure factors are calculated and compared to those obtained from electron diffraction data. The parameters are kept if a good agreement is reached, and the process runs for thousands of iterations to find a best fit of parameters.

Charge flipping [68, 69], a dual-space methods, is another method for solving structure from 3DED data. The key step is the charge flipping operation, where the negative peaks in the electron density function are set to positive. A new structure factor is then calculated from the new density and the new phases are kept while the amplitudes are replaced by the experimental values. The process is run for a repeated iteration until a converging solution is found.

Software packages such as SHELX [70, 71], SIR [72] and JANA [73] are the common programs used for structure solution from 3DED data. As most software are initially developed for X-ray crystallography, it is important to ensure that all the parameters are used for electrons, such as wavelength and scattering factors.

4.4.2.3 Structure refinement

An initial structure model can consist of only a part of atoms in the unit cell. Therefore, it is important to locate the missing atoms. As a set of structure factors can be calculated from the initial model, an electron potential map with coefficients of $|\boldsymbol{F}_0|$ can be obtained using the calculated phases. Practically, it is more useful to calculate difference maps using coefficients of $(|\boldsymbol{F}_0| - |\boldsymbol{F}_c|)$ with the calculated phases. In difference maps, an insufficient amount of electron potential generates peaks, which indicates missing atoms, as an example, while too much electron potential produces negative holes indicating the wrong assignment of atom types. In this way, new atoms can be located from the map and added to the model and this process is repeated until a satisfactory structural model has been obtained. Meanwhile, the structure factors from the models are calculated and compared with those experimentally observed. Least-square refinement procedures are used to minimize the difference between calculation and experiment similar to the refinement in X-ray diffraction.

Using kinematical approximation for 3DED data results in high *R* values when compared to those for X-ray data. This has been a concern for the 3DED field. However, recent studies on inorganic compounds have proved that the structural models obtained from 3DED are almost identical to those obtained by X-ray diffraction with average deviations of atomic positions by less than 0.1 Å [74–76]. This shows that the structural models obtained from 3DED data are indeed accurate and reliable. In addition, dynamical refinement [77] was developed, where the model intensities are calculated using the dynamical diffraction theory of electrons [78].

4.4.3 Applications of 3DED for studies of MOFs

With the development of 3DED techniques, *ab initio* structure determination of MOF nanocrystals can be performed as easy as single crystal X-ray diffraction (SCXRD) [79–81]. The first coordination polymer solved using electron diffraction data was the ζ-phase of Pigment Red [82]. Using a stepwise tilt-series, ED patterns of the needle-shaped crystals were collected. They are used to obtain the structural model after extracting the intensities. It reveals a specific molecular packing of Pigment Red (Figure 4.10). Because the small pigment crystals could only be acquired in a phase mixture, electron diffraction technique showed its unique advantage in this case.

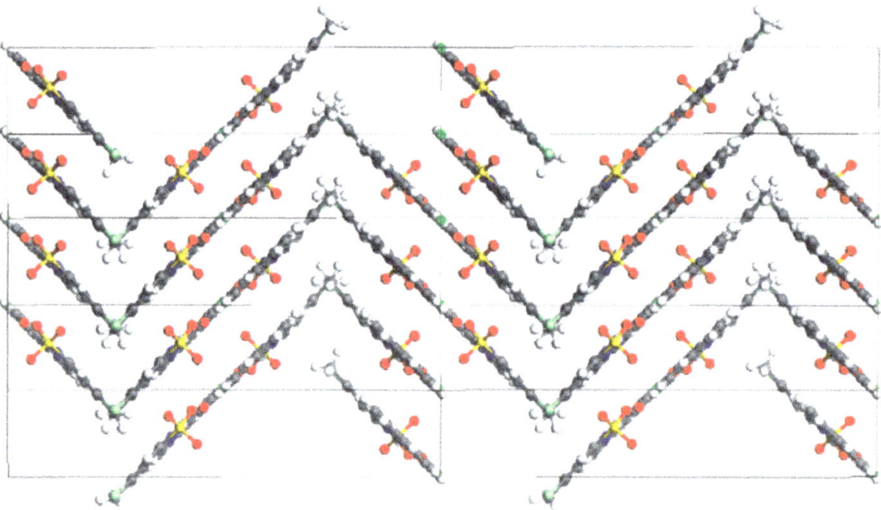

Figure 4.10: Structural models of ζ-phase Pigment Red [82] determined by using 3DED data. Reproduced from ref. [82], with permission from the American Chemical Society, copyright 2009.

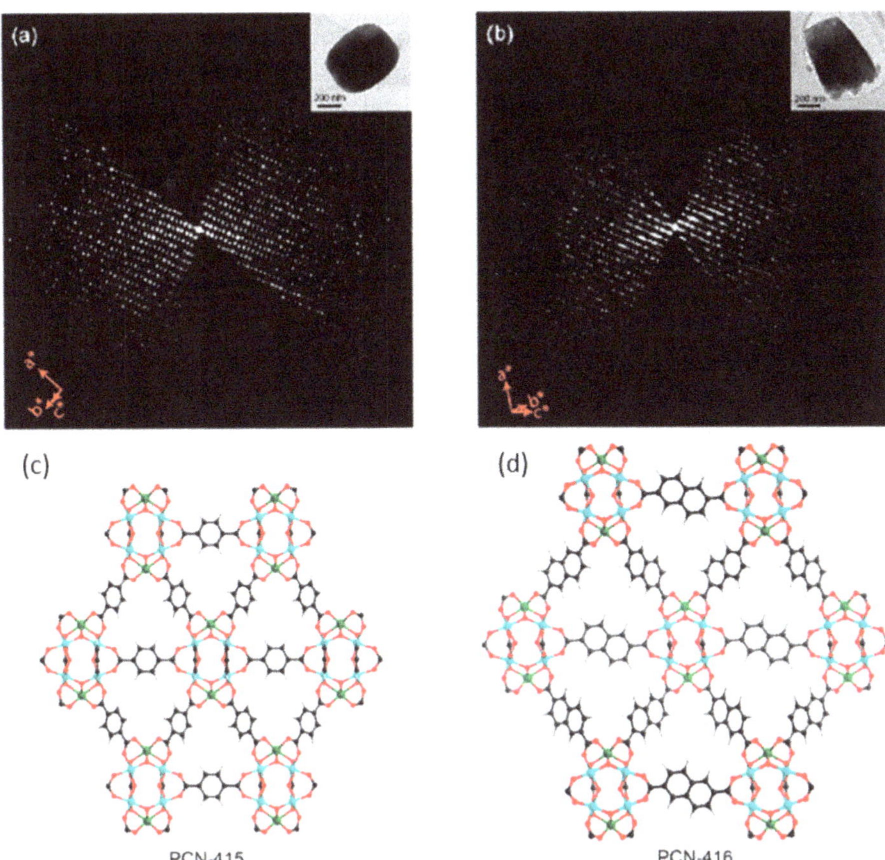

Figure 4.11: Reconstructed 3D reciprocal lattice of (a) PCN-415 and (b) PCN-416. (c) and (d) Structural models of PCN-415 and PCN-416 determined by 3DED.[83] Reproduced from ref. [83] with permission from the American Chemical Society, copyright 2018.

The structures of two isoreticular MOFs, PCN-415 and PCN-416, were successfully determined by using 3DED data at resolution of 0.75 and 1.05 Å, respectively (Figure 4.11) [83]. PCN-415 was synthesized by connecting the $[Ti_8Zr_2O_{12}(MeCOO)_{16}]$ clusters and 1,4-benzenedicarboxylate (BDC) linkers. Expanding BDC into 2,6-naphthalenedicarboxylate (NDC) gives rise to an isoreticular MOF, PCN-416. The structural models obtained from 3DED were compared with those refined against PXRD. The atomic positions refined by 3DED and PXRD showed a good agreement with a difference on average of 0.032 Å for Zr and Ti atoms and 0.071 Å for O and C atoms. The small deviation highlights the accuracy and reliability of the 3DED method. In addition, 3DED data enables precise locations of Ti and Zr in the mixed metal cluster, which is crucial for understanding the structure-property relationship and providing insights to the photoactivity of PCN-415 and PCN-416.

Furthermore, 3DED methods can be used to probe dynamic motions in MOFs [84]. Due to the high-quality 3DED data, the structures of two Zr-MOFs, MIL-140C and UiO-67, coexisting in a mixture, were determined *ab initio*. In addition, by refining the anisotropic displacement parameters (ADPs) the molecular motions of the linker molecules in both structures can be revealed (Figure 4.12). The linkers perform small amplitude liberations along the C-C axis. Interestingly, in MIL-140C, different degrees of motion of the same linker molecule can be identified depending on its local environment. It shows that 3DED can offer valuable information for the analysis of dynamic changes.

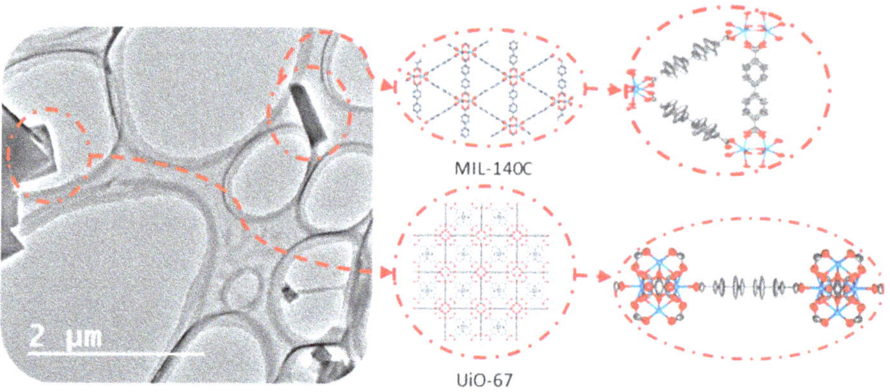

Figure 4.12: The structural and elliptical models of MIL-140C and UiO-67 which are determined by 3DED. Molecular motions of the linker molecules were revealed by using 3DED data. Reproduced from ref. [84] with permission from the American Chemical Society, copyright 2021.

In addition to structural analysis, by taking the advantage of the rapid data collection, 3DED can be used as a high-throughput analytical tool for discovering new materials, especially for those initially synthesized at a low concentration and in a phase mixture. A recent study reports the high-throughput single crystal analysis by 3DED leads to discovery of a MOF, ZIF-EC1, which shows a unique structure associated with electrocatalytic properties [85]. The material was discovered as a minor phase in the crystalline powder with the already known ZIF-C (Figure 4.13).

The X-ray and electron diffraction methods discussed above are both powerful techniques for structure determination, but both have their own limitations. Although usage of one technique is usually sufficient to employ in structure determination, in complicated circumstances, one technique alone is inadequate to solve the structure. Therefore, diverse techniques that provide complementary structural information can support each other to provide a complete successful structure determination. Although structure elucidation is well established in reticular chemistry, there is still

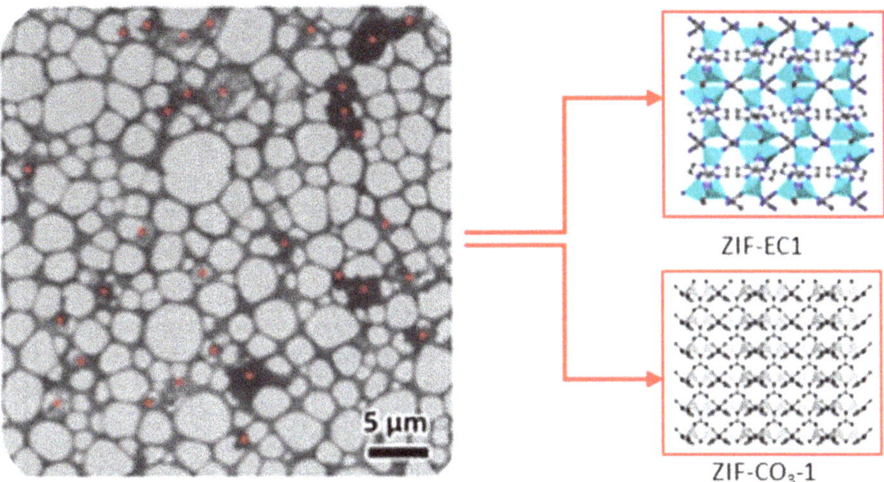

Figure 4.13: Examples of distinct MOF structures ZIF-EC1 and ZIF-C determined from phase mixtures by 3D ED. Reproduced from ref. [85] with permission from the John Wiley & Sons, Inc., copyright 2021.

much more to be explored in comprehending the formation of crystals and their respective mechanisms.

Bibliography

[1] Cheetham AK, Férey G, Loiseau T. Open-framework inorganic materials. Angew Chem Int Ed. 1999;38:3268–92.

[2] Parnham ER, Morris RE. Ionothermal synthesis of zeolites, metal-organic frameworks, and inorganic–organic hybrids. Acc Chem Res. 2007;40:1005–13.

[3] Venna SR, Jasinski JB, Carreon MA. Structural evolution of zeolitic imidazolate framework-8. J Am Chem Soc. 2010;132:18030–3.

[4] Usman, Ken AS et al. Downsizing metal-organic frameworks by bottom-up and top-down methods. NPG Asia Mater. 2020;12(1):1–18.

[5] Zhang B et al. Solvent determines the formation and properties of metal-organic frameworks. RSC Adv. 2015;5:37691–6.

[6] Cheng X et al. Size- and morphology-controlled NH_2-MIL-53(Al) prepared in DMF–water mixed solvents. Dalton Trans. 2013;42:13698–705.

[7] Schaate A, Roy P, Godt A, Lippke J, Waltz F, Wiebcke M, Behrens P. Modulated synthesis of Zr-based metal organic frameworks: from nano to single crystals. Chem Eur J. 2011;17:6643–51.

[8] Gandara F, Bennett TD. Crystallography of metal-organic frameworks. IUCrJ. 2014;1:563–70.

[9] Tsuruoka T, Furukawa S, Takashima Y, Yoshida K, Isoda S, Kitagawa S. Nanoporous nanorods fabricated by coordination modulation and oriented attachment growth. Angew Chem Int Ed. 2009;48:4739–43.

[10] Umemura A, Diring S, Furukawa S, Uehara H, Tsuruoka T, Kitagawa S. Morphology design of porous coordination polymer crystals by coordination modulation. J Am Chem Soc. 2011;133:15506–13.

[11] Martí-Rujas J. Structural elucidation of microcrystalline MOFs from powder X-ray diffraction. Dalton Trans. 2020;49:13897.
[12] Müller P. Crystal Structure Refinement: A Crystallographers Guide to SHELXL. New York: OUP Oxford; 2006.
[13] Bennett DW. Understanding Single-Crystal X-Ray Crystallography. Weinheim: Wiley-VCH; 2010.
[14] Clegg W. Crystal Structure Determination. New York: OUP Oxford; 1998.
[15] Pecharsky VK, Zevalij PY. Fundamentals of Powder Diffraction and Structural Characterization of Materials. Springer; 2009.
[16] Rietveld HM. J Appl Crystallogr. 1969;2:65–71.
[17] Young RA. The Rietveld Method. OUP; 1993.
[18] Harris KDM, Tremayne M, Kariuki BM. Contemporary advances in the use of powder X-ray diffraction for structure determination. Angew Chem Int Ed. 2001;40:1626–51.
[19] Harris KDM, Williams AP. Powder diffraction. In: Structure from Diffraction Methods. 1st ed. John Wiley & Sons, Ltd; 2014.
[20] Cavka JH, Jakobsen S, Olsbye U, Guillou N, Lamberti C, Bordiga S, Lillerud KP. A new zirconium inorganic building brick forming metal organic frameworks with exceptional stability. J Am Chem Soc. 2008;130:13850–1.
[21] Ga´ndara F, Uribe-Romo FJ, Britt DK, Furukawa H, Lei L, Cheng R, Duan X, O'Keeffe M, Yaghi OM. Porous, conductive metal-triazolates and their structural elucidation by the charge-flipping method. Chem Eur J. 2012;18:10595–601.
[22] Oszla´nyi G, Sütő A. The charge-flipping algorithm. Acta Crystallogr. 2008;A64:123–34.
[23] Mellot-Draznieks C, Newsam JM, Gorman AM, Freeman CM, Férey G. De novo prediction of inorganic structures developed through automated assembly of secondary building units. Angew Chem Int Ed. 2000;39:2270–5.
[24] Férey G, Serre C, Mellot-Draznieks C, Millange F, Surble´ S, Dutour J, Margiolaki I. A hybrid solid with giant pores prepared by a combination of targeted chemistry, simulation, and powder diffraction. Angew Chem Int Ed. 2004;43:6296–301.
[25] Mohideen MIH, Xiao B, Wheatley PS, McKinlay AC, Li Y, Slawin AMZ, Aldous DW, Cessford NF, Duren T, Zhao X, Gill R, Thomas KM, Griffin JM, Ashbrook SE, Morris RE. Protecting group and switchable pore-discriminating adsorption properties of a hydrophilic-hydrophobic metal-organic framework. Nat Chem. 2011;3:304–10.
[26] Infas Mohideen M, Allan PK, Chapman KW, Hriljacc JA, Morris RE. Ultrasound-driven preparation and pair distribution function-assisted structure solution of a copper-based layered coordination polymer. Dalton Trans. 2014;43:10438–42.
[27] Billinge SJL, Kanatzidis MG. Beyond crystallography: the study of disorder, nanocrystallinity and crystallographically challenged materials with pair distribution functions. Chem Commun. 2004;4:749–60.
[28] Egami T, Billinge SJL. Underneath the Bragg Peaks: Structural Analysis of Complex Materials. Oxford, UK, Boston: Pergamon, Kiddington; 2003.
[29] Xiao B, Byrne PJ, Wheatley PS, Wragg DS, Zhao X, Fletcher AJ, Thomas KM, Peters L, Evans JSO, Warren JE, Zhou W, Morris RE. Chemically blockable transformation and ultraselective low-pressure gas adsorption in a non-porous metal organic framework. Nat Chem. 2009;1:289–94.
[30] Mazaj M, Kaučič V, Zabukovec Logar N. Chemistry of metal-organic frameworks monitored by advanced X-ray diffraction and scattering techniques. Acta Chim Slov. 2016;63:440–58.
[31] Venna SR, Jasinski JB, Carreon MA. Structural evolution of zeolitic imidazolate framework-8. J Am Chem Soc. 2010;132:18030–3.
[32] Zhu M, Venna SR, Jasinski JB, Carreon MA. Room-temperature synthesis of ZIF-8: the coexistence of ZnO nanoneedles. Chem Mater. 2011;23:3590–2.

[33] Cravillon J, Nayuk R, Springer S, Feldhoff A, Huber K, Wiebcke M. Controlling zeolitic imidazolate framework nano- and microcrystal formation: insight into crystal growth by time-resolved in situ static light scattering. Chem Mater. 2011;23:2130–41.
[34] Cravillon J, Schröder CA, Nayuk R, Gummel J, Huber K, Wiebcke M. Fast nucleation and growth of ZIF-8 nanocrystals monitored by time-resolved in situ small-angle and wide-angle X-ray scattering. Angew Chem Int Ed. 2011;50:8067–71.
[35] Low ZX, Yao J, Liu Q, He M, Wang Z, Suresh AK, Bellare J, Wang H. Crystal transformation in zeolitic-imidazolate framework. Cryst Growth Des. 2014;14:6589–98.
[36] Friščić T, Halasz I, Beldon PJ, Belenguer AM, Adams F, Kimber SAJ, Honkimäki V, Dinnebier RE. In situ X-ray diffraction monitoring of a mechanochemical reaction reveals a unique topology metal-organic framework. Nat Chem. 2012;5:66–73.
[37] Millange F, Medina MI, Guillou N, Férey G, Golden KM, Walton RI. Time-resolved in situ diffraction study of the solvothermal crystallization of some prototypical metal-organic frameworks. Angew Chem Int Ed. 2010;49:763–6.
[38] Ragon F, Horcajada P, Chevreau H, Hwang YK, Lee UH, Miller SR, Devic T, Chang JS, Serre C. In situ energy-dispersive X-ray diffraction for the synthesis optimization and scale-up of the porous zirconium terephthalate UiO-66. Inorg Chem. 2014;53:2491–500.
[39] Reinsch H, Stock N. Formation and characterisation of Mn-MIL-100. CrystEngComm. 2013;15:544.
[40] Millange F, El Osta R, Medina ME, Walton RI. A time-resolved diffraction study of a window of stability in the synthesis of a copper carboxylate metal–organic framework. CrystEngComm. 2011;13:103.
[41] Huang Z, Grape ES, Li J, Inge AK, Zou X. 3D electron diffraction as an important technique for structure elucidation of metal-organic frameworks and covalent organic frameworks. Coord Chem Rev. 2021;427:213583.
[42] Huang Z, Willhammar T, Zou X. Three-dimensional electron diffraction for porous crystalline materials: structural determination and beyond. Chem Sci. 2021;12:1206–19.
[43] Gruene T, Holstein JJ, Clever GH, Keppler B. Establishing electron diffraction in chemical crystallography. Nat Rev Chem. 2021;5:660–8.
[44] Gemmi M, Mugnaioli E, Gorelik TE, Kolb U, Palatinus L, Boullay P, Hovmöller S, Abrahams JP. 3D electron diffraction: the nanocrystallography revolution. ACS Cent Sci. 2019;5:1315–29.
[45] Neumann W, Komrska J, Hofmeister H, Heydenreich J. Interpretation of the shape of electron diffraction spots from small polyhedral crystals by means of the crystal shape amplitude. Acta Crystallogr A. 1988;44:890–7.
[46] Huang Z, Ge M, Carraro F, Doonan C, Falcaro P, Zou X. Can 3D electron diffraction provide accurate atomic structures of metal-organic frameworks? Faraday Discuss. 2021;225:118–32.
[47] Palatinus L, Petříček V, Corrêa CA. Structure refinement using precession electron diffraction tomography and dynamical diffraction: theory and implementation. Acta Crystallogr A. 2015;71:235–44.
[48] Kolb U, Gorelik T, Kübel C, Otten MT, Hubert D. Towards automated diffraction tomography: Part I—Data acquisition. Ultramicroscopy. 2007;107:507–13.
[49] Boullay P, Palatinus L, Barrier N. Precession electron diffraction tomography for solving complex modulated structures: the case of $Bi_5Nb_3O_{15}$. Inorg Chem. 2013;52:6127–35.
[50] Zhang D, Oleynikov P, Hovmöller S, Zou X. Collecting 3D electron diffraction data by the rotation method. Z. Kristallogr. International Journal for Structural, Physical, and Chemical Aspects of Crystalline Materials Z Kristallogr. 2010;225:94–102.
[51] Wan W, Sun J, Su J, Hovmöller S, Zou X. Three-dimensional rotation electron diffraction: software RED for automated data collection and data processing. J Appl Crystallogr. 2013;46:1863–73.

[52] Gemmi M, Oleynikov P. Scanning reciprocal space for solving unknown structures: energy filtered diffraction tomography and rotation diffraction tomography methods. Z Kristallogr Cryst Mater. 2013;228:51–8.
[53] Shi D, Nannenga BL, Iadanza MG, Gonen T. Three-dimensional electron crystallography of protein microcrystals. eLife. 2013;2:e01345.
[54] Nederlof I, van Genderen E, Li YW, Abrahams JP. A medipix quantum area detector allows rotation electron diffraction data collection from submicrometre three-dimensional protein crystals. Acta Crystallogr D. 2013;69:1223–30.
[55] Nannenga BL, Shi D, Leslie AGW, Gonen T. High-resolution structure determination by continuous-rotation data collection in MicroED. Nat Methods. 2014;11:927–30.
[56] Gemmi M, La Placa MGI, Galanis AS, Rauch EF, Nicolopoulos S. Fast electron diffraction tomography. J Appl Crystallogr. 2015;48:718–27.
[57] Cichocka MO, Ångström J, Wang B, Zou X, Smeets S. High-throughput continuous rotation electron diffraction data acquisition via software automation. J Appl Crystallogr. 2018;51:1652–61.
[58] Kolb U, Gorelik T, Otten MT. Towards automated diffraction tomography. Part II—cell parameter determination. Ultramicroscopy. 2008;108:763–72.
[59] Vincent R, Midgley PA. Double conical beam-rocking system for measurement of integrated electron diffraction intensities. Ultramicroscopy. 1994;53:271–82.
[60] Smeets S, Wang B, Cichocka MO, Ångström J, Wan W. Instamatic. Zenodo 2019.
[61] Kolb U, Mugnaioli E, Gorelik TE. Automated electron diffraction tomography – a new tool for nano crystal structure analysis. Cryst Res Technol. 2011;46:542–54.
[62] Palatinus L, Brázda P, Jelínek M, Hrdá J, Steciuk G, Klementová M. Specifics of the data processing of precession electron diffraction tomography data and their implementation in the program PETS2.0. Acta Crystallogr B. 2019;75:512–22.
[63] Kabsch WXDS. Acta Crystallogr D. 2010;66:125–32.
[64] Kabsch W. Integration, scaling, space-group assignment and post-refinement. Acta Crystallogr D. 2010;66:133–44.
[65] Winter G, Waterman DG, Parkhurst JM, Brewster AS, Gildea RJ, Gerstel M, Fuentes-Montero L, Vollmar M, Michels-Clark T, Young ID, Sauter NK, Evans G. DIALS: implementation and evaluation of a new integration package. Acta Crystallogr D. 2018;74:85–97.
[66] Clabbers MTB, Gruene T, Parkhurst JM, Abrahams JP, Waterman DG. Electron diffraction data processing with DIALS. Acta Crystallogr D. 2018;74:506–18.
[67] Hauptman H. The direct methods of X-ray crystallography. Science. 1986;233:178–83.
[68] Oszlányi G, Sütő A. Ab initio structure solution by charge flipping. II. Use of weak reflections. Acta Crystallogr A. 2005;61:147–52.
[69] Oszlányi G, Sütő A. The charge flipping algorithm. Acta Crystallogr A. 2008;64:123–34.
[70] Sheldrick GM. A short history of SHELX. Acta Crystallogr A. 2008;64:112–22.
[71] Sheldrick GM. SHELXT – integrated space-group and crystal-structure determination. Acta Crystallogr A. 2015;71:3–8.
[72] Burla MC, Caliandro R, Carrozzini B, Cascarano GL, Cuocci C, Giacovazzo C, Mallamo M, Mazzone A, Polidori G. Crystal structure determination and refinement via SIR2014. J Appl Crystallogr. 2015;48:306–9.
[73] Petříček V, Dušek M, Palatinus L. Crystallographic computing system JANA2006: general features. Z Kristallogr Cryst Mater. 2014;229:345–52.
[74] Wang Y, Yang T, Xu H, Zou X, Wan W. On the quality of the continuous rotation electron diffraction data for accurate atomic structure determination of inorganic compounds. J Appl Crystallogr. 2018;51:1094–101.
[75] Huang Z, Ge M, Carraro F, Doonan CJ, Falcaro P, Zou X. Can 3D Electron Diffraction Provide Accurate Atomic Structures of Metal-Organic Frameworks?

[76] Roslova M, Smeets S, Wang B, Thersleff T, Xu H, Zou X. InsteaDMatic: towards cross-platform automated continuous rotation electron diffraction. J Appl Crystallogr. 2020;53:1217–24.
[77] Palatinus L, Brázda P, Boullay P, Perez O, Klementová M, Petit S, Eigner V, Zaarour M, Mintova S. Hydrogen positions in single nanocrystals revealed by electron diffraction. Science. 2017;355:166–9.
[78] Zuo JM, Spence JCH. Automated structure factor refinement from convergent-beam patterns. Ultramicroscopy. 1991;35:185–96.
[79] Portolés-Gil N, Lanza A, Aliaga-Alcalde N, Ayllón JA, Gemmi M, Mugnaioli E, López-Periago AM, Domingo C. Crystalline curcumin BioMOF obtained by precipitation in supercritical CO2 and structural determination by electron diffraction tomography. ACS Sustain Chem Eng. 2018;6:12309–19.
[80] Cichocka MO, Liang Z, Feng D, Back S, Siahrostami S, Wang X, Samperisi L, Sun Y, Xu H, Hedin N, Zheng H, Zou X, Zhou HC, Huang Z. A porphyrinic zirconium metal–organic framework for oxygen reduction reaction: tailoring the spacing between active-sites through chain-based inorganic building units. J Am Chem Soc. 2020;142:15386–95.
[81] Ge M, Zou X, Huang Z. Three-dimensional electron diffraction for structural analysis of beam-sensitive metal-organic frameworks. Crystals. 2021;11:263.
[82] Gorelik T, Schmidt MU, Brüning J, Beko S, Kolb U. Using electron diffraction to solve the crystal structure of a laked azo pigment. Cryst Growth Des. 2009;9:3898–903.
[83] Yuan S, Qin JS, Xu HQ, Su J, Rossi D, Chen Y, Zhang L, Lollar C, Wang Q, Jiang HL, Son DH, Xu H, Huang Z, Zou X, Zhou HC. $[Ti_8Zr_2O_{12}(COO)_{16}]$ cluster: an ideal inorganic building unit for photoactive metal-organic frameworks. ACS Cent Sci. 2018;4:105–11.
[84] Samperisi L, Jaworski A, Kaur G, Lillerud KP, Zou X, Huang Z. Probing molecular motions in metal–organic frameworks by three-dimensional electron diffraction. J Am Chem Soc. 2021;143:17947–52.
[85] Ge M, Wang Y, Carraro F, Liang W, Roostaeinia M, Siahrostami S, Proserpio DM, Doonan C, Falcaro P, Zheng H, Zou X, Huang Z. High-throughput electron diffraction reveals a hidden novel metal-organic framework for electrocatalysis. Angew Chem Int Ed. 2021;60:11391–7.

Ayalew H. Assen

5 Metal-organic frameworks for industrial gas separation

5.1 Introduction

In the chemical industry, the separation and purification of gases is a crucial yet challenging task representing 15–20 % of worldwide energy consumption. It is critical to have high-purity feedstocks and processes that lead to pure final products with minimum emissions of hazardous gases into the environment. Therefore, the development of systems and materials for effective separation and purification of raw materials and outputs is of the utmost importance. CO_2 capture from different sources, air separation, and splitting light hydrocarbon mixtures into their pure compounds are among the industrially relevant gas separation processes [1, 2]. However, these separations depend on energy-intensive cryogenic distillation technology (fractional distillation at very low temperatures) that operates based on differences in boiling points of the components to be separated [3]. There is a great interest to develop a more economical option that can complement or substitute the conventional distillation technique. In this regard, adsorptive separation, which is based on the selective binding affinity of gases to porous adsorbents or molecular size/shape differentiation of the components by pore apertures, is an active area of research. Using adsorption technology instead of distillation has potential to result in energy savings of up to 80 %. The forces that lead molecules to interact with the surface of adsorbents can be classified as either chemi or physisorption. Physisorption-based procedures are being more explored in most gas separations due to the considerable advantage in material regeneration that drastically alters the economics of such separation processes. To adsorb the gases to be separated, and hence to release from adsorbent materials, either pressure or temperature swing adsorption (PSA or TSA) setups or a combination thereof are utilized.

Adsorptive separation by porous physisorbents is achieved via different mechanisms. In the most prevalent separation mechanism (i. e., thermodynamic equilibrium separation), variations in adsorbate-surface and adsorbate packing interactions dictate the separation efficiency. It is advantageous when the surface of an adsorbent displays preferential adsorption of particular components over others. Alternatively, adsorptive separation can be achieved through either a diffusion-controlled kinetic separation mechanism or a molecular sieving effect. Kinetic separation is based on

Ayalew H. Assen, Technology Development Cell (TechCell), Applied Chemistry and Engineering Research Centre of Excellence (ACER CoE), Mohammed VI Polytechnic University (UM6P), Lot 660 – Hay Moulay Rachid, 43150, Ben Guerir, Morocco; and Department of Chemistry, College of Natural Science, Wollo University, Dessie, Ethiopia, e-mail: ayalew.assen@um6p.ma

https://doi.org/10.1515/9781501524721-005

varying diffusing rates that cause certain components to enter the pores and subsequently get adsorbed faster than others. In size or shape exclusion (the molecular sieving effect), only certain, distinct components of a gas mixture are allowed to enter the pores of an adsorbent to be adsorbed, while the remaining fractions are blocked from passing through the pore openings. Molecular sieving then requires an adsorbent with a cut-off aperture size to sieve one component from the others. Gas separation by total molecular exclusion is an ultimate target in the development of adsorbents for gas separation of isomers with identical physical and chemical characteristics, yet distinct sizes and/or shapes.

Conventional physisorbents such as zeolites, silica, and activated carbon have been employed in adsorption separation processes [4]. Alternative porous solids to separate gases are highly sought after due to the limitations associated with these traditional adsorbents. In the continued effort to develop advanced materials for gas separation, metal-organic frameworks (MOFs) have proven feasible as a viable option among different porous adsorbents. MOFs are made up of metal ions or clusters and well-coordinated organic linkers, which afford them with varied, ordered structures, customizable pore diameters and unrivaled surface areas [5, 6]. The possibility to tune a given MOF's structure and porosity attributes permits the materials' potential deployment for various gas separations. More importantly, the ease with which MOFs with certain functional groups can be constructed is an appealing property for adsorptive separation applications. Functional groups capable of selectively and reversibly binding certain gases can be purposefully placed on either the organic linker or the inorganic building block. When gas mixtures come into contact with adsorptive sites within the pores of a MOF, some of them can cluster together much closer than is normally conceivable in the gas phase leading to higher uptake capacities.

Over the past two decades, a significant amount of research has been carried out to investigate the application of MOFs in selective gas separation from different streams. Herein, we focus on MOF powders studied for selected important but challenging industrial gas separations, including air separation, separation of noble gases and light hydrocarbons. Molecular sizes of the gases to be separated and their adsorption characteristics are the main parameters that must be considered while designing MOFs for these gas separations. The physical parameters of the different gases to be discussed in this chapter are summarized in Table 5.1. The use of MOFs for CO_2 capture from different sources, removal of toxic gases, separation of liquid mixtures such as heavy hydrocarbons and the application of MOF-based membranes to separate industrially relevant gases are all of significant research interest, but are not covered in this chapter. These topics are discussed elsewhere [7–10] in literature and are detailed in other chapters of this book.

Because of their tunability that allows for precise control of several structural characteristics, including shape, size, polarity and functionality of pores, we can tailor MOFs to behave according to their intended purpose. MOF design strategies that drive gas separations include pore aperture engineering via the isoreticular chemistry

Table 5.1: Physical parameters of selected adsorbates [11, 12].

Gases	Normal boiling point (K)	Kinetic diameter (Å)	Polarizability ($\times 10^{25}$ cm^3)	Dipole moment ($\times 10^{18}$ esu cm)	Quadruple moment ($\times 10^{26}$ esu cm^2)
Nitrogen (N_2)	77.35	3.64–3.80	17.403	0.00	1.52
Oxygen (O_2)	90.17	3.467	15.812	0.00	0.39
Helium (He)	4.30	2.551	2.04956	0.00	0.00
Neon (Ne)	27.07	2.82	3.956	0.00	0.00
Argon (Ar)	87.27	3.542	16.411	0.00	0.00
Krypton (Kr)	119.74	3.655	24.844	0.00	0.00
Xenon (Xe)	165.01	4.047	40.44	0.00	0.00
Methane (CH_4)	111.66	3.758	25.93	0.00	0.00
Ethane (C_2H_6)	184.55	4.443	44.3–44.7	0.00	0.65
Ethylene (C_2H_4)	169.42	4.163	42.52	0.00	1.50
Acetylene (C_2H_2)	188.40	3.3	33.3–39.3	0.00	–
Propane (C_3H_8)	231.02	4.3–5.118	62.9–63.7	0.084	–
Propylene (C_3H_6)	225.46	4.678	62.6	0.366	–
n-Butane (*n*-C_4H_{10})	272.66	4.687	82	0.05	–
i-butane (*i*-C_4H_{10})	261.34	5.278	81.4–82.9	0.132	–
1-butene (1-C_4H_8)	266.92	4.46	81	0.359–0.438	–
cis-2-C_4H_8	276.87	4.94	82	0.30	–
trans-2-C_4H_8	274.03	4.31	81.82	0.00	–
Isobutene (*i*-C_4H_8)	266.25	4.84	80	0.50	–
1,3-butadiene (C_4H_6)	268.62	4.31	86.4	0.00	–

principle, framework interpenetration and adsorption/functional site control for enhancing specific MOF-gas interaction. Stimuli-dependent flexibility that induces gate-opening phenomena is another feature of MOFs that enable selective gas separations. The structural features of MOFs for realizing the right MOF for particular gas separation are summarized in Figure 5.1. It is noted that porous inorganic and carbon-based adsorbents (i. e., zeolites and activated carbons) currently used in industrial separation processes do not permit this level of control and degree of flexibility.

5.2 Air separation (nitrogen/oxygen separation)

Oxygen (O_2), in its purest form, is a much sought-after feedstock. It is used in various oxygenated-product manufacturing, combustion and industrial processes. The applications of O_2 include synthetic fuels manufacturing, waste remediation, onboard oxygen generation in aircraft, steel production and many other oxidative processes [3]. The importance of pure molecular O_2 is even critical worldwide to rescue patients from respiratory conditions caused by the spread of coronavirus. It is also a much-needed respiratory gas for other healthcare applications, including the treatment of acute and

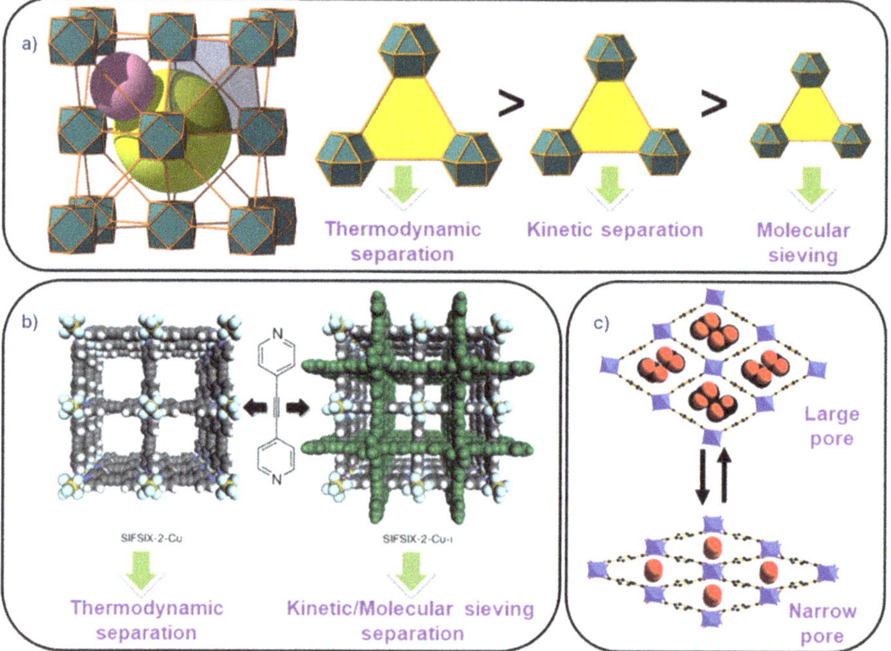

Figure 5.1: Structural features of MOFs exploited for gas separation: (a) Pore aperture engineering via the isoreticular chemistry principle, (b) framework interpenetration and (c) gate-opening/closing effect. Figure 5.1(b) is reproduced with permission from [13] Copyright © 2013, Nature Publishing Group.

chronic pulmonary disorders such as asthma and pneumonia. Nitrogen (N_2), on the other hand, is used to create inert atmospheres for system flushing/purging in industrial operations. It also allows for the safe storage and use of flammables while preventing combustibles from exploding. Furthermore, N_2 enhances the quality and shelf-life of air-sensitive items such as food, medicines and electronic devices. Pure N_2 is also required to produce NH_3, a valuable commodity in chemical industries and energy sectors. Due to the very low temperatures of its liquid form, N_2 is a common gas for cryogenic cooling and freezing. As these two important molecules are the main constituents in air, an efficient air separation material and method would boost the production of both pure N_2 and O_2.

Although O_2 is reasonably accessible from air as a mixture, current separation methods for obtaining pure O_2 from the air mixture (mainly O_2 and N_2) are prohibitively expensive. The energy-intensive cryogenic distillation process is often used to separate O_2 and N_2 from the air on an industrial scale [14]. Current O_2 generators focus on removing the more plentiful N_2 (78 % of the air) rather than collecting the needed O_2 (21 %). Very selective physisorbents have the potential to reverse this process and create highly pure O_2 with less energy.

Adsorbent-based devices, based on pressure swing adsorption (PSA) using zeolites, have also been deployed to construct cost-effective methods for N_2/O_2 separation at room temperature, particularly in portable medical devices [15]. Zeolites such as 13X, 4A, 5A, LiAgX, LiLSX and 10X are examples of adsorbents utilized in mobile oxygen concentrators to adsorb N_2 [15, 16]. However, for these devices, the zeolites employed were reported to have low adsorption uptakes and selectivities. While research in the field is still in its early stages, MOFs' tunability has shown immense promise to construct N_2/O_2 separators, with adsorption uptake and selectivity surpassing the commercially used LiX zeolite. The Cr(III)-based MIL-100(Cr), with coordinatively unsaturated Cr(III) sites, is one example of a MOF that thermodynamically captures N_2 over O_2 (Figure 5.2) [17]. Because zeolites are selective for N_2 (78 % of air) rather than O_2 (21 %), conventional O_2 concentrators filter out the greater amount of the air stream rather than the less prevalent, desired O_2. Using an oxygen-selective adsorbent material would lower the volume of air compressed in the concentrator, and hence the amount of energy required. The well-demonstrated tunable selectivity of MOFs towards gases of interest makes them potential candidates to capture the less abundant O_2.

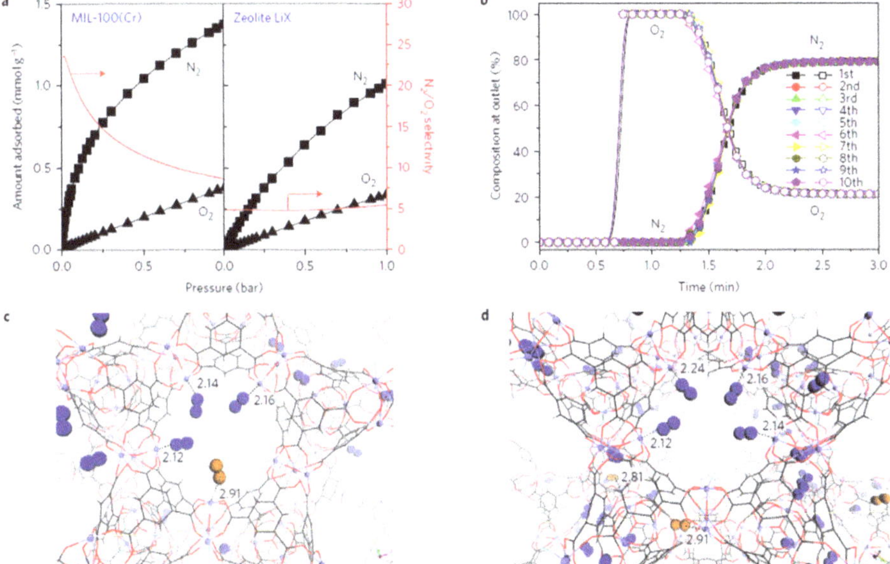

Figure 5.2: (a) Single-component N_2 and O_2 adsorption isotherms (black symbols) and IAST-predicted N_2/O_2 selectivities (red lines) at 293 K (molar composition 79/21) of MIL-100(Cr) and zeolite LiX; (b) Breakthrough curves of a binary 79/21 N_2/O_2 mixture for MIL-100(Cr) (0.41 g) at 298 K and 1 atm (flow rate = 5 $cm^3\ min^{-1}$) for 10 cycles; Local views of the snapshots extracted from the GCMC simulations performed for a N_2/O_2 mixture at 298 K and 0.1 bar, (c) and at 298 K and 1 bar, (d) in MIL-100(Cr). Reproduced with permission from ref. [17]. Copyright © 2016, Nature Publishing Group.

Because both molecules are quite similar in size, size or shape selective separation of N_2 and O_2 is a challenging process. MOFs with coordinatively unsaturated metal sites (CUSs) have attracted research interest in separating O_2 from N_2 due to variations in the bonding and electronic properties of the two molecules [18]. O_2 bonding can be side-on, bent or linear whereas N_2 bonding is invariably bent or linear [19]. The higher binding energy associated with the side-on bonding of O_2 leads to the selective O_2 adsorption by MOFs. The preferential adsorption of O_2 over N_2 by MOFs is illustrated by several well-known transition metal-based MOFs including HKUST-1 ($Cu_3(btc)_2$; btc^{3-} = 1,3,5-benzenetricarboxylate), $Cr_3(btc)_2$ (isostructural MOF to HKUST-1 with open Cr(II) coordination sites), M-MOF-74 (M_2(dobdc); H_2dobdc = 2,5-dihydroxyterephthalic acid, M = Co, Fe, Ni) and MIL-100 [19–22]. MOFs with redox-active CUSs are of considerable interest for O_2 over N_2 selective binding [22–25]. This is demonstrated by one of the most studied MOF platforms, M-MOF-74, which has coordinatively unsaturated M(II) centers. Fe-MOF-74 was shown to bind O_2 preferentially over N_2 with 18.2 wt% O_2 uptake at 211 K (Figure 5.3), which is equivalent to one adsorbed O_2 molecule per Fe(II) center [26]. Low-temperature Mossbauer and infrared spectra and Rietveld analysis of powder neutron diffraction data at 4 K provided evidence of a partial charge transfer from Fe(II) to O_2 at room temperature to form Fe(III) and O_2^{2-} upon adsorption. The reason for this was due to bonding of O_2 to Fe in a symmetric side-on mode with $d_{O\text{-}O}$ = 1.25(1) Å at low temperature and a slipping side-on mode with $d_{O\text{-}O}$ = 1.6(1) Å when oxidized at room temperature.

Metal centers of MOFs with variable oxidation states are not the only components targeted for favorable O_2 adsorption selectivity. The organic linkers of MOFs with electron donating moieties can also facilitate partial electron transfer and stronger O_2 binding. This is highlighted in the Co-triazolate and Co-pyrazolate isoreticular MOFs, Co-BTTri ($Co_3[(Co_4Cl)_3(BTTri)_8]_2$·DMF; H_3BTTri = 1,3,5-tri(1H-1,2,3-triazol-5-yl)benzene) and Co-BDTriP ($Co_3[(Co_4Cl)_3(BDTriP)_8]_2$·DMF; H_3BDTriP = 5,5′-(5-(1Hpyrazol-4-yl)-1,3-phenylene)bis(1H-1,2,3-triazole)) [27]. The more electron-donating pyrazolate group in Co-BDTriP led to higher O_2 affinities (Q_{st} = −47 kJ mol^{-1} at low loadings vs. −34 kJ mol^{-1} for Co-BTTri). The O_2 adducts in Co-BTTri are best characterized as Co(II)–dioxygen species with partial electron transfer, whereas the stronger binding sites in Co-BDTriP generate Co(III)–superoxo moieties. The electron-donating property of the bridging linkers also play a significant role in the O_2-framework binding and O_2/N_2 selectivity. For example, exchanging the -Cl^- groups of Co_2Cl_2(BBTA) (H_2BBTA = 1H,5H-benzo(1,2-d:4,5-d′)bistriazole) for -OH^- groups increased the strength of O_2 adsorption at the CUSs without increasing the N_2 affinity [28]. The increased electron-donating nature of the -OH^- bridging groups and the presence of H-bonding interactions between -OH^- and O_2 are ascribed to the strong O_2-$Co_2(OH)_2$(BBTA) binding.

Aside from a high density of CUSs, the degree of electron density within the frameworks is essential for the reversible sorption of O_2. This feature is demonstrated by O_2 adsorption in three Mn-MOFs, termed Mn_2(dobdc)$(DMF)_4$, $Mn_5(btac)_4(\mu_3\text{-}OH)_2(EtOH)_2$·DMF·3EtOH·3$H_2O$ (H_2btac = benzotriaole-5-carboxylic acid) and Mn_3(2,6-

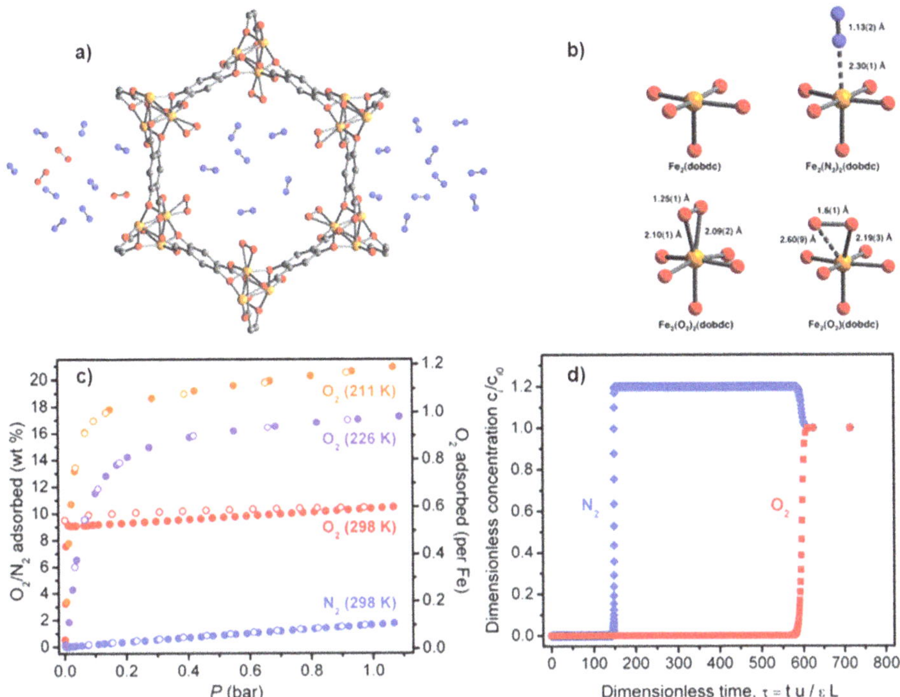

Figure 5.3: (a, b) First coordination spheres for the Fe(II) centers within Fe-MOF-74 and its O_2 and N_2 dosed variants, as determined from Rietveld analysis of neutron powder diffraction data, (c) Excess O_2 adsorption (filled symbol) and desorption (open symbol) isotherms collected for Fe-MOF-74 at different temperatures and N_2 isotherm at 298 K, (d) Calculated breakthrough curves during adsorption of simulated air (O_2:N_2 = 0.21:0.79) by Fe-MOF-74 at 211 K. Reproduced with permission from [26]. Copyright © 2011 American Chemical Society.

ndc)$_3$·4DMF (H_2ndc = 2,6-naphthalenedicarboxylic acid), with their channels decorated with coordinatively unsaturated Mn(II) sites [29]. Owing to the different electron densities at the Mn(II) sites in these three MOFs originating from their distinct coordination environments, differences in O_2 adsorption behavior were observed. The MOF with the highest electron density around the exposed Mn(II) center led to partial charge transfer from Mn(II) to the coordinated O_2 and showed reversible chemisorption of O_2. Too high or low electron density does not favor reversible O_2 binding.

In constructing MOFs that can selectively adsorb O_2 over N_2, early transition metal-based molecular building blocks are suitable for their more favorable interactions with O_2 than late transition metals [20]. This fact is exemplified by isostructural M-MIL-100 MOFs constructed from early (Sc) and middle (Fe) transition metal ion trinuclear oxo-clusters. In general, the binding energy of O_2 is substantially larger for early transition metals than it is for mid or late transition metals and the binding of N_2 is relatively weak for all transition metals. This general trend is also in agreement with experimental and simulation results for two distinct MOF structure variants, $M_3(btc)_2$ and $M_2(dobdc)$ (M = Fe, Co, Ni, Zn, Mg) [19, 21].

As demonstrated in the preceding discussions, enhancing uptake capacity and O_2/N_2 selectivity is possible by combining appropriate building blocks and tuning the MOF structure, characteristics of which are difficult to control in zeolites and other porous solid adsorbents. MOFs with CUSs and high porosities have potential for air separation via thermodynamic adsorption mechanisms. High porosities are useful to facilitate access to the metal sites. Although rare, ultramicroporous MOFs with pore dimensions that can distinguish the small difference in size between O_2 and N_2 have also shown potential for air separation, particularly at higher pressures [30]. The tailorability of MOFs also allows for future design of better-performing adsorbents that have a proper balance of O_2 and N_2 binding forces that are enough for the separation of the two molecules and regeneration of the adsorbents. In general, MOFs have the capability to generate pure O_2 at mild conditions (temperature and pressure), a promising potential to revolutionize the PSA process. However, the long-term use of such MOFs is challenged by the poor chemical stability of many framework structures and their deployment at the commercial level has yet to be seen.

5.3 Noble gas purification

Owing to their unique properties, such as being nearly nonreactive and non-flammable, noble gases like He, Ne, Ar, Kr and Xe at high purity levels, are ideal for a wide range of uses. These range from the creating inert atmospheres to welding, use as food preservatives, electronics, medical applications, leak detection and lighting and usage in fire extinguishers. However, their scarcity and expensive production and purification costs limit their uses to niche fields and they are very expensive and not commonly found in nature, frequently occurring at low quantities in the earth's atmosphere (5.22, 1.1 and 0.09 ppm for He, Kr and Xe, respectively) [31]. The most prevalent method of obtaining inert gases at the required purity is the prohibitively energy-intensive fractional distillation technique, which entails cooling gas mixtures to very low temperatures in multiple stages (e. g., to or below −186 °C in case of Ar). Adsorptive separation, which relies on pressure or temperature swing adsorption processes with advanced porous materials, is once again a promising alternative to circumvent the challenges faced by fractional distillation.

Among the noble gases, the separation of Xe and Kr has attracted tremendous attention over the past several decades. Although commonly used adsorbent materials such as zeolites (NaX and NaA) and activated carbon show promise, they are frequently characterized by limited adsorption capacity and selectivity [32, 33]. From the other potential adsorbents, MOFs are viewed as a viable alternative to improve uptake capacity and selectivity as they can be tailored to meet specific properties needed for such separation processes. Following the pioneering work by Mueller et al. in 2006 on developing a PSA process using HKUST-1 [34], energy-efficient MOF-based adsorptive Xe/Kr separation has shown considerable promise in the past decade resulting in uptake (up to 6.1 mmol g^{-1}) and selectivity (up to 22) at standard conditions [35]. Due to the higher polarizability of Xe compared with Kr, most MOFs are Xe selective under noncryogenic conditions. This property is exemplified by the Xe/Kr adsorptive separations performed on the well-known high surface area MOFs, MOF-5 (also called IRMOF-1 with a surface area of ~3400 $m^2 g^{-1}$), HKUST-1 (~1700 $m^2 g^{-1}$), the Zr-terephthalate MOF, UiO-66 (~1100 $m^2 g^{-1}$) [33, 34, 36]. A gas cylinder filled with these MOFs has significantly enhanced noble gas storage capacities compared with a cylinder without MOF when comparing at the same pressure.

In addition to accessible porosity, selectively strong adsorption sites on the surfaces are needed to have a pronounced impact on Xe/Kr selective separation. The well-known MOFs with polarizable CUSs and high surface areas, including the copper paddle wheel-based MOFs (HKUST-1 and MOF-505) and the M-dobdc-MOFs (also called M-MOF-74; M = Ni, Co, Zn, Mg) (surface area ~1000–1800 $m^2 g^{-1}$) are typical examples in this regard [37–40]. The relatively higher Xe/Kr selectivity of MOF-505 (9–10 for 20:80 Xe/Kr mixture at room temperature) than that of HKUST-1 (4.5 under similar conditions) is ascribed to a pore confinement effect. MOF-505 has relatively smaller pore sizes (4.8, 7.1, and 9.5 Å), while HKUST-1 contains small and large cavities of internal diameter ~13, 11, and 5 Å (Kinetic diameter of Xe is ~4.1 Å). The small octahedral pocket of HKUST-1 and the windows to the small pockets are the primary Xe binding sites [41]. Compared to HKUST-1, the higher density of CUSs in Ni-MOF-74 allowed higher Xe/Kr selectivity (~7) [33].

Noble gas inertness appears to limit the ability of MOFs to increase their binding affinity for these gases. Consequently, ultra-microporous MOFs that possess pore diameters that can perfectly fit a particular noble gas are employed to enhance MOF-gas binding energies. The rigid Co-squarate MOF, with a pore size (4.1 × 4.3 Å) comparable to the molecular dimension of Xe (~4.1 Å) and pore surface decorated with the highly polar -OH groups, is considered a representative example of such MOFs [42]. This MOF was shown to separate Xe from Kr with high Xe uptake capacity (~1.35 mmol g^{-1}) and Xe/Kr Ideal Adsorbed Solution Theory (IAST) selectivity (69.7) at low pressure (0.2 bar) and ambient temperature. In a second example, the microporous calcium-4,4'–sulfonyldibenzoate MOF (SBMOF-1, SB = Stony Brook) has a pore size of ca. 4.2 Å, a perfect fit to accommodate Xe atoms (kinetic diameter 4.1 Å) [43]. The MOF exhibited a high Xe adsorption affinity of 35 kJ mol^{-1} with Xe breakthrough uptake of 13.2 mmol g^{-1} and Xe/Kr selectivity of 16. The impact of smaller pores to enhance noble gas adsorption with Xe selectivity was also further illustrated by the small pore microporous Cd-MOF (pore diameter ~5.1 Å) constructed from the same V-shaped linker, 4,4'-sulfonyldibenzoate (SDB) [44]. The selective interaction of Xe with the framework was reflected in the higher Q_{st} of 35 kJ mol^{-1} than that calculated for Kr adsorption (16 kJ mol^{-1}). This Q_{st} is higher than the values for most of the bench-

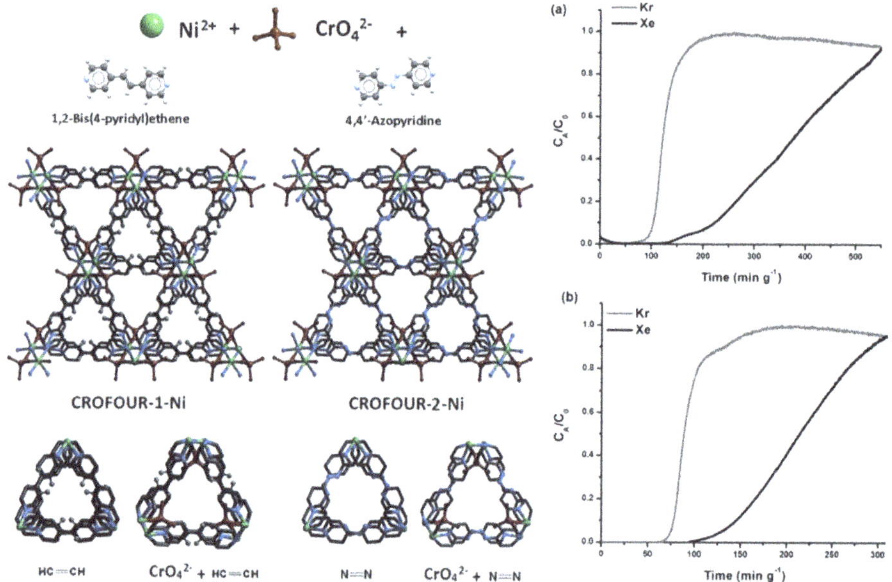

Figure 5.4: Schematic showing the assembly of CROFOUR-1-Ni and CROFOUR-2-Ni (left) and column breakthrough curves using Xe/Kr (20:80 gas mixture) at 298 K and 1 bar for CROFOUR-1-Ni (a) and CROFOUR-2-Ni (b). Reproduced with permission from [46]. Copyright © 2016 WILEY-VCH Verlag GmbH & Co. KGaA, Weinheim.

mark MOFs and other porous solids (Table 5.2). The hybrid isoreticular SIFSIX-3-M (M = Fe, Co, Ni, Cu, Zn) MOFs (channel size = <5 Å) is another example of ultramicroporous MOFs that have framework-Xe interaction with the Fe analog exhibiting a Q_{st} of 27.4 kJ mol^{-1} at zero loading of Xe [45]. These MOFs are assembled by pillaring SiF_6^{2-} anions with a pyrazine linker. Similarly, the hybrid ultra-microporous isoreticular CROFOUR-1-Ni and CROFOUR-2-Ni MOFs that are constructed by pillaring CrO_4^{2-} moieties with square grid sheets of $[Ni(L)_2]^{2+}$ (L = 1,2-bis(4-pyridyl)ethylene or 4,4'-azopyridine) (Figure 5.4), were reported to have a high Q_{st} (37.4 kJ mol^{-1} for CROFOUR-1-Ni and 30.5 kJ mol^{-1} for CROFOUR-2-Ni at low loading of Xe) and a Xe/Kr selectivity of 22 [46]. Other examples of small pore MOFs, designed by applying the isoreticular principle, have also been proven to exhibit MOF-noble gas interaction [47–49].

The flexibility of the framework is also important in selective MOF-noble gas interactions. For example, computational and experimental studies on Xe and Kr adsorption using SIFSIX-3-Ni revealed a rotational conformation change of pyrazine linkers upon Xe adsorption resulting in an inflection point in the isotherm. Though this was true for Xe adsorption, this behavior was not observed when applied for Kr adsorption [45]. Indeed, the pore flexibility phenomenon in SIFSIX-3-Ni allowed a greater Xe-MOF interaction and enhanced Xe/Kr selectivity. Furthermore, the flexible Zn-5-methyl-1H-tetrazole MOF, $[Zn(mtz)_2]$, also called USTA-49, is another example of a

Table 5.2: Representative examples of benchmark MOFs and other porous solids that have been explored for Xe/Kr adsorptive separation at ambient conditions.

Porous material	Metal/Ligand	BET area	Pore diameter (Å)	Xe/Kr uptake (mmol g^{-1})	Xe/Kr selectivity	Xe/Kr Q_{st} (kJ mol^{-1})	Ref.
Ni-MOF-74	Ni/2,5-dihydroxy-1,4-benzenedicarboxylate	950	11	3.83/1.69[a]	7.3	22/–	[39]
MOF-505	Cu/3,3′,5,5′-biphenyltetracarboxylate	1030	4.8, 7.1, 9.5	2.2/–	9-10	15/–	[37]
HKUST-1	Zn/1,3,5-benzenetricarboxylate	1710	13, 11, 5	3.18/1.92[a]	2.6	26/–	[37, 39]
SBMOF-1	Ca/4,4'–sulfonyldibenzoate	145	4.2	1.4/1.0	16	35/26	[43]
SBMOF-2	Ca/1,2,4,5-tetrakis(4-carboxyphenyl)-benzene	195	7 × 7	2.83/0.92	10	26/22	[50]
$Co_3(HCOO)_6$	Co/formate	300	5	1.97/0.77	22[b]/6[a]	28/22	[51]
MOF-Cu-H	Cu/5-(pyridin-3yl)-1,3-benzenedicarboxylate	868	6.4	3.19/1.6	15.8	33/26	[52]
Co-squarate	Co/squarate	95	4.1 × 4.3	1.35/–[c]	69.7[b]	44/36	[42]
CROFOUR-1-Ni	Cr & Ni/1,2-bis(4-pyridyl)ethylene	–		1.8/–	22[b]/19.8[a]	37/–	[46]
SIFSIX-3-Fe	Fe/SiF_6/Pyrazine	358	3.5-3.8	2.25/1.37	–	27/–	[45]
Ca-SINAP-1	Ca/1,1,2,2-tetra(4-carboxylphenyl)ethylene	143	5–11	2.8/1.5	10[b]	29/21	[53]
Noria	Porous organic cage ($C_{102}H_{96}O_{24}$)		5–7	1.5/0.7	9.4[b]	25–27	[54]
CC3 organic cage	Porous organic cage ($C_{72}H_{84}O_{12}$)	624	4.4	2.2/1.7	20.4[b]	31/23	[55]

[a]Breakthrough capacity/selectivity (20:80 Xe/Kr mixture); [b]IAST selectivity; [c]Uptake at 0.2 bar.

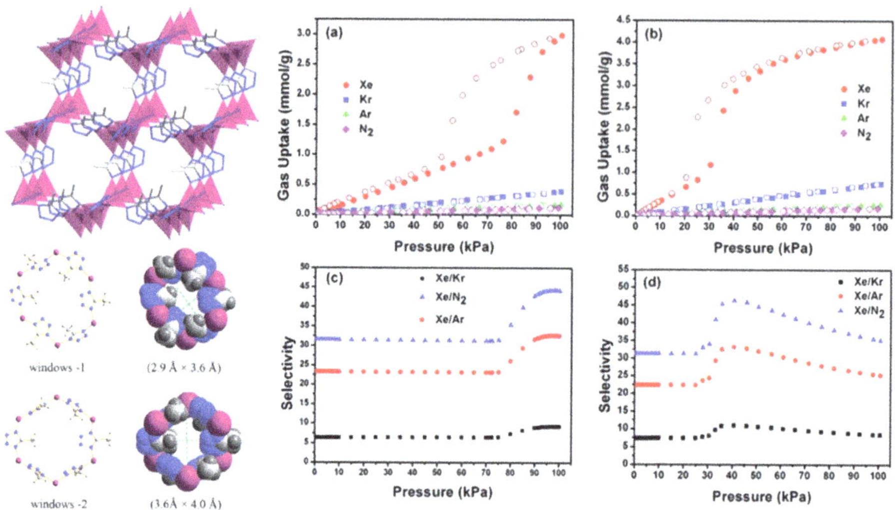

Figure 5.5: Crystal structure representation of UTSA-49 and its two different 6-membered ring windows (left). Adsorption (solid) and desorption (open) isotherms of noble gases on UTSA-49 at 298 K (a) and 273 K (b) and predicted adsorption selectivities of the MOF for 1:1 Xe/Kr, Xe/Ar and Xe/N_2 mixtures at 298 K (c) and 273 K (d). Reproduced with permission from ref. [56]. Copyright © 2015 The Royal Society of Chemistry.

MOF that was shown to have similar Xe adsorption behavior (Figure 5.5) [56]. Despite the small pore aperture sizes (2.9 × 3.6 Å and 3.6 × 4 Å) relative to the kinetic diameter of Xe (~4.1 Å), the pore flexibility of USTA-49 permitted the diffusion of a significant amount of Xe (2.7 mmol g^{-1} at 1 bar and 298 K) with very low Kr uptake (0.38 mmol g^{-1} under similar conditions) leading to high Xe/Kr selectivity (15.5). ZIF-8, ($Zn(mIM)_2$, mIM = 2-methylimidazolate) is also among those limited MOF examples that have displayed Xe/Kr separation capability via a gate-opening sorption mechanism [57].

Although the number of examples are rare, Kr-selective MOFs are alternatives for Xe/Kr separation. In this regard, the two-fold interpenetrated flexible MOF, [Cu(hfipbb)(H_2hfipbb)$_{0.5}$] (termed FMOFCu, hfipbb = 4,4'-hexafluoroisopropylidene-bisbenzoate]), is among the rare examples for inverse Kr/Xe separation [58]. This MOF, with a pore opening of 3.5 Å, possesses a gating effect at temperatures <273 K, thereby restricting diffusion of the larger Xe atoms while allowing the smaller Kr atoms to pass through. This phenomenon permits the kinetically driven splitting of Xe/Kr mixture via inverse Kr/Xe selectivity. The Ca-squarate ultra-microporous MOF, UTSA-280 or [$Ca(C_4O_4)(H_2O)$], with one-dimensional rigid channels of 3.8 Å, is another example of the rare Kr-selective MOFs (Kr/Xe selectivity of 72.1 with Kr uptake of 1.48 mmol g^{-1} from GCMC simulation) [59].

Aside from the commonly explored Xe/Kr separation mainly from spent nuclear fuels, separation of Xe from exhaled anesthetic gas mixtures and removal of the toxic radon are other industrially important separations [60, 61]. Given the tunability of

MOFs, there remains significant space for improving separation efficiency in the purification of noble gases.

5.4 Light hydrocarbons separation

Light hydrocarbon gases, methane and other C_{2-4} alkanes/alkenes, are energy sources employed in combustion engines and fuel cells. The ability to separate methane at ambient conditions, for instance, saves energy costs in natural gas upgrading and purification of C_{2-4} hydrocarbon feedstocks for further transformation. In their high purity forms, many light olefins (e. g., ethylene, propylene and butadiene) and other C_2 and C_3 hydrocarbons are also essential raw materials in the chemical industry and are useful precursors for a wide range of goods, including acetic acid and polymers (e. g., rubbers and plastics). As they are derivatives of crude oil fractionations, they exist as mixtures. For example, natural gas contains C_{2-4} hydrocarbon impurities in addition to its major component, methane (>90 %). Therefore, separating the hydrocarbon mixtures into their pure phase compounds is essential to get high-value feedstock for the chemical sector and cleaner fuels that cause little CO_2 emissions (e. g., CH_4). As a result, one of the petrochemical industry's essential processes is the separation of light hydrocarbon mixtures. The energy and cost-intensive cryogenic distillation is predominantly employed to achieve this separation. A lot of effort is being made to replace it with adsorption-based technology [1]. The application of MOFs for separating light hydrocarbon gases (C_{1-4} hydrocarbons) is discussed in the following sections.

5.4.1 Separation of methane from C_{2-4} hydrocarbons

MOFs can separate methane from other light hydrocarbon gases with C_2, C_3 and C_4 chains at ambient temperature and pressure conditions [62]. The separation of methane from longer hydrocarbon chains is relatively easier than separating C_{2-4} fractions. Most MOFs are more selective to longer chain hydrocarbons than methane due to enhanced van der Waals interactions between the larger molecules and the pore surface (Figure 5.6) [63]. The smaller molecular surface of shorter alkanes, on the other hand, reduces the strength of van der Waals interactions and as the length of the hydrocarbon chain increases, the enthalpy and entropy of adsorption become more negative. As a result, MOFs show a superior affinity for adsorption of longer alkane chains with higher isosteric heat of adsorption for C_{2-4} hydrocarbons than methane. This adsorption behavior was proven in different examples of MOFs, including the iconic MOF-5 (Q_{st} = −10.6 and −23.6 kJ mol^{-1} for CH_4 and n-C_4H_{10}, respectively) and HKUST-1 (Q_{st} = −12.0 and −29.6 kJ mol^{-1} for methane and n-butane, respectively) [64]. However, applying this concept to separate C_{2-4} hydrocarbon mixtures into their

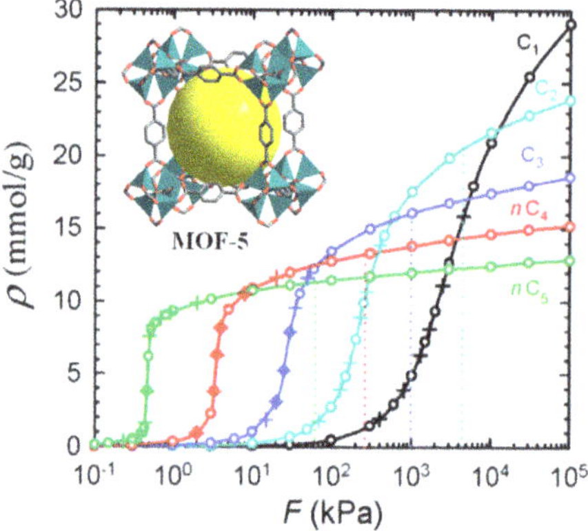

Figure 5.6: Adsorption isotherms of linear alkanes CH_4, $H_3C\text{-}CH_3$, $H_3C\text{-}CH_2\text{-}CH_3$, $H_3C\text{-}(CH_2)_2\text{-}CH_3$, and $H_3C\text{-}(CH_2)_3\text{-}CH_3$ on MOF-5 at 300 K. Reproduced with permission from [63]. Copyright © 2006 American Chemical Society.

pure phases is challenging due to the similar MOF-guest interactions among the C_{2-4} hydrocarbons.

Adsorption behavior and selectivity depend on different structural factors of MOFs. For isoreticular MOFs made from linkers of varying lengths, adsorption capacity is mainly associated with pore size, with increased linker length resulting in a reduced selectivity [65]. For isoreticular MOFs constructed from the same length linkers, the MOF-hydrocarbon interaction is related to the number of carbon atoms present in the linker, with enhanced binding energy being realized for MOFs having larger aromatic backbones. The *n*-butane/methane selectivity in IRMOFs, isoreticular to MOF-5, illustrates this fact [66]. Precise structural control of MOF pore size and functionality via isoreticulation, framework interpretation and decorating the interior surface with appropriate functional groups is considered useful to separate such mixtures by kinetics, thermodynamics or molecular sieving [7]. For example, the high density of CUSs in M-MOF-74 (M = Fe, Mg, Co) was demonstrated to permit the efficient separation of the 6-component $CH_4/C_2H_2/C_2H_4/C_2H_6/C_3H_6/C_3H_8$ mixture into their pure constituents [67]. The pore geometry also plays an important role in enhancing the adsorptive interaction through C-H···π and C-H···O interactions. The hydrocarbon separation by two microporous Ca-based MOFs possessing 1D open channels, SBMOF-1 and SBMOF-2, illustrates the role of pore geometry. The MOFs separate C_2H_6 from CH_4 with C_2H_6/CH_4 selectivity = ~74 for SBMOF-1 [68].

Flexibility, which enables guest-dependent breathing or gate-opening and closing behavior, is the other feature of MOFs that has been exploited for selective splitting of C_n hydrocarbon fractions from mixtures. The observed S-shaped (Type IV or V) alkane sorption isotherms in flexible MOFs are often used for the selective separation of hydrocarbons. The impact of flexibility is exemplified by MIL-53(Cr), which shows stepwise adsorption isotherm profiles at distinct pressures for $n\text{-}C_3H_8$ and longer fractions, but not for smaller hydrocarbon (CH_4 and C_2H_6) isotherms [69]. The mesh-adjustable molecular sieves (MAMS-1), formulated as $Ni_8(5\text{-BBDC})_6(\mu_3\text{-OH})_4$; BBDC = 5-*tert*-butyl-1,3-benzenedicarboxylate, is another MOF example that was demonstrated to have temperature-dependent molecular gate-opening behavior [70]. The framework's flexibility to endure a thermally induced reversible phase change between a narrow pore and a wide pore phase with a variation of the channel diameter between 2.9 and 5.0 Å leads to selective CH_4/C_2H_6, C_2H_6/C_3H_6 or $CH_4/C_2H_6/C_3H_6/C_4H_{10}$ mixture separations.

5.4.2 Light alkene/alkane separation

Light alkenes, such as ethene (C_2H_4) and propene (C_3H_6), are often prepared via hydrocarbon cracking and dehydrogenation of the corresponding alkanes. This is the reason for why some alkane byproducts exist as an impurity in the products. Light alkenes and alkanes have very close physical and chemical properties (i. e., polarizability, shape/size, dipole or quadrupole moments and boiling points/volatility), which make the separation of light alkene/alkane mixtures (C_2H_4/C_3H_6 and C_3H_6/C_3H_8) one of the petrochemical industry's most energy-consuming processes [71]. The commonly used distillation technique demands several steps and a high reflux ratio in addition to its high pressures and cryogenic temperature requirements (as low as −160 °C). In recent years, MOFs have shown immense promise for the adsorptive separation of light alkane/alkene mixtures with the potential to produce high purity hydrocarbons that are used in polymerization or other chemical transformation processes. This section will highlight MOFs that have shown satisfactory performance in separating C_2H_6/C_2H_4 and C_3H_8/C_3H_6 mixtures.

Thermodynamic equilibrium-driven separation is the most explored strategy in separating light olefin and paraffin mixtures using MOFs. Many MOFs have shown preferential adsorption of the olefin component due to adsorption sites within their building units, such as CUSs and H-bonding sites. Olefin selective binding was demonstrated in the separation of C_2H_6/C_2H_4 and C_3H_8/C_3H_6 mixtures using MOFs such as HKUST-1, M-MOF-74 (M = Mg, Co, Fe, Zn, Co, Ni, Mn), NOTT-300 and MIL-100(Fe). These MOFs showed selective adsorption of olefin components due to $\pi \cdots \pi$ stacking or strong interactions of the *p*-orbitals (π electrons) in olefins with the CUSs and/or the -OH functional groups tethered to the backbones of the frameworks [72–78]. Although rare, MOFs that exhibit preferential paraffin adsorption have also been reported [79–83]. The isoreticular principle serving as a useful guide to assembling those MOFs. For example, the zinc-azolate framework (MAF-49), constructed from Zn^{2+} and bis(5-amino-1H-1,2,4-triazol-3-yl)methane, was demonstrated to be C_2H_6-selective in the adsorption of C_2H_6/C_2H_4 mixture [79]. The inner pore wall structure of the MOF, particularly the spatial arrangements of H-bond acceptors that allow the formation of more C–H···N H-bonds with C_2H_6 than C_2H_4 (6 *vs.* 4), is attributed to the

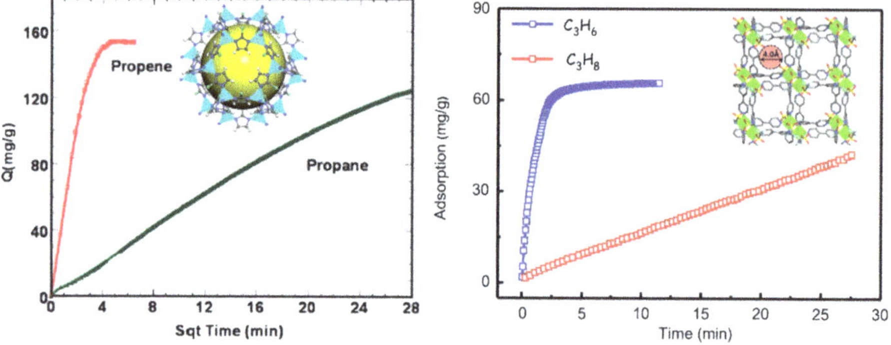

Figure 5.7: Crystal structure representations and kinetic adsorption profiles of ZIF-8 at 30 °C (left) and ELM-12 at 25 °C (right). Reproduced with permission from references [84, 85]. Copyright © 2009 American Chemical Society. Copyright © 2018 Elsevier B.V.

C_2H_6 selectivity. Regeneration is the major drawback in such separations as a high temperature is needed to separate the strongly adsorbed components. However, the alkane-selective MOFs could be considered practically advantageous as it is relatively easier to recover the alkane from the MOF sorbent.

MOFs for olefin/paraffin separation by diffusion or kinetic control have also been reported. ZIF-8 and ELM-12 ($[Cu(bpy)_2(OTf)_2]$; bpy = 4,4′-bipyridine, OTf^- = trifluoromethanesulfonate), are examples that were demonstrated to be suitable for C_3H_8/C_3H_6 kinetic separation (Figure 5.7) [84, 85]. The small pore aperture sizes of ZIF-8 (~3.4 Å) and ELM-12 (~4 Å) are the cause of the different diffusion rates of propylene and propane (size: 6.5 × 4.0 × 3.8 Å for C_3H_6 and 6.8 × 4.2 × 3.8 Å for C_3H_8). Pore flexibility also has a substantial role in the kinetically driven separation by the microporous ZIF-8. The synergistic effect of kinetics and thermodynamics has also been exploited for selective C_3H_8/C_3H_6 separation using MOFs. The recently reported pillared-layered isostructural $M(aip)(bpy)_{0.5}$ MOFs (M = Co, Ni and Zn; aip = 5-aminoisophthalic acid) is a representative example [86]. Precise control of pore apertures via the isoreticular principle is especially important to develop MOFs for kinetically-driven olefin/paraffin separations [87]. This strategy can be illustrated by the isoreticular series of rare-earth (RE)-based **ftw**-MOFs. The microporous isostructural RE-abtc-**ftw**-MOFs, $RE_6(OH)_8(abtc)_3(H_2O)_6(DMA)_2$ (RE = Y(III) or Tb(III), abtc = 3,3′,5,5′-azobenzene-tetracarboxylate; DMA = dimethylammonium), showed promising performance for kinetically controlled propane/propylene sieving due to fine-tuned aperture sizes [88, 89].

In the extreme case of kinetic separation, light olefins and paraffins can be separated by a molecular sieving mechanism (i. e., shape/size exclusion effect), if the kinetic diameter of at least one of the gas mixture components is smaller than the pore apertures. For example, the M-gallate-MOFs ($M(C_7O_5H_4)$; M= Ni, Mg, Co) with suitable pore aperture sizes (3.47 × 4.85, 3.56 × 4.84, 3.69 × 4.95 $Å^2$ for Ni, Mg and Co-gallate)

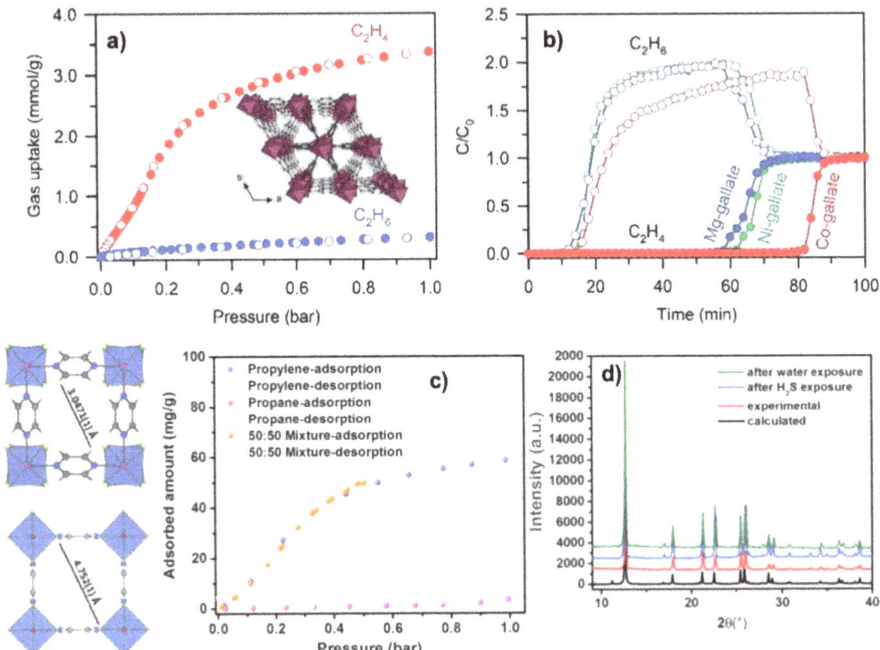

Figure 5.8: (a) Single-component C_2H_4 and C_2H_6 adsorption isotherms on Co-gallate at 298 K, (b) Breakthrough curves of M-gallate for the equimolar C_2H_4/C_2H_6 mixture at 273 K and 1 bar (flow rate = 0.5 mL min^{-1}), (c) Pure gas and equimolar C_3H_6/C_3H_8 mixture isotherms of NbOFFIVE-1-Ni at 298 K and (d) Breakthrough curves of equimolar C_3H_6/C_3H_8 mixture (4 mL min^{-1} flow) at 298 K and 1 bar. Reproduced with permission from [90, 92]. Copyright © 2016, American Association for the Advancement of Science. Copyright © 2018 Wiley-VCH Verlag GmbH & Co. KGaA, Weinheim.

(Figure 5.8) and the ultramicroporous Ca-squarate MOF, $[Ca(C_4O_4)(H_2O)]$ (termed UTSA-280), are among those MOFs that discriminate C_2H_4 (3.28 × 4.18 × 4.84 Å^3) and C_2H_6 (3.81 × 4.08 × 4.82 Å^3) via molecular sieving effect [90, 91]. The rigid 1D channels (3.2 × 4.5 Å^2 and 3.8 × 3.8 Å^2) of the MOF, which are in between the minimum cross-sectional areas of the smaller C_2H_4 and the larger C_2H_6, permit the size/shape selective C_2H_4/C_2H_6 separation at ambient conditions. For MOF-based propane/propylene separation by molecular sieving effect, the pillared-layered NbOFFIVE-1-Ni MOF ($Ni(Pyrazine)_2(NbOF_5)$) has shown huge potential to exclude the adsorption of propane while allowing propylene molecules to diffuse through the pores (Figure 5.8) [92]. The contracted aperture size of this MOF is attributed to the total sieving of C_3H_8 from C_3H_6 at ambient temperature and pressure. The channel size ranges from 3.04 to 4.75 Å, depending on the tilting angle of pyrazine.

The guest-dependent gate-opening effect in some flexible MOFs, which allows the adsorption and release of a particular component at a given gate-opening pressure, has played a significant role in developing MOFs that are suitable for light olefin/paraffin separation. ZIF-7 (Zn(BIM)2, BIM = benzimidazolate) exemplifies this gating effect.

This MOF preferentially adsorbs paraffins over olefins due to the different gate opening effects during the interactions of the pore opening with the hydrocarbons that have slightly different shapes [93, 94]. The separation of light olefin/paraffin mixtures due to different gate-opening behavior upon interaction of the gaseous components with MOFs has also been demonstrated in RPM3-Zn, $Zn_2(bpdc)_2(bpee)$ (bpdc = 4,4'-biphenyldicarboxylate; bpee = 1,2-bipyriylethylene) [95]. The critical factors for this are the right topology and appropriate building blocks to impart the desired shape, size and rigidity in the sought after pore aperture. In general, significant progress was made in developing MOF-based olefin/paraffin separators, however, there are still some issues, such as low adsorption capacity, which require the development of new and improved MOF adsorbents.

5.4.3 Light alkene/alkyne separation

This section will highlight MOFs that have been demonstrated as efficient for separating acetylene/ethylene and propyne/propylene mixtures by different mechanisms. C_2H_4, mainly produced from naphtha and ethane steam cracking, usually contains a trace amount of C_2H_2 impurity that can poison Ziegler–Natta catalysts during C_2H_4 polymerizations. Removal of the trace C_2H_2 side product is then important to produce polymer-grade ethene. The activated forms of the mixed-metal M'MOFs, termed $Zn_3(CDC)_3[Cu(SalPycy)]$ (M'MOF-3a), $Cd_3(BDC)_3[Cu(SalPyMeCam)]$ (M'MOF-4a), $Zn_3(CDC)_3[Cu(SalPyMeCam)]$ (M'MOF-5a), $Cd_3(BDC)_3[Cu(SalPytBuCy)]$ (M'MOF-6a) and $Zn_3(CDC)_3[Cu(SalPytBuCy)]$ (M'MOF-7a); CDC = 1,4-cyclohexanedicarboxylate and BDC = 1,4-benzenedicarboxylate, are among the first examples of MOFs that were demonstrated selective for C_2H_2/C_2H_4 adsorptive separation [96, 97]. The flexible linker employed and the resulting small micropores of the MOFs are important features to kinetically discriminate C_2H_2 and C_2H_4 with high selectivity (25.5 for M'MOF-3a at 195 K). However, the practical application of M'MOFs is limited by their very low C_2H_2 uptake. Highly porous MOFs, such as M-MOF-74 (M = Fe, Co, Mg) decorated with a high density of CUSs and the Al-based MOF, NOTT-300 ($[Al_2(OH)_2(L)]$; L^{4-} = biphenyl-3,3',5,5'-tetracarboxylate), were explored to circumvent the low C_2H_2 uptake problem [67, 74, 77]. These frameworks possess polyfunctional groups that allow cooperative H-bonding, $\pi \cdots \pi$ stacking, and intermolecular dipole interactions in the binding of C_2H_2 and C_2H_4 within the MOFs. Although such forces allow the hydrocarbons' discrimination via the thermodynamic mechanism, the more open structures of the MOFs resulted in lower C_2H_2/C_2H_4 selectivity.

MOFs have been designed to balance the uptake and selectivity "trade-off". The microporous MOF UTSA-100 ($[Cu(atbdc)]$; H_2atbdc = 5-(5-amino-1H-tetrazol-1-yl)-1,3-benzenedicarboxylic acid) is an example that improved adsorption selectivity without sacrificing uptake capacity [98]. The synergistic effects of the narrow pore apertures (3.3 Å) with suitable pore diameter (4.0 Å) and the weak acid-base interactions

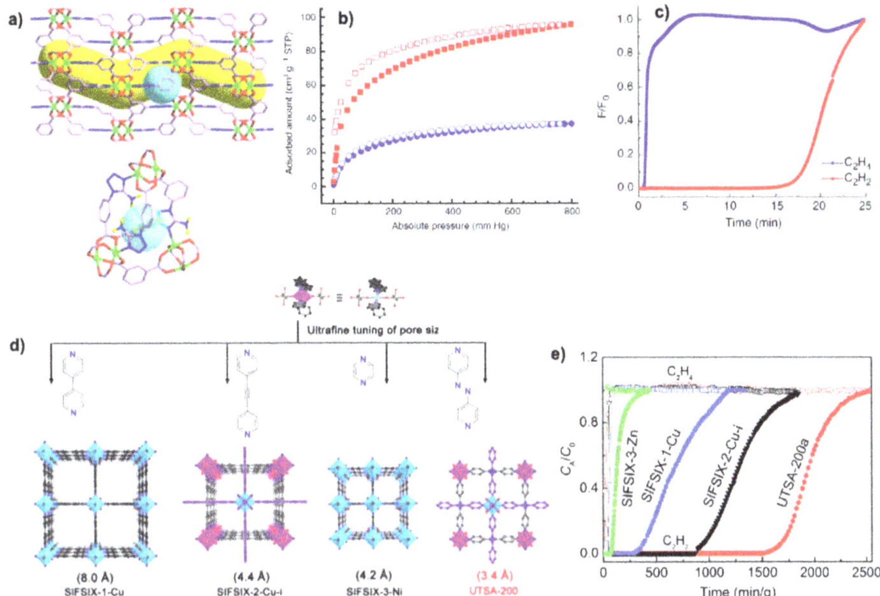

Figure 5.9: (a) The channel and cage (~4.0 Å diameter) structures of UTSA-100, (b) The C_2H_2 sits right at the small cage connecting two adjacent channels, (c) Single component C_2H_2 (red) and C_2H_4 (blue) adsorption isotherms at 296 K, (c) Column breakthrough curve for C_2H_2/C_2H_4 (1/99 mixture) on UTSA-100 at 296 K, (d) Isoreticular SIFSIX materials and (e) Column breakthrough curves for C_2H_2/C_2H_4 separations on some isoreticular SIFSIX MOFs at 298 K and 1.0 bar. Reproduced with permission from references [98–100]. Copyright © 2017 WILEY-VCH Verlag GmbH & Co. KGaA, Weinheim. Copyright © 2018 Wiley-VCH Verlag GmbH & Co. KGaA, Weinheim.

between -NH_2 functional groups on the linkers and C_2H_2 molecules permitted high C_2H_2 uptake and C_2H_2/C_2H_4 selectivity (10.72) (Figure 5.9).

The tunable $(SiF_6)^{2-}$-pillared SIFSIX MOF platform reported an impressive C_2H_2/C_2H_4 separation performance with balanced uptake and selectivity [99, 101–103]. The possibility to precisely control the pore and window sizes of SIFSIX MOFs and the presence of strongly basic SiF_6^{2-} sites lead to the specific adsorption of C_2H_2 with high adsorption capacity and selectivity at ambient conditions (Figure 5.9). The van der Waals interaction of C_2H_2 with the organic linkers is an additional driving force for selective adsorption. Facile pore size control in this MOF platform is achieved via either interpenetration or changing the length and/or functionality of the organic linker. Moreover, the SiF_6^{2-} sites enforce H-bonding interactions with the weakly acidic C_2H_2 molecules. The assembly of SIFSIX-14-Cu-i (also named as UTSA-200), by replacing the 4,4′-dipyridylacetylene (dpa, 9.6 Å) of the isoreticular SIFSIX-2-Cu-I MOF with the shorter 4,4′-azopyridine (azpy, 9.0 Å), led to the contraction of the 1D channel from 4.4 Å for SIFSIX-2-Cu-i to <4.0 Å for SIFSIX-14-Cu-i [99]. The ultrafinely-tuned aperture size led to highly selective C_2H_2/C_2H_4 separation by size exclusion of C_2H_4 (kinetic diameter of 4.2 Å) from C_2H_2 (kinetic diameter of 3.3 Å). This permitted the removal of trace C_2H_2 from C_2H_2/C_2H_4 mixture without sacrificing the adsorption capacity. These examples illustrate the power of isoreticular chemistry to facilitate the design of other highly selective MOFs for C_2H_2/C_2H_4 separation.

Propylene (C_3H_6), mainly produced by steam cracking of petroleum fractions, also contains trace amounts of propyne (C_3H_4) impurity. An efficient separation technique is needed to remove the ultralow impurity content and produce polymer-grade C_3H_6. MOFs have also shown promising potential for this challenging but indispensable separation in recent years. The anion (SiF_6^{2-} = SIFSIX, $NbOF_5^{2-}$ = NbOFFIVE)-pillared isoreticular SIFSIX series (e. g., SIFSIX-1-Cu, SIFSIX-2-Cu-i, SIFSIX-3-Ni, SIFSIX-3-Zn, NBOFFIVE-1-Ni, ZJUT-1 (SIFSIX-3-Ni but based on 2-aminopyrazine linker), and UTSA-200) are among the most explored MOFs for the separation of these two molecules of very similar size molecular size (C_3H_4: 4.16×4.01×6.51 Å^3 and C_3H_6: 5.25×4.16×6.44 Å^3) [100, 104, 105]. The precisely controlled apertures permits the size/shape selective kinetic separation of the two molecules with high C_3H_4 uptake and acceptable uptake and selectivity "trade-off" depending on the SIFSIX material that is used (Figure 5.10).

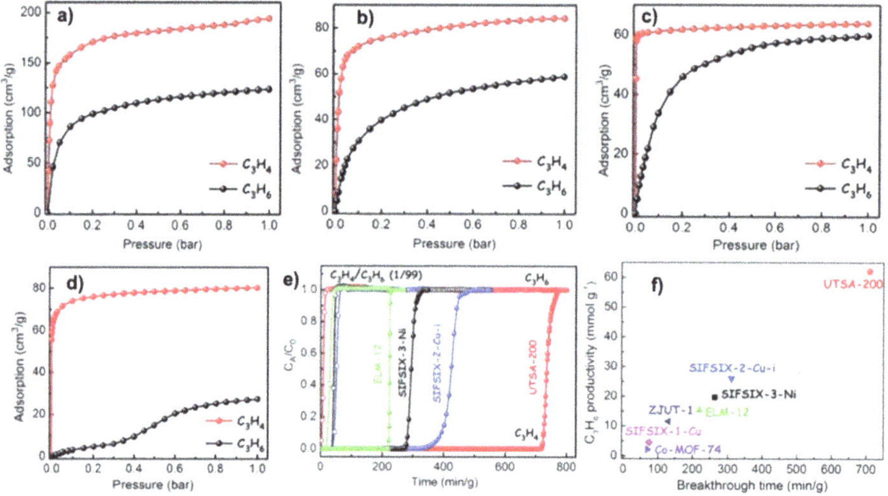

Figure 5.10: Single component C_3H_4 and C_3H_6 adsorption isotherms for SIFSIX MOFs: SIFSIX-1-Cu (a) SIFSIX-2-Cu-I, (b) SIFSIX-3-Ni, (c) UTSA-200, (d) at 298 K; Breakthrough curves for 1/99 (*v*/*v*) mixture (flow rate = 2.0 mL min^{-1}) at 298 K and 1 bar (e) and comparison of C_3H_6 productivity from C_3H_4/C_3H_6 mixtures of the indicated MOFs, with C_3H_4 concentration less than 1 ppm (f). Reproduced with permission from [100]. Copyright © 2018 Wiley-VCH Verlag GmbH & Co. KGaA, Weinheim.

The flexible-robust MOF, ELM-12, is another example of high C_3H_4 uptake and C_3H_4/C_3H_6 selectivity (IAST selectivity for 1/99 C_3H_4/C_3H_6 mixture of 84) [106]. The strong binding of C_3H_4 with the polar OTf^- groups and the dumbbell-shaped pores of the MOF play a significant role in separating the two molecules by ELM-12. However, it is worth mentioning that the use of a MOF as an adsorbing material to separate C_3H_4 from C_3H_6 mixtures is still in its early stages with just a few publications to date. Further practice of isoreticular chemistry could help to realize MOFs that separate the molecules via a full sieving effect.

5.4.4 C_4 hydrocarbon separation

C_4 hydrocarbons including *n*-butane (*n*-C_4H_{10}), isobutane (*i*-C_4H_{10}), butene isomers (1-C_4H_8, *i*-C_4H_8, *cis*-C_4H_8, *trans*-C_4H_8) and 1,3-butadiene (1,3-C_4H_6) are the other important light hydrocarbons used, in their high purity forms, either as fuel or feedstock to produce synthetic rubbers or other fine-chemicals. During their industrial production (i. e., catalytic or steam cracking and methanol-to-olefin processes) a mixture of the C_4 fractions is always produced, which demonstrates the need for highly efficient separations [107]. Efforts are being made to replace the conventional energy-intensive cryogenic fractional distillation with adsorption-based systems. More importantly, the practice of isoreticulation in MOF crystal chemistry has attracted attention in the development of adsorbents for the selective separation of C_4 hydrocarbons [108].

A considerable amount of literature has been published to apply MOFs for C_4 olefins separation through thermodynamic mechanisms. In this regard, MOFs with high density of CUSs, such as HKUST-1 ($Cu_3(btc)_2$, btc^{3-} = benzene-1,3,5-tricarboxylate) and M-MOF-74 or M_2(dobdc) (M = Mg, Mn, Fe, Co, Ni, Cu, Zn and $dobdc^{2-}$ = 2,5-dioxido-benzene-1,4-dicarboxylate), to enhance MOF-guest interaction via π-complexation have been explored [109–112]. However, the equilibrium-based separation is less efficient due to negligible selectivity and the high energy (heat) requirement to regenerate the MOF. Adsorbents that rely on nonequilibrium adsorption mechanisms, such as guest responsive structural flexibilities or gate-opening effects (e. g., the triply-interpenetrated flexible MOF SD-65 and Zn-BTM ($Zn_2(btm)_2$; H_2btm = bis(5-methyl-1H-1,2,4-triazol-3-yl)methane)), have also been reported [113–115]. More importantly, MOFs capable of shape and size-selective molecular recognition [e. g., the anion-pillared interpenetrated SIFSIX (GeFSIX-2-Cu-i, NbFSIX-2-Cu-i, GeFSIX-3-Ni and GeFSIX-14-Cu-i) and M-gallate (M = Ni, Mg, Co) MOFs] were also investigated and afforded promising results [116, 117]. The RE-based fumarate-**fcu**-MOF, $|(CH_3)_2NH_2|_2[RE_6(\mu_3\text{-}OH)_8(fum)_6(H_2O)_6]$ (RE = Y or Tb), with a triangular window size of ~4.7 Å, assembled via isoreticulation in RE-**fcu**-MOFs to tune the triangular pore aperture size also allowed the fractionation of C_4 olefins by kinetic effect [118]. The triangle window is the only way to access the MOF's octahedral and tetrahedral cages. The optimal aperture size for size and/or shape-selective recognition of the C_4 isomers (kinetics diameters in the range of 4.3–5.0 Å [119]) is attributed to the resulting separation performance. Given the tunability of MOFs, there is still much to be explored in the search for suitable MOF adsorbents capable of C_4 olefins splitting with high efficiency. As for other gas separations, the compromise of adsorption capacity for selectivity is the main challenge in developing MOFs for splitting C_4 olefins. This could increase the number of adsorption-desorption cyclic operations to be performed for a given separation. Chemically- and thermally-stable MOFs with big cages, but small apertures, are potential solutions for this challenge. The selection of a suitable network topology, the search for appropriate organic and inorganic molecular building blocks (MBBs) for the target net, and the discovery of appropriate reaction pathways to assemble the MOF from the MBBs are the main challenges to overcome.

Fuel upgrading by separating linear and branched C_4 paraffins is another critical task in the petrochemical industry. The process helps to produce high efficacy fuels in refineries, namely access to a high-octane number (RON) gasoline and/or a high cetane number diesel. The cost-effective separation of *n*-butane from isobutane also offers other potential benefits for practical industrial applications. For example, it is useful to produce LPG (liquefied petroleum gas) and refrigerant grade isobutane (also known by its commercial name, R600a). A significant number of reported work is available on paraffin/isoparrafin separation using MOFs with most focusing on C_{5+} liquid phases [7, 62]. The isoreticular RE-**fcu**-MOFs (RE = Y(III), Eu(III), Tb(III)) are among the few MOFs that possess tunable apertures for steric adsorptive separation of light hydrocarbon gas mixtures [108]. The most-efficient separation of the linear and branched C_4 paraffins (*n*-butane/isobutane mixture) was obtained using RE-fum-**fcu**-MOF. The separation by full molecular exclusion is attributed to the window size of ~4.7 Å, which is between the sizes of *n*-butane and isobutane. Controlling the number of defects in the isostructural MOF, Zr-fum-**fcu**-MOF, also allows the tuning of the triangular aperture for butane/isobutane separation. Depending on the defect level, the separation relies on either kinetic or full molecular exclusion mechanism [120]. While these research works have illuminated the potential of MOFs for linear/branched butane isomers, there is still a need to explore further in the search for a MOF that can separate the isomers by complete molecular sieving effect without compromising adsorption capacity.

5.5 Summary

MOFs provide novel avenues to pursue solutions to industrial gas separation processes due to their structural tunability. The modular synthesis to fit an application, from pore size to binding forces, give MOFs unparalleled gas separation performances that outperform traditional gas separation materials such as zeolites, silica and activated carbons. Among the different industrial gas separation applications, air separation, splitting of noble gases and fractionation of light hydrocarbon mixtures were given due attention in this chapter. An overview of the research conducted using MOFs to separate those industrially relevant gases was highlighted. The features of MOFs and their advantages in gas separations by different mechanisms (e. g., adjustment of the pore size for size/shape exclusion or kinetic separation, tailoring the pore wall functionality for thermodynamic equilibrium-based separation, controlling the gate-opening effect in flexible MOFs, and a combination thereof) were underscored. However, given the expensive organic and inorganic precursors of this class of materials and the limited stability for most of the resulting frameworks, MOFs are better suited to more specialized separations than bulk ones at the current stage. For example, MOFs can be utilized for efficient separation of O_2 from air for mobile oxygen concentrators.

There is currently no clear answer to whether MOFs are the solution for large-scale industrial gas separations for economic reasons. For instance, deploying MOFs for industrial-scale air separation to produce a large volume of N_2 for ammonia production is currently unlikely. MOFs could be used for efficient and cost-effective capture of toxic gases as components of personal protective equipment. On the contrary, deploying MOFs to capture pollutants from large point sources such as gas emissions from chemical production stacks or chimneys are mostly like not cost-efficient. The development of alternative large-scale and greener synthetic routes (e. g., mechanochemical synthesis) that can help to minimize MOF production costs might be considered a potential solution. Therefore, the main challenges ahead is the development of chemically and thermally stable MOFs at a reasonable cost that can be used over several cyclic adsorption-desorption measurements. The development of MOFs with high adsorption capacity without compromising selectivity is the other challenge. However, the construction of MOFs with little or no capacity and selectivity "trade-offs" is still very difficult. Despite the less energy-intensive nature of MOF-based separations, the purity level is also usually less than the result obtained by cryogenic separations. When these and other related challenges are overcome, the usage of MOFs for various bulk industrial separations will be right around the corner. At this stage, testing of different MOF adsorbents on a pilot scale could greatly promote the materials in the market.

Bibliography

[1] Sholl DS, Lively RP. Seven chemical separations to change the world. Nature. 2016;532:435–7.
[2] Ouikhalfan M, Lakbita O, Delhali A, Assen AH, Belmabkhout Y. Toward net-zero emission fertilizers industry: greenhouse gas emission analyses and decarbonization solutions. Energy Fuels. 2022;36:4198–223.
[3] Council NR. Separation Technologies for the Industries of the Future. Washington, DC: The National Academies Press; 1998.
[4] Padin J, Rege SU, Yang RT, Cheng LS. Molecular sieve sorbents for kinetic separation of propane/propylene. Chem Eng Sci. 2000;55:4525–35.
[5] Moumen E, Assen AH, Adil K, Belmabkhout Y. Versatility vs stability. Are the assets of metal–organic frameworks deployable in aqueous acidic and basic media? Coord Chem Rev. 2021;443:214020.

[6] Ubaid S, Assen AH, Alezi D, Cairns A, Eddaoudi M, Belmabkhout Y. Evaluating the high-pressure volumetric CH_4, H_2, and CO_2 storage properties of denser-version isostructural soc-metal–organic frameworks. J Chem Eng Data. 2022;67:1732–42.
[7] Adil K, Belmabkhout Y, Pillai RS, Cadiau A, Bhatt PM, Assen AH, Maurin G, Eddaoudi M. Gas/vapour separation using ultra-microporous metal–organic frameworks: insights into the structure/separation relationship. Chem Soc Rev. 2017;46:3402–30.
[8] Assen AH, Belmabkhout Y, Adil K, Lachehab A, Hassoune H, Aggarwal H. Advances on CO_2 storage. Synthetic porous solids, mineralization and alternative solutions. Chem Eng J. 2021;419:129569.
[9] Yaghi OM, Kalmutzki MJ, Diercks CS. Liquid- and gas-phase separation in MOFs. In: Introduction to Reticular Chemistry. 2019. p. 365–93.
[10] Adil K, Świrk K, Zaki A, Assen AH, Delahay G, Belmabkhout Y, Cadiau A. Perspectives in adsorptive and catalytic mitigations of NOx using metal–organic frameworks. Energy Fuels. 2022.
[11] Gehre M, Guo Z, Rothenberg G, Tanase S. Sustainable separations of C_4-hydrocarbons by using microporous materials. ChemSusChem. 2017;10:3947–63.
[12] Li JR, Kuppler RJ, Zhou HC. Selective gas adsorption and separation in metal–organic frameworks. Chem Soc Rev. 2009;38:1477–504.
[13] Nugent P, Belmabkhout Y, Burd SD, Cairns AJ, Luebke R, Forrest K, Pham T, Ma S, Space B, Wojtas L, Eddaoudi M, Zaworotko MJ. Porous materials with optimal adsorption thermodynamics and kinetics for CO_2 separation. Nature. 2013;495:80–4.
[14] Oxygen. In: Greenwood NN, Earnshaw A, editors. Chemistry of the Elements. 2nd ed. Oxford: Butterworth-Heinemann; 1997. p. 600–44.
[15] Santos JC, Cruz P, Regala T, Magalhães FD, Mendes A. High-purity oxygen production by pressure swing adsorption. Ind Eng Chem Res. 2007;46:591–9.
[16] Rege SU, Yang RT. Limits for air separation by adsorption with LiX zeolite. Ind Eng Chem Res. 1997;36:5358–65.
[17] Yoon JW, Chang H, Lee SJ, Hwang YK, Hong DY, Lee SK, Lee JS, Jang S, Yoon TU, Kwac K, Jung Y, Pillai RS, Faucher F, Vimont A, Daturi M, Férey G, Serre C, Maurin G, Bae YS, Chang JS. Selective nitrogen capture by porous hybrid materials containing accessible transition metal ion sites. Nat Mater. 2017;16:526–31.
[18] Zhou C, Cao L, Wei S, Zhang Q, Chen L. A first principles study of gas adsorption on charged CuBTC. Comput Theor Chem. 2011;976:153–60.
[19] Parkes MV, Sava Gallis DF, Greathouse JA, Nenoff TM. Effect of Metal in $M_3(btc)_2$ and $M_2(dobdc)$ MOFs for O_2/N_2 Separations: A Combined Density Functional Theory and Experimental Study. J Phys Chem C. 2015;119:6556–67.
[20] Sava Gallis DF, Chapman KW, Rodriguez MA, Greathouse JA, Parkes MV, Nenoff TM. Selective O_2 sorption at ambient temperatures via node distortions in Sc-MIL-100. Chem Mater. 2016;28:3327–36.
[21] Sava Gallis DF, Parkes MV, Greathouse JA, Zhang X, Nenoff TM. Enhanced O2 selectivity versus N2 by partial metal substitution in cu-BTC. Chem Mater. 2015;27:2018–25.
[22] Murray LJ, Dinca M, Yano J, Chavan S, Bordiga S, Brown CM, Long JR. Highly-selective and reversible O_2 binding in $Cr_3(1,3,5\text{-benzenetricarboxylate})_2$. J Am Chem Soc. 2010;132:7856–7.
[23] Southon PD, Price DJ, Nielsen PK, McKenzie CJ, Kepert CJ. Reversible and selective O_2 chemisorption in a porous metal–organic host material. J Am Chem Soc. 2011;133:10885–91.
[24] Reed DA, Xiao DJ, Jiang HZH, Chakarawet K, Oktawiec J, Long JR. Biomimetic O_2 adsorption in an iron metal–organic framework for air separation. Chem Sci. 2020;11:1698–702.
[25] Gopalsamy K, Babarao R. Heterometallic metal organic frameworks for air separation: a computational study. Ind Eng Chem Res. 2020;59:15718–31.

[26] Bloch ED, Murray LJ, Queen WL, Chavan S, Maximoff SN, Bigi JP, Krishna R, Peterson VK, Grandjean F, Long GJ, Smit B, Bordiga S, Brown CM, Long JR. Selective binding of O_2 over N_2 in a redox–active metal–organic framework with open iron(II) coordination sites. J Am Chem Soc. 2011;133:14814–22.
[27] Xiao DJ, Gonzalez MI, Darago LE, Vogiatzis KD, Haldoupis E, Gagliardi L, Long JR. Selective, tunable O_2 binding in cobalt(II)–triazolate/pyrazolate metal–organic frameworks. J Am Chem Soc. 2016;138:7161–70.
[28] Rosen AS, Mian MR, Islamoglu T, Chen H, Farha OK, Notestein JM, Snurr RQ. Tuning the redox activity of metal–organic frameworks for enhanced, selective O_2 binding: design rules and ambient temperature O_2 chemisorption in a cobalt–triazolate framework. J Am Chem Soc. 2020;142:4317–28.
[29] Wang W, Yang J, Li L, Li J. Research on the adsorption of O_2 in metal–organic frameworks with open manganese(II) coordination sites. Funct Mater Lett. 2012;06:1350004.
[30] Tang Y, Wang X, Wen Y, Zhou X, Li Z. Oxygen-selective adsorption property of ultramicroporous MOF $Cu(Qc)_2$ for air separation. Ind Eng Chem Res. 2020;59:6219–25.
[31] Banerjee D, Cairns AJ, Liu J, Motkuri RK, Nune SK, Fernandez CA, Krishna R, Strachan DM, Thallapally PK. Potential of metal–organic frameworks for separation of xenon and krypton. Acc Chem Res. 2015;48:211–9.
[32] Jameson CJ, Jameson AK, Lim HM. Competitive adsorption of xenon and krypton in zeolite NaA: 129Xe nuclear magnetic resonance studies and grand canonical Monte Carlo simulations. J Chem Phys. 1997;107:4364–72.
[33] Thallapally PK, Grate JW, Motkuri RK. Facile xenon capture and release at room temperature using a metal–organic framework: a comparison with activated charcoal. Chem Commun. 2012;48:347–9.
[34] Mueller U, Schubert M, Teich F, Puetter H, Schierle-Arndt K, Pastré J. Metal–organic frameworks–prospective industrial applications. J Mater Chem. 2006;16:626–36.
[35] Banerjee D, Simon CM, Elsaidi SK, Haranczyk M, Thallapally PK. Xenon gas separation and storage using metal-organic frameworks. Chem. 2018;4:466–94.
[36] Lee SJ, Yoon TU, Kim AR, Kim SY, Cho KH, Hwang YK, Yeon JW, Bae YS. Adsorptive separation of xenon/krypton mixtures using a zirconium-based metal-organic framework with high hydrothermal and radioactive stabilities. J Hazard Mater. 2016;320:513–20.
[37] Bae YS, Hauser BG, Colón YJ, Hupp JT, Farha OK, Snurr RQ. High xenon/krypton selectivity in a metal-organic framework with small pores and strong adsorption sites. Microporous Mesoporous Mater. 2013;169:176–9.
[38] Lee SJ, Kim KC, Yoon TU, Kim MB, Bae YS. Selective dynamic separation of Xe and Kr in Co-MOF-74 through strong binding strength between Xe atom and unsaturated Co2+ site. Microporous Mesoporous Mater. 2016;236:284–91.
[39] Liu J, Thallapally PK, Strachan D. Metal–organic frameworks for removal of Xe and Kr from nuclear fuel reprocessing plants. Langmuir. 2012;28:11584–9.
[40] Perry JJ, Teich-McGoldrick SL, Meek ST, Greathouse JA, Haranczyk M, Allendorf MD. Noble gas adsorption in metal–organic frameworks containing open metal sites. J Phys Chem C. 2014;118:11685–98.
[41] Hulvey Z, Lawler KV, Qiao Z, Zhou J, Fairen-Jimenez D, Snurr RQ, Ushakov SV, Navrotsky A, Brown CM, Forster PM. Noble gas adsorption in copper trimesate, HKUST-1: an experimental and computational study. J Phys Chem C. 2013;117:20116–26.
[42] Li L, Guo L, Zhang Z, Yang Q, Yang Y, Bao Z, Ren Q, Li J. A robust squarate-based metal–organic framework demonstrates record-high affinity and selectivity for xenon over krypton. J Am Chem Soc. 2019;141:9358–64.
[43] Banerjee D, Simon CM, Plonka AM, Motkuri RK, Liu J, Chen X, Smit B, Parise JB, Haranczyk M, Thallapally PK. Metal–organic framework with optimally selective xenon adsorption and separation. Nat Commun. 2016;7:ncomms11831.

[44] Banerjee D, Elsaidi SK, Thallapally PK. Xe adsorption and separation properties of a series of microporous metal–organic frameworks (MOFs) with V-shaped linkers. J Mater Chem A. 2017;5:16611–5.

[45] Elsaidi SK, Mohamed MH, Simon CM, Braun E, Pham T, Forrest KA, Xu W, Banerjee D, Space B, Zaworotko MJ, Thallapally PK. Effect of ring rotation upon gas adsorption in SIFSIX-3-M (M = Fe, Ni) pillared square grid networks. Chem Sci. 2017;8:2373–80.

[46] Mohamed MH, Elsaidi SK, Pham T, Forrest KA, Schaef HT, Hogan A, Wojtas L, Xu W, Space B, Zaworotko MJ, Thallapally PK. Hybrid ultra-microporous materials for selective xenon adsorption and separation. Angew Chem Int Ed. 2016;55:8285–9.

[47] Meek ST, Teich-McGoldrick SL, Perry JJ, Greathouse JA, Allendorf MD. Effects of polarizability on the adsorption of noble gases at low pressures in monohalogenated isoreticular metal–organic frameworks. J Phys Chem C. 2012;116:19765–72.

[48] Lee SJ, Kim S, Kim EJ, Kim M, Bae YS. Adsorptive separation of xenon/krypton mixtures using ligand controls in a zirconium-based metal-organic framework. Chem Eng J. 2018;335:345–51.

[49] Yan Z, Gong Y, Chen B, Wu X, liu Q, Cui L, Xiong S, Peng S. Methyl functionalized Zr-Fum MOF with enhanced Xenon adsorption and separation. Sep Purif Technol. 2020;239:116514.

[50] Chen X, Plonka AM, Banerjee D, Krishna R, Schaef HT, Ghose S, Thallapally PK, Parise JB. Direct observation of Xe and Kr adsorption in a Xe-selective microporous metal–organic framework. J Am Chem Soc. 2015;137:7007–10.

[51] Wang H, Yao K, Zhang Z, Jagiello J, Gong Q, Han Y, Li J. The first example of commensurate adsorption of atomic gas in a MOF and effective separation of xenon from other noble gases. Chem Sci. 2014;5:620–4.

[52] Xiong S, Gong Y, Hu S, Wu X, Li W, He Y, Chen B, Wang X. A microporous metal–organic framework with commensurate adsorption and highly selective separation of xenon. J Mater Chem A. 2018;6:4752–8.

[53] Wu XL, Li ZJ, Zhou H, Yang G, Liu XY, Qian N, Wang W, Zeng YS, Qian ZH, Chu XX, Liu W. Enhanced adsorption and separation of xenon over krypton via an unsaturated calcium center in a metal–organic framework. Inorg Chem. 2021;60:1506–12.

[54] Patil RS, Banerjee D, Simon CM, Atwood JL, Thallapally PK. Noria: a highly Xe-selective nanoporous organic solid. Chem Eur J. 2016;22:12618–23.

[55] Chen L, Reiss PS, Chong SY, Holden D, Jelfs KE, Hasell T, Little MA, Kewley A, Briggs ME, Stephenson A, Thomas KM, Armstrong JA, Bell J, Busto J, Noel R, Liu J, Strachan DM, Thallapally PK, Cooper AI. Separation of rare gases and chiral molecules by selective binding in porous organic cages. Nat Mater. 2014;13:954–60.

[56] Xiong S, Liu Q, Wang Q, Li W, Tang Y, Wang X, Hu S, Chen B. A flexible zinc tetrazolate framework exhibiting breathing behaviour on xenon adsorption and selective adsorption of xenon over other noble gases. J Mater Chem A. 2015;3:10747–52.

[57] Parkes MV, Staiger CL, Perry Iv JJ, Allendorf MD, Greathouse JA. Screening metal–organic frameworks for selective noble gas adsorption in air: effect of pore size and framework topology. Phys Chem Chem Phys. 2013;15:9093–106.

[58] Fernandez CA, Liu J, Thallapally PK, Strachan DM. Switching Kr/Xe selectivity with temperature in a metal–organic framework. J Am Chem Soc. 2012;134:9046–9.

[59] Xiong Xl, Chen Gh, Xiao St, Ouyang Yg, Li Hb, Wang Q. New discovery of metal–organic framework UTSA-280: ultrahigh adsorption selectivity of krypton over xenon. J Phys Chem C. 2020;124:14603–12.

[60] Abrahams BF, Dharma AD, Donnelly PS, Hudson TA, Kepert CJ, Robson R, Southon PD, White KF. Tunable porous coordination polymers for the capture, recovery and storage of inhalation anesthetics. Chem Eur J. 2017;23:7871–5.

[61] Zarabadi-Poor P, Marek R. In silico study of (Mn, Fe, Co, Ni, Zn)-BTC metal–organic frameworks for recovering xenon from exhaled anesthetic gas. ACS Sustain Chem Eng. 2018;6:15001–6.

[62] Bao Z, Chang G, Xing H, Krishna R, Ren Q, Chen B. Potential of microporous metal–organic frameworks for separation of hydrocarbon mixtures. Energy Environ Sci. 2016;9:3612–41.
[63] Jiang J, Sandler SI. Monte Carlo simulation for the adsorption and separation of linear and branched alkanes in IRMOF-1. Langmuir. 2006;22:5702–7.
[64] Farrusseng D, Daniel C, Gaudillère C, Ravon U, Schuurman Y, Mirodatos C, Dubbeldam D, Frost H, Snurr RQ. Heats of adsorption for seven gases in three metal–organic frameworks: systematic comparison of experiment and simulation. Langmuir. 2009;25:7383–8.
[65] Düren T, Snurr RQ. Assessment of isoreticular metal–organic frameworks for adsorption separations: a molecular simulation study of methane/n-butane mixtures. J Phys Chem B. 2004;108:15703–8.
[66] Düren T, Sarkisov L, Yaghi OM, Snurr RQ. Design of new materials for methane storage. Langmuir. 2004;20:2683–9.
[67] He Y, Krishna R, Chen B. Metal–organic frameworks with potential for energy-efficient adsorptive separation of light hydrocarbons. Energy Environ Sci. 2012;5:9107–20.
[68] Plonka AM, Chen X, Wang H, Krishna R, Dong X, Banerjee D, Woerner WR, Han Y, Li J, Light PJB. Hydrocarbon adsorption mechanisms in two calcium-based microporous metal organic frameworks. Chem Mater. 2016;28:1636–46.
[69] Trung TK, Trens P, Tanchoux N, Bourrelly S, Llewellyn PL, Loera-Serna S, Serre C, Loiseau T, Fajula F, Férey G. Hydrocarbon adsorption in the flexible metal organic frameworks MIL-53(Al, Cr). J Am Chem Soc. 2008;130:16926–32.
[70] Ma S, Sun D, Wang XS, Zhou HC. A mesh-adjustable molecular sieve for general use in gas separation. Angew Chem Int Ed. 2007;46:2458–62.
[71] Eldridge RB. Olefin/paraffin separation technology: a review. Ind Eng Chem Res. 1993;32:2208–12.
[72] Nicholson TM, Bhatia SK. Role of electrostatic effects in the pure component and binary adsorption of ethylene and ethane in Cu-tricarboxylate metal-organic frameworks. Adsorp Sci Technol. 2007;25:607–19.
[73] Lamia N, Jorge M, Granato MA, Almeida Paz FA, Chevreau H, Rodrigues AE. Adsorption of propane, propylene and isobutane on a metal–organic framework: molecular simulation and experiment. Chem Eng Sci. 2009;64:3246–59.
[74] Bloch Eric D, Queen Wendy L, Krishna R, Zadrozny Joseph M, Brown Craig M, Long Jeffrey R. Hydrocarbon separations in a metal-organic framework with open iron(II) coordination sites. Science. 2012;335:1606–10.
[75] Plaza MG, Ribeiro AM, Ferreira A, Santos JC, Hwang YK, Seo YK, Lee UH, Chang JS, Loureiro JM, Rodrigues AE. Separation of C_3/C_4 hydrocarbon mixtures by adsorption using a mesoporous iron MOF: MIL-100(Fe). Microporous Mesoporous Mater. 2012;153:178–90.
[76] Geier SJ, Mason JA, Bloch ED, Queen WL, Hudson MR, Brown CM, Long JR. Selective adsorption of ethylene over ethane and propylene over propane in the metal–organic frameworks M_2(dobdc) (M = Mg, Mn, Fe, Co, Ni, Zn). Chem Sci. 2013;4:2054–61.
[77] Yang S, Ramirez-Cuesta AJ, Newby R, Garcia-Sakai V, Manuel P, Callear SK, Campbell SI, Tang CC, Schröder M. Supramolecular binding and separation of hydrocarbons within a functionalized porous metal–organic framework. Nat Chem. 2015;7:121–9.
[78] Bachman JE, Kapelewski MT, Reed DA, Gonzalez MI, Long J. M_2(*m*-dobdc) (M = Mn, Fe, Co, Ni) metal–organic frameworks as highly selective, high-capacity adsorbents for olefin/paraffin separations. J Am Chem Soc. 2017;139:15363–70.
[79] Liao PQ, Zhang WX, Zhang JP, Chen XM. Efficient purification of ethene by an ethane-trapping metal-organic framework. Nat Commun. 2015;6:8697.
[80] Chen Y, Qiao Z, Wu H, Lv D, Shi R, Xia Q, Zhou J, Li Z. An ethane-trapping MOF PCN-250 for highly selective adsorption of ethane over ethylene. Chem Eng Sci. 2018;175:110–7.

[81] Hao HG, Zhao YF, Chen DM, Yu JM, Tan K, Ma S, Chabal Y, Zhang ZM, Dou JM, Xiao ZH, Day G, Zhou HC, Lu TB. Simultaneous trapping of C_2H_2 and C_2H_6 from a ternary mixture of $C_2H_2/C_2H_4/C_2H_6$ in a robust metal–organic framework for the purification of C_2H_4. Angew Chem Int Ed. 2018;57:16067–71.
[82] Lin RB, Wu H, Li L, Tang XL, Li Z, Gao J, Cui H, Zhou W, Chen B. Boosting ethane/ethylene separation within isoreticular ultramicroporous metal–organic frameworks. J Am Chem Soc. 2018;140:12940–6.
[83] Zhou P, Yue L, Wang X, Fan L, Chen DL, He Y. Improving ethane/ethylene separation performance of isoreticular metal–organic frameworks via substituent engineering. ACS Appl Mater Interfaces. 2021;13:54059–68.
[84] Li K, Olson DH, Seidel J, Emge TJ, Gong H, Zeng H, Li J. Zeolitic imidazolate frameworks for kinetic separation of propane and propene. J Am Chem Soc. 2009;131:10368–9.
[85] Li L, Lin RB, Wang X, Zhou W, Jia L, Li J, Chen B. Kinetic separation of propylene over propane in a microporous metal-organic framework. Chem Eng J. 2018;354:977–82.
[86] Chen Y, Wu H, Yu L, Tu S, Wu Y, Li Z, Xia Q. Separation of propylene and propane with pillar-layer metal–organic frameworks by exploiting thermodynamic-kinetic synergetic effect. Chem Eng J. 2022;431:133284.
[87] Lee CY, Bae YS, Jeong NC, Farha OK, Sarjeant AA, Stern CL, Nickias P, Snurr RQ, Hupp JT, Nguyen ST. Kinetic separation of propene and propane in metal–organic frameworks: controlling diffusion rates in plate-shaped crystals via tuning of pore apertures and crystallite aspect ratios. J Am Chem Soc. 2011;133:5228–31.
[88] Xue DX, Cadiau A, Weseliński ŁJ, Jiang H, Bhatt PM, Shkurenko A, Wojtas L, Zhijie C, Belmabkhout Y, Adil K, Eddaoudi M. Topology meets MOF chemistry for pore-aperture fine tuning: ftw-MOF platform for energy-efficient separations via adsorption kinetics or molecular sieving. Chem Commun. 2018;54:6404–7.
[89] Wang H, Dong X, Colombo V, Wang Q, Liu Y, Liu W, Wang XL, Huang XY, Proserpio DM, Sironi A, Han Y, Li J. Tailor-made microporous metal–organic frameworks for the full separation of propane from propylene through selective size exclusion. Adv Mater. 2018;30:1805088.
[90] Bao Z, Wang J, Zhang Z, Xing H, Yang Q, Yang Y, Wu H, Krishna R, Zhou W, Chen B, Ren Q. Molecular sieving of ethane from ethylene through the molecular cross-section size differentiation in gallate-based metal–organic frameworks. Angew Chem Int Ed. 2018;57:16020–5.
[91] Lin RB, Li L, Zhou HL, Wu H, He C, Li S, Krishna R, Li J, Zhou W, Chen B. Molecular sieving of ethylene from ethane using a rigid metal–organic framework. Nat Mater. 2018;17:1128–33.
[92] Cadiau A, Adil K, Bhatt PM, Belmabkhout Y, Eddaoudi M. A metal-organic framework–based splitter for separating propylene from propane. Science. 2016;353:137–40.
[93] van den Bergh J, Gücüyener C, Pidko EA, Hensen EJM, Gascon J, Kapteijn F. Understanding the anomalous alkane selectivity of ZIF-7 in the separation of light alkane/alkene mixtures. Chem Eur J. 2011;17:8832–40.
[94] Gücüyener C, van den Bergh J, Gascon J, Kapteijn F. Ethane/ethene separation turned on its head: selective ethane adsorption on the metal–organic framework ZIF-7 through a gate-opening mechanism. J Am Chem Soc. 2010;132:17704–6.
[95] Nijem N, Wu H, Canepa P, Marti A, Balkus KJ, Thonhauser T, Li J, Chabal YJ. Tuning the gate opening pressure of metal–organic frameworks (MOFs) for the selective separation of hydrocarbons. J Am Chem Soc. 2012;134:15201–4.
[96] Xiang SC, Zhang Z, Zhao CG, Hong K, Zhao X, Ding DR, Xie MH, Wu CD, Das MC, Gill R, Thomas KM, Chen B. Rationally tuned micropores within enantiopure metal-organic frameworks for highly selective separation of acetylene and ethylene. Nat Commun. 2011;2:204.

[97] Das MC, Guo Q, He Y, Kim J, Zhao CG, Hong K, Xiang S, Zhang Z, Thomas KM, Krishna R, Chen B. Interplay of metalloligand and organic ligand to tune micropores within isostructural mixed-metal organic frameworks (M′MOFs) for their highly selective separation of chiral and achiral small molecules. J Am Chem Soc. 2012;134:8703–10.
[98] Hu TL, Wang H, Li B, Krishna R, Wu H, Zhou W, Zhao Y, Han Y, Wang X, Zhu W, Yao Z, Xiang S, Chen B. Microporous metal–organic framework with dual functionalities for highly efficient removal of acetylene from ethylene/acetylene mixtures. Nat Commun. 2015;6:7328.
[99] Li B, Cui X, O'Nolan D, Wen HM, Jiang M, Krishna R, Wu H, Lin RB, Chen YS, Yuan D, Xing H, Zhou W, Ren Q, Qian G, Zaworotko MJ, Chen B. An ideal molecular sieve for acetylene removal from ethylene with record selectivity and productivity. Adv Mater. 2017;29:1704210.
[100] Li L, Wen HM, He C, Lin RB, Krishna R, Wu H, Zhou W, Li J, Li B, Chen B. A metal–organic framework with suitable pore size and specific functional sites for the removal of trace propyne from propylene. Angew Chem Int Ed. 2018;57:15183–8.
[101] Cui X, Chen K, Xing H, Yang Q, Krishna R, Bao Z, Wu H, Zhou W, Dong X, Han Y, Li B, Ren Q, Zaworotko MJ, Chen B. Pore chemistry and size control in hybrid porous materials for acetylene capture from ethylene. Science. 2016;353:141–4.
[102] Lin RB, Li L, Wu H, Arman H, Li B, Lin RG, Zhou W, Chen B. Optimized separation of acetylene from carbon dioxide and ethylene in a microporous material. J Am Chem Soc. 2017;139:8022–8.
[103] Nolan DO', Madden DG, Kumar A, Chen KJ, Pham T, Forrest KA, Patyk-Kazmierczak E, Yang QY, Murray CA, Tang CC, Space B, Zaworotko MJ. Impact of partial interpenetration in a hybrid ultramicroporous material on C_2H_2/C_2H_4 separation performance. Chem Commun. 2018;54:3488–91.
[104] Yang L, Cui X, Yang Q, Qian S, Wu H, Bao Z, Zhang Z, Ren Q, Zhou W, Chen B, Xing H. A single-molecule propyne trap: highly efficient removal of propyne from propylene with anion-pillared ultramicroporous materials. Adv Mater. 2018;30:1705374.
[105] Wen HM, Li L, Lin RB, Li B, Hu B, Zhou W, Hu J, Chen B. Fine-tuning of nano-traps in a stable metal–organic framework for highly efficient removal of propyne from propylene. J Mater Chem A. 2018;6:6931–7.
[106] Li L, Lin RB, Krishna R, Wang X, Li B, Wu H, Li J, Zhou J, Chen B. Flexible–robust metal–organic framework for efficient removal of propyne from propylene. J Am Chem Soc. 2017;139:7733–6.
[107] Bender M. An overview of industrial processes for the production of olefins – C_4 hydrocarbons. ChemBioEng Rev. 2014;1:136–47.
[108] Assen AH, Belmabkhout Y, Adil K, Bhatt PM, Xue DX, Jiang H, Eddaoudi M. Ultra-tuning of the rare-Earth fcu-MOF aperture size for selective molecular exclusion of branched paraffins. Angew Chem Int Ed. 2015;54:14353–8.
[109] Hartmann M, Kunz S, Himsl D, Tangermann O, Ernst S, Wagener A. Adsorptive separation of isobutene and isobutane on $Cu_3(BTC)_2$. Langmuir. 2008;24:8634–42.
[110] Kim H, Jung Y. Can metal–organic framework separate 1-butene from butene isomers? J Phys Chem Lett. 2014;5:440–6.
[111] Luna-Triguero A, Vicent-Luna JM, Poursaeidesfahani A, Vlugt TJH, Sánchez-de-Armas R, Gómez-Álvarez P, Calero S. Improving olefin purification using metal organic frameworks with open metal sites. ACS Appl Mater Interfaces. 2018;10:16911–7.
[112] Barnett BR, Parker ST, Paley MV, Gonzalez MI, Biggins N, Oktawiec J, Long JR. Thermodynamic separation of 1-butene from 2-butene in metal–organic frameworks with open metal sites. J Am Chem Soc. 2019;141:18325–33.
[113] Liao PQ, Huang NY, Zhang WX, Zhang JP, Chen XM. Controlling guest conformation for efficient purification of butadiene. Science. 2017;356:1193–6.
[114] Kishida K, Okumura Y, Watanabe Y, Mukoyoshi M, Bracco S, Comotti A, Sozzani P, Horike S, Kitagawa S. Recognition of 1,3-butadiene by a porous coordination polymer. Angew Chem Int Ed. 2016;55:13784–8.

[115] Lange M, Kobalz M, Bergmann J, Lässig D, Lincke J, Möllmer J, Möller A, Hofmann J, Krautscheid H, Staudt R, Gläser R. Structural flexibility of a copper-based metal–organic framework: sorption of C_4-hydrocarbons and in situ XRD. J Mater Chem A. 2014;2:8075–85.

[116] Zhang Z, Yang Q, Cui X, Yang L, Bao Z, Ren Q, Xing H. Sorting of C_4 olefins with interpenetrated hybrid ultramicroporous materials by combining molecular recognition and size-sieving. Angew Chem Int Ed. 2017;56:16282–7.

[117] Chen J, Wang J, Guo L, Li L, Yang Q, Zhang Z, Yang Y, Bao Z, Ren Q. Adsorptive separation of geometric isomers of 2-butene on gallate-based metal–organic frameworks. ACS Appl Mater Interfaces. 2020;12:9609–16.

[118] Assen AH, Virdis T, De Moor W, Moussa A, Eddaoudi M, Baron G, Denayer JFM, Belmabkhout Y. Kinetic separation of C_4 olefins using Y-fum-fcu-MOF with ultra-fine-tuned aperture size. Chem Eng J. 2021;413:127388.

[119] Ihmels EC, Fischer K, Gmehling J. Thermodynamic properties of the butenes: Part I. Experimental densities, vapor pressures, and critical points. Fluid Phase Equilib. 2005;228–229:155–71.

[120] Chen Z, Feng L, Liu L, Bhatt PM, Adil K, Emwas AH, Assen AH, Belmabkhout Y, Han Y, Eddaoudi M. Enhanced separation of butane isomers via defect control in a fumarate/zirconium-based metal organic framework. Langmuir. 2018;34:14546–51.

Bassem Al-Maythalony, Ala'a Al-Ghourani, Ying Siew Khoo, and Woei Jye Lau

6 Development of MOF-based membranes for gas and water separation

6.1 Introduction

The principles of reticular chemistry enable the construction of a special class of materials known as metal-organic frameworks (MOFs). MOFs are built from metal atoms or clusters connected by di-, tri- or multi-topic organic ligands, called linkers. The metal nodes and the linkers come together in the form of an extended crystalline structure that often possesses uniform and permanent porosity. The applications of MOFs are expanding tremendously owing to their design possibility, mild synthesis conditions, crystalline nature, permanent porosity, thermal stability and chemical stability in addition to their tunable pore sizes through pre and post-synthetic functionalization. MOFs are promising candidates to be applied in the areas of gas separation and water treatment. The use of MOFs in these applications propelled the field of reticular chemistry forward in the form of synthesizing new MOFs and modifying already existing MOFs, developing a high level of control over morphology, improving strategies for attaching MOF particles to surrounding matrices, and fabricating thin films on various supports. The MOFs that are suitable for separation do not require high porosity that, in many cases, comes at the expense of cost, rather, candidate MOFs suitable for separation should have a pore size in the range of the targeted molecules, in addition to chemical and thermal stability, ease for potential large-scale production and reasonable production cost.

For water and gas treatment applications, MOFs have been fabricated as: (i) fixed-beds for capturing certain components in a given feed stream. The efficiency of fixed beds in separation processes depends on the thermodynamic affinity between the targeted component(s) and the material. Separation via fixed-bed requires frequent activation for reuse after the saturation capacity of the active substance is reached and (ii) membranes to separate different components based on pore accessibility, rejection, as well as affinity, velocity and degree of interaction between given guest molecules and active sites. This form of separation is a continuous process that does not require frequent activation.

MOF-based membranes used in gas separation and water purification are fabricated as hollow-fiber, tubular and/or planar membranes. The most common fabri-

Bassem Al-Maythalony, Ala'a Al-Ghourani, Materials Discovery Research Unit, Advanced Research Centre, Royal Scientific Society (RSS), Amman 11941, Jordan, e-mail: bassem.maythalony@rss.jo
Ying Siew Khoo, Woei Jye Lau, School of Chemical and Energy Engineering, Faculty of Engineering, Universiti Teknologi Malaysia, 81310 Skudai, Johor, Malaysia

https://doi.org/10.1515/9781501524721-006

cation methods reported in the literature for examining MOF-based membranes are categorized as: (i) membranes with a continuous MOF layer at the top of a substrate. These types of membranes can be fabricated as two-dimensional thin films attached to a mechanically-stable support and (ii) composite membranes with MOF particles imbedded within the membrane matrix, known as mixed matrix membranes (MMMs). This type of membrane can be fabricated in the form of hollow-fiber, tubular and planar membranes because they gain stability from the mechanical robustness and resilience of the matrix.

This chapter elaborates on the different aspects of MOF-based membranes in gas separation and water purification applications. The discussion includes fabrication strategies, separation techniques, recent innovations in the field and challenges encountered and associated with the practical implementation of MOF-based membranes taking into consideration industrial and scientific perspectives along with the potential for expansion in the future.

6.2 MOF membrane fabrication strategies

The fabrication of MOF membranes comes with various difficulties. The challenges in pure MOF membranes as thin films can be summarized in: (i) preparing defect-free thin films, because of a low tendency of MOF particles for intergrowth into a continuous layer, (ii) chemical stability, (iii) thermal stability, (iv) mechanical robustness and (v) the high-throughput production. The crystalline nature of MOFs makes them brittle materials that cannot withstand extreme conditions, such as high temperature, high pressure and corrosive gases. Therefore, the successful implementation of MOFs in membrane science typically requires using a mechanically stable support and ensuring the satisfactory attachment of the MOF to the support.

On the other hand, the use of MOFs as fillers within a polymeric matrix to form MOF-based MMMs solves many challenges associated with MOF-based membranes, such as: (i) fabrication (i. e., the MMM with MOF filler allows for the exploration of different MOFs, even those that are not good for achieving particle intergrowth), (ii) the fabrication of mechanically stable membranes with robustness gained from the polymer mechanical resilience and thermal stability and (iii) the use of polymeric matrix facilitated improved permselectivity properties.

6.2.1 Methods for the fabrication of pure MOF membranes

This section discusses different strategies utilized in fabricating pure MOF membranes. These techniques are solvothermal and hydrothermal synthesis, microwave reactions, liquid phase epitaxy, secondary growth, self-assembly monolayer and spray drier.

6.2.1.1 Solvothermal and hydrothermal techniques

Hydrothermal and solvothermal methods are the simplest strategies for MOF fabrication. The two procedures are very similar, with one major difference being that water is used as the solvent in the hydrothermal method, while organic solvents are used in the solvothermal method. Solvothermal is the most common and effective method for the synthesis of MOFs. However, due to environmental sustainability and cost reduction, research continues to focus on developing hydrothermal synthesis strategies for the sake of greener synthesis methods. Solvothermal synthesis allows the use of a wide range of organic solvents, which tunes the reaction media via different polarities to carry out the reaction at higher temperatures due to the wider range of solvent boiling points. On the other hand, hydrothermal synthesis allows limited choices of the polarity of the reaction media and temperatures because of the relatively low boiling point. The fabrication of MOF membranes using solvothermal and hydrothermal strategies can be summarized in the dissolving of the metal precursor and the linker at specified conditions of concentration and pH in the presence of a substrate/support. This is followed by allowing the components to react for a certain period of time at a specified temperature to form the MOF structure. MOF particles grow and become heavier until they are deposited on the support by gravity or a chemical attraction toward the support.

The challenges associated with the use of solvothermal and hydrothermal methods for thin-film fabrication are the limited control of crystal growth, the irregularity of the particle size, the absence of chemical bonding between the thin film and the support (i. e., the reaction solution must be in direct contact with the target support), and the limited preference for the particle deposition direction that follows the direction of gravity. Additionally, the reliance on gravity to guide the orientation of grown particles' deposition restricts the layout of the substrate at the bottom of the reaction vessel.

It is worth noting that solvothermal and hydrothermal methods have been successfully implemented, separately and in combination with other synthesis techniques, in the fabrication of different challenging MOF membranes on various types of supports, such as alumina, titania and silica, among others. For example, a solvothermal growth technique was used to grow MOF-5 particles on the modified α-alumina support after forming MOF-5 seeds using a microwave-induced thermal deposition synthesis technique [1]. A **sod**-ZMOF membrane was successfully fabricated on a porous alumina substrate using the solvothermal method [2]. In this example, $In(NO_3)\cdot 6H_2O$ and 4,5-imidazole-dicarboxylic acid were reacted in N,N-dimethylformamide (DMF) in the presence of a ceramic support made of α-alumina placed in the bottom of the reaction vessel [2]. The reaction led to the formation of a continuous 36 μm-thick **sod**-ZMOF membrane with perfect crystal intergrowth [2]. The resulting membrane was examined in the separation of industrial gases and ex-

Figure 6.1: Zeolitic metal-organic framework (ZMOF) thin-film membrane. Solvothermal synthesis usually shows high density but poor crystal orientation.

hibited preferential permeation of CO_2 over the smaller H_2 gas and other industrially important gases, such as N_2 and CH_4 (Figure 6.1) [2].

6.2.1.2 Microwave induced synthesis

The microwave-induced MOF synthesis strategy was introduced by Jeong and coworkers [1, 3]. It attracted attention in the fabrication of MOF membranes because of its reduced reaction time, high scale production, narrow dispersity, facile control on morphology and facilitated secondary and seeded growth [4].

Microwave synthesis techniques are used in the fabrication of a certain MOF as continuous layers either directly or as a secondary growth technique by using a pre-modified support. The microwave synthesis technique, as secondary growth technique, was successfully utilized in the fabrication MOF-5 thin film on a premodified alumina support [1, 4, 5].

Microwave-assisted synthesis offers a powerful strategy for growing various continuous and defect-free MOF membranes (e. g., MOF-5 [1], ZIF-8 [1] ZIF-7 [6] and CAU-10-H) [7]. The microwave-induced method facilitated the production of complicated structures that are built from mixed linkers as in ZIF-78 [8]. Other MOFs that were made into membranes using the microwave reaction technique are UiO-66-NH_2[9] and NH_2-MIL-125(Ti) [10].

6.2.1.3 Layer-by-layer liquid phase epitaxy

Layer-by-layer liquid phase epitaxy (LBL) is a process in which a highly ordered crystalline MOF layer is grown on the substrate with a very controlled growth [11]. The fabrication process, in general, can be summarized in conditioning of the support followed by subsequent dip-coating in solutions containing metals and linkers with

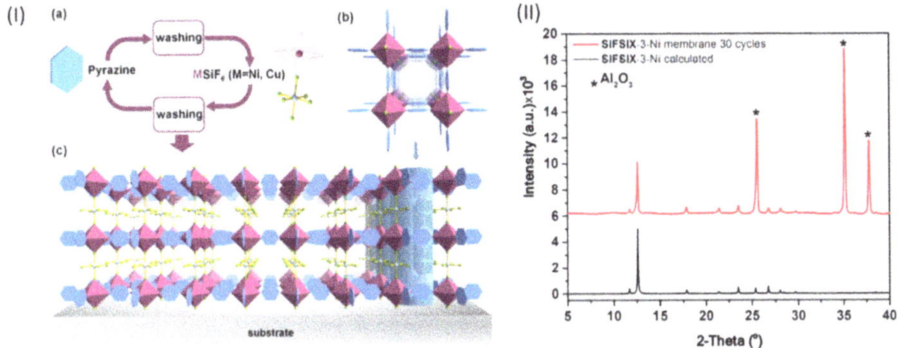

Figure 6.2: (I) (a) The liquid-phase epitaxy (LPE) method applied for the growth of SIFSIX-3-M thin film, (b) view of SIFSIX-3-M accessible channel, (c) scheme of SIFSIX-3-M membrane grown on Al_2O_3 substrate, while (II) shows the XRD patterns of fabricated SIFSIX-3-Ni membrane thin-film on Al_2O_3 support grown by means of the LPE method (red) and calculated (black) [18].

washing happening between every immersion; this cycle is repeated to the assigned number. There are two main factors that control the number of cycles normally assigned; (i) ensuring complete interlayer growth to avoid defects in the resulting thin film, and (ii) achieving the targeted thickness of thin film, which is normally proportional to the number of cycles (Figure 6.2). LBL is also used in the self-assembly monolayer (SAM) MOF-based membrane fabrication method [12]. LBL has many advantages compared to other methods of fabrication, such as (i) high control over the growth orientation, (ii) high control of thickness, (iii) improved covalent attachment between the multicrystalline MOF layer and the support, (iv) wide applicability to various supports, including polymeric, ceramic, gold and silicon supports, (v) facilitated high throughput support coating and (vi) facilitated intergrowth of challenging MOFs that normally suffer from very limited control of their intergrowth; such as the zirconium MOFs, UiO-66 [13]. LBL was successfully utilized in the growth of various MOF membranes, including HKUST-1 [14], ZIF-67 [15], ZIF-8 [15, 16], Ni-MOF-74 [17], UiO-66 and SIFSIX-3 [18]. Using LBL, a continuous membrane was fabricated demonstrating the successful confinement of HKUST-1 as a thin film on nanochannels of the support made from polyethylene terephthalate (np-PET) [19]. The resulting HKUST-1/np-PET showed a clear cut-off of CO_2 permeation versus other industrial gases and olefins [19].

6.2.1.4 Self-assembled monolayer

Self-assembled monolayer (SAM) is a process in which organic molecules are assembled spontaneously on a support. Normally, long-chain organic molecules with -SH, -COOH, $-NH_2$ functionalities are covalently attached to the support and arrange themselves on the support. These chains chemically bind MOF particles on the functionalized support to form a MOF thin film. Using the SAM method in the fabrication of MOF thin films was found to be useful in directing the orientation of the MOF crystals' growth. The surface modified by using SAM technique normally subjected to another MOF growth technique to fabricate continuous thin films. Several MOF thin films were successfully fabricated using the SAM method, such as UiO-66 [20], MOF-5 [21] and HKUST-1 [22]. The SAM method was successfully implemented for selective positioning of MIL-101 and UiO-66 at substrate surfaces in a patterned fashion [23]. In this example, an alkyne chain with carboxylic acid functionality allowed for the self-assembly of the MOFs on a silicon support. The modified support was used in the solvothermal reaction synthesis to form the patterned MOF layer [23].

6.2.1.5 Secondary growth

Several MOFs have been synthesized using secondary growth techniques of the MOF layer on a support that was seeded with nano-size MOF particles. The secondary growth can follow various synthetic methods, such as hydrothermal, solvothermal or microwave synthesis. The application of secondary growth synthesis requires preparing the MOF nano-seeds that are disseminated on the support. The modified support is transferred to another reaction solution and subjected to a crystal growth method to form MOF particles grown as thin films selectively on the modified sites of the support. This method has the advantage of improving the MOF-support attachment, adjusting the MOF growth orientation, and directing the growth location on the support resulting from the selective formation of a continuous layer on the targeted sites of the support. Using secondary growth, Jeong et al. successfully produced a continuous pure MOF-5 thin film [1]. The authors reported the synthesis of nano-seeds of MOF-5 using a microwave reaction that is then further subjected to solvothermal reaction conditions to produce a continuous layer of MOF-5 [1]. This is followed by several reports in which a secondary growth technique was used to grow MOF-5 thin films [24, 25]. Secondary growth methods were used by Gascon et al. to achieve a better attachment between the HKUST-1 and a porous alumina substrate [26]. The seeds of the MOFs were spread on the substrate by spin-coating, followed by secondary growth of HKUST-1 layer using solvothermal conditions [26]. Additionally, the secondary growth method has successfully been used for the controlled growth of hierarchical MOF-on-MOF architectures. This particular example demonstrated the use of secondary growth for easy stacking of different MOF layers to introduce different unique chemical functionality within the generated MOF layers [27].

6.2.1.6 Electrochemical reaction

The electrochemical fabrication technique provides milder conditions, lower energy and better control of growth for MOF membrane fabrication in the sub-5 μm range [18]. Successful electrochemical synthesis has been reported in the synthesis of a HKUST thin film that was deposited on a Cu mesh [28, 29]. Recently, Eddaoudi et al. reported an electrochemical directed-assembly strategy to fabricate polycrystalline MOF membranes that were examined in the separation of hydrocarbons [30].

6.2.1.7 Spray-drying synthesis

The spray dryer method is a feasible industrial synthesis method because it offers a rapid synthesis of crystalline MOF powders from solution, increased product yield, and shorter preparation time [31]. Additionally, spray drying allows facile engineering of the crystal growth through the use of nonsolvent to prompt MOF crystallization upon attachment to the support. A simultaneous spray-drying synthesis method is used for rapid synthesis and shaping of the MOF morphology and to strengthen the MOF particle attachments to the support [32]. This has successfully been used in the synthesis of γ-CD-MOFs [31], UiO-66, MIL-100 and $[Ni_8(OH)_4(H_2O)_2(L)_6]_n$ (where L = 1H-pyrazole-4-carboxylic acid) [32].

6.2.2 Methods for the fabrication of MOF-modified polymeric membranes

In this section, a brief overview is provided of the typical membrane fabrication methods commonly used by membrane scientists in synthesizing MOF-modified polymeric membranes. These techniques are phase inversion, surface coating, electrospinning and interfacial polymerization.

6.2.2.1 Phase inversion technique

The phase inversion technique was first reported in the late 1950s by Loeb and Sourirajan to fabricate asymmetric membranes for reverse osmosis (RO) applications [33]. Because of its simplicity and effectiveness in producing membranes with a wide range of properties (from porous to dense structures), the phase inversion technique has been widely used not only for lab-scale membrane fabrication but also for commercial membrane production. Prior to the MOF-modified membrane fabrication, a doped solution composed of a specific polymer, solvent, targeted MOF and additive (e. g., salt and secondary polymer) is prepared. The doped formulation (e. g., type of polymer, type of solvent, loading of MOF) can be varied depending on the membrane application. The MOF-modified membrane is then developed through either employing a casting technique (for producing a flat sheet membrane) or via a spinning technique (for producing a hollow fiber membrane). A large amount of water is used to initiate phase inversion. Once the membrane is formed, different post-treatments, including heat treatment and glycerol preservation, are carried out to realize the required membrane surface properties.

6.2.2.2 Surface coating

Compared to the phase inversion technique, surface coating offers a unique advantage as it does not alter the bulk properties of the membrane. Typically, surface coating is performed on the existing membrane surface to alter its top layer's properties to achieve desired characteristics. Several coating techniques can be applied to develop a MOF-coated membrane. These include dip-, spin-, spray- and layer-by-layer (LbL) coating. Dip-coating, for instance, is the easiest method to be performed. LbL method can produce nm-scale thickness during the deposition process, and the layer properties are tuneable by the number of sequential adsorption steps. A major drawback of surface-coating is the possible formation of a relatively dense layer atop the existing membrane leading to reduced membrane water flux [34].

6.2.2.3 Electrospinning

Electrospinning is another membrane production method that can form a highly porous nanofiber membrane. This property is particularly important for the adsorptive MOF-modified membranes as the existence of an extremely high surface area can lead to the effective removal of heavy metal ions. Electrospinning uses electric force to draw charged threads of a polymer solution containing MOF to form a fiber mat with diameters on the order of nanometers. During the electrospinning process, parameters such as polymer solution properties (e. g., viscosity, conductivity and surface tension), humidity, temperature, the distance between capillary tip and collector and voltage can impact the final characteristics of the membranes produced [35].

6.2.2.4 Interfacial polymerization

Interfacial polymerization (IP) is a technique involving the cross-linking between two different types of active monomers on the surface of the microporous substrate. This technique is commercially used to fabricate polyamide (PA) thin-film composite (TFC) membranes for industrial nanofiltration (NF)/reverse osmosis (RO)/forward osmosis

(FO) processes. Researchers use this technique to introduce nanomaterials into the PA selective layer to address the drawbacks of the typical TFC membranes, such as flux/rejection trade-offs and poor antifouling properties. A MOF can be introduced into either an aqueous amine solution or acyl chloride organic solution during the membrane fabrication. However, most researchers prefer to disperse nanomaterials/nanoparticles into an aqueous phase as they can easily achieve a more homogenous solution to minimize the particle agglomeration in the PA matrix of the thin-film nanocomposite (TFN) membranes [36].

6.3 MOF-based membranes for separation applications

6.3.1 MOF-based membranes for gas separation

Membrane technology plays a major role in the downstream processes of various important and industrially challenging separation and purification of CO_2, CH_4, H_2, N_2, O_2, olefins, paraffins, dehumidification and H_2S removal. Polymer-based membranes, in many cases, provide high selectivity, while suffering from low permeability impacting their industrial implementation. On the other hand, membranes made from zeolites offer higher permeability but they suffer from limited ranges of porosity and reduction of selectivity when exposed to moisture [37, 38]. The reactivation of zeolite membranes requires energy-intensive regeneration conditions [39]. MOF-based membranes are attractive for their ability to perform selective separations of complicated gas mixtures. For the sake of improved gas separation performance in both selectivity and permeability, the field has witnessed a large number of reports on various MOFs syntheses, structural modifications, morphological control and incorporation into different types of membrane configurations.

Over the past 28 years, the number of publications on MOF-based membranes for gas separation has exponentially increased. These publications are distributed between pure MOF membranes and MMMs, as shown in Figure 6.3. Since 2009 the number of filed patents on MOFs for gas separation has increased annually, showing a greater interest in the technological implementation of MOF-based membranes for gas separation applications, as demonstrated in Figure 6.4.

The performance of the membrane in separation is a combination of the permeance (P) and the selectivity (S). In general, the increase of permeability comes at the expense of a decrease in selectivity, and vice versa. This trade-off between P and S is expressed in the upper bound curve represented by Robson et al. that is calculated for various binary gas mixtures [40].

Several MOFs have been used in gas separation using membrane technology. For example, IRMOFs, many ZIFs, MIL compounds, SIFSIX, UiO-66 and –67 and MOF-74,

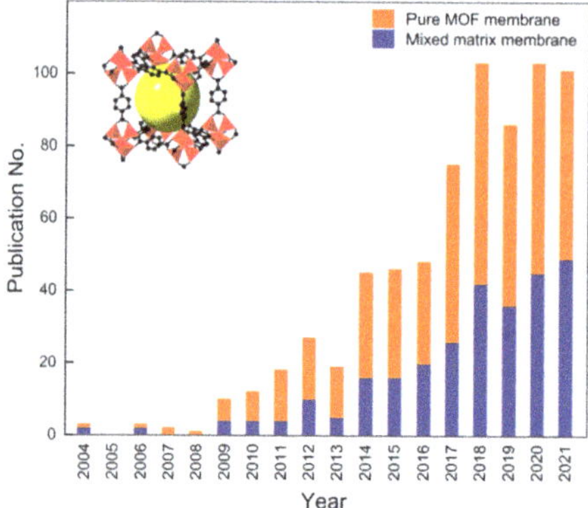

Figure 6.3: Histograms showing the exponential increase in the number of publications on MOF-based membranes from 2004 until 2021. Every data column is the sum of the reported membranes of pure MOF and MMMs.

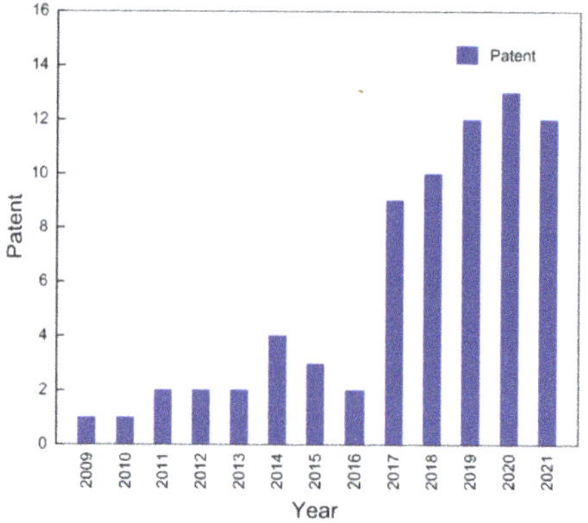

Figure 6.4: Histograms showing the exponential increase of registered patents on MOF-based membranes. Every data column is the sum of the reported membranes of pure MO and MMMs.

among others. Some of the MOF materials have been successfully synthesized as pure MOF films, while due to the complexity of the production process of the pure MOF membrane have been investigated as a filler in the MMMs.

The presented information in Tables 6.1 to 5 summarizes the separation performance of MOF-based membranes, which are higher than the upper-bound curve in both the MMMs and the pure MOF membranes. The selection criteria of the MOFs presented here is to ensure that the performance of the membrane exceeds the upper bound curve. The score criterion, as defined by Qian et al., was used in the evaluation of performance based on reported permeability and selectivity measurement [5]. The score of each membrane is calculated from the shortest distance between the membrane location on the permeability and selectivity plot and the upper bound curve for

Table 6.1: The CO_2/CH_4 separation on MOF-based MMMs.

MOF	Loading (wt%)	Polymer	Permeance (CO_2)	α (CO_2/CH_4)	Ref.
Y-fum-fcu-MOF	20	6FDA-DAM	1050.0	136.4	[41]
UiO-66-NH_2	20	PIM-1	12498.0	392	[42]
MOF-74	20	PIM-1	19000,0	700	[43]
NUS-8	2	PIM-1	6462	217	[44]
ZIF-71	30	PIM-1	3458.6	97.1	[45]
PEI-g-ZIF-8	20	PVAm	1790	40.7	[46]
UiO-66	14	6FDA-DAM	1912	30.8	[47]
ZIF-7-NH_2	30	Cross-linked poly(ethyleneoxide)	212	62.4	[48]
KAUST-7	33	6FDA-Durene	1030	33.2	[49]
ZIF-90	15	6FDA-DAM	720	36.9	[50]
ZIF-67	20	PIM-1	5181	16.8	[51]
HKUST-1-IL	20	6FDA-Durene	1101.6	29.3	[52]

Table 6.2: The H_2/CH_4 separation on MOF-based MMMs.

MOF	Loading (wt%)	Polymer	Permeance (H_2)	α (H_2/CH_4)	Ref.
MOF-74	20	PIM-1	11469	10.3	[43]
ZIF-8	30	TB-6FDA-PI	1858	35.7	[46]
UiO-66-NH_2	10	PIM-1	2641	25.9	[53]
UiO-66(Ti)	13.3	PIM-1	8380	6	[54]
ZIF-11	30	PEI-Ultem1000	8.7	18125	[55]
ZIF-71	30	6FDA-Durene	4533	56	[56]
ZIF-90	50	6FDA-TP polyimide	179	99.4	[57]

Table 6.3: The CO_2/CH_4 separation on membrane made from MOF.

MOF	Score (CO_2; GPU)	α (CO_2/CH_4)	Ref.
IRMOF-1	7612	328	[58]
ZIF-8	163	181.1	[59]
HKUST-1	4190	12.5	[60]

a given gas mixture. Tables 6.1 and 6.2 present a list of the best performing MOF-based MMMs in the separation of CO_2/CH_4 and H_2/CH_4, while Tables 6.3, 6.4 and 6.5 represent the best performing MOFs that have been used as membranes in the separation of CO_2 from CH_4, N_2 and H_2, respectively.

Table 6.4: The CO_2/N_2 separation on membrane made from MOF.

MOF	Score (CO_2; GPU)	α (CO_2/N_2)	Ref.
IRMOF-1	614	410	[58]
HKUST-1	4190	13.2	[60]

Table 6.5: The H_2/CO_2 separation on membrane made from MOF.

MOF	Q (H_2; GPU)	α (H_2/CO_2)	Ref.
MAMS-1	800	266.7	[61]
ZIF-8/GO	240	406	[62]
$Zn_2(bIM)_3$	3136.2	129.6	[63]
ZIF-95	7340	35	[64]
NH_2-MIL-53	5930	31	[65]
ZIF-9	22179.1	14.9	[66]
$Cu_3(BTC)_2$	287463.0	4.7	[67]
$Zn(BDC)(TED)_{0.5}$	11791.1	16.9	[68]
NH_2-Mg-MOF-74	244.9	87.5	[69]
$Co_2(bIM)_4$	597	58.5	[70]
ZIF-9–67 hybrid	41965.4	8.9	[71]
ZIF-7	7027	18.4	[72]
CAU-1-NH	4480.3	21.4	[73]
ZIF-100	188.2	77.8	[74]
JUC-150	546	39	[75]
ZIF-90	884	21.6	[76]
ZIF-L	582.4	24.4	[77]
UiO-67	3046.6	8.7	[78]
MOF-5	32855.4	2.8	[79]
MIL-96(Al)	1582.1	8.7	[80]
CAU-1	576.5	13.2	[81]
NH_2-MIL-125(Ti)	133.5	24.8	[82]
$[Ni_2(L\text{-}asp)_2(bpe)]\cdot(G)$	121.9	22.6	[83]
$Cu(bipy)_2(SiF_6)$	896.1	7.8	[84]
ZIF-22	603	8.5	[85]
ZIF-78	307	11	[86]

6.3.2 MOF-based membranes for water purification

6.3.2.1 Effect of MOF incorporation on membrane water flux and rejection

MOF nanomaterials with high flexibility, tunable porosity, large surface area and high compatibility with polymer matrices have made them suitable for membrane fabrication and/or surface modification for use in water treatment processes [87]. For example, Xiao et al. [88] reported an outstanding improvement in membrane flux af-

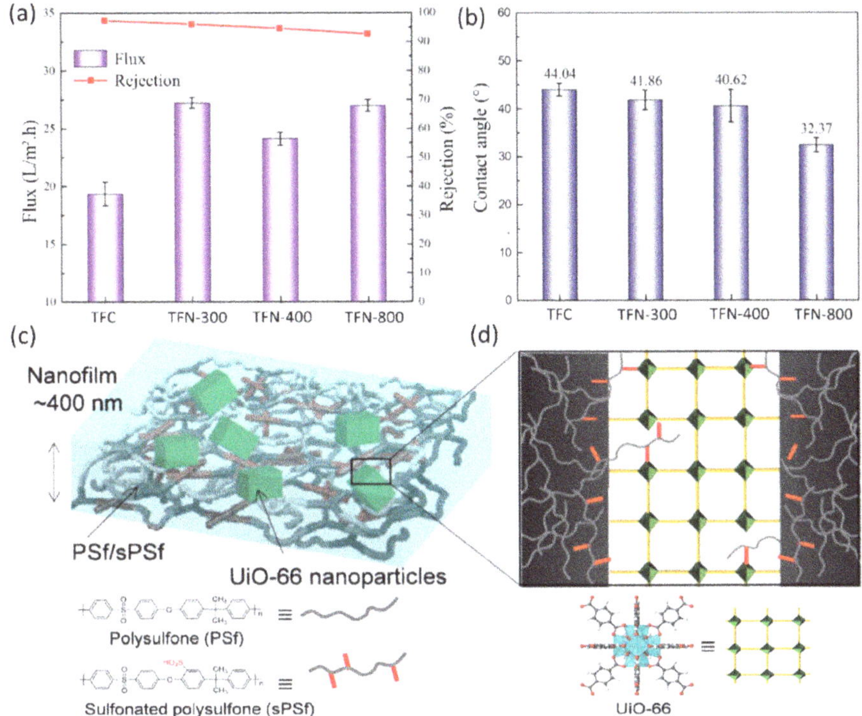

Figure 6.5: (a) Water contact angle and (b) membrane filtration performance of pristine and 0.1 w/v % CAU-1 incorporated TFN membrane (Note: TFN-300, TFN-400 and TFN-800 correspond to the TFN membrane incorporating CAU-1 with a size of 300, 400 and 800 nm, respectively) [88], (c) schematic illustration of mixed matrix nanofilm containing UiO-66 nanoparticles[90] and (d) PSf/sPSf polymer chains that show the local binding interactions between the UiO-66 framework and the sPSf chains [90].

ter CAU-1 was incorporated within the polyamide (PA) selective layer of a nanofiltration membrane. Figure 6.5a shows the impacts of CAU-1 with different sizes (300, 400 and 800 nm) on the membrane water flux and its Na_2SO_4 rejection performance when tested at 0.4 MPa using a 1000 ppm salt solution. By introducing 0.1 wt/v% CAU-1 (dispersed in a trimesoyl chloride (TMC)/*n*-hexane organic solution) into the membrane matrix (regardless of particle sizes), all the resultant TFN membranes exhibited significantly higher water flux compared to the bare TFC membrane. In particular, the membrane with 300 nm sized CAU-1 particles showed the highest water flux among the studied membranes, achieving 40 % higher water flux than that of the bare TFC membrane with a minimum drop in salt selectivity (decreased from ca. 97 to 96 %). The notable difference in water flux may be due to the presence of CAU-1, which provides unique transport channels for water molecules, allowing it to pass through the membrane at reduced resistance. In addition, the existence of small amount of CAU-1 in the selective layer can render the membrane more hydrophilic as evidenced by the water contact angle decreasing from ca. 44 to 41.9°, as shown in Figure 6.5b. This also partially contributed to flux improvement. Since CAU-1 is positively charged, the authors speculated that its introduction during the IP process might affect the diffusion of piperazine (PIP) monomer from the aqueous phase to the hexane/TMC organic phase, which eventually formed a thinner PA layer that was favorable for water transport. Although the TFN membranes comprised of 400 and 800 nm CAU-1 particles were also able to achieve higher water productivity compared to the bare TFC membrane, both membranes suffered from a much lower salt removal rate (down to 93 %) owing to possible surface defects formed within the PA matrix during the incorporation of the larger-sized CAU-1. When silicon dioxide (SiO_2)[89] and titanium dioxide (TiO_2) [89] are used for TFN membrane fabrication, the particle size of both materials is in fact very small (<50 nm) to minimize the surface defects due to severe particle agglomeration/aggregation. It is noted that the typical thickness of a PA layer is between 100 and 500 nm, and thus, large particles are not recommended as the thin layer cannot accommodate larger sizes when introduced.

Liu et al. [90] incorporated UiO-66 into a PSf/sulfonic polysulfone (sPSf) membrane matrix. Unlike conventional MMMs with a thickness of 100–150 microns, the MMM developed by Liu et al. [90] had nm-level thickness (~400 nm) as illustrated in Figure 6.5c,d. The aperture size of UiO-66 (~6 Å), which is in the range of hydrated ions (6.6–10.5 Å) and water molecules (~2.8 Å), offers great potential to selectively allow water molecules to pass through while retaining hydrated ions to diffuse through it. Based on the findings, it was found that the high-performance MMM can be produced at a UiO-66/polymer weight ratio of 15 (15 wt%) and this membrane exhibited an increased water permeance of 0.52 $L\,m^{-2}\,h^{-1}\,bar^{-1}$, which was much better compared to the pristine membrane (0.026 $L\,m^{-2}\,h^{-1}\,bar^{-1}$). The MMM also showed satisfactory performance in removing SO_4^{2-} ions and phenol red (molecular weight: 354.4 g mol^{-1}) with rejection maintained at least 96 % when tested using a 20 mmol L^{-1} feed solution composed of NaCl, Na_2SO_4 and phenol red at 30 bar. However, it must be pointed out that although the nanofilm demonstrated better water flux upon the incorporation of UiO-66, its water production rate is far below the industrial standard of NF membranes that can produce water at the rate of 5–20 $L\,m^{-2}\,h^{-1}\,bar^{-1}$. This is due to the extremely dense structure of the nanofilm, even though its layers are thin.

Aside from incorporating a MOF within the PA layer of composite membranes or embedding it within the asymmetric structure of MMMs, coating of a given MOF on the existing membrane surface is also a strategy for enhancing membrane filtration performance in terms of flux and rejection. Wang et al. [91] demonstrated the coating of metal azolate framework-4, SOD-[$Zn(mim)_2$] (MAF-4), on the polyethersulfone (PES) membrane surface via *in situ* growth followed by epitaxial growth of SOD-metal azolate framework-7, [$Zn(mtz)_2$] (MAF-7). Although successful from a fabrication perspective, the hydrophobic nature of MAF-4 compromised the water molecules' transportation. To enhance the membrane hydrophilicity, hydrophilic MAF-7 was applied and grown epitaxially on the MAF-4 modified membrane [92]. The results showed that

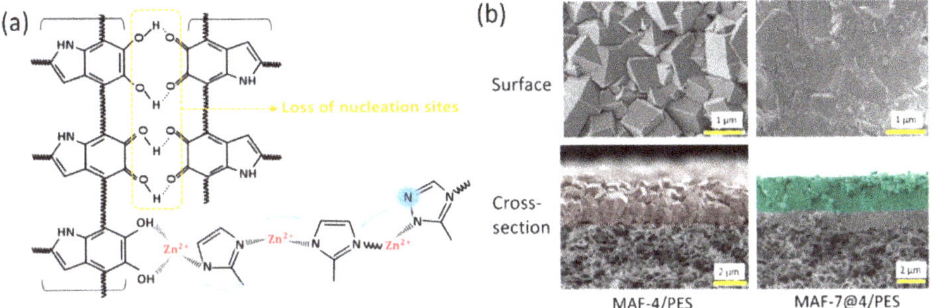

Figure 6.6: (a) Schematic diagram of MOF growth mechanism on the Zn-doped linkage layer and (b) the SEM images of MAF-4/PES and MAF-7@4/PES membranes [91].

the presence of the MAF-7@4/PES coating layer enhanced the PES membrane surface hydrophilicity (with the water contact angle decreasing from 70 to 45°) and played a significant role in improving water flux and salt rejection. Figure 6.6a presents the MOF growth mechanism on the Zn-doped linkage layer. First, zinc ions blended with dopamine were adhered to the PES surface. The MAF-4 was then coated on the membrane surface via *in situ* growth followed by epitaxial growth of MAF-7. The resulted MAF-7@4/PES membrane was reported to exhibit much higher water permeability and NaCl rejection (1.24 $L\,m^{-2}\,h^{-1}\,MPa^{-1}$ and 98.7 %) compared to the MAF-4/PES membrane (0.72 $L\,m^{-2}\,h^{-1}\,MPa^{-1}$ and 97.1 %) when both were tested at 1.5 MPa using a 0.2 wt% NaCl feed solution. The better performance of the MAF-7@4/PES membrane was due to the presence of an additional hydrophilic MAF-7, which grew epitaxially along with the dense MAF-4 layer, offering unique water transport properties. As shown in Figure 6.6b, the PES membrane was completely covered by rhombic MAF-4 crystals (indicated in light brown), as evidenced in the MAF-4/PES membrane. After the growth of MAF-7 (indicated in green) atop the MAF-4 crystals (MAF-7@4/PES membrane), the membrane exhibited higher compactness and homogeneity, which tended to uniformize and reduce the grain boundary defects for the MOF layer formed on the PES surface.

Shu et al. [93] also modified the surface of a membrane (made of hydrolyzed polyacrylonitrile) by coating it with a thin polyethyleneimine (PEI) layer containing BUT-203 (2D MOF) at different loading (50 to 200 $mg\,L^{-1}$). It is reported that the 2D MOF was compatible with the polymer matrix and enhanced the membrane water permeability due to the high porosity of the MOF structure. The researchers found that the pure water permeability was increased from 14.83 to 146.33 $L\,m^{-2}\,h^{-1}\,bar^{-1}$ as the BUT-203 loading increased from zero to 200 $mg\,L^{-1}$. At an optimum BUT-203 loading of 100 $mg\,L^{-1}$, the coated membrane exhibited extraordinary solute separation efficiency toward anionic charged dyes, reaching 99.9 % and 99.7 % rejection against congo red (CR) and eriochome black T (EBT), respectively. These results were attributed to the high negative charge of the BUT-203-modified membrane (−67 mV) that offered a strong repulsive force toward the anionic dye molecules via electrostatic repulsion effect, leading to a higher rejection and better fouling resistance.

Table 6.6 summarizes the studies related to the utilization of MOFs in modifying the membrane properties for different processes, including UF, NF, RO and FO. In short, it is clear that membranes modified using MOFs, in general, exhibit better water flux and satisfactory rejection compared to the control membranes.

Table 6.6: Summary of different types of MOF-modified membranes for water applications.

Types of MOF-based membranes	[a]Nanomaterials	[a]Type	[a]Best TFN membrane performance (compared with control TFC membrane)	Ref.
TFC membrane	0.025–0.1 *wt/v* % MIL-101(Cr) in organic solution	RO	Optimum nanomaterial loading: 0.05 *wt/v* % MIL-101(Cr) Water permeability at 16 bar: 2.2 $L\,m^{-2}\,h^{-1}\,bar^{-1}$ (TFN) vs 1.5 $L\,m^{-2}\,h^{-1}\,bar^{-1}$ (TFC) NaCl rejection (2000 ppm): remained at same level (>99 %)	[94]
	0.3 wt% ZIF-8 at different size (50, 150, 400 nm) in organic solution	RO	Optimum nanomaterial size: 50 nm ZIF-8 Water permeability at 20 bar: 1.85 $L\,m^{-2}\,h^{-1}\,bar^{-1}$ (TFN) vs 1.15 $L\,m^{-2}\,h^{-1}\,bar^{-1}$ (TFC) NaCl rejection (2000 ppm): 99 % (TFN) vs. 95 % (TFC) Other property: TFN membrane showed excellent fouling resistance with higher normalized flux (>0.90) compared to TFC (~0.65) after tested using 100 ppm BSA.	[95]
	0.05–0.3 wt% MIL-125 in organic solution	RO	Optimum nanomaterial loading: 0.15 wt% MIL-125 Water permeability at 20.7 bar: 4.11 $L\,m^{-2}\,h^{-1}\,bar^{-1}$ (TFN) vs 3.02 $L\,m^{-2}\,h^{-1}\,bar^{-1}$ (TFC) NaCl rejection (2000 ppm): Both membranes showed similar rejection (~98 %)	[96]
	0.05–0.2 wt/v % Zirconiumv (IV)-carboxylate UiO-66 in organic solution	FO	Optimum nanomaterial loading: 0.1 wt% UiO-66 Draw solution: 2M NaCl; Feed solution: DI water Pure water flux at osmotic pressure: 27 $L\,m^{-2}\,h^{-1}$ (TFN) vs 22.8 $L\,m^{-2}\,h^{-1}$ (TFC) NaCl rejection: Both membranes showed similar rejection (95 %)	[97]

Table 6.6 (continued)

Types of MOF-based membranes	[a]Nanomaterials	[a]Type	[a]Best TFN membrane performance (compared with control TFC membrane)	Ref.
	0.05–0.2 wt/v % PSS-modified ZIF-8 in aqueous solution	NF	Optimum nanomaterial loading: 0.1 wt/v % PSS-modified ZIF-8 Pure water permeability at 4 bar: 14.9 L/m^2.h.bar (TFN) vs 6.94 L m^{-2} h^{-1} bar^{-1} (TFC) Na_2SO_4 rejection (1000 ppm): Both membranes showed similar rejection (~93 %) Other property: TFN membrane showed excellent water permeability (14.9 L m^{-2} h^{-1} bar^{-1}) than that of TFC membrane (6.94 L m^{-2} h^{-1} bar^{-1}) after subjected to 500 ppm RB5 test.	[98]
Mixed-matrix membrane	0.1–1.0 wt% Ce(III) in PES/PVP/DMAc casting solution	NF	Optimum nanomaterial loading: 0.1 wt% Ce (III) Pure water permeability at 3 bar: 7.07 L m^{-2} h^{-1} bar^{-1} (TFN) vs. 4.90 L m^{-2} h^{-1} bar^{-1} (TFC) Salt rejection: *n/a* Other property: TFN membrane showed excellent dye rejection (99 %) than that of TFC membrane (86 %) after subjected to 30 mg L^{-1} Direct Red 16 dye test.	[99]
	2.5–20 wt% MIL-125(Ti) in PVDF casting solution	UF	Optimum nanomaterial loading: 10 wt% MIL-125(Ti) Permeate permeability at 4 bar: 64.3 L m^{-2} h^{-1} bar^{-1} (TFN) vs ~2 L m^{-2} h^{-1} bar^{-1} (control) RhB dye rejection: maintained	[100]
	5–50 wt% UiO-66 in PSf/DMAc casting solution	NF	Optimum nanomaterial loading: 15 wt% UiO-66 Water permeability at 30 bar: 0.52 L m^{-2} h^{-1} bar^{-1} (TFN) vs 0.026 L m^{-2} h^{-1} bar^{-1} (TFC) Na_2SO_4 rejection (20 mM): 94 % (TFN) vs. 96 % (TFC)	[90]

Table 6.6 (continued)

Types of MOF-based membranes	[a]Nanomaterials	[a]Type	[a]Best TFN membrane performance (compared with control TFC membrane)	Ref.
	Basolite A100, Basolite F300, Basolite C300 (1.0 wt% MOF)	FO	Optimum nanomaterial type: Basolite C300 Draw solution: 0.5 M $MgCl_2$; Feed solution: DI water Pure water flux at osmotic pressure: ~110 $L/m^2.h$ (TFN) vs. 78.1 $L\,m^{-2}\,h^{-1}$ (TFC) Salt rejection: *n/a*	[101]
Surface-coated membrane	50–200 mg L^{-1} BUT-203	NF	Optimum loading: 100 mg L^{-1} BUT-203 Pure water permeability at 1 bar: ~100 $L/m^2.h.bar$ (TFN) vs. 14.83 $L\,m^{-2}\,h^{-1}\,bar^{-1}$ (TFC) Methylene blue rejection (100 mg L^{-1}): Both were maintained at 98 %.	[93]
	0.05–0.2 wt% PDA/ZIF-8	NF	Optimum loading: 0.1 wt% PDA/ZIF-8 Pure water permeability at 5 bar: 3.37 $L\,m^{-2}\,h^{-1}\,bar^{-1}$ (TFN) vs. 2.03 $L\,m^{-2}\,h^{-1}\,bar^{-1}$ (TFC) $MgCl_2$ rejection (10 mM): 93 % (TFN) vs. 94.4 % (TFC) Other property: TFN membrane showed excellent frr (73 %) compared to TFC (97 %) after tested using 1 g L^{-1} SA and 1 mM $CaCl_2$	[102]
	TFC-PS-Ag/MOF and TFC-BPA-Ag/MOF (0.05 wt% Ag/MOF)	FO	Optimum membrane: TFC-PS-Ag/MOF Draw solution: 2M NaCl; Feed solution: DI water Water flux at osmotic pressure: 25.5 $L\,m^{-2}\,h^{-1}$ (TFN) vs ~12.5 $L\,m^{-2}\,h^{-1}$ (TFC) NaCl rejection: n. a. Other property: TFN membrane exhibited significant anti-biofouling properties (62 % of E-coli reduced) than TFC membrane (5 % of E-coli reduced)	[103]

[a]Abbreviations: Basolite A100 (MIL 53 or aluminium terephthalate); Basolite C300 (copper benzene-1,3,5-tricarboxylate); Basolite F300 (iron benzene-1,3,5-tricarboxylate); BPA (3-bromopropionic acid); BUT-203 ($Cu(NDC(SO_3)_2)_{1/2}(DPE)\,(H_2O)$); $CaCl_2$ (calcium chloride); DMAc (dimethylacetamide); FRR (flux recovery rate); *n/a* (not available); PDA (polydopamine); PES (polyethersulfone); PSf (polysulfone); PSS (poly(sodium 4-styrenesulfonate)); PVDF (polyvinylidene fluoride); PVP (polyvinylpyrrolidone); SA (sodium alginate).

6.3.2.2 Organic/Inorganic fouling resistance

Besides the intrinsic water flux and rejection capacity of membranes, their resistance against different types of surface fouling is critical when the membranes are used for water applications. A membrane with greater antifouling properties can reduce operating cost through the need for minimum cleaning frequency and extended lifespan of the membrane.

Aiming to improve the antifouling of TFC membranes for RO processes, Aljundi et al. [104] embedded 0.1–0.8 wt% ZIF-8 within a PA selective layer made of 2 wt% *m*-phenylenediamine (MPD) and 0.15 wt% trimesoyl-chloride (TMC) during the interfacial polymerization process. The authors reported that 0.4 wt% was the best ZIF-8 loading to produce the RO membrane that achieved promising antifouling characteristics. By subjecting the membranes to a fouling solution composed of 2000 ppm NaCl and 100 ppm bovine serum albumin (BSA) for 4 h, the ZIF-8-modified RO membrane suffered from a much lower flux decline (13 % drop from the initial water flux) compared to the bare TFC RO membrane (53 %). Such data clearly demonstrated the positive role of ZIF-8 in reducing the interaction between foulants and the membrane surface. More importantly, the ZIF-8-modified RO membrane also displayed higher pure water flux ($2.3\,L\,m^{-2}\,h^{-1}\,bar^{-1}$) than that of the bare RO membrane ($1.1\,L\,m^{-2}\,h^{-1}\,bar^{-1}$) without compromising salt rejection during filtration of 2000 ppm NaCl aqueous solution at 15 bar.

In general, the improvement in membrane hydrophilicity can play an important role in enhancing membrane antifouling properties [34]. Wang et al. [91] reported that the antifouling property of a PES membrane against $0.5\,g\,L^{-1}$ BSA solution could be enhanced upon surface coating with either MAF-7 or MAF-7@4 nanomaterials (see Figure 6.6). The *in situ* growth of MAF-4 layer on the PES substrate followed by the epitaxial growth of MAF-7 on the MAF-4 layer could lead to the lowest water contact angle (45°), which was related to the highest surface hydrophilicity. Moreover, the surface of MAF-4/PES membrane also became smoother after the addition of MAF-7, offering the additional feature of improving fouling resistance. The superior surface hydrophilicity of MAF-7@4/PES membrane coupled with its smoother surface had made it suffer from minimum flux deterioration (22.4 %) at the end of the filtration study, as shown in Figure 6.7a. The bare PES, MAF-4/PES and MAF-7/PES membranes recorded ~80 %, 75.6 % and 66.2 % flux decline, respectively. Neither the rough MAF-7/PES nor the hydrophobic MAF-4/PES membranes were as competent as the MAF-7@4/PES membrane in mitigating the BSA deposition on the membrane surface. The fluorescence distribution of BSA foulants on the MAF-7@4/PES membrane was significantly lower with less intensity than the bare PES and other two MAF-modified membranes (Figure 6.7b). This observation suggested that the MAF-7@4/PES membrane with a high degree of surface hydrophilicity and smoother morphology could offer a good solution to overcome the issue of organic fouling.

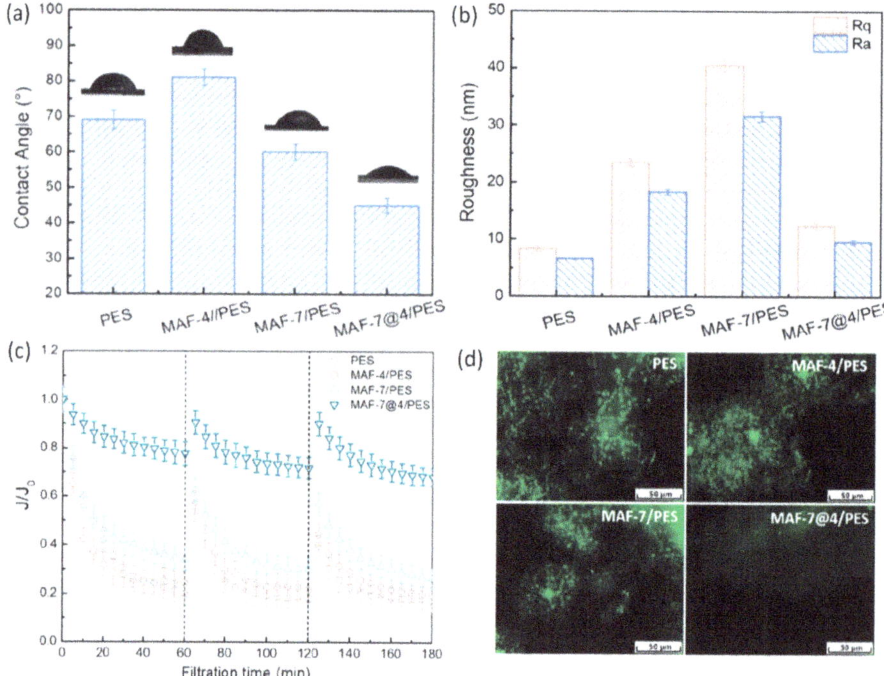

Figure 6.7: (a) normalized flux of membrane for cyclic filtration (0.5 g L^{-1} BSA solution for 3 h) with DI water cleaning performed at 60 and 120 min and (b) epifluorescence microscopic images showing the fluorescein-labeled BSA (FITC-BSA) deposited onto the membrane surfaces after being used for 3 h filtration of 0.5 g L^{-1} BSA solution [91].

Mohammadnezhad et al. [99] showed that the incorporation of 0.1 wt% nanocrystalline Ce(III) MOF within a PES membrane improves the antifouling properties of the membrane. The improved performance of the MMM is attributed to its significantly lower surface roughness (S_a = 9.23 nm) compared to the bare membrane (S_a = 43.86 nm) and enhanced membrane hydrophilicity with the water contact angle being reduced from 63.2 to 57.2°. These properties play an important role in minimizing the foulants from being absorbed in the cavities and grooves of the membrane. Moreover, the PES membrane embedded with 0.1 wt% Ce(III) MOF increased the water flux, recording a 44 % higher value than the control PES membrane. The antifouling ability of the MOF-modified membranes, however, attenuated to a 69.51 % flux recovery rate (FRR) when a much higher loading of the Ce(III) MOF was introduced (0.5 wt.%). The excessive addition of Ce(III) MOF leads to particle agglomeration within the membrane matrix, leading to increased surface roughness and pore dimension.

6.3.2.3 Antibacterial resistance

Aside from demonstrating the positive impacts on the membrane water flux/rejection as well as antifouling characteristics, some researchers also reported the additional roles of MOFs in alleviating bacterial growth on membranes after being incorporated within a membrane matrix [105]. A silver-based MOF was previously reported to demonstrate significant biocidal properties attributed to the silver clusters surrounded by the organic ligands [106]. Firouzjaei et al. [106] successfully synthesized graphene oxide-silver-based MOF (GO-Ag-MOF) and used it to develop an anti-biofouling TFN membrane. Figure 6.8a illustrates the interaction between these three nanomaterials to form hybrid GO-Ag-MOF. The superoxide anions, oxidative stress and sharp edge of GO nanosheets are also important for biocidal activity [107]. When the membranes were exposed to *E. coli* suspensions for 3 h, the results (shown in Figure 6.8b–e) indicated that the membrane embedded with GO-Ag-MOF exhibited the highest antimicrobial strength, with a 96 % extirpation rate being achieved with the Ag-MOF-embedded (80 %) membrane and GO-embedded (66 %) membrane being followed thereafter. The bare TFC membrane (control) did not show any biocidal activity as it only achieved 2 % dead bacteria. The findings clearly demonstrated the enhanced antimicrobial properties of GO-Ag-MOF-embedded membranes compared to other three studied membranes, owing to not only the roles of GO and Ag in killing the bacteria but also the improvements in membrane surface hydrophilicity (reduced from 71 to 54°) and surface negative charge increasing from −52 to −63 mV at pH = 7. The latter two properties provided a hydration layer and repulsion force against the

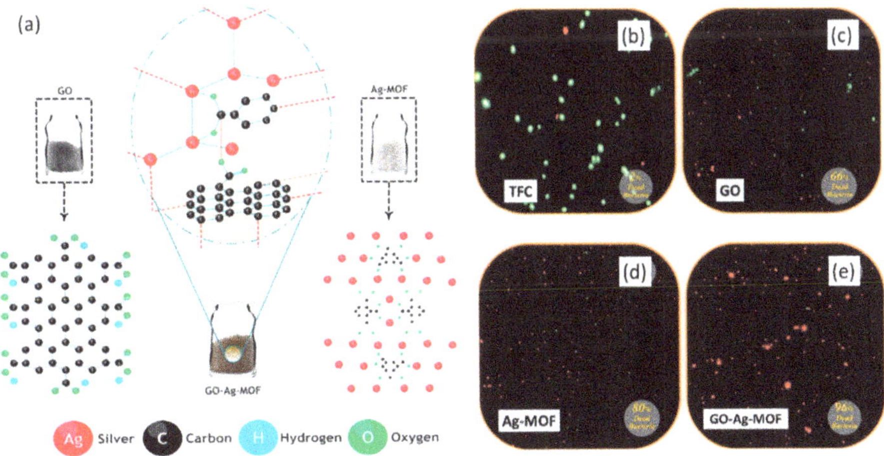

Figure 6.8: (a) Schematic diagram of GO-Ag-MOF interaction and (b–d) Fluorescence images of the membrane with and without nanomaterials incorporation after incubation with *E. coli* (Note: green and red represent live and dead bacteria, respectively) [106].

negatively charged *E. coli* bacteria, thus decreasing its interaction with the membrane surface.

Pejman et al. [103] functionalized the TFC membrane surface using an Ag-based MOF nanomaterial combined with 3-bromopropionic acid (BPA) (carboxyl-containing zwitterions) via surface grafting with the aim to develop an anti-biofouling TFC membrane. The COO-functional group of BPA is important as it can potentially demonstrate a higher antibacterial capability. The antimicrobial properties of membranes were determined by two methods, plate counting method and confocal microscopy as shown in Figure 6.9a. The membranes were examined after being incubated in *E. coli* suspensions at 37 °C for 24 h. By using the cultured-based pour plating method (Figure 6.9a), the membrane modified with the Ag-based MOF, together with BPA (TFC-BPA-MOF), was likely to yield 100 % die-off of *E.coli* compared to the control (100 % alive) and the TFC-BPA (10 % dead) membranes. The fluorescence images (Figure 6.9a) further confirmed the viable and non-viable bacteria cells on the different membrane samples. The results clearly revealed that the ratio of living-to-dead cells was approximately 100 % for the control membrane, which was in good agreement with the heterotrophic plate count. The TFC-BPA and TFC-BPA-MOF membranes yielded 14 % and 100 % dead bacteria cells, respectively. These findings clearly suggested that the TFC-BPA-MOF membrane showed significantly better anti-microbial behavior compared to the other two membranes. This can be explained by the fact that the presence of 2-methylimidazole in the structure of Ag-MOF tended to form an even stronger hydration layer on the membrane surface [108] and the Ag-MOF itself can create sustainable antibacterial activity [109].

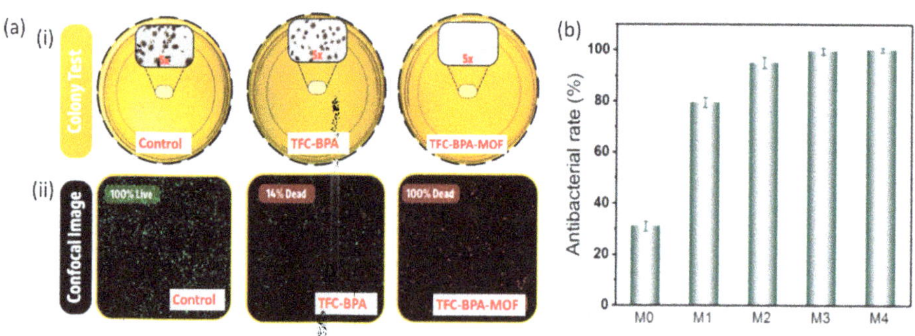

Figure 6.9: (a) Antimicrobial activity tests of control and TFC-BPA and TFC-BPA-MOF membranes after tested with *E. coli* suspensions for 24 h at 37 °C using heterotrophic plate count (i) and confocal microscopy (ii)[103] and (b) antimicrobial activity of MIL-125(Ti)-incorporated membranes via plate counting after subjected to *E. coli* suspensions for 12 h at 37 °C (Note: M0, M1, M2, M3, M4 and M5 represent the membrane incorporated with MIL-125(Ti) loadings of zero, 2.5, 5, 10 and 20 wt%, respectively) [100].

Zhou et al. [100] attempted to incorporate the highly photocatalytic MIL-125(Ti) into a polyvinylidene fluoride (PVDF) membrane to form a MMM with the goal to improve antibacterial activity. Photons of natural light can stimulate the MIL-125(Ti) to initiate photocatalysis. Thus, abundant substances with strong oxidizing properties such as superoxide ions (O_2^-), hydroxyl radicals (•OH) and hydrogen peroxide (H_2O_2) can be generated after a photocatalytic reaction. As presented in Figure 6.9b, the antimicrobial properties of MIL-125(Ti)-modified membranes were enhanced from 79.5 to 100 % by increasing MIL-125(Ti) from 2.5 (M1) to 20 wt% (M4) after being tested with *E. coli* for 12 h. Meanwhile, the control membrane (M0) yielded only a 31.1 % antibacterial rate under the same conditions. The improved surface properties can be explained by the strong oxidizing compounds endowed by MIL-125(Ti) that react with the organic matter of *E. coli* to form water and carbon dioxide.

6.3.2.4 Adsorption behavior

MOFs are also very useful for heavy metal adsorption as they have a very large surface area (up to 10,000 $m^2 g^{-1}$), highly-controlled porous structures, flexibility in tuning with other metallic or nonmetallic elements and open crystal structure. Lately, some MOFs have emerged as potential materials to remove different types of heavy metal ions via adsorption mechanisms. These include Zr-based MOF for As(V) [110] and Hg(II) removal [111], Zn-MOF-74 for As(III) and As(V) removal [112] and MIL-125 for Pb(II) and Hg(II) separation[113]. Furthermore, several reviews have shown the potential use of different MOFs for a wide range of heavy metal ions separation [114, 115] However, compared to the number of publications related to MOFs for heavy metal removal, the number of papers reporting MOF-modified membranes for adsorptive removal of heavy metal ions is significantly less. In 2018, Efome et al. [116] developed an adsorptive nanofiber membrane by co-electrospinning of hydrophilic polyacrylonitrile (PAN) and Zr-based MOF-808 solution and used for Zn(II) and Cd(II) ion removal. The authors reported a MOF loading of up to 20 wt% in PAN nanofiber membranes and the resultant membranes were selective for heavy metal ions removal. The maximum adsorption capacities that the developed adsorptive nanofiber membrane could achieve were 287.06 and 225.05 $mg\,g^{-1}$ for Zn(II) and Cd(II), respectively. With respect to water flux, the developed membrane could treat the Zn- or Cd-contaminated water solution at a rate of approximately 345 $L\,m^{-2}\,h^{-1}$ at an operating pressure of only 0.4 bar. This adsorptive membrane offers an alternative solution for rapid and efficient wastewater treatment, but more research is needed to optimize its properties in terms of nanomaterial concentration, permeance and adsorptive capacity.

Using MOF-loaded PAN nanofiber membrane, Chen et al. [117] reported that the strong adsorption sites of MOF-808 allowed the PAN membrane to adsorb effectively various heavy metal ions in a short time. Since industrial wastewater usually contains

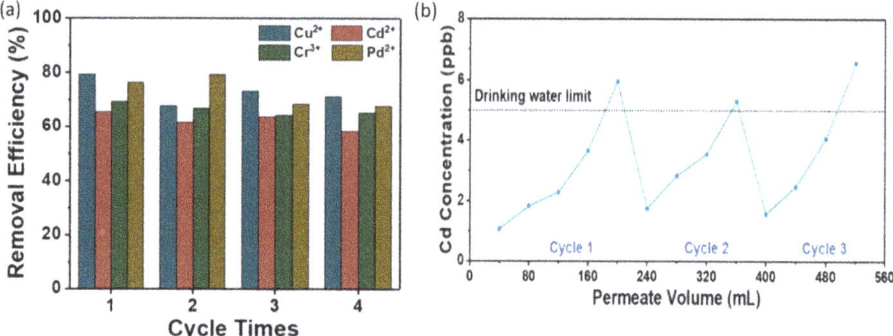

Figure 6.10: (a) The removal efficiency of the membrane containing 60 wt% MOF (labeled as PME-60) against various heavy metal ions via adsorption (Note: The membrane was washed with EDTA-2NA solution after each cycle, and a total of 4 cycles was performed on the same membrane) and (b) cycle test of membrane's adsorption efficiency of very low concentrations of the Cd(II) ion in emulsion [117].

a variety of heavy metal ions, the authors simulated an oil-in-water solution containing four different ions (e. g., Cu(II), Cd(II), Cr(III) and Pd(II) at 10 ppm each) and used it to evaluate the membrane performance. As shown in Figure 6.10a, the membrane demonstrated the removal rates in the order of Cu(II) > Pd(II) > Cr(III) > Cd(II). The variation in the ion adsorption efficiencies indicated that the developed membrane was more selective toward Cu(II) and Pd(II) compared to the Cr(III) and Cd(II). This may be related to the electronegativity of the ions and their coordination to the carboxyl groups (–COO–) in MOF-808-EDTA. At the end of the experiment (4th cycle), the removal efficiency of the membrane decreased by an average of 10 %, suggesting that the membrane could effectively be reused after proper cleaning. The authors further reported that the membrane can handle emulsions containing 100 ppb Cd(II) ion by producing the permeate with and ion concentration below the prescribed concentration for EU standards for drinking water (Figure 6.10b), provided that proper cleaning was carried out to retrieve the membrane adsorption capacity.

Other MOF-functionalized nanofiber membranes that have been previously documented to exhibit promising adsorption capacity against selected heavy metal ions (under optimum conditions) include UiO-66-NH_2/PAN/chitosan membrane (Cd(II) capacity: 415.6 mg g^{-1}; Pb(II) capacity: 441.2 mg g^{-1})[118] and UiO-66-$(COOH)_2$/PAN membrane (Tb(III) capacity: 214.1 mg g^{-1}; Eu(III) capacity: 191.9 mg g^{-1}) [119]. In addition to heavy metal ion removal, Ali et al. [120] used ZIF-8-modified membranes for adsorptive removal of dyes. The authors introduced different loading of ZIF-8 (5–20 wt%) into polydimethoxysilane (PDMS) solution followed by coating it atop the surface of P84 porous support. When the ZIF-8 loading was increased up to 20 wt%, the dye rejection of the membrane increased from 65 to 87 % when tested at 10 bar. The authors also claimed that the outstanding dye rejection performance of ZIF-8-

modified membrane could be due to the unique pore structure of MOF that only allows water molecules to pass through while retaining dyes. In addition, the authors also explained that dye molecules could be adsorbed onto the ZIF-8 surface, leading to a better dye removal rate.

Wang et al. [121] incorporated NH_2-UiO-66 and MIL-100(Cr) into the ultrahigh-molecular-weight polyethylene (UHMWPE) to yield a MMM via thermally induced phase separation-hot pressing method. The authors believed that the special adsorption sites (basic side chains in NH_2-UiO-66 and open metal sites in MIL-100(Cr)) of both MOFs could play an effective role in removing dye molecules. The results showed that the changes in membrane surface charge upon MOF incorporation affected the adsorption behavior toward dye molecules. At the optimum MOF loading, the resultant positively charged NH_2-UiO-66-incorporated membrane (16 mV at pH = 7) displayed exceptional adsorption behavior toward the negatively charged dye molecules, showing a high removal rate (99 %) and water permeability (115.9 $L\,m^{-2}\,h^{-1}\,bar^{-1}$) after being tested using 100 ppm orange G solution at 0.2 MPa. Meanwhile, the membrane with negatively charged MIL-100(Cr) showed higher separation efficiency for the positively charged dye molecules (100 ppm). This membrane was able to remove crystal violet (99 % removal), rhodamine B (99.2 %) and methylene blue (99.2 %) with fluxes recorded between 108 and 120 $L\,m^{-2}\,h^{-1}\,bar^{-1}$.

6.4 Challenges associated with MOF-based membrane fabrication and performance limitation

6.4.1 Gas applications

Despite the great promise for performance improvement in gas separation by using MOF-based membranes, major challenges must be considered for successful implementation in industry. The synthesis strategy is an important challenge that needs to be addressed to ensure proper MOF particle intergrowth and phase purity. Control over the degree of defects is another aspect that needs to be addressed because defects provide a free way for mixture component penetration through the membrane layer. Defects are generated in the form of pinholes, cracks and grain boundaries between the MOF particles and between the MOF particles and the support. The selection of MOF, matrix or composite as well as phase purity, compatibility between MOF and support, and the implemented synthesis method are responsible for various defects because different phases will have different interactions, stabilities and diffusion properties. The dynamic properties of MOF materials, such as structure flexibility and breathing, as in MIL-53, is another aspect that can influence the performance of MOF-based membranes because it impacts molecular sieving properties and results in lower selectivity.

The judicious selection of the MOF, matrix and fabrication method can help to avoid defect problems. However, nonavoidable defects can be treated with dense, inert polymeric materials, such as polydimethylsiloxane (PDMS). Additionally, mixing different MOFs with polymeric matrices in the form of MMMs is useful for addressing some complications related to the fabrication of MOF-based membranes. Many reported MMMs retain superior thermal stability, flexibility and mechanical resilience compared to pure MOF thin films. But the use of the polymeric matrix in the composite membrane is associated with some drawbacks, such as low permeability, which in many cases is an intrinsic property of the polymer itself, membrane stability, aging, plasticizing and swelling. Aging factors influence membrane performance over time and upon exposure to elevated pressures and temperatures, while plasticizing and swelling are typically observed when separating condensable biogas and industrial gases. Aging, plasticizing and swelling reduce permeability and/or selectivity over time which adversely affect membrane performance.

Using MOFs as fillers in MMMs is associated with some challenges related to the synthesis of the MOFs in the nanoscale with narrow particle size dispersity. But also, the high loading of MOF filler in the polymeric matrix, in many cases, is associated with the creation of defects and influences the mechanical flexibility through the increased brittleness of the membrane. The high MOF loading also increases the chance of forming defects from interfacial due to a lack of compatibility between the MOF and polymer. In summary, for the industrial implementation of polymer-based membranes, the following points need to be addressed: (i) the separation of pure gas mixture differs from the performance obtained on a real gas mixture due to the presence of pollutants such as moisture, H_2S, NO_x, and SO_x, (ii) durability, (iii) mechanical robustness, (iv) scale-up of membrane production, (v) the capital cost of membrane production, (vi) membranes must be manufactured in a capillary shape to maximize exposed area versus unit volume, and finally (vii) the compatibility between the MOF and the surrounding matrix needs improvement to minimize the chance of defects from grain boundaries.

6.4.2 Water applications

One of the main challenges encountered during membrane fabrication for water applications is the agglomeration of MOFs in the membrane matrix (either top surface or cross-section). The agglomeration of MOFs in the PA selective layer of the TFC membrane is more critical as the skin layer is very thin (a few hundred nanometers), and any large particle clusters are not able to be well accommodated within the thin layer. Any minor defects in the top selective layer leads to a reduction in salt rejection. As reported in many studies, large amounts of particles should be prevented during TFC or

TFN membrane fabrication. For a MMM made via the phase inversion technique, large particle clumps are normally present on the membrane surface and cross-sectional view (under scanning electron microscopy examination) even though extra effort is undertaken in ensuring dispersion of the MOFs in the dope solution prior to use. The uneven distribution of MOFs in the membrane matrix adversely affects membrane structure integrity and lowers its mechanical property. Although many researchers in recent years have demonstrated the potential of electrospun nanofiber membrane for heavy metal removal, there remain major challenges related to MOF-nanofiber membrane fabrication. These include the speed of fabrication, which is significantly slower than the conventionally-made membranes, a high chance of needle blockage during electrospinning, and largely limited to flat sheet membrane making.

Furthermore, it must be pointed out that there is yet any commercial nanofiber membrane currently on the market for water applications. Regarding membrane performance, the literature reveals that little attention is paid to the long-term stability of the developed MOF-modified membranes, despite the fact that the membrane performance stability is one of the key aspects to convincing industrial adoption. There is a high chance the MOF-modified membrane might experience performance instability (in terms of water flux and solute rejection) due to the poor compatibility between MOF and membrane matrix, leading to the leaching of MOF from the membrane. The leaching of MOFs from adsorptive membranes might be more significant as such membranes must to go through frequent chemical washing at high pH to retrieve their adsorptive capacity. The possible degradation of polymeric membranes might cause the embedded MOF to leach out, affecting the membrane performance and the permeate quality. Another issue associated with the data reported by researchers is that most of the membrane performances are, in fact, lower compared to the commercial membranes on the market. Although many studies reported the good performance of the MOF-modified membranes for water applications (compared to their respective control membrane), the data are not very convincing to compete with commercial membranes. Therefore, it is strongly recommended that a commercial membrane be tested along with the developed MOF-modified membrane in order to provide a better insight into the properties of the membrane developed.

6.5 Conclusion and future directions

Membrane technology is an attractive alternative to the current unsustainable and conventional gas separation and water purification methods. With the discovery of more than 100,000 structures, MOFs offer a wide range of diverse properties with the potential to improve viability that may exceed the performance reported by previously reported films. The adjustability of MOF apertures through linker exchange and modulation allows for improved selectivity through enhanced diffusion. The difference in kinetic diameter allows selective diffusion of small molecules and hinders diffusion of larger molecules, resulting in the generation of diffusion selectivity for H_2 versus CO_2 in several ZIFs. On the other hand, modifying the thermodynamic affinity between MOFs and guest molecules by various strategies, such as creating coordinatively unsaturated metal sites and adding various functional groups, allows for improving selectivity by modifying the adsorption of guest molecules on the host framework. Simultaneously adjusting solubility and diffusion properties can provide cooperative features for improved separation. A wide range of MOF-based membrane fabrication strategies, the diverse choices of supports and MOF synthesis methods definitely facilitate the process of the exploration of different MOFs in membrane applications. They provide a means for overcoming complications in the industrial implementation of MOF-based membranes and addressing the challenge of the capital cost of the membranes in gas separation and water purification applications.

Bibliography

[1] Yoo Y, Lai Z, Jeong HK. Fabrication of MOF-5 membranes using microwave-induced rapid seeding and solvothermal secondary growth. Microporous Mesoporous Mater. 2009;123:100–6.

[2] Al-Maythalony BA, Shekhah O, Swaidan R, Belmabkhout Y, Pinnau I, Eddaoudi M. Quest for anionic MOF membranes: continuous sod-ZMOF membrane with CO_2 adsorption-driven selectivity. J Am Chem Soc. 2015;137:1754–7.

[3] Yoo Y, Jeong HK. Rapid fabrication of metal organic framework thin films using microwave-induced thermal deposition. Chem Commun. 2008;21:2441–3.

[4] Liu Y, Hori A, Kusaka S, Hosono N, Li M, Guo A, Du D, Li Y, Yang W, Ma Y. Microwave-assisted hydrothermal synthesis of [Al(OH) (1, 4-NDC)] membranes with superior separation performances. Chem Asian J. 2019;14:2072–6.

[5] Qian Q, Asinger PA, Lee MJ, Han G, Mizrahi Rodriguez K, Lin S, Benedetti FM, Wu AX, Chi WS, Smith ZP. MOF-based membranes for gas separations. Chem Rev. 2020;120:8161–266.

[6] Li YS, Liang FY, Bux H, Feldhoff A, Yang WS, Caro J. Molecular sieve membrane: supported metal–organic framework with high hydrogen selectivity. Angew Chem. 2010;122:558–61.

[7] Jin H, Mo K, Wen F, Li Y. Preparation and pervaporation performance of CAU-10-H MOF membranes. J Membr Sci. 2019;577:129–36.

[8] Hillman F, Brito J, Jeong HK. Rapid one-pot microwave synthesis of mixed-linker hybrid zeolitic-imidazolate framework membranes for tunable gas separations. ACS Appl Mater Interfaces. 2018;10:5586–93.

[9] Huang A, Wan L, Caro J. Microwave-assisted synthesis of well-shaped UiO-66-NH_2 with high CO_2 adsorption capacity. Mater Res Bull. 2018;98:308–13.

[10] Sun Y, Song C, Guo X, Hong S, Choi J, Liu Y. Microstructural optimization of NH_2-MIL-125 membranes with superior H_2/CO_2 separation performance by innovating metal sources and heating modes. J Membr Sci. 2020;616:118615.

[11] Liu B, Fischer RA. Liquid-phase epitaxy of metal organic framework thin films. Sci China Chem. 2011;54:1851–66.

[12] Kang Z, Ding J, Fan L, Xue M, Zhang D, Gao L, Qiu S. Preparation of a MOF membrane with 3-aminopropyltriethoxysilane as covalent linker for xylene isomers separation. Inorg Chem Commun. 2013;30:74–8.

[13] Liu X, Demir NK, Wu Z, Li K. Highly water-stable zirconium metal–organic framework UiO-66 membranes supported on alumina hollow fibers for desalination. J Am Chem Soc. 2015;137:6999–7002.
[14] Nan J, Dong X, Wang W, Jin W, Xu N. Step-by-step seeding procedure for preparing HKUST-1 membrane on porous α-alumina support. Langmuir. 2011;27:4309–12.
[15] Knebel A, Wulfert-Holzmann P, Friebe S, Pavel J, Strauß I, Mundstock A, Steinbach F, Caro J. Hierarchical nanostructures of metal-organic frameworks applied in gas separating ZIF-8-on-ZIF-67 membranes. Eur J Chem. 2018;24:5728–33.
[16] Shekhah O, Eddaoudi M. The liquid phase epitaxy method for the construction of oriented ZIF-8 thin films with controlled growth on functionalized surfaces. Chem Commun. 2013;49:10079–81.
[17] Lee DJ, Li Q, Kim H, Lee K. Preparation of Ni-MOF-74 membrane for CO2 separation by layer-by-layer seeding technique. Microporous Mesoporous Mater. 2012;163:169–77.
[18] Chernikova V, Shekhah O, Belmabkhout Y, Eddaoudi M. Nanoporous fluorinated metal–organic framework-based membranes for CO_2 capture. ACS Appl Nano Mater. 2020;3:6432–9.
[19] Usman M, Ali M, Al-Maythalony BA, Ghanem AS, Saadi OW, Ali M, Jafar Mazumder MA, Abdel-Azeim S, Habib MA, Yamani ZH, Ensinger W. Highly efficient permeation and separation of gases with metal–organic frameworks confined in polymeric nanochannels. ACS Appl Mater Interfaces. 2020;12:49992–50001.
[20] Lu G, Cui C, Zhang W, Liu Y, Huo F. Synthesis and self-assembly of monodispersed metal-organic framework microcrystals. Chem Asian J. 2013;8:69–72.
[21] Conato MT, Jacobson AJ. Control of nucleation and crystal growth kinetics of MOF-5 on functionalized gold surfaces. Microporous Mesoporous Mater. 2013;175:107–15.
[22] Ji H, Hwang S, Kim K, Kim C, Jeong NC. Direct in situ conversion of metals into metal–organic frameworks: a strategy for the rapid growth of MOF films on metal substrates. ACS Appl Mater Interfaces. 2016;8:32414–20.
[23] Semrau AL, Pujar SP, Stanley PM, Wannapaiboon S, Albada B, Zuilhof H, Fischer RA. Selective positioning of nanosized metal–organic framework particles at patterned substrate surfaces. Chem Mater. 2020;32:9954–63.
[24] Zhao Z, Li Z, Lin J. Secondary growth synthesis of MOF-5 membranes by dip-coating nano-sized MOF-5 seeds. Huagong Xuebao/CIESC J. 2011;62:507–14.
[25] Zhao Z, Ma X, Li Z, Lin Y. Synthesis, characterization and gas transport properties of MOF-5 membranes. J Membr Sci. 2011;382:82–90.
[26] Gascon J, Aguado S, Kapteijn F. Manufacture of dense coatings of $Cu_3(BTC)_2$ (HKUST-1) on α-alumina. Microporous Mesoporous Mater. 2008;113:132–8.
[27] Gu Y, Wu Yn, Li L, Chen W, Li F, Kitagawa S. Controllable modular growth of hierarchical MOF-on-MOF architectures. Angew Chem. 2017;129:15864–8.
[28] Schäfer P, van der Veen MA, Domke KF. Unraveling a two-step oxidation mechanism in electrochemical Cu-MOF synthesis. Chem Commun. 2016;52:4722–5.
[29] Van Assche TR, Desmet G, Ameloot R, De Vos DE, Terryn H, Denayer JF. Electrochemical synthesis of thin HKUST-1 layers on copper mesh. Microporous Mesoporous Mater. 2012;158:209–13.
[30] Zhou S, Shekhah O, Jia J, Czaban-Jóźwiak J, Bhatt PM, Ramírez A, Gascon J, Eddaoudi M. Electrochemical synthesis of continuous metal–organic framework membranes for separation of hydrocarbons. Nat Energy. 2021;6:882–91.
[31] Tse JY, Kadota K, Nakajima T, Uchiyama H, Tanaka S, Tozuka Y. Crystalline rearranged CD-MOF particles obtained via spray-drying synthesis applied to inhalable formulations with high drug loading. Cryst Growth Des. 2021.

[32] Garzon-Tovar L, Cano-Sarabia M, Carné-Sánchez A, Carbonell C, Imaz I, Maspoch D. A spray-drying continuous-flow method for simultaneous synthesis and shaping of microspherical high nuclearity MOF beads. React Chem Eng. 2016;1:533–9.
[33] Cohen Y, Glater J. A tribute to Sidney Loeb—The pioneer of reverse osmosis desalination research. Desalin Water Treat. 2010;15:222–7.
[34] Khoo YS, Lau WJ, Liang YY, Yusof N, Ismail AF. Surface modification of PA layer of TFC membranes: does it effective for performance improvement? J Ind Eng Chem. 2021;102:271–92.
[35] Ismail AF, Hilal N, Jaafar J, Wright C. Nanofiber Membranes for Medical, Environmental, and Energy Applications. CRC press; 2019.
[36] Lau W, Gray S, Matsuura T, Emadzadeh D, Chen JP, Ismail A. A review on polyamide thin film nanocomposite (TFN) membranes: history, applications, challenges and approaches. Water Res. 2015;80:306–24.
[37] Khulbe K, Matsuura T, Feng C, Ismail A. Recent development on the effect of water/moisture on the performance of zeolite membrane and MMMs containing zeolite for gas separation; review. RSC Adv. 2016;6:42943–61.
[38] Fernández-Barquín A, Rea R, Venturi D, Giacinti-Baschetti M, De Angelis MG, Casado-Coterillo C, Irabien Á. Effect of relative humidity on the gas transport properties of zeolite A/PTMSP mixed matrix membranes. RSC Adv. 2018;8:3536–46.
[39] Li Y, Li L, Yu J. Applications of zeolites in sustainable chemistry. Chem. 2017;3:928–49.
[40] Robeson LM. The upper bound revisited. J Membr Sci. 2008;320:390–400.
[41] Liu Y, Liu G, Zhang C, Qiu W, Yi S, Chernikova V, Chen Z, Belmabkhout Y, Shekhah O, Eddaoudi M. Enhanced CO_2/CH_4 separation performance of a mixed matrix membrane based on tailored MOF-polymer formulations. Adv Sci. 2018;5:1800982.
[42] Tien-Binh N, Rodrigue D, Kaliaguine S. In-situ cross interface linking of PIM-1 polymer and UiO-66-NH_2 for outstanding gas separation and physical aging control. J Membr Sci. 2018;548:429–38.
[43] Tien-Binh N, Vinh-Thang H, Chen XY, Rodrigue D, Kaliaguine S. Crosslinked MOF-polymer to enhance gas separation of mixed matrix membranes. J Membr Sci. 2016;520:941–50.
[44] Cheng Y, Tavares SR, Doherty CM, Ying Y, Sarnello E, Maurin G, Hill MR, Li T, Zhao D. Enhanced polymer crystallinity in mixed-matrix membranes induced by metal–organic framework nanosheets for efficient CO_2 capture. ACS Appl Mater Interfaces. 2018;10:43095–103.
[45] Hao L, Liao KS, Chung TS. Photo-oxidative PIM-1 based mixed matrix membranes with superior gas separation performance. J Mater Chem A. 2015;3:17273–81.
[46] Gao Y, Qiao Z, Zhao S, Wang Z, Wang J. In situ synthesis of polymer grafted ZIFs and application in mixed matrix membrane for CO_2 separation. J Mater Chem A. 2018;6:3151–61.
[47] Ahmad MZ, Navarro M, Lhotka M, Zornoza B, Téllez C, de Vos WM, Benes NE, Konnertz NM, Visser T, Semino R. Enhanced gas separation performance of 6FDA-DAM based mixed matrix membranes by incorporating MOF UiO-66 and its derivatives. J Membr Sci. 2018;558:64–77.
[48] Xiang L, Sheng L, Wang C, Zhang L, Pan Y, Li Y. Amino-functionalized ZIF-7 nanocrystals: improved intrinsic separation ability and interfacial compatibility in mixed-matrix membranes for CO2/CH4 separation. Adv Mater. 2017;29:1606999.
[49] Chen K, Xu K, Xiang L, Dong X, Han Y, Wang C, Sun LB, Pan Y. Enhanced CO_2/CH_4 separation performance of mixed-matrix membranes through dispersion of sorption-selective MOF nanocrystals. J Membr Sci. 2018;563:360–70.
[50] Bae TH, Lee JS, Qiu W, Koros WJ, Jones CW, Nair S. A high-performance gas-separation membrane-containing submicrometer-sized metal–organic framework crystals. Angew Chem. 2010;122:10059–62.
[51] Wu X, Liu W, Wu H, Zong X, Yang L, Wu Y, Ren Y, Shi C, Wang S, Jiang Z. Nanoporous ZIF-67 embedded polymers of intrinsic microporosity membranes with enhanced gas separation performance. J Membr Sci. 2018;548:309–18.

[52] Lin R, Ge L, Diao H, Rudolph V, Zhu Z. Ionic liquids as the MOFs/polymer interfacial binder for efficient membrane separation. ACS Appl Mater Interfaces. 2016;8:32041–9.
[53] Ghalei B, Sakurai K, Kinoshita Y, Wakimoto K, Isfahani AP, Song Q, Doitomi K, Furukawa S, Hirao H, Kusuda H. Enhanced selectivity in mixed matrix membranes for CO_2 capture through efficient dispersion of amine-functionalized MOF nanoparticles. Nat Energy. 2017;2:1–9.
[54] Smith SJ, Ladewig BP, Hill AJ, Lau CH, Hill MR. Post-synthetic Ti exchanged UiO-66 metal-organic frameworks that deliver exceptional gas permeability in mixed matrix membranes. Sci Rep. 2015;5:1–6.
[55] Boroğlu MŞ. Structural characterization and gas permeation properties of polyetherimide (PEI)/zeolitic imidazolate (ZIF-11) mixed matrix membranes. J Turk Chem Soc A Chem. 2016;3:183–205.
[56] Japip S, Wang H, Xiao Y, Chung TS. Highly permeable zeolitic imidazolate framework (ZIF)-71 nano-particles enhanced polyimide membranes for gas separation. J Membr Sci. 2014;467:162–74.
[57] Zhang Q, Luo S, Weidman JR, Guo R. Preparation and gas separation performance of mixed-matrix membranes based on triptycene-containing polyimide and zeolite imidazole framework (ZIF-90). Polymer. 2017;131:209–16.
[58] Rui Z, James JB, Kasik A, Lin Y. Metal-organic framework membrane process for high purity CO_2 production. AIChE J. 2016;62:3836–41.
[59] Yeo ZY, Chai SP, Zhu PW, Mohamed AR. Development of a hybrid membrane through coupling of high selectivity zeolite T on ZIF-8 intermediate layer and its performance in carbon dioxide and methane gas separation. Microporous Mesoporous Mater. 2014;196:79–88.
[60] Usman M, Ali M, Al-Maythalony BA, Ghanem AS, Saadi OW, Ali M, Jafar Mazumder MA, Abdel-Azeim S, Habib MA, Yamani ZH. Highly efficient permeation and separation of gases with metal–organic frameworks confined in polymeric nanochannels. ACS Appl Mater Interfaces. 2020;12:49992–50001.
[61] Wang X, Chi C, Zhang K, Qian Y, Gupta KM, Kang Z, Jiang J, Zhao D. Reversed thermo-switchable molecular sieving membranes composed of two-dimensional metal-organic nanosheets for gas separation. Nat Commun. 2017;8:1–10.
[62] Wang X, Chi C, Tao J, Peng Y, Ying S, Qian Y, Dong J, Hu Z, Gu Y, Zhao D. Improving the hydrogen selectivity of graphene oxide membranes by reducing non-selective pores with intergrown ZIF-8 crystals. Chem Commun. 2016;52:8087–90.
[63] Peng Y, Li Y, Ban Y, Yang W. Two-dimensional metal–organic framework nanosheets for membrane-based gas separation. Angew Chem. 2017;129:9889–93.
[64] Kanehashi S, Kishida M, Kidesaki T, Shindo R, Sato S, Miyakoshi T, Nagai K. CO_2 separation properties of a glassy aromatic polyimide composite membranes containing high-content 1-butyl-3-methylimidazolium bis (trifluoromethylsulfonyl) imide ionic liquid. J Membr Sci. 2013;430:211–22.
[65] Zhang F, Zou X, Gao X, Fan S, Sun F, Ren H, Zhu G. Hydrogen selective NH_2-MIL-53 (Al) MOF membranes with high permeability. Adv Funct Mater. 2012;22:3583–90.
[66] Huang Y, Liu D, Liu Z, Zhong C. Synthesis of zeolitic imidazolate framework membrane using temperature-switching synthesis strategy for gas separation. Ind Eng Chem Res. 2016;55:7164–70.
[67] Li W, Yang Z, Zhang G, Fan Z, Meng Q, Shen C, Gao C. Stiff metal–organic framework–polyacrylonitrile hollow fiber composite membranes with high gas permeability. J Mater Chem A. 2014;2:2110–8.
[68] Huang A, Chen Y, Liu Q, Wang N, Jiang J, Caro J. Synthesis of highly hydrophobic and permselective metal–organic framework Zn(BDC) $(TED)_{0.5}$ membranes for H_2/CO_2 separation. J Membr Sci. 2014;454:126–32.

[69] Wang N, Mundstock A, Liu Y, Huang A, Caro J. Amine-modified Mg-MOF-74/CPO-27-Mg membrane with enhanced H_2/CO_2 separation. Chem Eng Sci. 2015;124:27–36.
[70] Nian P, Liu H, Zhang X. Bottom-up fabrication of two-dimensional Co-based zeolitic imidazolate framework tubular membranes consisting of nanosheets by vapor phase transformation of Co-based gel for H_2/CO_2 separation. J Membr Sci. 2019;573:200–9.
[71] Zhang C, Xiao Y, Liu D, Yang Q, Zhong C. A hybrid zeolitic imidazolate framework membrane by mixed-linker synthesis for efficient CO_2 capture. Chem Commun. 2013;49:600–2.
[72] Li W, Meng Q, Li X, Zhang C, Fan Z, Zhang G. Non-activation ZnO array as a buffering layer to fabricate strongly adhesive metal–organic framework/PVDF hollow fiber membranes. Chem Commun. 2014;50:9711–3.
[73] Kong C, Du H, Chen L, Chen B. Nanoscale MOF/organosilica membranes on tubular ceramic substrates for highly selective gas separation. Energy Environ Sci. 2017;10:1812–9.
[74] Wang N, Liu Y, Qiao Z, Diestel L, Zhou J, Huang A, Caro J. Polydopamine-based synthesis of a zeolite imidazolate framework ZIF-100 membrane with high H_2/CO_2 selectivity. J Mater Chem A. 2015;3:4722–8.
[75] Kang Z, Xue M, Fan L, Huang L, Guo L, Wei G, Chen B, Qiu S. Highly selective sieving of small gas molecules by using an ultra-microporous metal–organic framework membrane. Energy Environ Sci. 2014;7:4053–60.
[76] Huang A, Wang N, Kong C, Caro J. Organosilica-functionalized zeolitic imidazolate framework ZIF-90 membrane with high gas-separation performance. Angew Chem. 2012;124:10703–7.
[77] Zhong Z, Yao J, Chen R, Low Z, He M, Liu JZ, Wang H. Oriented two-dimensional zeolitic imidazolate framework-L. membranes and their gas permeation properties. J Mater Chem A. 2015;3:15715–22.
[78] Knebel A, Sundermann L, Mohmeyer A, Strauß I, Friebe S, Behrens P, Caro JR. Azobenzene guest molecules as light-switchable CO_2 valves in an ultrathin UiO-67 membrane. Chem Mater. 2017;29:3111–7.
[79] Li JR, Kuppler RJ, Zhou HC. Selective gas adsorption and separation in metal–organic frameworks. Chem Soc Rev. 2009;38:1477–504.
[80] Knebel A, Friebe S, Bigall NC, Benzaqui M, Serre C, Caro JR. Comparative study of MIL-96 (Al) as continuous metal–organic frameworks layer and mixed-matrix membrane. ACS Appl Mater Interfaces. 2016;8:7536–44.
[81] Zhou S, Zou X, Sun F, Ren H, Liu J, Zhang F, Zhao N, Zhu G. Development of hydrogen-selective CAU-1 MOF membranes for hydrogen purification by 'dual-metal-source' approach. Int J Hydrog Energy. 2013;38:5338–47.
[82] Sun Y, Liu Y, Caro J, Guo X, Song C, Liu Y. In-plane epitaxial growth of highly C-oriented NH2-MIL-125 (Ti) membranes with superior H_2/CO_2 selectivity. Angew Chem. 2018;130:16320–5.
[83] Kang Z, Fan L, Wang S, Sun D, Xue M, Qiu S. In situ confinement of free linkers within a stable MOF membrane for highly improved gas separation properties. CrystEngComm. 2017;19:1601–6.
[84] Fan S, Sun F, Xie J, Guo J, Zhang L, Wang C, Pan Q, Zhu G. Facile synthesis of a continuous thin $Cu(bipy)_2(SiF_6)$ membrane with selectivity towards hydrogen. J Mater Chem A. 2013;1:11438–42.
[85] Huang A, Bux H, Steinbach F, Caro J. Molecular-sieve membrane with hydrogen permselectivity: ZIF-22 in LTA topology prepared with 3-aminopropyltriethoxysilane as covalent linker. Angew Chem. 2010;122:5078–81.
[86] Dong X, Huang K, Liu S, Ren R, Jin W, Lin Y. Synthesis of zeolitic imidazolate framework-78 molecular-sieve membrane: defect formation and elimination. J Mater Chem. 2012;22:19222–7.

[87] Jun BM, Al-Hamadani YA, Son A, Park CM, Jang M, Jang A, Kim NC, Yoon Y. Applications of metal-organic framework based membranes in water purification: a review. Sep Purif Technol. 2020;247:116947.
[88] Xiao S, Huo X, Tong Y, Cheng C, Yu S, Tan X. Improvement of thin-film nanocomposite (TFN) membrane performance by CAU-1 with low charge and small size. Sep Purif Technol. 2021:118467.
[89] Jin L, Shi W, Yu S, Yi X, Sun N, Ma C, Liu Y. Preparation and characterization of a novel PA-SiO2 nanofiltration membrane for raw water treatment. Desalination. 2012;298:34–41.
[90] Liu TY, Yuan HG, Liu YY, Ren D, Su YC, Wang X. Metal–organic framework nanocomposite thin films with interfacial bindings and self-standing robustness for high water flux and enhanced ion selectivity. ACS Nano. 2018;12:9253–65.
[91] Wang Z, Qi J, Lu X, Jiang H, Wang P, He M, Ma J. Epitaxially grown MOF membranes with photocatalytic bactericidal activity for biofouling mitigation in desalination. J Membr Sci. 2021;630:119327.
[92] Liang W, Xu H, Carraro F, Maddigan NK, Li Q, Bell SG, Huang DM, Tarzia A, Solomon MB, Amenitsch H. Enhanced activity of enzymes encapsulated in hydrophilic metal–organic frameworks. J Am Chem Soc. 2019;141:2348–55.
[93] Shu L, Xie LH, Meng Y, Liu T, Zhao C, Li JR. A thin and high loading two-dimensional MOF nanosheet based mixed-matrix membrane for high permeance nanofiltration. J Membr Sci. 2020;603:118049.
[94] Xu Y, Gao X, Wang X, Wang Q, Ji Z, Wang X, Wu T, Gao C. Highly and stably water permeable thin film nanocomposite membranes doped with MIL-101(Cr) nanoparticles for reverse osmosis application. Materials. 2016;9:870.
[95] Wang F, Zheng T, Xiong R, Wang P, Ma J. Strong improvement of reverse osmosis polyamide membrane performance by addition of ZIF-8 nanoparticles: effect of particle size and dispersion in selective layer. Chemosphere. 2019;233:524–31.
[96] Kadhom M, Hu W, Deng B. Thin film nanocomposite membrane filled with metal-organic frameworks UiO-66 and MIL-125 nanoparticles for water desalination. Membranes. 2017;7:31.
[97] Ma D, Peh SB, Han G, Chen SB. Thin-film nanocomposite (TFN) membranes incorporated with super-hydrophilic metal–organic framework (MOF) UiO-66: toward enhancement of water flux and salt rejection. ACS Appl Mater Interfaces. 2017;9:7523–34.
[98] Zhu J, Qin L, Uliana A, Hou J, Wang J, Zhang Y, Li X, Yuan S, Li J, Tian M. Elevated performance of thin film nanocomposite membranes enabled by modified hydrophilic MOFs for nanofiltration. ACS Appl Mater Interfaces. 2017;9:1975–86.
[99] Mohammadnezhad F, Feyzi M, Zinadini S. A novel Ce-MOF/PES mixed matrix membrane; synthesis, characterization and antifouling evaluation. J Ind Eng Chem. 2019;71:99–111.
[100] Zhou S, Gao J, Zhu J, Peng D, Zhang Y, Zhang Y. Self-cleaning, antibacterial mixed matrix membranes enabled by photocatalyst Ti-MOFs for efficient dye removal. J Membr Sci. 2020;610:118219.
[101] Lee JY, She Q, Huo F, Tang CY. Metal–organic framework-based porous matrix membranes for improving mass transfer in forward osmosis membranes. J Membr Sci. 2015;492:392–9.
[102] Fu W, Chen J, Li C, Jiang L, Qiu M, Li X, Wang Y, Cui L. Enhanced flux and fouling resistance forward osmosis membrane based on a hydrogel/MOF hybrid selective layer. J Colloid Interface Sci. 2021;585:158–66.
[103] Pejman M, Firouzjaei MD, Aktij SA, Das P, Zolghadr E, Jafarian H, Shamsabadi AA, Elliott M, Esfahani MR, Sangermano M. Improved antifouling and antibacterial properties of forward osmosis membranes through surface modification with zwitterions and silver-based metal organic frameworks. J Membr Sci. 2020;611:118352.
[104] Aljundi IH. Desalination characteristics of TFN-RO membrane incorporated with ZIF-8 nanoparticles. Desalination. 2017;420:12–20.

[105] Firouzjaei MD, Shamsabadi AA, Sharifian GhM, Rahimpour A, Soroush M. A novel nanocomposite with superior antibacterial activity: a silver-based metal organic framework embellished with graphene oxide. Adv Mater Interfaces. 2018;5:1701365.

[106] Firouzjaei MD, Shamsabadi AA, Aktij SA, Seyedpour SF, Sharifian GhM, Rahimpour A, Esfahani MR, Ulbricht M, Soroush M. Exploiting synergetic effects of graphene oxide and a silver-based metal–organic framework to enhance antifouling and anti-biofouling properties of thin-film nanocomposite membranes. ACS Appl Mater Interfaces. 2018;10:42967–78.

[107] Seabra AB, Paula AJ, de Lima R, Alves OL, Durán N. Nanotoxicity of graphene and graphene oxide. Chem Res Toxicol. 2014;27:159–68.

[108] Shen L, Zhang X, Zuo J, Wang Y. Performance enhancement of TFC FO membranes with polyethyleneimine modification and post-treatment. J Membr Sci. 2017;534:46–58.

[109] Seyedpour SF, Rahimpour A, Najafpour G. Facile in-situ assembly of silver-based MOFs to surface functionalization of TFC membrane: A. novel approach toward long-lasting biofouling mitigation. J Membr Sci. 2019;573:257–69.

[110] Assaad N, Sabeh G, Hmadeh M. Defect control in Zr-based metal–organic framework nanoparticles for arsenic removal from water. ACS Appl Nano Mater. 2020;3:8997–9008.

[111] Ding L, Luo X, Shao P, Yang J, Sun D. Thiol-functionalized Zr-based metal–organic framework for capture of Hg(II) through a proton exchange reaction. ACS Sustain Chem Eng. 2018;6:8494–502.

[112] Yu W, Luo M, Yang Y, Wu H, Huang W, Zeng K, Luo F. Metal-organic framework (MOF) showing both ultrahigh As(V.) and As(III) removal from aqueous solution. J Solid State Chem. 2019;269:264–70.

[113] Awad FS, AbouZeid KM, El-Maaty WMA, El-Wakil AM, El-Shall MS. Efficient removal of heavy metals from polluted water with high selectivity for mercury(II) by 2-imino-4-thiobiuret–partially reduced graphene oxide (IT-PRGO). ACS Appl Mater Interfaces. 2017;9:34230–42.

[114] Ru J, Wang X, Wang F, Cui X, Du X, Lu X. UiO series of metal-organic frameworks composites as advanced sorbents for the removal of heavy metal ions: synthesis, applications and adsorption mechanism. Ecotoxicol Environ Saf. 2021;208:111577.

[115] Haldar D, Duarah P, Purkait MK. MOFs for the treatment of arsenic, fluoride and iron contaminated drinking water: a review. Chemosphere. 2020;251:126388.

[116] Efome JE, Rana D, Matsuura T, Lan CQ. Insight studies on metal-organic framework nanofibrous membrane adsorption and activation for heavy metal ions removal from aqueous solution. ACS Appl Mater Interfaces. 2018;10:18619–29.

[117] Chen X, Chen D, Li N, Xu Q, Li H, He J, Lu J. Modified-MOF-808-loaded polyacrylonitrile membrane for highly efficient, simultaneous emulsion separation and heavy metal ion removal. ACS Appl Mater Interfaces. 2020;12:39227–35.

[118] Jamshidifard S, Koushkbaghi S, Hosseini S, Rezaei S, Karamipour A, Irani M. Incorporation of UiO-66-NH_2 MOF into the PAN/chitosan nanofibers for adsorption and membrane filtration of Pb(II), Cd(II) and Cr(VI) ions from aqueous solutions. J Hazard Mater. 2019;368:10–20.

[119] Hua W, Zhang T, Wang M, Zhu Y, Wang X. Hierarchically structural PAN/UiO-66-(COOH) 2 nanofibrous membranes for effective recovery of terbium(III) and europium(III) ions and their photoluminescence performances. Chem Eng J. 2019;370:729–41.

[120] Ali M, Aslam M, Khan A, Gilani MA, Khan AL. Mixed matrix membranes incorporated with sonication-assisted ZIF-8 nanofillers for hazardous wastewater treatment. Environ Sci Pollut Res. 2019;26:35913–23.

[121] Wang H, Zhao S, Liu Y, Yao R, Wang X, Cao Y, Ma D, Zou M, Cao A, Feng X. Membrane adsorbers with ultrahigh metal-organic framework loading for high flux separations. Nat Commun. 2019;10:1–9.

James Kegere and Ha L. Nguyen

7 Metal-organic frameworks for CO_2 capture and conversion

7.1 Introduction

Over the past decades, there has been an exponential increase in the use of fossil fuels due to energy demands needed for the industrialization around the world. Our excessive use of fossil fuels has led to a drastic increase in CO_2 emission level, accounting for 60 % of the total CO_2 emissions from all sources. The excess CO_2 released into the atmosphere is currently close to 36 GT per year, which causes global problems including climate change and CO_2 poisoning in marine and terrestrial environments. Furthermore, CO_2 emissions will continue to rise to 500 ppm by 2050 and with this will come many more negative effects [1, 2]. Researchers are currently under pressure to develop technologies that would minimize the impact of needless CO_2 emissions as well as to reduce the actual emissions at point sources [3–9]. If those technologies remain undeveloped, continued CO_2 emission will result in rising sea levels, increased water acidity and drastic weather changes (e. g., change in rainfall patterns). In the fullness of time, these destructive effects will cause the earth to become uninhabitable and many of the endangered, yet vital species will become extinct from their ecosystems.

Recent CO_2 mitigation measures have centered around the use of chemical adsorbents, in which amines (e. g., ethanolamine) are conventionally applied in the capture and storage of CO_2. However, due to the prohibitive cost and risks associated with the use of these volatile compounds, intensive global research is being pursued into finding alternative materials (porous solid adsorbents) having superior uptake performance and low cost. Such materials have presented a superior adsorption capacity (>4.4 wt.%), longer durability and reusability [4–6, 9]. Concurrent efforts to improve the porosity and active surfaces of such adsorbents have gained attention since these properties help improve CO_2 capture and storage in adsorbents. Moreover, these active surfaces and pores provide more accessible surface area for further CO_2 conversion that yield useful products from an otherwise harmful gas [4–6]. Some of the widely used porous solid adsorbents include metal oxides owing to their basic and ionic nature. Oxides of calcium and magnesium are especially popular in this regard due to the ease of forming metal carbonates in the presence of CO_2, which can enhance the

James Kegere, Department of Chemistry, United Arab Emirates University, Al-Ain 15551, United Arab Emirates
Ha L. Nguyen, Department of Chemistry and Berkeley Global Science Institute, University of California Berkeley Berkeley, CA 94720, United States; and Joint UAEU–UC Berkeley Laboratories for Materials Innovations, United Arab Emirates University, Al-Ain 15551, United Arab Emirates, e-mail: nguyen.lh@berkeley.edu

https://doi.org/10.1515/9781501524721-007

CO_2 uptake. The adsorption process is governed by a complex interplay between thermodynamics and kinetics. Considering these two factors is key for most CO_2 capture applications, but one factor (kinetics) is less critical than the other as illustrated in the use of calcium oxides for CO_2 capture.

Other commonly used and explored adsorbent materials include ionic liquids in matrices, silicates, polymers, activated carbon and zeolites given their porosity and enhanced surface area for CO_2 capture and storage in addition to their reusability and durability. Zeolites are, however, could be less desirable due to their hydrophilic nature associated with extensive drying step, thus cost penalty when dealing with moisture-containing flue gas. Other less hydrophilic porous materials are sufficient for use; however, they are faced with the challenge of poor selectivity toward CO_2.

Metal-organic frameworks (MOFs) have gained an unmatched reputation as game changers in the field of gas adsorption, storage, separation and conversion. Interesting structural properties of MOFs are their tunable pore sizes, large surface areas, hydrophobicity controlled by building block selection, stable in various conditions, low regeneration costs and high thermal stability. In general, CO_2 uptake requires high surface area and narrow pore size for effective retention of adsorbates—a property that is easily achieved using MOFs due, in larger part, to their designable structures [1, 6, 10, 11]. Captured CO_2 is especially useful in multiple fields ranging from engineering to food and beverages. It can also be converted into fuels and chemical feedstocks for use in various applications, all of which make CO_2 uptake and conversion a priority in materials research in the coming decade. Herein, we dedicate this chapter to providing a deep understanding of how to use MOFs for CO_2 capture by discussing lessons from designed MOFs that have been applied to this end. Along with this, we pay close attention to the conversion of CO_2 through photo and electrochemical reduction pathways. Finally, we highlight strategies to enhance MOFs' activity in the photo and electrochemical reduction of CO_2.

7.2 Conventional materials used in CO_2 capture

The dire need for ameliorating CO_2 emission has prompted increased research into discovering materials with intrinsic abilities to capture, store and transform CO_2. Developing materials bearing such ideal properties requires careful examination of the different parameters that lead to their industrial use, including selectivity of the material towards CO_2, especially as it pertains to separating CO_2 from flue gas, and minimization of the energy penalty of regenerating the material after CO_2 capture and release. Another important parameter is the nature of the interaction/affinity of the material toward CO_2. Strong binding between CO_2 and the surface of the material means a high energy requirement for desorption, while weak binding properties leads to a low energy penalty. We shall now outline the properties of conventional adsorbents used in CO_2 capture [5–10].

7.2.1 Aqueous alkanolamine absorbents

Alkanolamines have a long history of use for capture CO_2, and surprisingly, they remain the state-of-the-art materials due to the unique affinity of amines toward CO_2 [1, 2, 5, 13, 14]. Amines in alkanolamines execute nucleophilic attack on the carbon atom of a given CO_2 molecule to form the C–N bond in carbamate or bicarbonate species depending on the nature of the amine functionality. To improve the affinity of alkanolamines toward CO_2, amines can be substituted, but the strong interaction that exists between CO_2 and the amines will come at the expense of a high regeneration energy to break the resulting C–N bond formation. Among the variety of alkanolamines, monoethanolamine is the most efficient adsorbent material known to date. Others are diethanolamine, triethanolamine, 2-amino-2-methyl-1-propanol and N-methyldiethanolamine (MDEA), all of which have been studied in the CO_2 capture and sequestration technology. MDEA is usually dissolved in water and its reaction with CO_2 results in the formation of an anionic carbamate and an ammonium cation. Although alkanolamines possess superior CO_2 capture properties, they have limitations that easily negate their performance, which include instability of the alkanolamines in aqueous media and corrosive properties that effect the reactor vessels. The former effects the regeneration, and hence, the lifespan of the absorbent [13, 14]. Solid porous materials, amine supported materials were, therefore, developed to address these concerns and as a result, diverse materials including zeolites, activated carbons, and MOFs have been studied for gas capture and separation [17, 18].

7.2.2 Activated carbons

Activated carbons are a popular class of adsorbents that are prepared by pyrolysis of various carbonaceous materials such as biomass, resins, [3, 7, 14–16]. Compared to zeolites, activated carbons have a lower enthalpy of CO_2 adsorption and this is due to the uniform electric potential on their surfaces rendering them less effective in CO_2 capture at low pressure. However, at higher pressure, owing to the high surface area of these materials, the adsorption potential for CO_2 is enhanced. This explains the reason for activated carbons being used for CO_2 sorption in high-pressure settings [14–16].

7.2.3 Zeolites

Zeolites are hydrated aluminosilicate frameworks that are fully ordered or partially disordered and possess interesting chemical and physical properties. Zeolites are popular sorbents that have been applied in various chemical and environmental fields [15–19]. In CO_2 sorption, zeolites have commanded significant attention owing to their unique properties that include extensive and narrow pore dimensions, large surface area, low density, high stability and sometimes structural flexibility. One other property of zeolitic materials is the rich presence of cations in the structure, which increases electrostatic interactions with CO_2. There are diverse types of zeolites classified as types A, X, Y and ZSM-5 [16]. NaX and NaA zeolites are the two commonly explored forms when pursuing CO_2 sorption [15]. It is easy synthesize zeolites as most of the starting materials are obtained from industrial waste, such as coal fly ash. Aquino et al., for instance, used coal fly ash to synthesize NaX and NaA zeolites and subsequently examined these materials for CO_2 sorption, which showed significant uptake capacity [15]. Unlike activated carbons however, zeolites are limited in performance in the presence of steam and require elevated temperature for regeneration, consequently leading to lowered efficiency [16].

7.3 MOFs in CO_2 uptake

The presence of a narrow pore distribution under low temperature facilitates the uptake of CO_2 [1, 5, 7, 8, 14, 20–22]. It is important to understand the interaction mechanism of CO_2 and the sorbent since this assist in the design of better MOF materials for CO_2 capture. Several reports have attributed enhanced CO_2 capture by MOFs that bear open metal sites and/or polar functional groups. Studies to ascertain the contributing factors of CO_2 uptake revealed that attractive forces also come from O-atoms within the pores whose interaction with CO_2 through electrostatic forces creates a strong binding affinity of CO_2 molecules on the pore walls [20–22]. Second, CO_2–CO_2 interaction is promoted by small micropores especially those materials having pores less than 0.15 nm. MOFs composed of linkers with –COOH, O = C, and/or –OH can be advantageous for CO_2 sorption. For example, studies conducted by Cai *et al.* on the adsorption of CO_2 on M-OH functionalized MOFs revealed the enhanced adsorption of CO_2 with the narrowing of pores albeit with reduced adsorption kinetics, and this was attributed to the strong interaction between the MOF framework and CO_2 molecules [20]. Furthermore, it was established bimetallic MOF frameworks further increased the interaction between the MOF, leading to high uptake capacity at room temperature and low pressure as compared to the single metal centered MOF that required high energy for the desorption [20]. Especially, Ni-OH functionalized Zn-MOF exhibited superior CO_2 sorption at room temperature due to enabling thermodynamic driving force that promotes interaction between the MOF and CO_2 (Figure 7.1a) [20].

Density functional theory (DFT) is a powerful calculation method to reveal the CO_2 uptake mechanism in MOFs. Hou et al. [21] studied the influence of Lewis acid

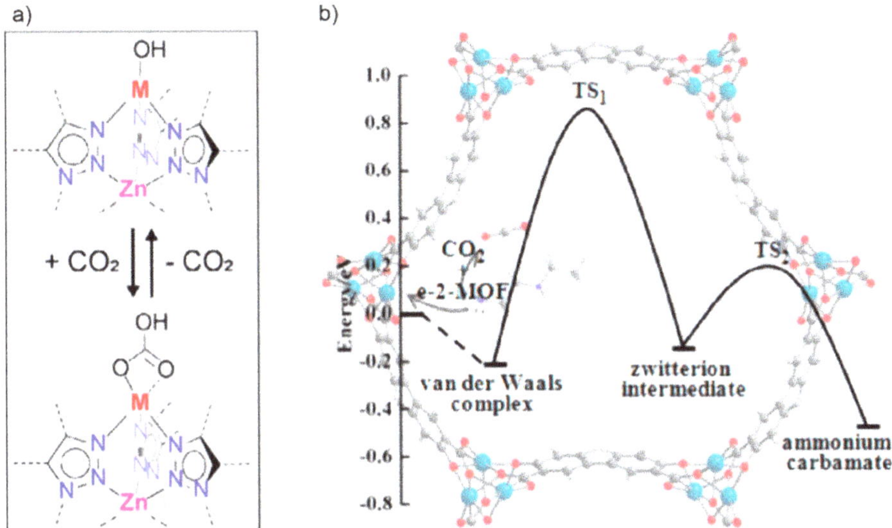

Figure 7.1: Mechanism of CO_2 uptake in MOFs having (a) bimetallic building units and (b) amine-based functionalities. Reproduced with permission from [20] and [22]. Copyright 2020, American Chemical Society.

and base on CO_2 sorption in MOFs DFT and Grand-Canonical Monte Carlo simulation. This work was carried on M-MOF-74 (M = Mg and Zn) [21]. The authors determined that CO_2 adsorption into the MOF structure was a result of strong Lewis acid/base interaction between O atoms in the CO_2 molecules and the metal centers in the MOF. Furthermore, Zhang et al. [22] reported similar findings using corrected DFT-D3. Particularly, the authors studied the interaction of CO_2 in N-ethyl ethylenediamine (e-2) factionalized MOF-74 (M_2(dobpdc), M = Mg, Sc and Zn, dobpdc = 4,4′-dioxidobiphenyl-3,3′-dicarboxylate) to fully understand the detailed mechanism of CO_2 adsorption in MOFs including all involved pathways (intermediates and transition states). In their findings, CO_2 interacted with M_2(dobpdc) via e-2 functional groups through physisorption generating C–N bond that resulted into a zwitterion acting as an intermediate for insertion of CO_2 molecules into the M–N bond (Figure 7.1b) [22].

Apart from these, the electrostatic interactions that increase CO_2 uptake play an important role which was explained by Mulliken et al. Especially, by using charge distribution calculations the authors showed that O atoms in –COOH have a negative charge distribution of ca. −0.345 to −0.370 e, which provided a strong electronegativity while C- and H-atoms provided electropositivity. Combining these electrostatic interactions creates strong attractive forces that bind CO_2 molecules to the internal pore surfaces [1, 8, 14]. Considering this, CO_2 uptake in MOFs may go through a similar pathway.

7.3.1 CO_2 capture processes

Prior to the discovery of MOFs, aqueous alkanolamine absorbents were widely applied in natural gas upgrading industry for CO_2 separation and uptake, as discussed in the previous section. Amine groups in each alkanolamine react with C-atoms forming N–C bonds through nucleophilic attack to subsequently generate carbamate or bicarbonate compounds. These materials are limited in their performance by factors such as thermal instability, corrosiveness and a high energy penalty of regeneration.

MOFs are a class of porous materials made up of metal nodes interconnected by organic linkers through strong bonds [23–27]. The well-defined structure of the MOFs being designed and tailored suits them perfectly not only for CO_2 capture and storage but adsorption of other small molecules [23]. Most MOFs are stable when exposed to organic solvents and many of them retain their crystal structures in the presence of aqueous media. Another interesting aspect of MOFs is the tunability of the structure [18–21], which is in contrast with other porous materials (e. g., zeolites and activated carbons). Hence, MOFs are categorized by both their dynamic/flexible or rigid structures. Each of these categories has been tailored for applications that require specific characteristics [24].

CO_2 capture technology exists in three processes: precombustion, post-combustion and oxy-fuel combustion CO_2. The precombustion method converts fossil fuels into CO and H_2 (syngas) that is followed by water-gas shift reactions to generate CO_2 at high partial pressures shortly before combustion. Pre-combustion capture has the relative advantage of higher component concentration and pressure, which lowers the energy penalty to a value half of what post-combustion consumes. Separation of CO_2/H_2 is easier than CO_2/N_2 in precombustion than in post-combustion processes. Oxyfuel combustion requires pure oxygen for the combustion and production of the syngas that is comprised of CO_2, H_2O and trace amounts of NO_x and SO_x, among others. In this process, it is easier to condense water vapor and harvest CO_2 thus obviating CO_2 capture and separation costs. However, the cost of filtering air to get pure O_2 is high hindering the applicability of this technology [24].

Post-combustion capture is the most "mature" process, in which CO_2, as a portion of flue gas generated after combusting carbon-based fuels, is selectively captured. Over 60 % of electricity in the United States, for example, is produced from fossil fueled power plants, which are solely responsible for anthropogenic CO_2. To reduce the CO_2 emission, selectively capturing CO_2 from flue-gas stream is of immense importance and currently has received most consideration from the advanced materials research community (including those in reticular chemistry). The major challenge in post-combustion CO_2 capture is to selectively capture CO_2 over the other gases given that the concentration of CO_2 in the flue gas is low (15 % CO_2, ~75 % N_2, and small fractions of H_2O, O_2, NO_x and SO_x), leading to a relatively low driving force (i. e., high energy input is required) [25]. In post-combustion capture, exhaust gas with partial pressures of 0.13–0.16 bar at 40–60 °C is fed to the MOF-based packed bed reactor.

CO_2 is captured on the internal pore walls of the MOF by van der Waals forces along with covalent bonding. The excess gas after saturation is desorbed through temperature or pressure swing processes. To achieve working capacities for various MOFs that are comparable to amines as shown in Figure 7.2, certain MOFs must be brought to an elevated temperature (200 °C) for regeneration. For example, Mg-MOF-74 shows the highest CO_2 capture working capacity (5.87 mmol g^{-1} at ambient temperature) while USTA-16 (USTA = University of Texas at San Antonio) exhibits higher selectivity for CO_2 over N_2. This makes USTA-16 the most economically viable material for industrial scale application [25].

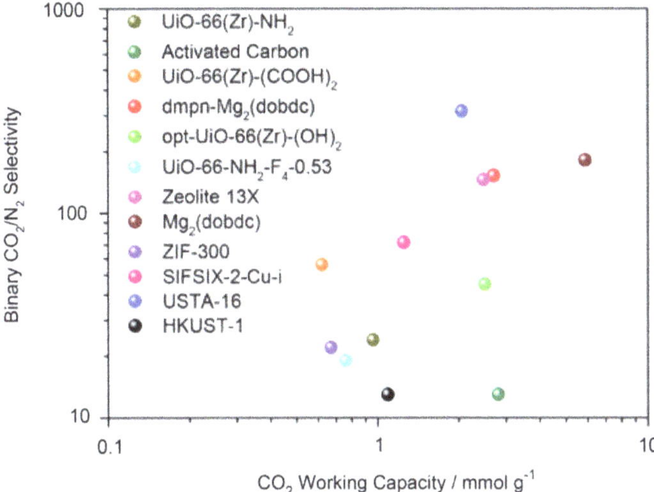

Figure 7.2: Selectivity in terms of working capacity of various MOFs for post-combustion CO_2 sequestration. Reproduced with permission from [25]. Copyright 2020, Elsevier.

While oxy-fuel combustion is suitable for newly built power plants, precombustion has been reported to be effective in gasification plants. On the other hand, post-combustion is more reliable for retrofitting existing power plants since combustion occurs in coal powered plant boilers. For selective capture of CO_2 relying on MOFs, CO_2 input conditions such as partial pressures (usually at low pressure of 0.15 bar) and CO_2 concentration are crucial factors. To enhance the CO_2 uptake affinity, the presence of polar functional groups lining the pores of MOFs coupled with coordinatively unsaturated sites (CUSs) or incorporated entities in specific pore confinement plays an essential role in boosting the CO_2 uptake performance, well beyond other porous materials [17, 19, 20].

7.3.2 Designs of MOFs for CO_2 capture

The ability to capture CO_2 lies in the structure stability and evacuation of guest molecules. MOFs have intrinsic structural ability to be activated to remove solvent guest molecules that create open adsorptive metal sites (Figure 7.3). These sites provide high affinity to CO_2 molecules [24]. In MOFs, CUSs exist at the pore walls of the structure. An iconic example of which is Cu-MOF [18, 19] based on Cu(−COO)$_4$ clusters or secondary building units (SBUs). Other CUS-rich MOFs include MIL-100, MIL-101 [25] (MIL = Materials of Institute Lavoisier), among many others [28]. The metal sites interact with CO_2 molecules through strong electrostatic forces, typical of Lewis acid behavior. This interaction results in improved CO_2 capture.

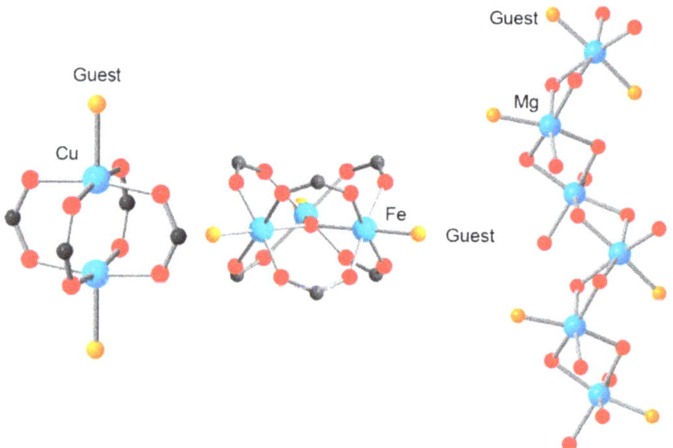

Figure 7.3: Representation of typical metal clusters in MOFs bearing open metal sites. Atom colors: M (Cu, Fe or Mg), blue; C, black; O, red and Guest, orange. H atoms omitted for clarity.

A study reported by Snurr et al. demonstrated that equilibrium CO_2/CH_4 selectivity by MOFs having CUSs outperformed those without [29]. This finding further revealed the biased selectivity of CO_2 over CH_4 as per the mechanism explained earlier. Most isoreticular MOFs containing Mg, Fe, Co, Mn, Ni, Cu and Zn with CUSs exhibit high equilibrium selectivity toward CO_2 and this happens mostly at low pressure, which is ideal for flue gas separation [17, 18]. MOF-74 has so far been the most effective in CO_2 capture while USTA-16 has shown the highest selectivity of CO_2 over other gases [25]. Metal sites in MOFs behave like Lewis acid centers and are the preferred sites that CO_2 highly interacts with. High selectivity compared to other MOFs has made USTA-16 a commercially viable technology in CO_2 adsorption technology. There are, however, efforts to improve on the selective efficiency in other MOFs, which includes increasing the density of CUSs, doping, and controlling pore size and volume. These factors have been at the center of research in MOF synthesis for gas adsorption and separation technology. According to research findings by Rajagopalan et al. [30], who studied the temperature swing adsorption on MOFs, it was revealed that interaction between negatively charged O-atoms on CO_2 and metal centers with low electronegativity in MOFs at low partial pressure created a strong attraction and selective adsorption of CO_2 [30]. Adsorption of CO_2 in MOFs normally occurs along with other gases, and the challenge would then be how to separate the mixed components after capture. This could be overcome by creating CUSs with low electronegative metals. One example is the metalation of MOF-74 with Mg reported by Long et al. [31]. Their findings indicated that inclusion of Mg into MOF-74 led to the interaction between Mg ions and hydroxyl ions in the MOF, thus creating a synergistic enhancement of CO_2 uptake in the MOF channels. It has been established those synthetic methods that enhance saturation with CUSs improved the stability of MOFs in CO_2 sorption. Wang et al. reported the enhanced performance of a Ni-based MOF with a trinuclear SBU structure with high stability and CO_2 uptake of up to 37.57 $cm^3\ g^{-1}$ and selectivity over N_2 [32]. Lanthanide-based MOFs are another group of MOFs that have proven effective in high CO_2 selectivity. For example, an Eu-MOF reported by Liao et al. [33] exhibited interesting CO_2 uptake (with a composition selectivity of 15/85 CO_2/N_2 at 273 K and 1 atm.). Further to increasing the efficiency of CO_2 capture in open metal sites, post-synthetic treatment can be considered. For example, $HCu[(Cu_4Cl)_3(BTTri)_8]$ (Cu-BTTri, BTTri = 1,3,5-benzenetristriazolate) MOF with CUSs (Figure 7.4a) is coordinated by ethylenediamine (en), and the resulting en-functionalized MOF exhibits remarkably high affinity for CO_2 (Figure 7.4b) at low pressures and low initial heat of adsorption [26]. Similarly, acrylamide and metal salts are used to functionalize and increase CO_2 uptake in many MOFs [28, 29, 34, 35]. For instance, metal salts such as $Cu(BF_4)_2$ in Al(OH)(bpydc) (bpydc = 2,2′-bipyridine-5,5′-dicarboxylate) are a proven enhancer of CO_2 uptake in

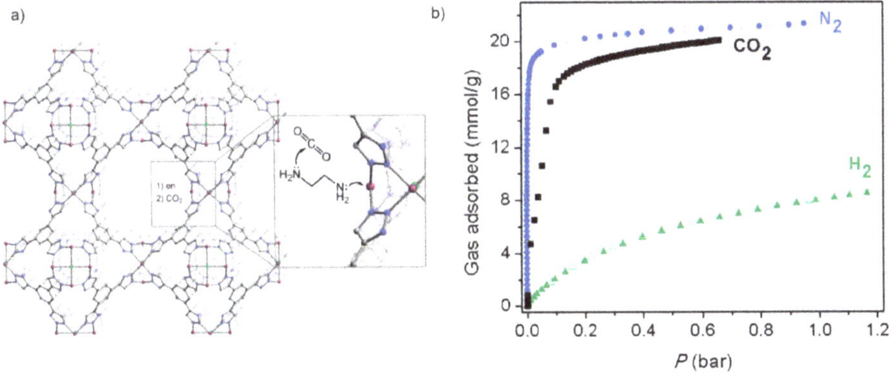

Figure 7.4: Crystal structure of Cu-BTTri (a) presenting Cu CUSs which are used to stitch with ethylenediamine to form en-functionalized Cu-BTTri (b). Atom colors: Cu, purple; C, grey; N, blue and Cl, green. H atoms omitted for clarity. Reproduced with permission from [26]. Copyright 2009, American Chemical Society.

MOFs by increasing the electric dipole on the surface of the MOF [34]. The heat of adsorption can be increased from 23 to 30 kJ mol^{-1} with isotherm data collected at 298, 303 and 308 K [28].

Another method of enhancing CO_2 capture is to create basicity in the MOF, which, in turn, enhances the interaction between the framework and CO_2 molecules on the acidic sites. This can be achieved through incorporation of amines and using amine-based linkers in the synthesis, for instance, in using amine-functionalized terephthalic acids [17, 18, 27]. Doping of the MOF with the amines in the pores is another strategy to modify MOFs with basic compounds that lead to high uptake capacity and selective adsorption of CO_2 [30, 31]. Studies involving doping of MOFs with P- and N-heteroatoms have also proven effective in enhancing CO_2 capture [36]. This is because P- and N-heteroatoms enhance the polarity on the MOF surfaces, which facilitates interaction with CO_2 molecules [36].

Other surface property-altering methods, such as direct polymerization, have been reported. Forgan et al. altered the surface chemistry of UiO-66-NH_2 (UiO = University of Oslo) by direct polymerization, which produced a two-fold increase in CO_2 uptake as compared to the parent MOF [17, 18, 20–25, 27]. UiO-66-NH_2 post-synthetically modified with dianhydrides results in anhydrides, which then are reactive with diamines. The resulting imidized surface creates a layer of polyimides on the surface of the MOF. Polymerized MOF surfaces are characterized by high stability as well as enhanced diffusivity and selectivity toward CO_2 [18, 28]. Nugent et al. [39] reported a new MOF functionalization technique that involves introducing inorganic ions with narrow pores into SIFSIX-3-Zn for improved selectivity toward CO_2. The narrow pores in the MOF afforded superior selectivity toward CO_2 even at a low concentration of 10 % CO_2 in a mixture with N_2 in both wet and dry conditions. In general, functionalizing MOFs with narrow pore materials is a promising technique for gas separation [25, 28].

7.3.2.1 Heterocyclic linkers

Introduction of heterocyclic linkers containing N-, and/or O- atoms in the UiO family has been discovered to increase both selectivity and adsorption capacity (Figure 7.5) [40]. These materials that incorporate N-atoms (Zr-BPYDC; BPYDC = 2,2′-bipyridine-5,5′-dicarboxylate) [40], S-atoms (Zr-BTDC; BTDC = 2,2′-bithiophene-5,5′-dicarboxylic) and O-atoms (Zr-BFDC; BFDC = 2,2′-bifuran-5,5′-dicarboxylic) [40] proved the abovementioned while their parent MOF, UiO-67, showed lower adsorption capacity. The study further revealed from this family of heterolinkers that the O-based linker in Zr-BFDC had the highest CO_2 adsorption capacity. Linker functionalization in MOFs has been proven to be useful technique for improving performance of MOFs in CO_2 uptake. The improved CO_2 uptake is rooted in the increased metal-linker interactions resulting in a strong bond that offers stability and higher CO_2 adsorption capacity for the MOF. However, this host-guest interaction is not without

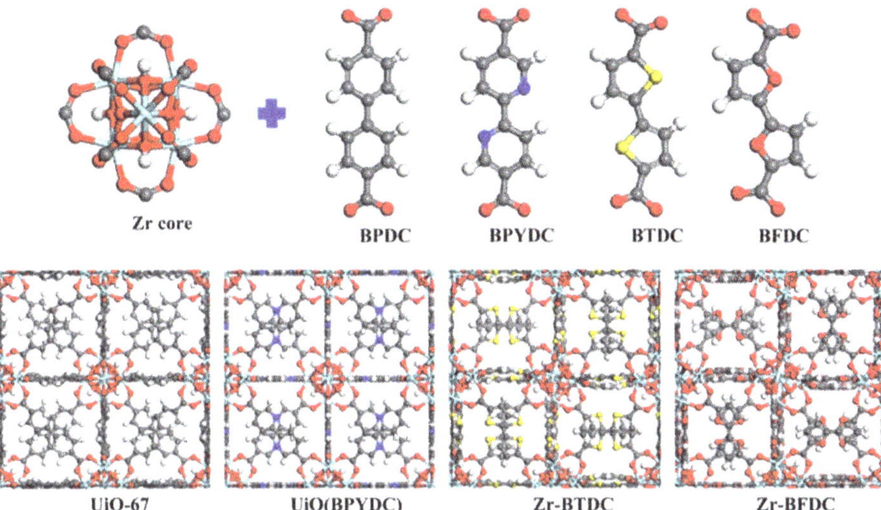

Figure 7.5: Combination of Zr-oxide cluster and various functionalized linkers to form UiO-66-type structures—a strategy to enhance CO_2 uptake capacity. Atom colors: Zr, green, C, grey; O, red; S, yellow and H, white. Reproduced with permission from [40]. Copyright 2018, Elsevier.

its downside. Particularly, introduction of heterolinkers in MOF structures eventually leads to reduced pore volumes of the respective MOF given that functionalized linkers occupy part of the available pore space, thereby resulting into lower CO_2 uptake [40]. Therefore, rational designs must be taken into account when introducing any guest molecules into the pores of MOFs for their beneficial aspects and should not overshadow their potential negative impact on pore space volume [38, 41].

7.3.2.2 Amino and other functional groups

Cadmium-based MOFs, such as $\{[Cd(BIPA)(IPA)]\cdot DMF\}_n$, $\{[Cd(BIPA)(HIPA)]\cdot DMF\}_n$, and $\{[Cd(BIPA)(NIPA)]\cdot 2H_2O\}_n$ (BIPA = bis(4-(1H-imidazol-1-yl)phenyl)amine), IPA = isophthalic acid, DMF = N,N'-dimethylformamide, HIPA = 5-hydroxyisophthalic acid and NIPA = 5-nitroisophthalic acid) [42] were synthesized using amine-based linkers along with phenol, hydroxyl and nitro groups, respectively [25]. These MOFs exhibited high selectivity and affinity for CO_2 adsorption. This is owed to the reduced pore space and functionalization of the MOFs with 'pro' CO_2 groups. This property has made these Cd(II) MOFs equally suitable for metal detection [42]. He et al. synthesized a MOF, termed ZJNU-98 (ZJNU = Zhejiang Normal University; $[Cu_2(L)(H_2O)_2]\cdot 5DMF\cdot H_2O)$, using an aminopyridine-functionalized diisophthalate linker (H_4L, 5,5′-(pyridin-3-amine-2,5-diyl)diisophthalic acid, Figure 7.6) [43]. The authors subsequently investigated the CO_2 uptake using this MOF, which revealed a significantly superior per-

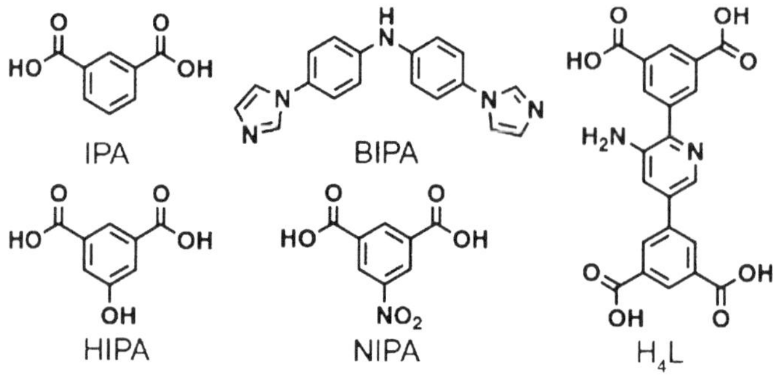

Figure 7.6: Chemical structures of linkers used to construct MOFs for enhanced CO_2 uptake capacity.

formance compared with the parent MOF, NOTT-101 (pyridinic N-modified; NOTT = University of Nottingham) constructed from a nonamine linker [43]. Similarly, Mutyala et al. incorporated MIL-101(Cr) into polyethylimine (PEI) and the results showed high CO_2 uptake due to $-NH_2$ groups that chemically react with CO_2 to produce carbamate [9]. Another study further revealed a positive correlation between the amount of amine incorporated into the framework and the amount of CO_2 adsorbed. For example, 70 % MIL-101(Cr) presence in PEI yielded 3.81 mmol g^{-1}, which is 4 times higher than that of the parent MIL-101(Cr) [9].

Generally, functionalized MOFs usually have higher affinity toward CO_2 than their parent, pristine MOFs. However, functionalization of MOFs should be preferably conducted before the MOF synthesis through modification of the linkers and/or metals nodes (on the linkers). The reason for this is that post-synthetic modifications may provide minor impact in terms of conversion and properties of the resulting MOFs, except for the case in which the creation of active metal centers is carried out using post-synthetic metalation [26].

7.3.2.3 Fluorination of MOFs

Like amines, fluoride groups in MOFs have an immense influence on CO_2 selectivity and uptake capacity [25]. Inspiration for fluorination was conceived to not only curb CO_2 emissions in industry, but also to capture H_2S, which is another very harmful gas. Reports have shown enhanced selection of CO_2 and H_2S over CH_4 in a mixed gas system. For example, SIFSIX-3-Ni [44] (with a formula of $[M(pyz)_4SiF_6]$, M = Fe or Ni, pyz = pyrazine) (Figure 7.7a) and NbOFFIVE-1-Ni [37] are particularly suited for CO_2/H_2S adsorption as they show strong selectivity with values greater than 1, while AIFFIVE-1-Ni exhibited similar selectivity with more robustness under different temperature, concentration and pressure conditions [44]. The fluorination of these MOFs increased their selectivity toward CO_2 and is one of the most effective methods to synthesize MOFs for CO_2 capture purposes (Figure 7.7b) [18, 34].

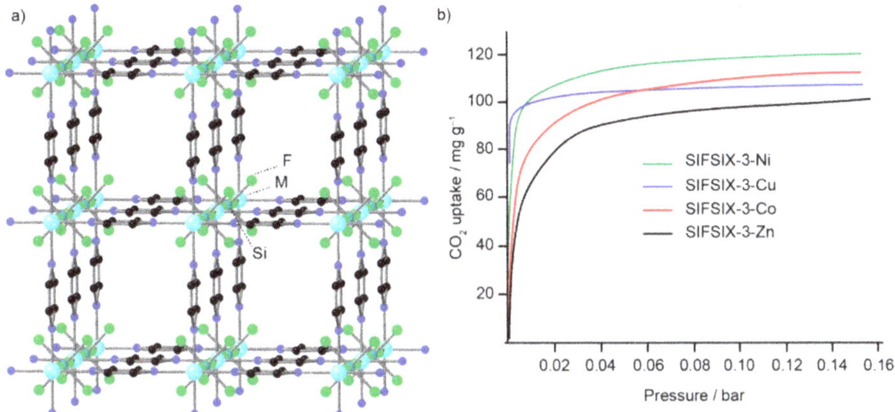

Figure 7.7: (a) Crystal structure of SIFSIX-3-M (M = Ni, Zn, Cu and Co) and (b) CO_2 uptake isotherms at 298 K of SIFSIX-3-M. Atom colors: M, light blue; Si, dark green; C, black; N, navy blue and F, bright green. H atoms are omitted for clarity.

7.3.2.4 Ozonolysis of MOFs

There is a rising need to modify the pore structure in MOFs from microporous to mesoporous to increase their capacity for adsorption and storage of CO_2. Ozonolysis is one of the main methods used to clip linkers with nonterminal olefins within a given MOF structure [25]. This post-synthetic step is only made successful in MOFs whose reactivities and structural properties are similar in many reaction conditions, but different in an ozone environment. One example is the ozonolysis of Zr-FCU [45], which was conducted by synthesizing the MOF composed of 4,4′-azobenzene dicarboxylic acid (H_2azo) and 4,4′-stilbenzene dicarboxylic acid (H_2sti) as linkers [45]. This MOF was ozonolyzed by ozone under $ZnCl_4$, DMF, HCl and L-proline. The MOF synthesized using H_2sti had its porosity lost and surface area sacrificed whereas the H_2azo-based Zr-FCU MOF maintained its porous structure after exposure to ozone. These findings indicated that the H_2sti-based MOF underwent ozonolysis [45]. Interestingly, even though it lost some pore structure and surface area, successful transformation from microporous to mesoporous structure and concomitant increase of pore volume by over 40 % was obtained. It should be noted that increasing the pore volume of MOFs is important for the sorption of CO_2 at high pressure with moderate selectivity and may enhance the recyclability of the MOF for long term use [45].

7.3.3 MOF membranes for CO_2 capture

Membrane technology has attracted attention in gas capture and separation processes owing to its environmental friendliness, low cost and energy efficiency. The feed stream in membranes is driven by a difference in pressures between the two sides of the membranes [18, 36]. The performance of membranes is bolstered by using MOF materials with well-defined pore structures. Polymers are currently the ideal host materials in membrane systems given their ease of fabrication and low cost. However, with the advances in MOF-based and MOF/fiber composite membranes that capitalize on their microporous, well-defined and tunable pores and surface properties, the future of membrane technology will undergo significant improvement. Using pure MOFs for membranes has also been reported with promising results [25]. ZIF-7 ($Zn_2(bim)_4$, bim = benzimidazole) is one example that has been so far evaluated by Peng et al. [46]. The authors synthesized $Zn_2(bim)_4$ that was then fabricated on an Al_2O_3 support with thickness of 1–2 nm. The highly crystalline MOF nanosheet membrane achieved H_2 diffusivity as well as high H_2/CO_2 selectivity (from 53–291) [25]. Moreover, hybrid systems containing MOFs and fibers exhibited even better performance according to recent reports. Studies conducted on the MIL family of MOFs with fibers have provided new insights into these two systems in what is now known as MOF-led mixed-matrix membranes (MMMs) [47]. MIL-53(Al) has been utilized owing to its high porosity, chemical stability and unique breathing properties. It was found to be ideal to produce MMMs for CO_2 capture and separation [47]. Although the technology is still in infancy and further studies will be required to fully understand other aspects.

7.3.4 MOF regeneration

An emphasis in our previous discussion has centered around superior CO_2 selectivity and uptake capacity. While these two properties are behind many research programs, there is an additional critical dimension for an adsorbent material to be reusable as use of a new material for every cycle would be economically and environmentally untenable. A material in this regard should adsorb and desorb the adsorbate under certain conditions, such as lowered pressure and inert gas purging. Numerous studies have been conducted to simultaneously adsorb and desorb CO_2/H_2S from a CH_4 rich environment in a single step [37]. Belmabkhout et al. used a fluorinated MOF and NbOFFIVE-1-Ni for CO_2 sorption and the findings displayed high stability of the MOFs under hydrolytic conditions as well as high recyclability [37]. The adsorption capacity for the fluorinated MOFs revealed remarkable consistency during many desorption cycles. Liu et al. reported a study in which they introduced $-CH_3$ functional groups into $Zn(BDC)(DMBPY)_{0.5}\cdot(DMF)_{0.5}(H_2O)_{0.5}$ and $Zn(NDC)(DMBPY)_{0.5}\cdot(DMF)_2$ (H_2BDC = 1,4-benzenedicarboxylic acid, DMBPY = 2,2′-dimethyl-4,4′-bipyridine and H_2NDC = 2,6-naphthalenedicarboxylic acid). These MOFs presented proper interaction with CO_2 in addition to limited interpenetration and structural stability upon unloading [48]. Although the MOFs had lower pore volume, the methyl-functionalized MOFs exhibited superior CO_2 sorption capacity thanks to the H-bond interaction between the methyl functional groups and the CO_2 molecules. Importantly, that H-bond interaction is weak enough to avoid structural collapse upon the desorption [48]. MOF recyclability and reusability are also as important as other parameters such as surface area, tunability and porosity. The shift from liquid-based CO_2 sorbent materials was mainly to overcome the short-lived performance and high energy consumption during regeneration of aqueous amines [41, 42]. For MOF-based solid CO_2 sorbent materials, to bring MOFs to industrial scale applications that make economic sense, recyclability and reusability of the MOF must be carefully studied to ensure the two parameters are achieved without loss of adsorption capacity [49]. Recyclability and reusability studies can be done through studying the adsorption and desorption behavior and narrowing down the number of cycles (adsorption and desorption) that a given MOF undergoes while maintaining its other attractive structural properties [49].

7.4 MOFs for CO_2 conversion

Photocatalysis, photoelectrochemical catalysis, thermochemical catalysis, biological and electrochemical catalysis are all candidates processes to carry-out CO_2 conversion to provide user-friendly, value-added products from CO_2 [38–41]. In this section, we discuss the chemistry of MOFs for photocatalytic and electrocatalytic conversion of CO_2.

7.4.1 CO_2 photocatalytic conversion

Photocatalytic conversion of CO_2 into useful products is an effective strategy for solving the CO_2 emission problem. In this technology, semiconducting materials are used. The problem facing photocatalysts in the uphill reaction of CO_2 conversion relies on low solar light efficiency due to wide bandgaps, fast recombination of photoelectrons and holes, as well as low CO_2 adsorption capacity. Transition metals are effective photocatalysts due to high light-heat conversion efficiency and serve as inspiration for photocatalytic conversion [12, 31, 38–41]. Although knowledge of the detailed mechanism behind CO_2 photocatalysis has not been fully elucidated, recent studies have given important clues to the chemistry of the CO_2 photoreduction pathway [46–54]. The following six equations [55] provide a summarized detail of what is known of the photocatalytic process (Table 7.1). A salient feature of photocatalysis is the generation of unstable intermediates and varied products that are dependent on the reaction routes [55].

MOFs possess superior surface area, accessible reactive metal centers and, more importantly, low and tunable bandgap energy, all of which are essential factors for this application. These properties have endowed MOFs with effective catalytic conversion

Table 7.1: CO_2 reduction reactions and their corresponding potentials.

CO_2 reduction reaction	Redox potential vs. normal hydrogen electrode (NHE)
$CO_2 + e^- \rightarrow CO_2^-$	E = −1.90 V
$CO_2 + H^+ + 2e^- \rightarrow HCO^{2-}$	E = −0.49 V
$CO_2 + 2H^+ + 2e^- \rightarrow CO + H_2O$	E = −0.53 V
$CO_2 + 4H^+ + 4e^- \rightarrow HCHO + H_2O$	E = −0.48 V
$CO_2 + 6H^+ + 6e^- \rightarrow CH_3OH + H_2O$	E = −0.38 V
$CO_2 + 8H^+ + 8e^- \rightarrow CH_4 + 2H_2O$	E = −0.24 V
$2H^+ + 2e^- \rightarrow H_2$	E = −0.41 V
$H_2O \rightarrow 1/2O_2 + 2H^+ + 2e^-$	E = 0.82 V

of CO_2 to useful products through photoreduction, hydrogenation and cycloaddition [19, 31, 38–41, 43–55]. Typically, in the photoreduction of CO_2, electrons are generated by ultraviolet (UV) or visible light irradiation. The excited electrons then migrate from the valance band (VB) to the conduction band (CB) leaving holes behind. Electrons in the CB subsequently interact with CO_2 to reduce to the relevant product(s) (Figure 7.8). Uniform distribution of pore space and metal active centers highly favor electron movement and access to photoreduction sites (i. e., metal centers) where the reaction with CO_2 takes place. This is made easy by conjugation of the MOF framework backbone, which permits uptake, circulation and eventual reduction of CO_2 into potential carbon-neutral fuels. Parallel to the photo-reduction in the CB is the oxidation of water in the VB that releases O_2. Finally, electrons and holes recombine, which suppresses the photoreduction process. MOFs can photocatalyze the reduction of CO_2 with a high conversion efficiency (>90 %). Zhang et al. reported a high CO_2 photocatalytic conversion efficiency along with high recyclability of CO_2 to methane for the very

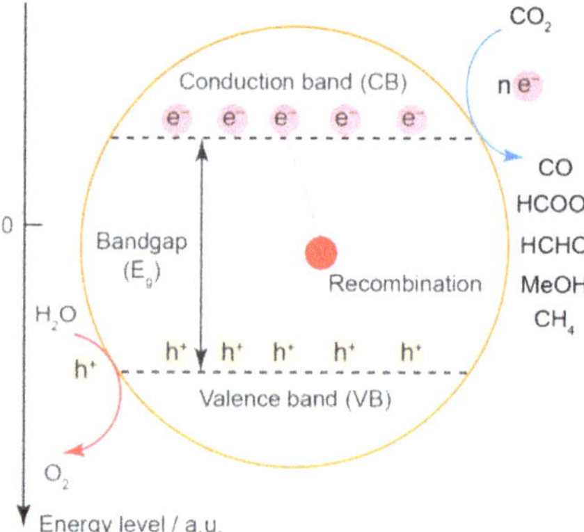

Figure 7.8: Fundamental process underlying CO_2 photoreduction catalyzed by a semiconductor.

first time [51]. They used UiO-68-NHC (NHC = N-heterocyclic carbene), in which an imidazolium bromide-based linker had been incorporated. The catalysis in this case occurred through transfer of hydrogen from silanes to CO_2 thereby ensuring a high catalytic activity [51].

Bandgap engineering has been the center of structure design in photocatalytic MOFs used for the photoreduction of CO_2. Two iconic examples of this concept in MOF chemistry will be discussed in detail. The first example belongs to MIL-125 and its variant MIL-125-NH_2 (Figure 7.9a). MIL-125 adopts a face-centered cubic (**fcu**) structure that is comprised of 8-membered ring Ti SBUs and BDC linkers [59]. Due to the weak conjugation of the linker backbone, MIL-125 only absorbs UV light with a large bandgap energy (3.6 eV). However, when functionalized with amino group to form MIL-125-NH_2, the resulting MOF material has a much smaller bandgap (2.65 eV) and can photocatalyze CO_2 to $HCOO^-$ under visible light irradiation [60]. The enhanced photoreduction of CO_2 was attributed to linker-to-metal charge transfer behavior. A similar study conducted on Fe-MOFs by Sun et al. presented predictable results in the ability to reduce CO_2 to CO using amino-functionalized Fe-MOFs (i. e., amino-based versions of MIL-101(Fe), MIL-53(Fe), MIL-88B(Fe)) [61]. Interestingly, applying the same strategy toward UiO-66 to produce UiO-66-NH_2 can help promote not only CO_2 photoreduction [62], but also the hydrogen evolution reaction (HER) [63]. In general, amino groups facilitate linker-to-metal charge transfer behavior, which in turn, enhances the pho-

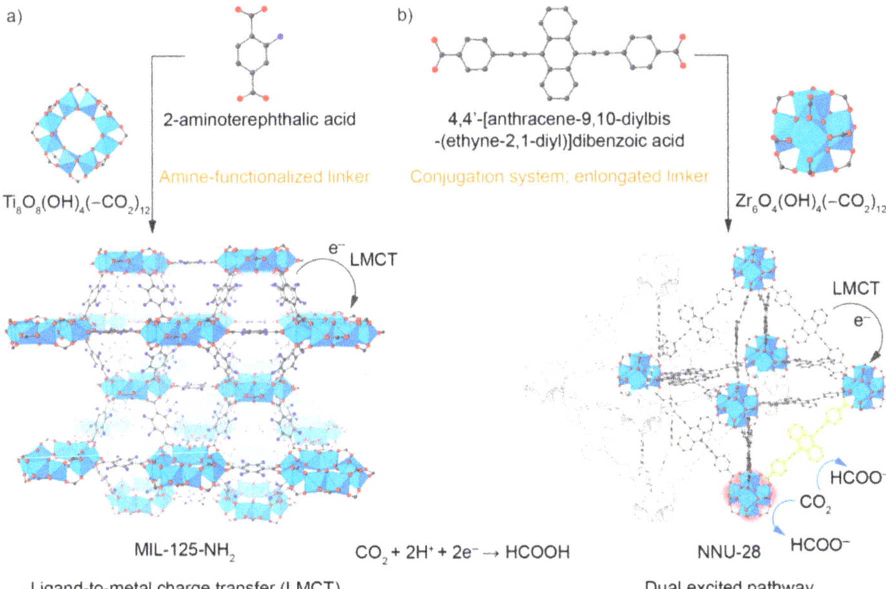

Figure 7.9: Bandgap engineering of MIL-125-NH_2 (a) and NNU-28 (b) through amino-functionalized entities and the highly conjugation system, respectively. Atom colors: M, blue polyhedra; C, gray; O, red and N, blue. H atoms omitted for clarity.

tocatalytic properties of a chosen MOF. The second example is NNU-28 (NNU = Northeast Normal University) [64]. This MOF is composed of Zr-oxo SBUs and the linker integrates a naphthyl unit that ensures strong conjugation of the MOF backbone. As a result, NNU-28 reduced CO_2 to formate via a dual pathway mechanism (Figure 7.9b) with a high production yield (183.3 μmol h^{-1} $\mu mol_{MOF}{}^{-1}$). This is the benchmark performance for CO_2 to formate photoreduction among Zr-based MOFs [53].

Photocatalysts can absorb light energy only when the bandgap energy of the incident light equals or is above the bandgap energy of the photocatalyst. Bandgap energy is an integral component of all photocatalysts as it dictates the pathway of reaction and eventual products. Most importantly, it plays a vital role in multiple proton and electron transfer (MPET) for multiple CO_2 photoreduction pathways [54, 55]. Recombination of electrons and holes is common through straightforward electron-hole pairing. Therefore, to minimize this recombination that precludes premature termination of the CO_2 photoreduction, various research efforts have been carried-out to engineer a given MOF with more complicated electron-hole recombination pathways, thereby providing photoexcited electrons a longer life to carry-out reduction. Bulk phase and surface recombination of photogenerated electrons and holes increases the photocatalytic efficiency in MOFs. Once fast recombination cannot occur, electrons with high enough reducing power reduce CO_2 into various products while holes with high oxidizing ability can oxidize water [41, 42, 56–67]. Many factors contribute to the effectiveness of the photocatalysis, keys among which include the light trapping ability, efficiency of the generated electrons and the surface properties of MOFs [66–68].

Porphyrin-like pigments have been widely used in CO_2 photoreduction due to their ability to minimize fast recombination of photoexcited electrons and holes. To have MOFs capable of prolonging the lifetime of photoelectrons, porphyrin-functionalized linkers can be used as organic linking units. Jiang et al. used PCN-222 (PCN = porous coordination network; Figure 7.10) as a photocatalyst for the photoreduction of CO_2 [69]. Bearing a mesoporous structure, which is an important feature to maximize the exposed tetracarboxyphenylporphyrin linker (TCPP), PCN-222 strongly absorbed visible light from 200 nm to 800 nm. The long-lived excited electrons were facilitated by this MOF (≈11 ns) because of TCPP units that enable an electron trap state. This means the detrapped time needed to release electrons from a deep state is long enough to overcome the fast recombination of the excited electrons and holes. This enhances the photocatalytic activity of PCN-222 producing a total yield of 30 μmol $HCOO^-$ after 10 h in the presence of triethanolamine (TEOA) as a sacrificial agent.

Using porphyrin derivatives to synthesize MOFs has yielded many emerging MOF photocatalysts such as PMOF-1 (PMOF = porphyrin-based MOF) [70], Zr-PP-1 [71] and PCN-601 [72]. The active metal sites in these MOFs not only enhance the lifetime of photoelectrons, but also have an impactful role on the C–O cleavage of CO_2. PMOF-1 was metalated with Rh and the resulting Rh-PMOF-1 exhibited sufficient photocatalytic reduction of CO_2 to $HCOO^-$. The $HCOO^-$ yield was 6.1 $\mu mol_{MOF}{}^{-1}$ within 18 h irradiation under visible light and the turnover frequency (TOF) was 0.34 h^{-1}. The high activity of

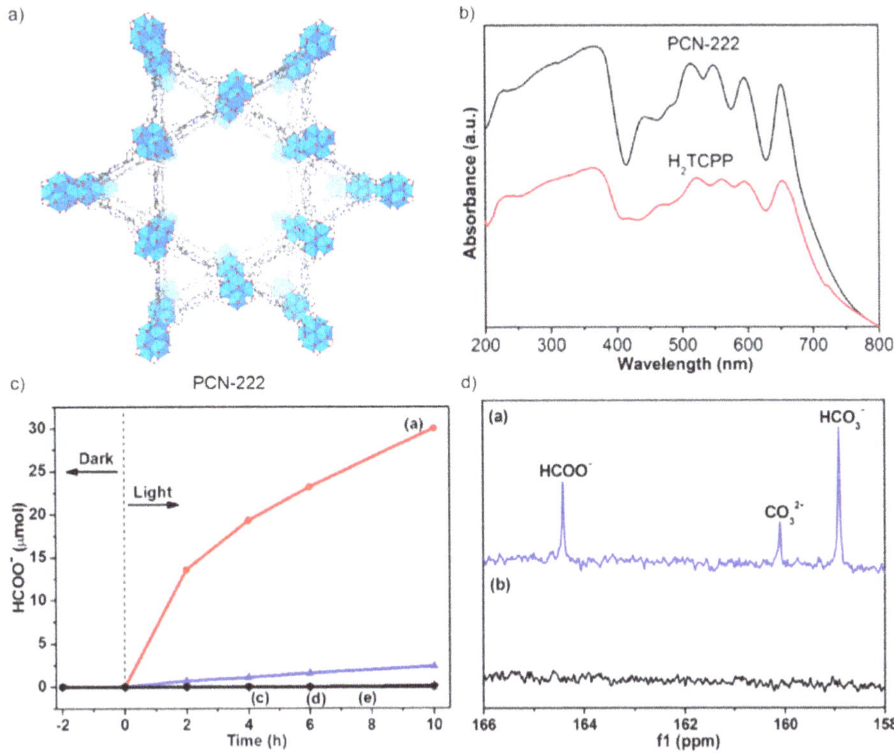

Figure 7.10: PCN-222 comprising of TCPP linking units and Zr-oxide SBUs (a) has a strong visible light absorption, (b) which dictates the high photocatalytic reduction of CO_2 to format (c and d). Inset of c) and d): (a) PCN-222, (b) TCPP linker, (c) no PCN-222, (d) no TEOA and (e) no CO_2. Atom colors: Zr, blue polyhedral; C, grey; O, red and N, blue. H atoms omitted for clarity. Reproduced with permission from [69]. Copyright 2015, American Chemical Society.

Rh-PMOF-1 was attributed to the distinct Rh CUSs on the linkers [70]. Similarly, Zr-PP-1 coupled orbitals between its Co-porphyrin entities and Zr SBUs, which, in turn, generated a delocalized π-conjugated network. This was the main reason for the electron migration from the linker to the catalytic active centers (i. e., Co sites) that subsequently photocatalyzed CO_2 to CO with a production rate of 14 μmol g^{-1} h^{-1} [71]. The reaction was carried out under visible light irradiation in the presence of TEOA and the MOF could be recycled for 45 h. Interestingly, in the case of PCN-601, the Ni-oxo SBUs play an essential role in being both mediators and active sites for multielectron transfer, thus facilitating the CO_2 to CH_4 photoreduction. More importantly, the holes in PCN-601 oxidize water to peroxide to make this photoreduction free of sacrificial agents [72].

MOF structures can be tuned to allow accessibility to all metal sites, more important heteroatoms, as well as linker sites [55]. This improves the electron-hole separation efficiency and eventually the photocatalytic performance. Additionally, MOFs with a highly conjugated backbone play a key role in facilitating the movement of

Table 7.2: Emerging MOFs and MOF-based materials for CO_2 photoreduction.

MOF	Product(s)	Product yield	Reaction condition	Ref.
MOF **4** (a UiO-67 variant)	CO	5 CO-TONS[a]	UV–Vis, in MeCN[b] and TEA[c] as a sacrificial agent within 5 h	[73]
PCN-222	$HCOO^-$	30 µmol	UV–Vis, in CH_3CN and TEOA[d] as sacrificial agent within 10 h	[69]
NNU-28	$HCOO^-$	26.4 µmol	UV–Vis, in CH_3CN and TEOA as sacrificial agent within 10 h	[64]
$Zr_6O_4(OH)_4(L)\cdot$ 6DMF	$HCOO^-$	19.6 µmol	UV–Vis, in CH_3CN and TEOA as sacrificial agent within 12 h	[74]
Ag-Re_n-MOF (a modified UiO-67)	CO	1.5CO-TON (10^{-2} h)	UV–Vis, in MeCN and TEA as sacrificial agent within 48 h	[75]
UiO-66-CrCAT (a modified UiO-66)	HCOOH	51.73 ± 2.64 µmol	UV–Vis, in MeCN and TEOA containing BNAH[e] (0.1 M) within 6 h	[76]
MOF-525-Co	CO and CH_4	2.42 mmol CO and 0.42 mmol CH_4	UV–Vis, TEOA as sacrificial agent within 6 h	[77]
Co-ZIF-9	CO	85.6 µmol/h	UV–Vis, in MeCN, H_2O and TEOA as sacrificial agent within 3 h	[78]
Zn/PMOF	CH_4	10.43 mmol	UV–Vis, in MeCN/H_2O/TEOA within 4 h	[79]
Co-ZIF-9/TiO_2	CH_4	140 ppm	UV–Vis, carried-out on a glass reactor with deionized H_2O as reducing agent within 8 h	[80]
Cu-TiO_2/ZIF-8	CO and CH_3OH	29.7 ± 1.3 ppm CO and 31.3 ± 3.4 ppm CH_3OH	UV–Vis, in DMA[f] and TEOA as sacrificial agent within 5 h	[81]
$CsPbBr_3$@ZIF-8	CO and CH_4	1.2 µmol CO and 5 µmol CH_4	UV–Vis, gaseous reaction mode using sealed Pyrex bottle heated up to 250 °C under vacuum, within 3 h	[82]
$CsPbBr_3$@ZIF-67	CO and CH_4	2 µmol CO and 10 µmol CH_4	UV–Vis, gaseous reaction mode using sealed Pyrex bottle heated up to 250 °C under vacuum, within 3 h	[82]
NH_2-MIL-125	$HCOO^-$	8.14 µmol	UV–Vis, in MeCN and TEOA as a sacrificial agent, within 10 h.	[60]

[a]CO-TONS: evolved CO molecules per catalytic site, [b]MeCN: acetonitrile, [c]TEA: triethylamine, [d]TEOA: triethanolamine, [e]BNAH: 1-benzyl-1,4-dihydronicotinamide, [f]DMA: N,N-dimethylacetamide.

electrons to active sites and avail channels for reactant adsorption [55]. Combination of these features improves charge transfer efficiency, limits bulk phase photo-induced electron-hole recombination and maximizes utilization of solar energy. Table 7.2 lists emerging MOFs and MOF-based photocatalysts applied for the photoreduction of CO_2.

7.4.2 Kinetics of CO_2 photoreduction

It is worth mentioning that CO_2 is a linear molecule with a net dipole of zero well known for its extreme stability ($\Delta H = \approx 800\,\text{kJ mol}^{-1}$), which presents challenges to any form of transformation. Photocatalysis offers an opportunity to take advantage of the linear structure of CO_2 through bending and creating a dipole moment that facilitates further chemical reaction [41, 42, 72–78]. Understanding atomic reactions that occur at the surface of a photocatalyst are essential for designing efficient photocatalysts and near accurate development of kinetic models. There are two kinetic models to consider: (i) microkinetic and (ii) intrinsic kinetics. The microkinetic model deals with fundamental atomic-scale studies that exclude heat and mass transfer processes. Advances in computational quantum mechanics have powered microkinetic modeling of photocatalytic behavior with density functional theory (DFT) being the most predominant method used. The DFT method provides detailed information on reaction steps that occur on the photocatalyst surface [55–57]. In these studies, several photoreduction mechanisms have been proposed, which include the formaldehyde, carbene and glyoxal route. Characterization techniques, such as scanning electron microscopy and UV-vis for bandgap calculation carried out either before or after the reaction, do not provide a detailed picture of the reactants and reaction mechanisms involved [17, 18, 55]. However, diffuse reflectance infrared Fourier transform spectroscopy, photoluminescence, Raman spectroscopy and near-ambient pressure X-ray photo-electron spectroscopy provide detailed information about the chemical reaction steps involved on the photocatalytic surface [55]. The two most common intermediates during decomposition of CO_2 on the surface of photocatalysts is carbonate and bicarbonates. Carbonates are the most dominant in photocatalysis and recognizing the process of forming them, along with the poisoning bidentate carbonates, aids in fully elucidating the kinetic model of the photocatalytic reaction [55]. Reaction kinetics for CO_2 photocatalysis is often affected by temperature, pressure, and light transport and intensity, the latter of which is the most impactful factor among them. Although limited studies have been conducted to fully understand these, varying light irradiance during the photoreduction experiment has showed a mixture of products and intermediates. Pressure and temperature are associated with mass transport at the photocatalytic surface and do affect the reaction kinetics in almost equal measure [17, 18, 55]. For determination of the intrinsic kinetic model, the distribution of light in the photocatalyst and the light energy absorbed are important whereas for nonphoto differential photocatalysis, light dependence is influenced by the position of the photocatalyst in the photoreactor. According to Liu et al. [84], there is a direct relationship between the concentration of light and photoreduction of CO_2. Additionally, doping of heterojunction photocatalysts changed the internal electric field of the P–N junction, and hence increased the photocatalytic conversion of CO_2 [84]. This proves the synergistic impact of composites in photocatalysis in the presence of concentrated photoenergy. In the study, the double-edged system ensured that the products of CO_2

reduction were employed in hydrogen production [84]. This work also revealed that concentration of CO_2 and pH of the reaction medium are factors of critical influence to the half-reduction in photocatalytic reactions and directly impacts the selectivity and efficiency of the process. The authors discovered that higher negative potential levels off with high pH and, at pH > 10, CO_3^{2-} is predominant while pH = 3 is characterized with production of H_2CO_3 [84]. Interestingly, at pH = 3, CO_2 conversion was characterized by total organic content while, at pH > 10, CO_2 was converted to formic acid and formaldehyde [84].

7.4.3 Electrocatalytic conversion of CO_2

The CO_2 electrocatalytic reduction (CER), like that of CO_2 photoreduction, is a multi-electron transfer and proton coupled reaction that converts CO_2 into various products through different intermediate compounds [79–91]. Typically, the reaction involves a diffusion-reaction conduction process in an electrolyte where dissolved reactants interact with a solid electrode. This process primarily requires high electric and ionic conductivity apart from a high surface area. This is to allow proper interaction between different reaction phases involved [19, 76–79].

CO_2 electrocatalysis occurs on the cathode where the MOF is employed to reduce CO_2 into various products. There are three processes that take place in order: adsorbing CO_2 on the surface of cathode (i. e., the MOF), transferring the charge throughout the structure of the MOF to react with CO_2 molecules and desorbing the intermediates and products off of the MOF surface [92]. Electrocatalytic reactions are normally limited by a high energy barrier between the reactants and intermediates, which means that the choice of the electrocatalyst is crucial in lowering this energy bar and inducing low energy intermediates [86]. Electrochemical and electrocatalytic conversion of CO_2 are two processes that are currently at the center stage for the generation of clean fuels and mitigation of greenhouse gases [93]. This process is carried-out by an up to 14 electron exchange process that takes place in two half equations as shown in Table 7.3.

In general, CER will occur at potentials more negative than the standard potentials listed in Table 7.3. The difference between the reaction potential and the standard one is overpotential. Ideally, it is desirable to minimize the overpotential since it is related to two aspects: (i) the high potential (sluggish kinetics) to form bent CO_2^- and (ii) the energy consumption of the electrolyte.

It is worth mentioning that CUSs in MOFs act as Lewis active sites and are mainly responsible for the CER. Additionally, the presence of homogenously segregated active sites on the molecular- and atomic-scale derived from the inorganic and organic units in MOFs have made MOFs suitable candidates for electrocatalytic conversion of CO_2 [19, 76, 78, 81–83]. Moreover, MOFs have a confining effect in that they limit evolution

Table 7.3: Standard electrochemical potentials for CO_2 reduction [93].

CO_2 reduction reaction	Potential vs. NHE
$CO_2 + 2H^+ + 2e^- \rightarrow CO + H_2O$	−0.106
$2CO_2 + 2H^+ + 2e^- \rightarrow H_2C_2O_4$	−0.500
$CO_2 + 2H^+ + 2e^- \rightarrow HCOOH + H_2O$	−0.250
$CO_2 + 4H^+ + 4e^- \rightarrow CH_2O + 2H_2O$	−0.070
$CO_2 + 4H^+ + 4e^- \rightarrow C + 2H_2O$	0.210
$CO_2 + 6H^+ + 6e^- \rightarrow CH_3OH + H_2O$	0.016
$CO_2 + 8H^+ + 8e^- \rightarrow CH_4 + 2H_2O$	0.169
$CO_2 + 12H^+ + 12e^- \rightarrow C_2H_4 + 4H_2O$	0.064
$CO_2 + 14H^+ + 14e^- \rightarrow C_2H_6 + 4H_2O$	0.084

of the hydrogen reaction (i. e., HER) while enhancing selectivity toward CO_2 reduction products [32]. Many reports have demonstrated increased efficiency in catalytic conversion of CO_2 into specific products while limiting HER. For example, MIL-53(Al) increased the efficiency by 40 % due to the confinement effect of Al [47]. To capitalize on the confinement effect for increasing the catalytic efficiency in MOFs, electron donating linkers have been used to dope various MOFs. This increases the electron transfer kinetics on the surface and active centers of the MOFs [19, 83, 84, 87].

In terms of designing a MOF structure suitable for CER, metal sites play a significant role. It is ideal that the metal active centers in the MOF have a high binding energy with the adsorbed CO_2 molecules and/or intermediates, but not so strong such that intermediates, or products can easily be desorbed. Due to the requirement that charge-carriers need to be mobilized straightforwardly, MOF electrocatalysts should also be electrically conductive. It is obvious that MOFs are widely composed of carboxylate-metal linkages, so the electrical conductivity of MOFs is typically not high enough to meet the requirements of CER. Therefore, designing MOF structures that are electrically conductive is imperative to enhance CER performance [97].

The design of conductive MOFs, as mentioned above, relies on careful selections of metal SBUs and organic linking units. It is important to enhance the accessibility of CO_2 molecules to the open framework—more active sites can then be fully exploited with the more accessible sites being better. To this end, a 3D structure of metal-catecholate was synthesized by Yaghi et al. [98]. The combination of organic linkers bearing a high charge transfer feature (cobalt phthalocyanine-2,3,9,10,16,17,23,24-octaol, CoPc) and single metal sites acting as accessible active sites in an anionic 3D MOF formulated as $[Fe_6(OH_2)_4(CoPc)_3]^{6-}$ (MOF-1992, Figure 7.11a) generated a satisfactory electrocatalyst. MOF-1992 employed on the cathode was able to reduce CO_2 to CO in water at a current density of −16.5 mA (Figure 7.11b). The overpotential was −0.52 V and the Faradaic efficiency (FE) was 80 %. In terms of stability, MOF-1992 could electrocatalyze the CO_2 reduction for 6 h, reaching a TON of 5800 [98].

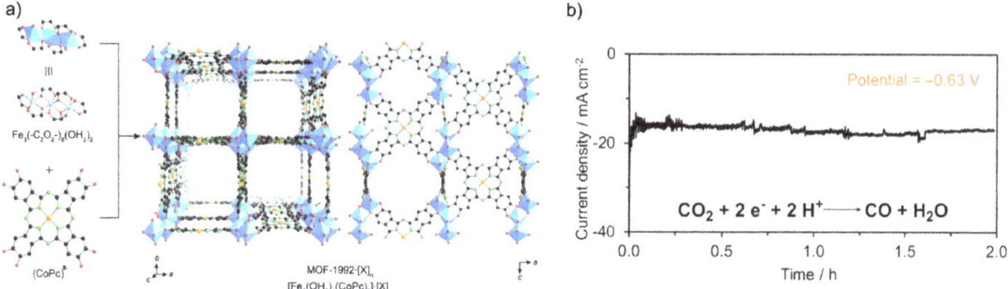

Figure 7.11: Crystal structure of MOF-1992 displaying 3D interconnected pores, which increase the accessible active metal sites (a) and the chronoamperometry at a fixed potential of −0.63 V vs. reversible hydrogen electrode (RHE) of MOF-1992 (b). Atom colors: Fe, blue polyhedra; Co, orange; C, grey; O, red and N, green. H atoms are omitted for clarity. Reproduced with permission from [98]. Copyright 2019, American Chemical Society.

Apart from the number of active sites, the intrinsic catalytic property of an individual active center is crucial in determining the high performance of MOF electrocatalysts. This is the reason why enhancing the number of accessible active sites as described above in the case of MOF-1992 plays an important role in enhancing the electrocatalytic CO_2 reduction. Additionally, nanostructures of MOFs having large surface area and a highly conjugated backbone that facilitates charge transfer is the target of structure designs [90]. The active sites of MOFs that are embedded on the linkers to form MOF electrocatalysts can be Co, Fe and Cu. As reported by Yang et al. [99], a thin-film porphyrin-based MOF, formulated as $Al_2(OH)_2$TCPP-Co and comprised of Co-TCPP linkers acting as the electrocatalytic entities, was synthesized and was demonstrated to achieve good CER activity. CO_2 was reduced into CO with a FE of 76 % selectivity and TON of 1400 at −0.7 V vs reversible hydrogen electrode (RHE). By employing TCPP-based linkers to construct MOF structures, researchers also found high electrocatalytic activity for Fe_MOF-525 [100], PCN-222(Fe) [101] and NNU-15 [100]. Particularly, NNU-15, with Co active sites coordinating with OH^- and [5,10,15,20-tetra(4-imidazol-1-yl)phenyl]porphyrin linker, was able to electrocatalyze CO_2 to CO with an excellent FE of 99.2 % at −0.6 V vs. RHE. The activity of these MOF electrocatalysts is attributed to the metal-based porphyrin-like linking units. In other words, the organic building blocks in MOFs can be modified to enhance CER activity of the resulting MOFs. This modification will also help create the catalytic sites even though the host MOF is inactive for CER. To this end, cocatalysts can be incorporated into MOF structures as reported by Lin et al. [102], Sun et al. [89] and others [89–100].

ZIF-8, for example, can be used as an electrocatalytic host to convert CO_2 to CO and its performance was heavily boosted by doping the MOF with 1,10-phenanthroline (ZIF-A-LD) [105]. This, according to the authors, is attributed to improved charge transfer as well as formation of *COOH with the help of the donor linker. The result was even better when doped with other electron-donating fillers to form (ZIF-A-LD/CB) [105]. Using a Schiff base reaction, Yang et al. were able to incorporate Co(II) phthalocyanine into ZIF-90 to form CoTAPc–ZIF-90 [106] with subsequent tests showing superior electron transfer and ultimately many folds improvement in electrocatalysis of CO_2 [106]. Yan et al. exploited the catalytic properties of Cu_2O by incorporating it in Cu-MOF to form Cu_2O@Cu-MOF [106]. Electrocatalytic studies on the MOF for different reaction products revealed the synergistic effect of combining two catalytically active systems with an overall FE of 76 % and a FE of 63 % for CH_4 [106]. Furthermore, Albo et al. [107] combined two MOFs (HKUST-1 and CAU-17) using porous carbon as a support for electrocatalytic conversion of CO_2 to different alcohols [107]. Their findings further proved the nexus between MOF composition and the level of alcohol production as well electron charge density with selectivity [107]. There are, however, areas of interest such as reaction mechanism and the role of metal centers in the reduction of CO_2 along with the effect of electron confinement in MOFs [100–105]. After all these investigations, it is of paramount necessity to test the performance of electrocatalytic MOFs when included in membranes for realizing continuous gas-phase CO_2 reduction processes [81, 82, 84, 98, 108]. Finally, electron-donating and -receiving centers were both incorporated into MOF structures composed of Zn-ε-keggin clusters (ε-$PMo_8VMo_4VIO_{40}Zn_4$) with M-TCPP linkers (M = Co, Fe, Ni and Zn) (Figure 7.12) [109]. The uniquely oriented electronic transportation channel of these PMOFs, formed by polyoxometalate and metalloporphyrin units, is a critical factor for facilitating electron transfer, thus enhancing CER performance. Co-PMOF was the best electrocatalyst, reducing CO_2 to CO with an excellent FE (98.7 %) and a high TOF of 1656 h^{-1} at −0.8 V vs. RHE.

7.4.4 MOF composites in photocatalytic and electrocatalytic conversion of CO_2

MOFs in conjunction with other materials have showed encouraging performance in the conversion of CO_2 through increase in charge separation. With MOF composites formed by coupling MOFs with different semiconductors, the charge separation can even be more efficient [19, 78, 79, 82]. For instance, TiO_2/MOF composites have been studied extensively in photocatalysis and electrocatalysis. TiO_2/HKUST-1 is one such example whereby a composite MOF was created by introducing TiO_2 into HKUST-1 [110]. While the surface area of the MOF was decreased, the loading capacity that translated into gas adsorption capacity was increased. Another feature of the composite system was the inhibition of electron-hole recombination [110]. Furthermore, TiO_2/ZIF-8 composite has also been reported and compared to the parent, pristine ZIF-8 [111]. TiO_2/ZIF-8 exhibited a two-fold higher in catalytic conversion of CO_2 while maintaining a lower energy penalty. Recently, TiO_2 was loaded into the mesopore of MIL-101 and functionalized MIL-101 [112]. The sample with 42 % TiO_2 in MIL-101-Cr-NO_2 presented excellent CO_2 photoreduction activity. Specifically, the CO_2 photoreduction was conducted without the need for a sacrificial agent and CO was produced with a high production rate of 12 mmol g^{-1} h^{-1}.

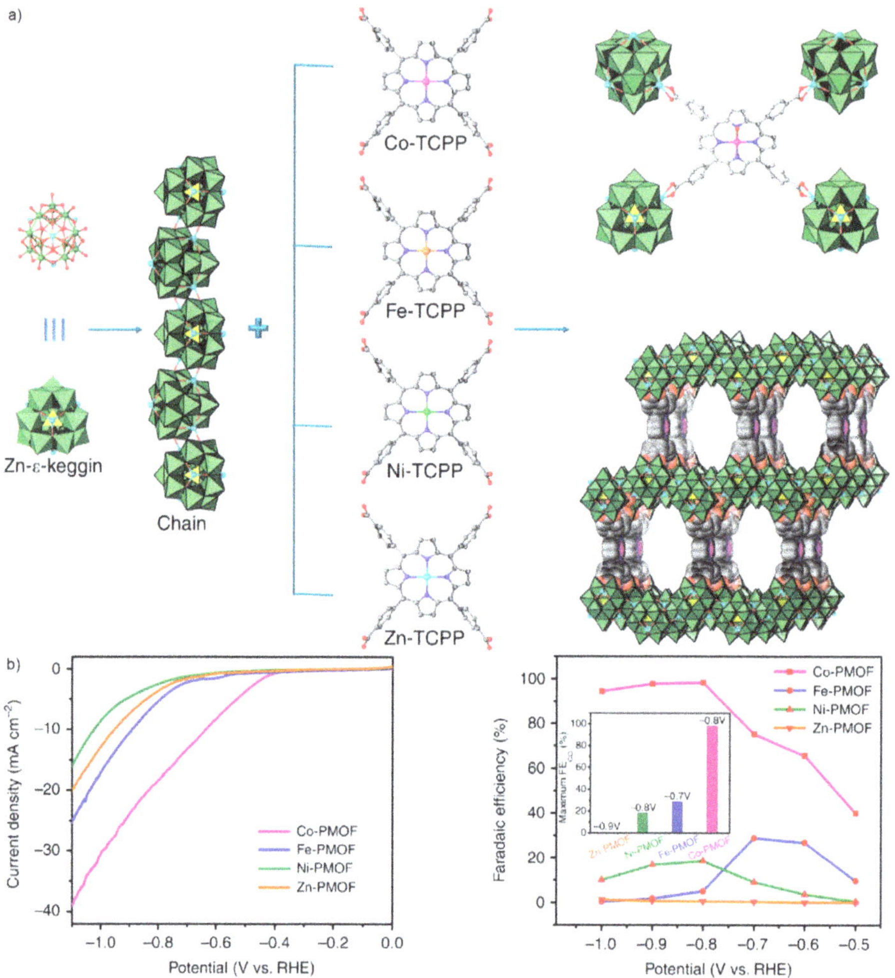

Figure 7.12: Strategy to construct excellent M-PMOF electrocatalysts (M = Co, Fe, Ni and Zn) adopting Zn-ε-keggin clusters and M-TCPP entities (a). Electrocatalytic property of M-PMOFs is characterized by current density and FE (b), which showed that Co-PMOF was the best among the isoreticular structures reported in this study. Atom colors: Zn-ε-keggin clusters, green polyhedra; Zn, light blue; Ni, bright green; Fe, orange; Co, pink; C, grey; O, red and N, blue. H atoms omitted for clarity.

Loading MOFs with novel metals results in a decrease of bandgap and concomitant enhancement of electron charge separation [111]. The photocatalytic activity of M-doped NH_2-MIL-125(Ti) MOFs, as a representative example, displayed a 62 % improvement in the synthesis of $HCOO^-$ and H_2 formation from CO_2 when doped with Au or Pt under visible light irradiation [111]. UiO-66 has been doped with carbon nitride nanosheets through self-assembly under visible light and their subsequent deployment in the conversion of CO_2 to CO was found to outperform the parent, pristine UiO-66 [111]. Other materials of high virtue in electro and photocatalysis are derived from MOF-electrode conjugation to increase FE. For example, a Re-based MOF and its highly oriented thin film was deposited onto a conductive fluorine-doped tin oxide (FTO) electrode via a chemical reaction method between tin oxide and hydrofluoric acid. The resulting composite afforded a superior FE of over 80 % with exceedingly high selectivity for the reduction of CO_2 to HCOOH [111]. Depositing porphyrin-based MOF-525 onto FTO glass has been employed in the electrochemical catalysis of CO_2 into a mixture of CO and H_2 with superior catalytic results [111].

7.4.5 Production cost of MOF-based systems for CO_2 capture and conversion

It is not currently feasible to commercialize MOF-based technology for capture of CO_2 due to their excessive cost of production even though designs to lower the cost of electricity by improving the hydrothermal MOF synthesis and increase CO_2 capture ability has been achieved to some extent [24, 25, 87]. Simply put, the overall cost, which includes MOF production, gas sorption and MOF regeneration, currently renders MOF materials economically unviable. In comparison with conventional adsorbents such as zeolites, the cost of raw materials for MOFs is high and this is attributed to a dearth of manufacturing industries for the nascent field of MOFs. Use of organic solvents for the solvothermal MOF synthesis increases the cost by close to 83 % and more research is needed to improve the viability of using aqueous and assisted liquid grinding routed in the synthesis. Preliminary findings show that these novel routes would reduce the production costs by 34–83 % [25]. On a positive note, MOFs have a great advantage over other materials, such as alkanolamines, by having a lower heat capacity value. Heat capacity is critical in the regeneration of adsorbent materials and, in the past, alkanolamines were deemed superior due to effective CO_2 capture properties yet were limited by their high heat capacity. For instance, the heat capacity of MOF-177 is significantly lower than most alkanolamines [113]. Alkanolamines when dissolved in water must be heated to elevated temperatures to desorb CO_2 because alkanolamine heat capacity values at room temperature are close to that of pure water and are multiple times higher than that of MOFs. Weime et al. reported a linear correlation between heat capacity and temperature and MOFs still possess a lower heat capacity and offer great relief in terms of energy costs [84]. One other important aspect to consider is the enthalpy of adsorption and research has so far revealed that due to the presence of CUSs in MOFs, a higher enthalpy of adsorption is expected as is evidenced by Mg-MOF-74. To minimize the impact of high adsorption enthalpy, Queen et al. studied analogs of Mg-MOF-74 to control the interaction between CO_2 and the metal sites [84]. Their findings showed that analogs of Zn, Fe-, Mn- and Co-MOF-74 are dynamically suited to CO_2 sorption compared to Mg-MOF-74 [84]. Partial substitution and impregnation of the MOF with mono- or single-ended amines to saturate metal sites is another proven method for lowering enthalpy of adsorption in each MOF [84]. It is therefore necessary to couple molecular characteristics with process economics to develop a tenable alternative to existing synthesis methods and MOF materials.

7.5 Conclusion

This chapter focused on fundamental MOF chemistry by which MOF materials can be used as platforms (pristine and modified structures) for CO_2 capture and conversion into various value-added products. It is obvious that MOFs hold great promise for selectively capturing CO_2 from flue gas with additional potential for being upgraded further for direct air capture of CO_2. Additionally, given that these materials bear porous, rigid and stable structures that can also be engineered to incorporate active entities, MOFs are maturing as photo- or electrocatalysts for CO_2 conversion. Although significant efforts are needed for enhancing the selectivity and production yield, MOFs remain of the most interesting class of materials for the reduction of CO_2.

Bibliography

[1] Alami AH, Abu Hawili A, Tawalbeh M et al. Materials and logistics for carbon dioxide capture, storage and utilization. Sci Total Environ. 2020;717:137221.
[2] Nazir G, Rehman A, Park SJ. Sustainable N-doped hierarchical porous carbons as efficient CO2 adsorbents and high-performance supercapacitor electrodes. J CO_2 Util. 2020;42:101326.
[3] Omer RM, Omer RM, Al-Tikrity ETB, El-Hiti GA et al. Porous aromatic melamine Schiff bases as highly efficient media for carbon dioxide storage. Processes. 2020;8:17.
[4] Scott V, Gilfillan S, Markusson N et al. Last chance for carbon capture and storage. Nat Clim Change. 2013;3:105–11.
[5] Mohamed MG, Zhang X, Mansoure TH et al. Hypercrosslinked porous organic polymers based on tetraphenylanthraquinone for CO_2 uptake and high-performance supercapacitor. Polymer. 2020;205:122857.
[6] Pal A, Uddin K, Saha BB et al. A benchmark for CO_2 uptake onto newly synthesized biomass-derived activated carbons. Appl Energy. 2020;264:114720.
[7] Song X, Peng C, Ciais P et al. Nitrogen addition increased CO_2 uptake more than non-CO_2 greenhouse gases emissions in a Moso bamboo forest. Sci Adv. 2020;6:eaaw5790.
[8] Li Y, Li W, Cheng Z et al. Improved CO_2 uptake and supercapacitive energy storage using heteroatom-rich porous carbons derived from conjugated microporous polyaminoanthraquinone networks. ChemNanoMat. 2020;6:58–63.
[9] Mutyala S, Jonnalagadda M, Mitta H, Gundeboyina R. CO_2 capture and adsorption kinetic study of amine-modified MIL-101(Cr). Chem Eng Res Des. 2019;143:241–8.
[10] Li J, Zhao D, Liu J et al. Covalent organic frameworks: a promising materials platform for photocatalytic CO_2 reductions. Molecules. 2020;25:2425.
[11] Huang X, Zhang YB. Reticular materials for electrochemical reduction of CO_2. Coord Chem Rev. 2021;427:213564.

[12] Senftle TP, Carter EA. The holy grail: chemistry enabling an economically viable CO_2 capture, utilization, and storage strategy. Acc Chem Res. 2017;50:472–5.
[13] Chowdhury FA, Goto K, Yamada H et al. A screening study of alcohol solvents for alkanolamine-based CO_2 capture. Int J Greenh Gas Control. 2020;99:103081.
[14] Sumida K, Rogow DL, Mason JA et al. Carbon dioxide capture in metal–organic frameworks. Chem Rev. 2012;112:724–81.
[15] Ma X, Yang Y, Wu Q et al. Underlying CO_2 mechanism of CO_2 uptake onto biomass-based porous carbons: Do adsorbents capture chiefly through narrow micropores. Fuel. 2020;282:118727.
[16] Bakhtyari A, Mofarahi M, Lee C-H. CO_2 adsorption by conventional and nanosized zeolites. Elsevier Inc. 2020;00:193–228.
[17] Choi HJ, Jo D, Min JG, Hong SB. The origin of selective adsorption of CO2 on merlinoite zeolites. Angew Chem Int Ed. 2021;60:4307–14.
[18] Kumar S, Srivastava R, Koh J. Utilization of zeolites as CO_2 capturing agents: advances and future perspectives. J CO_2 Util. 2020;41:101251.
[19] Su F, Lu C, Kuo SC, Zeng W. Adsorption of CO_2 on amine-functionalized y-type zeolites. Energy Fuels. 2010;24:1441–8.
[20] Cai Z, Bien CE, Liu Q, Wade CR. Insights into CO_2 adsorption in M-OH functionalized MOFs. Chem Mater. 2020;32:4257–64.
[21] Hou XJ, He P, Li H, Wang X. Understanding the adsorption mechanism of C_2H_2, CO_2, and CH_4 in isostructural metal-organic frameworks with coordinatively unsaturated metal sites. J Phys Chem C. 2013;117:2824–34.
[22] Yang LM, Zhang H, Pan H, Ganz E. Atomistic level mechanism of CO_2 adsorption in N-ethylethylenediamine-functionalized M_2(dobpdc) metal-organic frameworks. Cryst Growth Des. 2020;20:6337–45.
[23] Trickett C, Helal A, Al-Maythalony B et al. The chemistry of metal–organic frameworks for CO_2 capture, regeneration and conversion. Nat Rev Mater. 2017;2:17045.
[24] Ge X, Ma S. CO_2 capture and separation of metal–organic frameworks. Mater Carbon Capture. 2020;2050:5–27.
[25] Younas M, Younas M, Rezakazemi M, Daud M et al. Recent progress and remaining challenges in post-combustion CO_2 capture using metal-organic frameworks (MOFs). Prog Energy Combust Sci. 2020;80:100849.
[26] Asgari M, Semino R, Schouwink PA et al. Understanding how ligand functionalization influences CO_2 and N_2 adsorption in a sodalite metal-organic framework. Chem Mater. 2020;32:1526–36.
[27] Li Z, Zhu Q. MOF-based materials for photo- and electrocatalytic CO_2 reduction. EnergyChem. 2020;00:100033.
[28] Kökçam-Demir Ü, Goldman A, Esrafili L et al. Coordinatively unsaturated metal sites (open metal sites) in metal-organic frameworks: design and applications. Chem Soc Rev. 2020;49:2751–98.
[29] Bae YS, Farha OK, Spokoyny AM et al. Carborane-based metal-organic frameworks as highly selective sorbents for CO_2 over methane. Chem Commun. 2008;00:4135–7.
[30] Rajagopalan AK, Rajendran A. The effect of nitrogen adsorption on vacuum swing adsorption based post-combustion CO2 capture. Int J Greenh Gas Control. 2018;78:437–47.
[31] Herm ZR, Swisher JA, Smit B, Krishna R, Long JR. Metal-organic frameworks as adsorbents for hydrogen purification and precombustion carbon dioxide capture. J Am Chem Soc. 2011;133:5664–7.
[32] Zou R, Guo W et al. A new microporous metal-organic framework with a novel trinuclear nickel cluster for selective CO_2 adsorption. Inorg Chem Commun. 2019;104:78–82.

[33] Liao M, Wei M, Mo JP et al. Acidity and Cd^{2+} fluorescent sensor and selective CO_2 adsorption by a water-stable Eu-MOF. Dalton Trans. 2019;48:4489–94.
[34] Bloch ED, Bloch ED, Britt D, Lee C et al. Metal insertion in a microporous metal-organic framework lined with 2, 2′-bipyridine. J Am Chem Soc. 2010;132:14382–4.
[35] Zheng B, Bai J, Duan J et al. Enhanced CO_2 binding affinity of a high-uptake rht-type metal-organic framework decorated with acylamide groups. J Am Chem Soc. 2011;133:748–51.
[36] Wang R, Mi JS, Dong XY et al. Creating a polar surface in carbon frameworks from single-source metal-organic frameworks for advanced CO_2 uptake and lithium-sulfur batteries. Chem Mater. 2019;31:4258–66.
[37] Belmabkhout Y, Guillerm V, Eddaoudi M. Low concentration CO_2 capture using physical adsorbents: are metal-organic frameworks becoming the new benchmark materials. Chem Eng J. 2016;296:386–97.
[38] Schoedel A, Ji Z, Yaghi OM. The role of metal–organic frameworks in a carbon-neutral energy cycle. Nat Energy. 2016;1:16034.
[39] Nugent P, Giannopoulou EG, Burd SD et al. Porous materials with optimal adsorption thermodynamics and kinetics for CO_2 separation. Nature. 2013;495:80–4.
[40] Hu J, Liu Y, Liu J et al. High CO2 adsorption capacities in UiO type MOFs comprising heterocyclic ligand. Microporous Mesoporous Mater. 2018;256:25–31.
[41] Zhu AW, Zhang C, Li Q et al. Selective reduction of CO_2 by conductive MOF nanosheets as an efficient co-catalyst under visible light illumination. Appl Catal B, Environ. 2018;238:339–45.
[42] Wang Z, Han L, Gao X et al. Three Cd(II) MOFs with di ff erent functional groups: selective CO_2 capture and metal ions detection. Inorg. Chem. 2018;00:2–9.
[43] He M, Xu T, Jiang Z et al. Incorporation of bifunctional aminopyridine into anNbO-type MOF for the markedly enhanced adsorption of CO_2 and C_2H_2 over CH_4. Inorg Chem Front. 2019;6:1177–83.
[44] Elsaidi SK, Mohamed MH, Simon CM et al. Effect of ring rotation upon gas adsorption in SIFSIX-3-M (M = Fe, Ni) pillared square grid networks. Chem Sci. 2017;8:2373–80.
[45] Guillerm V, Xu H, Albalad J et al. Postsynthetic selective ligand cleavage by solid-gas phase ozonolysis fuses micropores into mesopores in metal-organic frameworks. J Am Chem Soc. 2018;140:15022–30.
[46] Peng Y, Li Y, Ban Y et al. Metal-organic framework nanosheets as building blocks for molecular sieving membranes. Science. 2014;346:1356–9.
[47] Couck S, Denayer JFM, Baron GV et al. An amine-functionalized MIL-53 metal-organic framework with large separation power for CO_2 and CH_4. J Am Chem Soc. 2009;131:6326–7.
[48] Liu H, Zhao Y, Zhang Z et al. The effect of methyl functionalization on microporous metal-organic frameworks' capacity and binding energy for carbon dioxide adsorption. Adv Funct Mater. 2011;21:4754–62.
[49] Sanz-pérez ES, Dantas TCM, Arencibia A et al. Reuse and recycling of amine-functionalized silica materials for CO_2 adsorption. J Chem Eng. 2017;308:1021–33.
[50] Cao C, Shi Y, Xu H et al. A multifunctional MOFs as recyclable catalyst for fixation of CO_2 with aziridines or epoxides and luminescent probe of Cr(VI). Dalton Trans. 2018;47:4545–53.
[51] Venkata C, Raghava K, Harish VVN et al. Metal-organic frameworks (MOFs)-based efficient heterogeneous photocatalysts: synthesis, properties and its applications in photocatalytic hydrogen generation, CO_2 reduction and photodegradation of organic dyes. Int J Hydrog Energy. 2019;45:7656–79.
[52] Wang C, Wang X, Liu W. The synthesis strategies and photocatalytic performances of TiO_2/MOFs composites. Chem Eng J. 2019;391:123601.
[53] Li D, Kassymova M, Cai X et al. Photocatalytic CO_2 reduction over metal-organic framework-based materials. Coord Chem Rev. 2020;412:213262.

[54] Kong X, He T, Zhou J et al. In situ porphyrin substitution in a Zr(IV)-MOF for stability enhancement and photocatalytic CO_2 reduction. Small. 2021;17:2005357.
[55] Xu D, Liu W, Liu J et al. Recent advances in MOF-based nanocatalysts for photo-promoted CO_2 reduction applications. Catalysis. 2019;9:658.
[56] Alkhatib II, Garlisi C, Pagliaro M et al. Metal-organic frameworks for photocatalytic CO2 reduction under visible radiation: a review of strategies and applications. Catal Today. 2018;340:209–24.
[57] Mazari SA, Hossain N, Basirun WJ et al. Overview of catalytic conversion of CO_2 into fuels and chemicals using metal organic frameworks. Process Saf Environ Prot. 2021;146:67–92.
[58] Qin J, Xu P, Huang Y et al. High loading of Mn(ii)-metalated porphyrin in a MOF for photocatalytic CO_2 reduction in gas–solid conditions. Chem Commun. 2021;57:8468–71.
[59] Dan-Hardi M, Serre C, Frot T et al. A new photoactive crystalline highly porous titanium(IV) dicarboxylate. J Am Chem Soc. 2009;131:10857–9.
[60] Fu Y, Sun D, Chen Y et al. An amine-functionalized titanium metal–organic framework photocatalyst with visible-light-induced activity for CO_2 reduction. Angew Chem. 2012;125:3364–7.
[61] Wang D, Huang R, Liu W et al. Fe-based MOFs for photocatalytic CO2 reduction: role of coordination unsaturated sites and dual excitation pathways. ACS Catal. 2014;4:4254–60.
[62] Sun D, Fu Y, Liu W et al. Derivatives: towards a better understanding of photocatalysis on metal–organic frameworks. Chem Eur J. 2013;19:14279–85.
[63] Silva CG, Luz I, Llabrés I Xamena FX et al. Water stable Zr-benzenedicarboxylate metal-organic frameworks as photocatalysts for hydrogen generation. Chem - A Eur J. 2010;16:11133–8.
[64] Chen D, Xing H, Wang C, Su Z. Highly efficient visible-light-driven CO_2 reduction to formate by a new anthracene-based zirconium MOF via dual catalytic routes. J Mater Chem A. 2016;4:2657–62.
[65] Nguyen HL. Reticular materials for artificial photoreduction of CO_2. Adv Energy Mater. 2020;10:2002091.
[66] Thompson WA, Sanchez Fernandez E, Maroto-Valer MM. Review and analysis of CO_2 photoreduction kinetics. ACS Sustain Chem Eng. 2020;8:4677–92.
[67] Deng X, Qin Y, Hao M et al. MOF-253-supported Ru complex for photocatalytic CO_2 reduction by coupling with semidehydrogenation of 1,2,3,4-tetrahydroisoquinoline (THIQ). Inorg Chem. 2019;58:16574–80.
[68] Wu L, Mu Y, Guo X et al. Encapsulating perovskite quantum dots in iron-based metal–organic frameworks (MOFs) for efficient photocatalytic CO_2 reduction. Angew Chem. 2019;58:9491–5.
[69] Xu H, Hu J, Wang D et al. Visible-light photoreduction of CO_2 in a metal-organic framework: boosting electron-hole separation via electron trap states. J Am Chem Soc. 2015;137:13440–3.
[70] Liu J, Fan YZ, Li X et al. A porous rhodium(III)-porphyrin metal-organic framework as an efficient and selective photocatalyst for CO_2 reduction. Appl Catal B, Environ. 2018;231:173–81.
[71] Chen EX, Qiu M, Zhang YF et al. Acid and base resistant zirconium polyphenolate-metalloporphyrin scaffolds for efficient CO2 photoreduction. Adv Mater. 2018;30:1704388.
[72] Fang ZB, Liu TT, Liu J et al. Boosting interfacial charge-transfer kinetics for efficient overall CO_2 photoreduction via rational design of coordination spheres on metal-organic frameworks. J Am Chem Soc. 2020;142:12515–23.
[73] Wang C, Xie Z, Kathryn E et al. Doping metal-organic frameworks for water oxidation, carbon dioxide reduction, and organic photocatalysis. J Am Chem Soc. 2011;133:13445–54.
[74] Sun M, Yan S, Sun Y et al. Enhancement of visible-light-driven CO_2 reduction performance using an amine-functionalized zirconium metal–organic framework. Dalton Trans. 2018;47:909–15.

[75] Choi K, Kim D, Rungtaweevoranit B et al. Plasmon-enhanced photocatalytic CO conversion within metal-organic frameworks under visible light plasmon-enhanced photocatalytic CO_2 conversion within metal-organic frameworks under visible light. J Am Chem Soc. 2017;139:356–62.
[76] Lee Y, Kim S, Fei H et al. Photocatalytic CO_2 reduction using visible light by metal-monocatecholato species in a metal-organic framework. Chem Commun. 2015;51:16549–52.
[77] Zhang H, Wei J, Dong J et al. Efficient visible-light-driven carbon dioxide reduction by a single-atom implanted metal–organic framework. Angew Chem. 2016;128:14310–4.
[78] Wang S, Wang X. Environmental photocatalytic CO_2 reduction by CdS promoted with a zeolitic imidazolate framework. Appl Catal B, Environ. 2015;162:494–500.
[79] Sadeghi N, Sharifnia S, Arabi MS. A porphyrin-based metal organic framework for high rate photoreduction of CO_2 to CH_4 in gas phase. Biochem Pharmacol. 2016;16:450–7.
[80] Yan SC, Ouyang SX, Gao J et al. A room-temperature reactive-template route to mesoporous ZnGa2 O4 with improved photocatalytic activity in reduction of CO_2. Angew Chem Int Ed. 2010;122:6400–4.
[81] Maina JW, Schutz J, Grundy L et al. Inorganic nanoparticles/metal organic framework hybrid membrane reactors for efficient photocatalytic conversion of CO_2. ACS Appl Mater Interfaces. 2017;9:35010–7.
[82] Liao K, Yu J, Xu Y. Core shell CsPbBr3 zeolitic imidazolate framework nanocomposite for efficient photocatalytic CO_2 reduction. ACS Energy Lett. 2018;3(11):2656–62.
[83] Feng X, Pi Y, Song Y et al. Metal-organic frameworks significantly enhance photocatalytic hydrogen evolution and CO_2 reduction with Earth-abundant copper photosensitizers. J Am Chem Soc. 2020;142:690–5.
[84] Wieme J, Vandenbrande S, Lamaire A et al. Thermal engineering of metal–organic frameworks for adsorption applications: a molecular simulation perspective. ACS Appl Mater Interfaces. 2019;11:38697–707.
[85] Zhang W, Hu Y, Ma L et al. Progress and perspective of electrocatalytic CO_2 reduction for renewable carbonaceous fuels and chemicals. Adv Sci. 2018;5:1700275.
[86] Zhang L, Zhao ZJ, Gong J. Nanostructured materials for heterogeneous electrocatalytic CO_2 reduction and their related reaction mechanisms. Angew Chem Int Ed. 2017;56:11326–53.
[87] Al-Omari AA, Yamani ZH, Nguyen HL. Electrocatalytic CO_2 reduction: from homogeneous catalysts to heterogeneous-based reticular chemistry. Molecules. 2018;23:2835.
[88] Adegoke KA, Maxakato NW. Porous metal–organic framework (MOF)-based and MOF-derived electrocatalytic materials for energy conversion. Mater Today Energy. 2021;21:100816.
[89] Jinxuan Y, Yan G, Chenghuan G et al. Highly oriented MOF thin film-based electrocatalytic device for the reduction of CO_2 to CO exhibiting high faradic efficiency. J Mater Chem A. 2016;4:15320–6.
[90] Pettinari C, Tombesi A. Metal–organic frameworks for chemical conversion of carbon dioxide. MRS Energy Sustain. 2020;7:31.
[91] Shoaib S, Shah A, Najam T et al. Metal-organic framework-based electrocatalysts for CO_2 reduction. Small Struct. 2021;80:2100090.
[92] Mahmood A, Guo W, Tabassum H et al. Metal-organic framework-based nanomaterials for electrocatalysis. Adv Energy Mater. 2016;61:600423.
[93] Elhenawy M, Khraisheh M, AlMomani F et al. Metal-organic frameworks as a platform for CO_2 capture and chemical processes: adsorption, membrane separation, catalytic-conversion, and electrochemical reduction of CO_2. Catalysts. 2020;10:1293.
[94] Cui Q, Qin G, Wang W et al. Titania-modified silver electrocatalyst for selective CO_2 reduction to CH_3OH and CH_4 from DFT study. J Phys Chem C. 2017;121:16275–82.

[95] Zhao Y, Zheng L, Jiang D et al. Nanoengineering metal–organic framework-based materials for use in electrochemical CO_2 reduction reactions. Small. 2021;17:2006590.
[96] Zhang X, Zhang Y, Li Q et al. Highly efficient and durable aqueous electrocatalytic reduction of CO_2 to HCOOH with a novel Bismuth-MOF: experimental and DFT study. J Mater Chem A. 2020;8:9776–87.
[97] Yi J, Si D, Xie R et al. Conductive two-dimensional phthalocyanine-based metal–organic framework nanosheets for efficient electroreduction of CO_2. Angew Chem. 2021;133:17245–51.
[98] Matheu R, Gutierrez-Puebla E, Monge MA et al. Three-dimensional phthalocyanine metal-catecholates for high electrochemical carbon dioxide reduction. J Am Chem Soc. 2019;141:17081–5.
[99] Kornienko N, Zhao Y, Kley CS et al. Metal-organic frameworks for electrocatalytic reduction of carbon dioxide. J Am Chem Soc. 2015;137:14129–35.
[100] Hod I, Sampson MD, Deria P et al. Fe-porphyrin-based metal-organic framework films as high-surface concentration, heterogeneous catalysts for electrochemical reduction of CO_2. ACS Catal. 2015;5:6302–9.
[101] Dong BX, Qian SL, Bu FY et al. Electrochemical reduction of CO_2 to CO by a heterogeneous catalyst of Fe-porphyrin-based metal-organic framework. ACS Appl Energy Mater. 2018;1:4662–9.
[102] Guo Y, Shi W, Yang H et al. Cooperative stabilization of the [pyridinium-CO_2-Co] adduct on a metal-organic layer enhances electrocatalytic CO_2 reduction. J Am Chem Soc. 2019;141:17875–83.
[103] Zhou Y, Liu S, Gu Y, et al. In(III) Metal-Organic Framework Incorporated with Enzyme-Mimicking Nickel Bis(dithiolene) Ligand for Highly Selective CO_2 Electroreduction. J Am Chem Soc. 2021;143:14071–6.
[104] Huang NY, He H, Liu S et al. Electrostatic attraction-driven assembly of a metal-organic framework with a photosensitizer boosts photocatalytic CO_2 reduction to CO. J Am Chem Soc. 2021;143:17424–30.
[105] Dou S, Song J, Xi S et al. Boosting electrochemical CO_2 reduction on metal–organic frameworks via ligand doping. Angew Chem Int Ed. 2019;58:4041–5.
[106] Yang Z, Zhang X, Long C et al. Covalently anchoring cobalt phthalocyanine on zeolitic imidazolate frameworks for efficient carbon dioxide electroreduction. CrystEngComm. 2020;22:1619–24.
[107] Albo J, Perfecto-Irigaray M, Beobide G, Irabien A. Cu/Bi metal-organic framework-based systems for an enhanced electrochemical transformation of CO_2 to alcohols. J CO_2 Util. 2019;33:157–65.
[108] Li J, Wang Y, Chen Y et al. Metal–organic frameworks toward electrocatalytic applications. Appl Sci. 2019;9:2427.
[109] Wang YR, Huang Q, He CT et al. Oriented electron transmission in polyoxometalate-metalloporphyrin organic framework for highly selective electroreduction of CO_2. Nat Commun. 2018;9:4466.
[110] Di Credico B, Redaelli M, Bellardita M et al. Step-by-step growth of HKUST-1 on functionalized TiO_2 surface: an efficient material for CO_2 capture and solar photoreduction. Catalysts. 2018;8:353.
[111] Zou YH, Wang HN, Meng X et al. Self-assembly of TiO_2/ZIF-8 nanocomposites for varied photocatalytic CO_2 reduction with H_2O vapor induced by different synthetic methods. Nanoscale Adv. 2021;3:1455–63.
[112] Jiang Z, Xu X, Ma Y et al. Filling metal–organic framework mesopores with TiO_2 for CO2 photoreduction. Nature. 2020;586:549–54.
[113] Mason JA, Sumida K, Herm ZR et al. Evaluating metal-organic frameworks for post-combustion carbon dioxide capture via temperature swing adsorption. Energy Environ Sci. 2011;4:3030–40.

Thach N. Tu, Long H. Ngo, and Tung T. Nguyen

8 Metal-organic frameworks as adsorbents for onboard fuel storage

8.1 Fundaments of adsorption

Excess, total and absolute adsorption

During the process of adsorption, gas molecules are attracted by the surface to accumulate at a higher density than that would normally be present at the same temperature, T, and pressure, P. For adsorption on a two-dimensional surface, the strength of the interaction between the gas and surface will depend on the distance between the surface and gas molecules and become negligible at a certain distance—the Gibbs dividing surface. Since there are no interactions between the surface and gas molecules, only bulk or free gas molecules are present beyond the Gibbs dividing surface and, therefore, divide the total free volume into adsorbed and bulk regions. The absolute amount adsorbed, n_{abs}, is defined simply as the total number of molecules that are in the adsorbed region. Since the Gibbs dividing surface cannot be determined experimentally, all adsorption measurements provide excess adsorption, n_{ex}, which is defined as the difference between the absolute adsorption amount and the amount of bulk gas that would have been present in the adsorbed region, V_a, in the absence of a surface [1],

$$n_{\text{ex}} = n_{abs} - V_a \rho_{\text{bulk}}(P, T) \tag{8.1}$$

Since V_a also cannot be determined experimentally, estimation of n_{abs} from the measured excess adsorption becomes impossible. This led to proposing an approximation estimation, the total adsorption, n_{tot}, which includes all gas molecules within the pores of an adsorbent. Total adsorption can be calculated from the excess adsorption using equation (8.2), in which total pore volume, V_p, is determined from the amount of the adsorbed N_2 at the condensed P/P_0 at 77 K [2],

$$n_{\text{tot}} = n_{\text{ex}} + V_p \rho_{\text{bulk}}(P, T) \tag{8.2}$$

Acknowledgement: This study was financially supported by the Alexander von Humboldt Foundation.

Thach N. Tu, Nguyen Tat Thanh University, 300A Nguyen Tat Thanh Street, District 4, Ho Chi Minh City 755414, Vietnam; and Faculty of Chemistry and Pharmacy, University of Regensburg, Universitätsstraße 31, 93053 Regensburg, Germany, e-mail: tnthach@ntt.edu.vn
Long H. Ngo, Tung T. Nguyen, NTT Hi-Tech Institute, Nguyen Tat Thanh University, Ho Chi Minh City, Vietnam

https://doi.org/10.1515/9781501524721-008

Working capacity is the usable amount of methane defined as the difference of methane uptake between maximum adsorption operational pressure and the minimum operational desorption pressure of the engine (e. g., 5 bar in the case of CH_4). In adsorption, the amount of uptake is affected by temperature. This means that the deliverable capacity can be varied by alternating the desorption temperature while maintaining the same minimum operational desorption pressure at 5 bar. Since working capacity is the primary factor to evaluate the performance of the adsorbent, designed materials should have a large total uptake at maximum adsorption operational pressure and much lower total uptake at minimum operational desorption pressure to maximize the amount of usable fuel [3].

Isosteric heats of adsorption (Q_{st}) is the average binding energy as a function of the amount of adsorbed guest molecules, n, at a specific surface coverage. Q_{st} is determined by using the Clausius–Clapeyron relation and is used commonly for analysis of the adsorbents. To calculate Q_{st} from this relation, the total adsorption data is often used as an approximation for absolute adsorption in heat of adsorption calculations [4],

$$-Q_{st} = RT^2 \left(\frac{\partial \ln P}{\partial T} \right)_n \tag{8.3}$$

8.2 Metal-organic frameworks for onboard natural gas storage

8.2.1 Natural gas for lower CO_2 emission

With an increasing use of fossil fuel needed to power the growth of the global economy, atmospheric carbon dioxide (CO_2) emissions are accelerating at an unprecedented rate. Specifically, atmospheric CO_2 from fossil fuel burning has reached a crisis level (>400 ppm) placing mankind in the difficult position of having to find a way to mitigate and adapt to environmental problems that arise from climate change. Consequently, there is an urgent need to develop alternative cleaner fuels that can replace traditional fossil fuels to lower the emission of CO_2 into the atmosphere.

Natural gas (NG), a type of earth-abundant fossil fuel mainly consisting of methane, represents an alternative/immediately deployable option due to the relatively lower CO, CO_2, SO_x and NO_x released per unit of energy when compared with other fossil fuels (e. g., gasoline) [3]. The utilization of NG for fuelling automobiles has been gaining widespread interest [5], however, the bottleneck preventing extensive use of the NG-based automobile engine is the low energy density of NG compared to that of gasoline (0.04 MJ L^{-1} *vs.* 32.4 MJ L^{-1}, respectively) [6]. To overcome this challenge, compressed natural gas (CNG) is used to store NG at ambient temperature, however, the system requires supercritical storage conditions (200 to 250 bar) to reach an energy density of 9 MJ L^{-1}, which is only 26 % that of gasoline. Furthermore, the cost of heavy and expensive storage vessels, multistage compression facilities and the raised concerns for risk of ignition make CNG less competitive with gasoline [2, 5, 7].

Adsorbed natural gas (ANG) has emerged as an alternative method to tackle the challenges encountered with CNG. By employing porous sorbent materials, this method allows NG to be stored at much lower pressures (below 100 bar) while maintaining a competitive storage capacity. Therefore, this technology reduces the risk of ignition and vessel costs by using lightweight, conformable fuel tanks that are permitted to be integrated into the limited space available within a small car [6]. The fast-paced development of ANG technology was further heightened by the U.S. Department of Energy's (DOE) 2012 research program, which targeted ambitious methane storage capacities of 350 cm^3 (STP) cm^{-3} or 0.5 g g^{-1} at room temperature [5].

8.2.2 The current status of metal-organic frameworks for methane storage

Metal-organic frameworks (MOFs) are a new type of highly porous material, whose structures are constructed from inorganic secondary building units (SBUs) and organic linkers being stitched together via strong coordinated bonds to form periodic frameworks [8]. Due to the modular nature of their design and synthesis, the pore size and chemical properties of MOFs can be tuned based on a targeted architecture [9, 10]. This has led to increasing interest in employing MOFs in catalysis [11], separations [12], storage [5, 13], proton conduction [14, 15] and controlled delivery of guest molecules [16, 17]. Among these applications, MOFs have been extensively used for methane storage (Table 8.1) [5, 13]. MOFs with coordinatively unsaturated metal sites (CUSs) in the structure, such as HKUST-1, Ni-MOF-74 and the **tbo/nbo**-type architecture MOF materials (i. e., PCN-14, NJU-Bai-43, NOTT-101a, UTSA-110a, UTSA-76, PCN-46a, ZJU-105a, UTSA-111a, NJU-Bai 19, Table 8.2), display high total volumetric methane storage capacities (Table 8.1). MOFs without CUSs, such as MOF-5, MOF-177, MOF-205 and MOF-210, with high surface areas and large pore sizes also have been shown to achieve satisfactory capacities at high pressures (Table 8.1). Although significant advances have been achieved in the field of methane storage by MOFs, the results are far less than the DOE 2012 target. Therefore, further work on the design and synthesis of MOFs that possess an optimized methane adsorption energy to achieve higher working capacities are highly sought. Furthermore, MOFs also suffer several limitations when applied for methane storage. For example, MOFs often experience a loss of capacity after compressing and have a high associated financial cost in their synthesis. All of these issues must be overcome before the practical NG storage using MOFs can be deemed practical.

Table 8.1: Total methane uptake and working capacity for the highest performing MOFs.

Material	Pore volume ($cm^3\ g^{-1}$)	BET surface area ($m^2\ g^{-1}$)[a]	Total uptake Volumetric ($cm^3\ cm^{-3}$)		Working capacity (5–65 bar) Volumetric ($cm^3\ cm^{-3}$)		Refs.
			35 bar	65 bar	35 bar	65 bar	
MOFs with coordinated unsaturated metal sites (CUS) in the structure							
Ni-MOF-74	0.51	1350	228	251	106	129	[18]
HKUST-1	0.78	1850	227	267	150	190	
Al-soc-MOF-1	2.3	5585	127	193	106	172	[19]
NU-1501-Al	2.90	7920	106	166	86	146	[20]
PCN-14	0.87	1753	195	230	122	157	[21]
NJU-Bai-43	1.22	3090	202	254	146	198	[22]
NOTT-101[b]	1.080	2805	194	237	138	181	[23]
NOTT-102[b]	1.268	3342	181	237	136	192	
UTSA-110[b]	1.263	3241	187	241	136	190	[24]
UTSA-111[b]	1.229	3252		234	129	183	[25]
NU-111	2.09	4930	138	206	111	179	[26]
UTSA-76	1.092	2820	211	257	151	197	[27]
MFM-115[b]	1.38	3394	186	238	138	191	[28]
MFM-112[b]	3800	1.62	162	125	218	181	
NJU-Bai-19	1.063	2803	205	246	144	185	[29]
PCN-46[b]	1.243	3224	184	233	141	190	[30]
ZJU-105[b]	1.037	2608	181	228	128	175	
MOFs without coordinated unsaturated metal sites (CUS) in the structure							
VNU-21	0.58	1440	142	182	101	140	[31]
MOF-5	1.38	3320	126	188	104	166	[32]
MOF-177	1.89	4500	122	193	102	175	
MOF-205	2.16	4460	120	183	101	164	
MOF-210	3.60	6240	82	143	70	131	
MOF-519	0.938	2400	200	260[a]	151	211[a]	[3]
MOF-520	1.277	3290	162	214	125	177	
MOF-905-NO_2	1.29	3380	132	185	107	160	[6]
MOF-905-Me_2	1.39	3640	138	192	111	165	
MOF-905	1.34	3490	145	206	120	175	
MOF-950	1.30	3440	145	195	109	159	
MAF-38	0.808	2022	226	263	150	187	[33]
Fe-ncb-ABDC	1.76	4475	–	197	–	172	[34]
ST-2	2.44	5172	117	184	98	165	[35]
MIL-53(Al)-OH	~0.68	1284	167	217	114	164	[36]
Co(bdp)	0.652	2911	161	203	–	197	[37]

[a] Brunauer–Emmett–Teller. [b] Poor reproducibility and likely overestimation due to the difficulty in controlling the composition [3].

8.2.3 Strategies for improving methane storage capacities of metal-organic frameworks

The pore geometry of adsorbents is an important factor in the adsorption of methane. For those porous materials having similar surface area, those with small pore size tend to interact stronger with guest molecules to enhance uptake capacity in the low-pressure range while the larger pore size tends to have weaker interactions and will exhibit a low uptake capacity at low pressure albeit with a better adsorption profile at higher pressure ranges. In the field of ANG storage, working capacity is the amount of methane storage capacity above 5 bar. Optimal materials typically possess a high surface area and suitable pore size to achieve high storage capacity at the maximum operating pressure while minimizing the adsorbed amount at a pressure below 5 bar. Previously, E. Wilmer et al. reported a computational approach to screen over 137,953 hypothetical MOFs for methane uptake at 35 bar and 298 K. This study concluded that the suitable pore size for methane adsorption in MOFs was between 4 and 8 Å, which is large enough to pack one or two methane molecules inside. For example, MAF-38, which possesses an ellipsis-shaped pore with a size of $9.0 \times 14.2\,Å^2$, Brunauer–Emmett–Teller (BET) surface area of $2022\,m^2\,g^{-1}$, and pore volume of $0.615\,cm^3\,cm^{-3}$ produced a storage (working capacity) of 226 (150), 263 (187) and 273 (197) cm^3(STP) cm^{-3} at 35, 65 and 80 bar and 298 K, respectively [33].

While the ideal pore size being between 4 and 8 Å is perhaps correct to improve the uptake capacity at low pressure, optimization criteria to maximize the working capacity, especially in the high-pressure range remains complicated due to the influence of different structural features. Indeed, MAF-38 adsorbs up to $76\,cm^3\,cm^{-3}$ at 5 bar and its adsorption isotherm is almost saturated at a pressure >65 bar making this MOF material less competitive to store methane at ANG pressures >65 bar. This MOF example points to the fact that there is a need to design larger pore size materials to minimize adsorbed CH_4 below 5 bar and increase the storage capacity at high pressure. The literature has shown several remarkable examples that demonstrate the successful design of ultrahigh surface area MOFs that have hierarchical pore size architectures composed of a large pore (<20 Å in diameter) and a small pore that work together to achieve high working capacity. The compound Al-soc-MOF-1 (Figure 8.1), which has a cubic-shaped pore size of 14.3 Å, connected windows of 6 Å, BET surface area of $5585\,m^2\,g^{-1}$ and pore volume of $2.3\,cm^3\,g^{-1}$ achieved a volumetric working capacity of 201 cm^3(STP) cm^{-3} at 80 bar and 298 K. Furthermore, due to its low crystal density, Al-soc-MOF-1 afforded a gravimetric uptake that was >700 $cm^3\,g^{-1}$ (0.5 g/g) at temperatures below 288 K and pressure of 80 bar, which reached the 2012 DOE CH_4 gravimetric uptake target [19]. The compound, MOF-520, with a pore size of 9.9 and 16.2 Å in diameter, BET surface area of $3290\,m^2\,g^{-1}$ and pore volume of $1.277\,cm^3\,g^{-1}$ reached a volumetric working capacity of 125 and 194 $cm^3\,cm^{-3}$ at 80 bar and 298 K, respectively [3]. It is noted that the parent MOF of MOF-520, MOF-519, displayed a volumetric working capacity of 230 $cm^3\,cm^{-3}$ at 80 bar and 298 K [3], however, this value

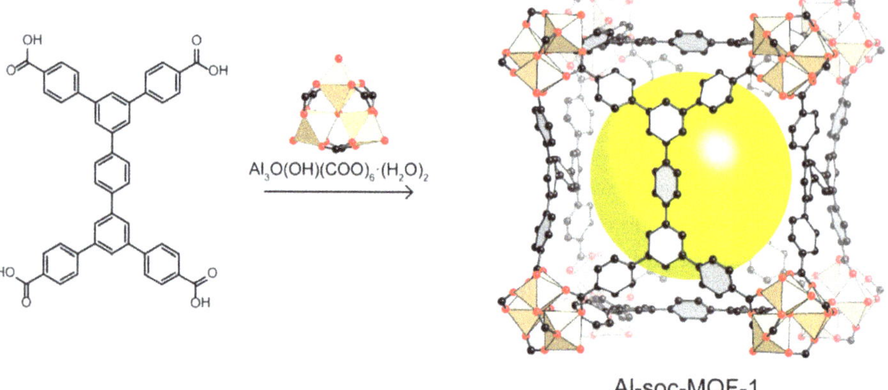

Figure 8.1: Structure of Al-soc-MOF-1 with small pore windows and large cubic pore size of 14.3 Å highlighted as the yellow ball. Atom colors: C, black; Al, light orange; O, red. H atoms are omitted for clarity [19].

has proven difficult to reproduce [6]. The compound MOF-905 with pore sizes of 6 and 18 Å in diameter, BET surface area of 3490 $m^2\,g^{-1}$ and pore volume of 1.34 $cm^3\,g^{-1}$ produced a volumetric working capacity of 120 and 203 $cm^3\,cm^{-3}$ at 35, 80 bar and 298 K, respectively. Interestingly, MOF-905 exhibited a much higher working capacity than that of the isoreticularly expanded structure, MOF-205, which has a much larger pore size [6]. Another noteworthy compound, ST-2, possesses a complicated distribution of pore diameters: (i) elongated polyhedral cages (inner diameter: 3.3 × 3.2 × 2.3 nm^3 and 2.3 × 2.3 × 1.5 nm^3), (ii) spherical icosahedral cages (inner diameter: 2.4 × 2.4 × 2.3 nm^3) and (iii) adamantane-like cages (inner diameter: 0.8 nm), BET surface area of 5172 $m^2\,g^{-1}$ and pore volume of 2.44 $cm^3\,g^{-1}$. With support from Grand Canonical Monte Carlo (GCMC) simulations, ST-2 was claimed to have a remarkable deliverable capacity of 289 cm^3 (STP) cm^{-3} at 298 K and 5–200 bar while an experimental volumetric working capacity of 187 cm^3 (STP) cm^{-3} is reached at 80 bar and 298 K [35].

8.2.4 Coordinately unsaturated metal sites (CUSs)

CUSs are a type of active site generated from the removal of volatile coordinated solvent ligands from the metal coordination sites during the material's activation process. These sites are introduced into MOFs by elaborated material design employing SBUs that are prone to the removal of axial solvent ligands from metal coordinative sites while maintaining the same structural architecture or by simple doping of the framework with metal centers [38]. The unsaturated nature of the metal ion's coordination sphere enables them to bind strongly with guest molecules.

8.2.5 CUSs in traditional MOFs

To date, the MOF-74 (or CPO-27, Figure 8.2) series is the most famous example for the design and application of CUSs in MOFs. The parent MOF-74 in this iso-structural series has a 1D channel with a pore size of 11 Å, BET surface area of ~1350 $m^2\ g^{-1}$, and most notably, a very high density of CUSs (~4.5 sites/nm^3) [39, 40]. Among the iso-structures of MOF-74 constructed from different metals, Ni-MOF-74 was found to give the highest methane uptakes of 228 and 251 $cm^3\ cm^{-3}$ at 35 and 65 bar, respectively, however, a low volumetric working capacity was calculated for this material (129 $cm^3\ cm^{-3}$) due to high uptake at a pressure below 5 bar [39]. Another case of a MOF with CUS is HKUST-1 (Figure 8.2), one of the first reported MOFs in the literature. HKUST-1 is composed of copper paddlewheel SBUs that are connected by 1,3,5-benzenetricarboxylate. The crystal structure is highlighted by three different types of pores that have diameters of approximately 5 Å (green), 10 Å (yellow), respectively [41]. Only the large cuboctahedral cage has CUSs that point to the pore. HKUST-1 is also a well-known example for use in methane storage with its total methane uptake at 298 K being 230 and 267 $cm^3\ cm^{-3}$ at 35 and 65 bar, respectively [18]. The volumetric working capacity of HKUST-1 is also remarkable with a value of 190 $cm^3\ cm^{-3}$ despite suffering a huge loss in capacity due to HKUST-1's high uptake at low pressure and near-saturation at pressures >65 bar.

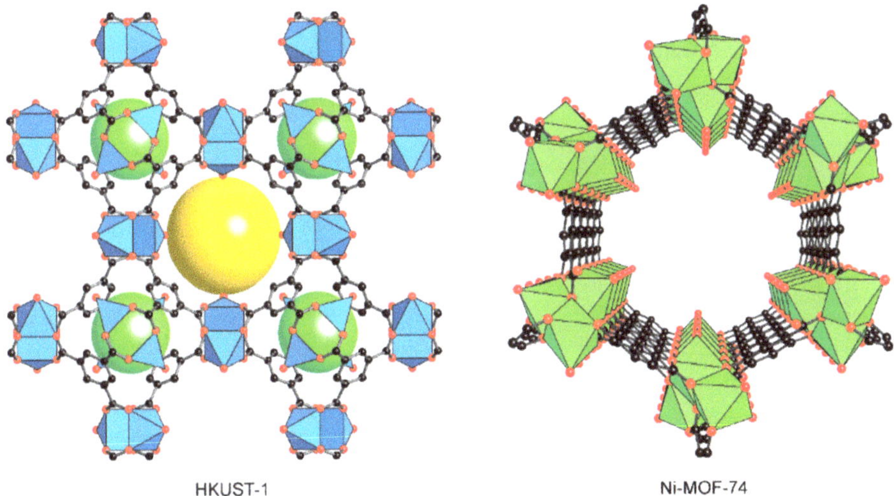

Figure 8.2: Crystal structures of HKUST-1 and MOF-74. Atom colors: C, black; Cu, blue polyhedral; Ni, green polyhedral; O, red. H atoms are omitted for clarity. Yellow and green balls provide an indication of the pore sizes and free space [18, 39].

8.2.6 CUSs in nbo-based MOFs

The **nbo**-based architecture has been widely employed to construct porous framework materials for methane storage. The first example of the **nbo**-structure is MOF-505, which is constructed from 3,3',5,5'-biphenyltetracarboxylate linker and paddle-wheel dicopper SBUs [42]. Following this, a number of iso-reticularly extended MOF-505 structures synthesized by linkers with a lengthened biphenyl core were also reported (Figure 8.3). The MOF compound, UTSA-76a, with a terphenyl core-based linker exhibits a methane adsorption capacity (working capacity) of 257 (197) cm^3 (STP) cm^{-3} at 298 K and 65 bar [27]. Further extending the **nbo**-based architecture by using increasingly longer linkers have led to materials with impressive uptake and working capacity. Several representative examples have attained methane adsorption capacities (working capacities) of 234 (183) $cm^3\,cm^{-3}$ for UTSA-111a [25], 237 (192) $cm^3\,cm^{-3}$ for NOTT-102a [23], 241 (190) $cm^3\,cm^{-3}$ for UTSA-110a,[24] and 254 (198) $cm^3\,cm^{-3}$ for NJU-Bai-43 at 65 bar and 298 K (Figure 8.3) [22].

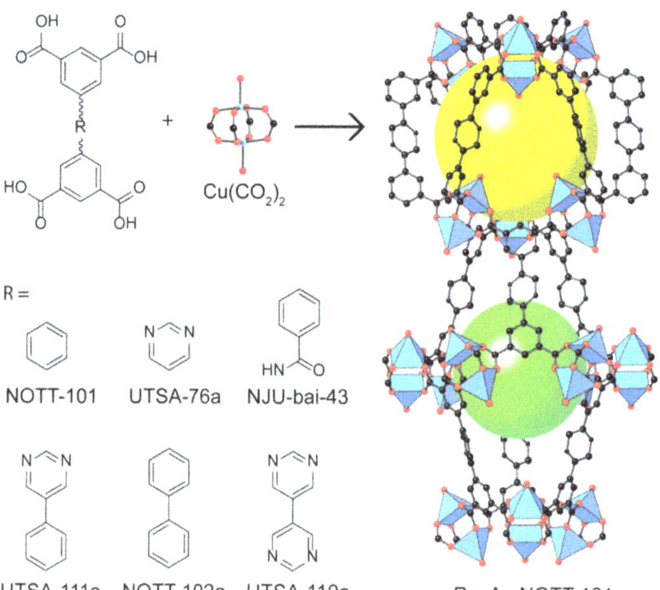

Figure 8.3: **nbo**-based MOFs constructed from different linkers and functionalities for optimizing methane uptake and working capacity. Atom colors: C, black; Cu, blue polyhedral; O, red. H atoms are omitted for clarity. Yellow and green balls provide an indication of the pore sizes and free space [22–25].

8.2.7 CUSs in rht-type copper–hexacarboxylate MOFs

The **rht**-based architecture has also been employed to construct porous framework materials for methane adsorption. This type of architecture employs C3-symmetrical hexacarboxylate linkers to connect 24 edges of a cuboctahedron. The resulting hierarchical structure has three different cages that can be categorized has having small to very large pore size [43]. The literature demonstrated the potential of **rht**-frameworks for methane storage; for example, MFM-112 and NU-111 have working capacities of 181 and 177 $cm^3\ cm^{-3}$ at 65 bar and 298 K (Figure 8.4), respectively [26, 28]. Although it is true that the presence of CUSs in MOFs enhance methane uptake capacity, this strategy suffers a drawback in that the strong binding energy of CUSs lead to a loss of capacity due to high uptake at a pressure below 5 bar. Furthermore, CUSs are easily coordinated by trace impurities in the NG resulting in reduced storage capacity during charge/recharge cycles. For these reasons, design strategies are needed beyond simply introducing CUSs to discover new MOFs for methane storage [31].

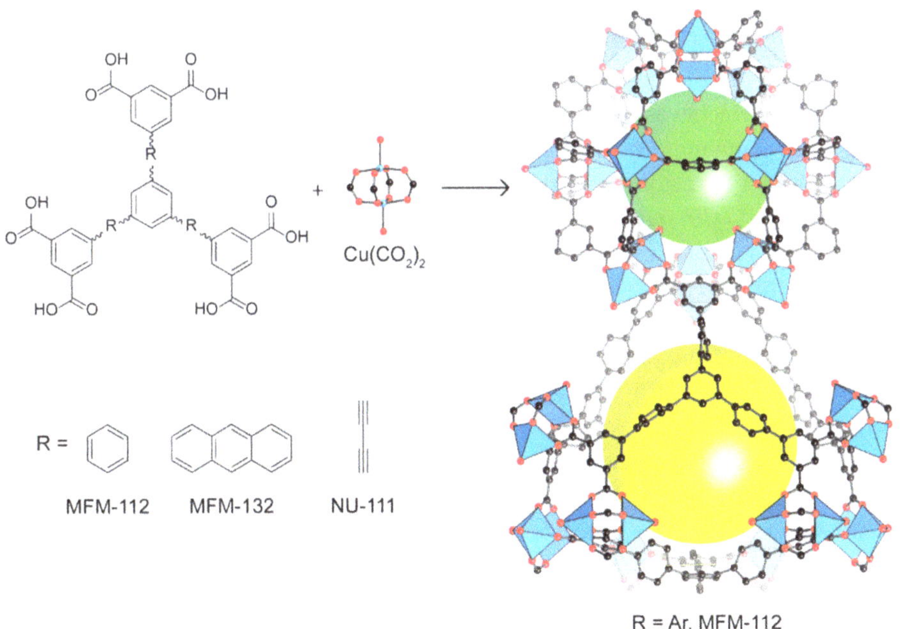

Figure 8.4: **rht**-based MOFs constructed from different for optimizing methane uptakes and working capacity. Atom colors: C, black; Cu, blue polyhedral; O, red. H atoms are omitted for clarity. Yellow and green balls provide an indication of the pore sizes and free space [26, 28].

8.2.8 Framework functionalization for enhancing the methane uptake capacity

The introduction of functional groups onto the framework backbone may alter the pore size, pore volume and methane heat of adsorption. Jiang et al. reported the influence of framework functionalities on the total volumetric methane uptake of MOF-905 at high pressure. Specifically, at 35 bar, MOF-905-Naph (i. e., MOF-905 whose linker contains a naphthalene unit) exhibits the highest total volumetric methane uptake (146 cm^3 cm^{-3}) among four MOF-905 materials, followed by MOF-905, MOF-905-Me_2 and MOF-905-NO_2. However, upon increasing the pressure to 80 bar, the benefits of framework functionalization disappear [6].

While functionalities introduced on the side chain of linkers may reduce the pore volume, which hinders the total uptake and working capacity at high pressure, linker design with functionalized groups stitched directly onto the linker's backbone led to resulting MOFs experiencing an improved total uptake and working capacity. Recent literature have reported promising examples of this strategy; for example, Chen et al. investigated the methane adsorption of four **nbo**-based materials constructed from terphenyl core linkers of N-contained rings (NOTT-101, ZJU-5, UTSA-75 and UTSA-76; Figure 8.3). The results indicated a significant increase in the total volumetric methane storage capacities for those MOFs with pyridine (ZJU-5), pyridazine (UTSA-75) and pyrimidine (UTSA-76) functionalities in the linker backbone when compared to NOTT-101, which was built from an unfunctionalized terphenyl-based linker. The pyrimidine-containing linker in UTSA-76a yielded significant enhancement of the total volumetric methane storage (working capacities) at 65 bar from 237 (181) cm^3 cm^{-3} in NOTT-101a to 257 (197) cm^3 cm^{-3} in UTSA-76a, respectively (Figure 8.3) [44]. Chen et al. further extended the UTSA-76a structure by employing a longer linker containing two pyrimidine rings in the backbone to synthesize USTA-110a [24, 25]. When assessed for its methane adsorption properties, USTA-110a displayed a slightly lower volumetric working capacity compared with that of UTSA-76a, yet USTA-110a achieved significantly higher gravimetric working capacity (Figure 8.3). Beyond pyrimidine functionalities, the integration of amide groups into the linker backbone has also proven favorable for enhancing methane adsorption. Bai et al. investigated the methane adsorption properties of three **nbo**-based compounds, NJU-Bai-41, –42 and –43, built from a amide-containing linker backbone [22]. This study revealed a considerable increase in the CH_4 working capacity of amide-constructed materials over the parent PCN-14 MOF material [21]. Notably, the compound NJU-Bai-43 reached a volumetric working capacity of 198 cm^3 cm^{-3} at 298 K and 65 bar, which is amongst the highest reported for MOFs under these conditions (Figure 8.3).

8.2.9 Flexible MOFs for enhancing working capacity

While tireless efforts have been undertaken to improve the working capacity of MOFs, a novel strategy has recently been proposed whereby flexible MOFs are used to minimize methane uptake at low pressure (<5 bar). Long et al. investigated the methane adsorption of two flexible MOFs, termed Co(bdp) and Fe(bdp) [37]. Both compounds were observed to feature sharp increasing/decreasing steps of adsorption and desorption in their methane sorption isotherms. Further structural analysis showed that both materials undergo a structural phase transition in response to CH_4 pressure and temperature. Specifically, Co(bdp) and Fe(bdp) remain in a closed-packing phase at low CH_4 pressures, therefore limiting the adsorbed amount of CH_4 at a pressure below 5 bar. When the pressure reached specific values, a sudden structural transformation occurred for both the materials from closed-packing to an open phase. This resulted in steep uptake of CH_4 at the structural transformation-inducing pressure (Figure 8.5). Due to this special property, both materials had high volumetric working capacities, in which the compound, Co(bdp), reached a value of 197 $cm^3\ cm^{-3}$ at 298 K and 65 bar. Remarkably, this behavior produced a greater working capacity than such capacities have been attained for classical, structurally-rigid MOF adsorbents. Zhao et al. adopted this strategy to design an aluminum-based MIL-53 analogue for methane storage. Their study demonstrated the successful design of carboxylate-based flexible MOFs capable of moving from the closed-packing phase to an open phase as trigged by adsorption pressure. Specifically, the compound MIL-53(Al)-OH synthesized from -OH-functionalized terephthalic acid attained a promising deliverable working capacity of 164 $cm^3\ cm^{-3}$ at 298 K and 65 bar [36].

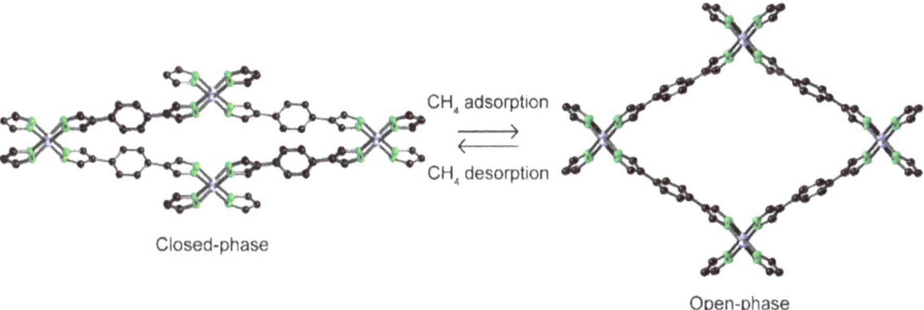

Figure 8.5: Phase transition in response to CH_4 pressures and temperature of Co(bdp). Atom colors: C, black; Co, light purple; N, green. H atoms are omitted for clarity [37].

8.2.10 Designing MOFs to balance volumetric and gravimetric working capacities

The 2012 DOE research program targets both volumetric and gravimetric storage capacities of the entire system. Their 2012 targeted working capacities were 350 $cm^3\ cm^{-3}$ and 0.5 $g\ g^{-1}$ at room temperature, which required that both size and weight requirements of the onboard tank being taken into consideration for achieving these targets. It is noted that a trade-off exists between gravimetric and volumetric capacities, in which large-pore-size MOFs (e. g., MOF-210, MOF-205, MOF-177, ST-2, NU-111) exhibit high gravimetric storage capacity while the small-pore-size MOFs (e. g., HKUST-1, MAF-38, UTST-76) are favored for higher volumetric storage capacity. This means that there remains a great challenge in providing satisfactory volumetric and gravimetric capacities within a single material.

Several MOF materials have been designed and synthesized to optimize the structural factors needed for realizing both high volumetric and gravimetric working capacities. For example, Al-soc-MOF-1, the first reported such compound, is highlighted by both a high volumetric and gravimetric working capacity of 201 cm^3(STP) cm^{-3} at 80 bar and 298 K and 624 $cm^3\ g^{-1}$ at 80 bar and 288 K, respectively. Chen et al. also extended the structure of NOTT-101 by adding pyrimidine rings to linker in the synthesis of USTA-110a and UTSA-111a. Both materials exhibited balanced improvement between their volumetric and gravimetric working capacities when compared to UTSA-76a (Figure 8.3) [24]. Farha et al. reported an Al-based MOF, termed, NU-1501-Al, that was shown to balance its volumetric and gravimetric working capacities. This compound was assembled from H_6PET-2 and a trimer Al-cluster to form an extended structure of **acs** topoloy with pore sizes of ~1.7 and 2.2 nm, BET surface area of 7920 $m^2\ g^{-1}$ and a pore volume of 2.93 $cm^3\ g^{-1}$ (Figure 8.6). NU-1501-Al realized methane working capacities (5 to 80 bar) of ~0.44 $g\ g^{-1}$ (174 cm^3 (STP) cm^{-3}) at 296 K and 0.54 $g\ g^{-1}$ (214 $cm^3\ cm^{-3}$) at 270 K, respectively [20]. Later on, Bai et al. reported the structural design and optimization of several Fe-based MOFs for balancing the volumetric/gravimetric trade-off, in which the compound Fe-ncb-ABDC demonstrated a combination of simultaneously high CH_4 gravimetric and volumetric working capacities of 0.302 (0.37) $g\ g^{-1}$ and 196 (240) $cm^3\ cm^{-3}$ at 298 (273) K and between 5 and 80 bar, respectively [34].

8.2.11 Sol-gel approach for the synthesis of monolithic MOFs, a promising direction toward the practical use of MOFs for methane storage

In practice, ANG storage requires that adsorbents be efficiently packed to minimize the dead volume between the polycrystalline powder particles. The traditional packing method relies on mechanical pressure to press the adsorbent (e. g., zeolites) into

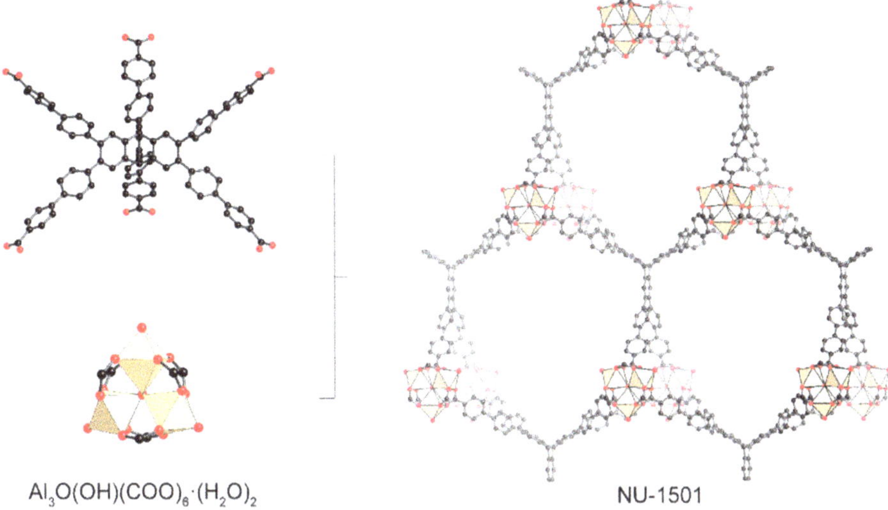

Figure 8.6: Structure of NU-1501-Al with small pore and large 1D channel. Atom colors: C, black; Al, light orange; O, red. H atoms are omitted for clarity [20].

desirable shapes and sizes. Unfortunately, most reported storage capacity values for MOFs are calculated using the single-crystal density of those respective MOFs without considering issues arising from different packing scenarios. This is unfortunate because the high mechanical pressure used in compression of the powder particles is unsuitable for MOFs, which experience partial collapse of the internal pore structure under high packing pressure [18]. For example, compressing HKUST-1 at 5 tons by mechanical pressure yielded a density of 1.1 g cm^{-3} which is significantly larger than the ideal crystal density of 0.883 g cm^{-3}. Furthermore, the total volumetric uptake value was ~50 % of the value for the ideal sample due to loss of porosity upon pressing. This situation highlights the challenge of densification and pelletization of MOFs for applications in industry [18], Recently, a new protocol for densification of MOFs without using mechanical pressure has been developed, in which monolithic MOFs were synthesized with a minimal amount of interstitial space. This method is derived from the nucleation of a large amount of discrete MOF nanoparticles, followed by gelation to form a colloidal network structure, and then finalized by mild drying [45]. Several successful examples have been reported, in which monolithic MOFs were employed for methane storage. As a representative example, the monolithic $_{mono}$HKUST-1 with a bulk density 1.06 g cm^{-3} exhibited CH_4 uptake of 259 cm^3 cm^{-3} at 65 bar and 298 K [46]. In a second representative example, centimeter-sized monoliths of UiO-66 were shown to achieve CH_4 uptake of 211 cm^3 cm^{-3} at 65 bar [47].

8.2.12 Perspective methane storage by MOFs

The future of MOF design and synthesis for methane storage may focus on several directions: (1) Improving volumetric and gravimetric working capacity as well as balancing between two factors. This can be achieved by the design of MOFs with suitable pore size, shape and functionalities to improve the uptake at high pressure while limiting the unusable at a pressure below 5 bar. Respecting the design of an ideal material for methane storage application, the MOF candidates may have to tether the optimized pore size with flexible structure when responding to CH_4 pressure to minimize methane uptake at low pressure (<5 bar); (2) Designed MOFs with better endurance against the poisonous trace-compound in natural gas as the accumulated amount of them over time can reduce the materials adsorption capacity. Respecting this issue, MOFs should be synthesized from the stable SBUs and have a minimizing amount of open metal sites in the backbone; (3) Designing materials that can almost maintain the same working capacity at varying adsorption temperatures. This feature is important and may become the critical point in the design of the adsorption material to work in climate conditions when the change in temperature is significant over time. Regarding this issue, the combination of strong methane adsorption sites into a flexible framework backbone when responding to CH_4 pressure is highly desired; (4) Further improving and extending the scope of monolith MOFs are needed for efficiently packing MOFs into the storage vessel without harmful to their intrinsic properties.

8.3 Metal-organic frameworks for onboard hydrogen storage

8.3.1 H_2 fuel for zero CO_2 emission

Hydrogen (H_2) is considered the ideal alternative energy source due in large part to its immense mass-energy content (142 MJ kg^{-1} compared to 45.8 MJ kg^{-1} of gasoline and 47.2 MJ kg^{-1} of natural gas), low atomic mass, widespread availability and zero-emission of CO_2 and other environmentally harmful by-products upon burning. Indeed, extensive utilization of hydrogen fuel is a critical pathway to pursue to reduce CO_2 emission [48]. Currently, 96 % of industrial hydrogen is produced by the steam reforming method [49] with the other 4 % being produced from electrolysis [50]. There are also several emerging technologies used for the production of pure hydrogen, which can be utilized for both short and long-term goals. These not only are based from wind, solar, geothermal and hydro sources, but also from biomass conversion, concentrated solar power, microorganism production and semiconductor use [51]. Despite the benefit, current technology for exploiting H_2 as a reliable energy source is suffering from two main bottlenecks: (i) the development of effective and durable

Table 8.2: The DOE targets for onboard hydrogen storage.

Storage parameter	Units	2020	2025	Ultimate
Total gravimetric (System Gravimetric Capacity)	$kWh\ kg^{-1}$	1.5	1.8	2.2
	$kg\ H_2\ kg^{-1}$	0.045	0.055	0.065
Total volumetric (System Volumetric Capacity)	$kWh\ L^{-1}$	1.0	1.3	1.7
	$kg\ H_2\ L^{-1}$	0.030	0.040	0.050

hydrogen fuel cells and (ii) the achievement of an equal or higher energy storage density with that of gasoline (0.01188 MJ L^{-1} *vs.* 32.4 MJ L^{-1} at ambient temperature and pressure). According to the DOE's 2025 program, the target system uptake is aimed to be 5.5 wt% with total volumetric and gravimetric capacity target values of 0.040 kg $H_2\ L^{-1}$ system and 0.055 kg $H_2\ kg^{-1}$ system, respectively (Table 8.2) [52].

The traditional method for H_2 storage involves the utilization of high-pressure vessels (typically at 350 and 700 bar or ~5000 and ~10,000 psi, respectively) with the hydrogen density stored in compressed form strongly dependant on the storage pressure. The volumetric density of the stored H_2 is 7.8 kg $H_2\ m^{-3}$ at 10 MPa and 20 ºC, which increases to 39 kg $H_2\ m^{-3}$ at an elevated pressure of ca. 69 MPa [53]. To achieve high density, reinforced materials are necessary for the storage vessels, such as carbon fiber and glass fiber, but these are costly to manufacture. An alternative to high-pressure compression is H_2 storage by porous adsorbent materials based on physical adsorption via intermolecular interactions, which is attractive due to increased safety and ease of use [54]. Carbonaceous materials, such as carbon nanotubes (CNTs), activated carbon (AC), graphene, as well as zeolites and MOFs, have been physically and chemically treated to achieve high H_2 storage capacities.

Carbon materials that possess low atomic weight and have a microporous nature can adsorb H_2 via van der Waal's interactions between the H_2 gas molecules and their internal surface. Although these materials can store up to 4.5 wt% of H_2 at 77 K, the highest measured H_2 storage value is <1 wt% at room temperature even at high pressure [55]. For example, at room temperature and a H_2 pressure of 45 bar, the H_2 storage capacity is <0.18 wt% in graphite nanofibers (GNFs) and multiwalled nanotubes (MWNTs). Only purified single-walled nanotubes (SWNTs) display slightly higher H_2 uptake of ca. 0.63 wt% [56]. At present, a recorded hydrogen uptake higher than 1 wt% was measured for activated carbon (ca. 1.06 wt% at room temperature and 15 bar H_2 pressure). This material prepared from biomass with KOH activation, which was attributed to the large stable surface area and pore size distribution [57].

Zeolites are another class of porous materials that have been investigated for H_2 storage, however, this class of inorganic material is only able to store negligible amounts of H_2 at room temperature or at temperatures >200 ºC. The highest uptake value achieved by zeolites was in NaA, which realized 0.28 and 0.30 wt% at room temperature and 270 ºC, respectively. At cryogenic temperatures (77 K and 15 bar), elevated H_2 storage capacities of 1.54, 1.79 and 1.81 wt.% could be reached by NaA, NaX and NaY zeolites, respectively [58]. Zeolites display diverse H_2 uptake behavior depending on both the framework structure and the nature of the cations present. Calcium-exchanged zeolite X exhibited an enhanced gravimetric hydrogen storage capacity of 2.19 wt% at 15 bar and 77 K with a BET surface area of 669 $m^2\ g^{-1}$ [59]. The limitation in storage capacity of these traditional materials has led to a continued exploration of creating new materials that can storage appreciable quantities of H_2 at ambient temperature.

8.3.2 The current status of MOFs for hydrogen storage

MOFs have considerable potential as a class of next-generation materials for H_2 storage owing to their ultrahigh surface area values, tuneable pore size and shape and internal surface polarity [70]. In addition, MOFs have inherent benefits for H_2 storage because of their reversible uptake and high-rate hydrogen adsorption process owing to the physical adsorption mechanism that occurs [81]. Additionally, their pore size and framework topology can be designed based on the principles of reticular chemistry [10] to obtain high-surface-area materials with significantly enhanced H_2 adsorption properties. Indeed, it is all of these unique characteristics of MOFs that make them great candidates for H_2 storage applications (Table 8.3) [82, 83]. Historically, the first example of utilizing MOFs for H_2 storage was reported in 2003 [70] in which, a MOF material, termed MOF-5, whose composition is based on the di-topic 1,4-benzenedicarboxylic acid linker (H_2BDC) and the tetranuclear $Zn_4O(CO_2)_6$ metal cluster (Figure 8.7), was shown to realize a H_2 gas uptake of up to 4.5 wt% at 78 K and 1.0 wt% at room temperature and 20 bar. Since then, a wide range of MOFs have been investigated, designed and modified to achieve higher values of H_2 uptake capacity.

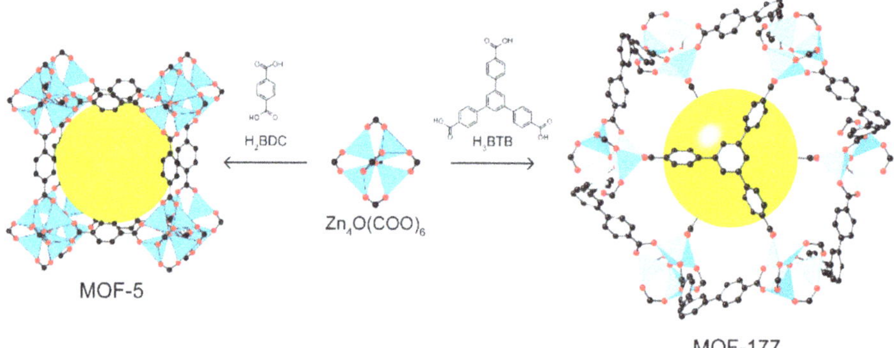

Figure 8.7: Utilization of di-topic and tri-topic linkers in the synthesis of MOF-5 and MOF-177 from the tetranuclear $Zn_4O(CO_2)_6$ SBU. Atom colors: C, black; Zn, light blue; O, red. H atoms are omitted for clarity. Yellow balls provide an indication of the pore sizes and free space [71, 72].

Table 8.3: Hydrogen uptake for the best performing materials.

Material	Storage Condition Temp. (K) / Press (bar)	BET surface area ($m^2\ g^{-1}$)	H_2 adsorption capacity (wt%)	Ref.
AC (Maxsorb)	77/30	3306	5.70	[60]
AC (Maxsorb)	303/100	3306	0.67	[61]
AC (AX-21)	77/60	2745	10.80	[62]
AC (KOH-treated)	77/20	2770	6.20	[63]
SWCNT	133/0.4	–	5–10	[64]
CNT	273–295/1.0	290–800	≤1.0	[65, 66]
CNT (K-doped)	343/1.0	130 (specific)	14.00	[67]
MWCNT (Ni-doped)	– (electrochemical)	(electrochemical)	0.75	[68]
CNF (N-doped)	298/100	870 (specific)	2.00	[69]
MOF-5	78/20	2500-300	4.50	[70]
	298/20		1.00	[70]
IRMOF-8	298/10	1801	2.00	[71, 72]
MOF-177	78/70	4600	7.50	[73]
	298/100		0.62	[74, 75]
NU-100	77/56	6143	10.00	[74, 75]
NU-109	77/45	7010	8.30	
NU-110	77/45	7140	8.82	[74, 75]
	298/180		0.57	
MOF-399	77/56	7157	9.02	[75]
	298/140		0.46	
Cr-MIL-53	77/16	1020	3.10	[76]
Al-MIL-53	77/16	1026	3.80	[76]
Cu-MOF-5	77/65	1154	3.60	[77]
	298/65		0.35	
MOF-210	77/80	6240	17.60	[77]
	298/80		2.70	
Be-MOF	77/1.0	4030	1.60	[78]
	298/95		2.30	
COF-1	77/1.0	711	1.70	[79]
	77/70		3.80	
COF-5	77/1.0	1590	0.10	[79]
	77/80		3.40	
COF-102	77/1.0	3620	0.50	[79, 80]
	77/100		10.00	

One of the highest H_2 uptake capacities (7.5 wt%) was reported at 77 K and 70 bar using MOF-177, which is comprised of the same tetranuclear $Zn_4O(CO_2)_6$ clusters linked by the tri-topic 1,3,5-benzenetribenzene carboxylic acid (BTB) linker [71, 72]. Another early example witnessed the employment of an extended hexa-topic carboxylic acid linker consisting of multi-ethynylene units for the construction of an **rht**-topological MOF, termed NU-100. This material is highlighted by a high BET surface area of 6143 $m^2\ g^{-1}$ leading to a high H_2 uptake capacity of 10.0 wt% at 77 K and 56 bar [73–75]. Other extended MOF materials based on the **rht** topology, NU-109 and NU-110 (BET surface areas of 7010 $m^2\ g^{-1}$ and 7140 $m^2\ g^{-1}$, respectively), have also shown high H_2 capacities of 8.30 and 8.82 wt% at 77 K and 45 bar, respectively. Recently, a MOF with ultrahigh porosity, NU-1501-M (Figure 8.6, M = Al, Fe), with an impressive surface area (gravimetric BET area of 7310 $m^2\ g^{-1}$ and a volumetric BET surface area of 2060 $m^2\ cm^{-3}$) was demonstrated to have a high H_2 capacity (14.0 wt%, 46.2 g L^{-1}) under a combined temperature and pressure swing (77 K and 100 bar to 160 K and 5 bar) [20]. These results indicate that underlying structural properties and pore sizes of MOFs can be fine-tuned by changing the SBUs and linkers to realize H_2 uptake capacities that approach DOE targets.

Aside from expanding the linkers, controlling binding energy through the modification of active metal sites is considered a well-developed strategy for enhancing the H_2 uptake capacity of MOFs. Many metals (e. g., Li, Na, K, Mg, Ca, Be, Ti, Pt, Pd, Cu, Fe, Co, Ni, Zn) have been used as clusters or for doping within preformed MOFs (Table 8.4) [104]. For example, MIL-53 and its analogues, prepared from H_2BDC and various transition metals displayed different H_2 uptake capacities. Specifically, Cr-MIL-53 and Al-MIL-53 realized H_2 uptakes of 3.1 wt% and 3.8 wt% at 77 K and 16 bar, respectively [76]. The compound, $Be_{12}(OH)_{12}(BTB)_4$ (Be-MOF), synthesized from the lightweight Be element adsorb 2.3 wt% H_2 at 298 K and 95 bar [78]. These results indicate that the H_2 adsorption capability of MOFs is directly impacted by different types of metal ions.

Table 8.4: Isosteric heats of hydrogen adsorption by MOFs.

Nonopen metal site			**Open metal site**		
Compounds	**Q_{st}/kJ mol^{-1}**	**Ref.**	**Compounds**	**Q_{st}/kJ mol^{-1}**	**Ref.**
MOF-5, IRMOF-1	3.8	[77]	MOF-177	4.4	[84]
SNU-70	4.88	[85]	MIL-102	6	[86]
PCN-610/NU-100	4.42	[87]	MIL-100	6.3	[88]
$Be_{12}(OH)_{12}(BTB)_4$	5.5	[78]	UMCM-150	7.3	[89]
NOTT-400	5.96	[90]	$Cr_3(BTC)_2$	7.4	[91]
IRMOF-8	6.1	[92]	CPO-27-Zn	8.3	[93]
UMCM-2	6.4	[94]	MIL-101	10	[88]
NOTT-401	6.65	[90]	MOF-74(Mg)	10.3	[95]
SNU-77H	7.05	[96]	Mn_2(m-dobdc)	10.5	[97]
MOF-5 interpenetrated	7.6	[98]	SNU-5	11.6	[84]
SNU-6	7.74	[99]	Co_2(m-dobdc)	12.1	[97]
MOF-646	7.8	[100]	Ni_2(m-dobdc)	13.7	[97]
sod-ZMOF	8.4	[101]	SNU-15’	15.1	[102]
rho-ZMOF	8.7	[101]			
TUDMOF-2	9.5	[103]			

The presence of CUSs also play an important role in controlling H_2 gas uptake. For example, HKUST-1, with open metal sites prepared from copper(II) ions and benzene-1,3,5-tricarboxylate (BTC), displayed a BET surface area of up to 2,000 $m^2\ g^{-1}$ [105] and a maximum H_2 uptake of 3.6 wt% at 77 K and 0.35 wt% at room temperature and 65 bar [77]. The MOF, M_2(m-dobdc) (M = Co, Ni; m-dobdc^{4-} = 4,6-dioxido-1,3-benzenedicarboxylate), and its MOF-74 isomer, M_2(dobdc) (M = Co, Ni; dobdc^{4-} = 1,4-dioxido-1,3-benzenedicarboxylate), with CUSs represent some of the most promising adsorbents for H_2 uptake at near-ambient temperatures. Specifically, Ni_2(m-dobdc) exhibited a volumetric working capacity between 100 and 5 bar of 11.0 g L^{-1} at 25 °C and 23.0 g L^{-1} with a temperature swing between –75 and 25 °C [106]. A partially desolvated Mn-based MOF, constructed from a BTT linker (BTT^{3-} = 1,3,5-benzenetristetrazolate), yielded excellent a total H_2 uptake of 6.9 wt% at 70 K and 90 bar and a record-high isosteric heat of adsorption of 10.1 kJ mol^{-1}, which is directly attributed to the binding of H_2 molecules to the coordinatively unsaturated Mn(II) centers within the framework [107]. The replacement of Mn(II) with Cu(II) in a sodalite-type MOF resulted in a fully desolvated compound leading to an enhanced density of exposed coordination sites for strong H_2 binding [108].

8.3.3 H_2 storage at cryogenic temperatures—effect of pore size, pore volume and surface area

Numerous publications have confirmed a linear relationship between H_2 adsorption capacity at 77 K and surface area [58, 109–112], which suggests employing materials with higher surface area to improve hydrogen adsorption in MOFs at 77 K. In general, H_2 adsorption capacity at low pressure (1 bar) and a temperature of 77 K is greatly affected by surface areas ranging from 100 to 2000 $m^2\ g^{-1}$; however, surface areas greater than 2000 $m^2\ g^{-1}$ have shown no appreciably impact on the amount of H_2 adsorbed under the same conditions. At low H_2 pressure and cryogenic temperature, H_2 molecules are likely to predominantly bind to the thermodynamically favorable sites in MOFs, which have the highest hydrogen affinity, therefore, the uptake capacity in this region can be enhanced by applying materials with strong binding sites, such as CUSs, catenation, or functionalization of linkers [96].

It is also widely known that surface area and pore volume have a significant effect on H_2 adsorption at high pressure. H_2 adsorption in the high-pressure range exhibits a fairly linear relationship with surface area with the capacity for high-pressure H_2 uptake at 77 K increasing upon increasing the surface area [113]. MOF-5 (or IRMOF-1) and its iso-reticular extended structures (IRMOF = isoreticular metal-organic framework), whose pore size and surface area are controlled by the length of the linkers, were shown to be prototypical examples of high surface area MOFs as H_2 gas adsorptive materials (Figure 8.8). In the IRMOF series, materials constructed from longer linkers have higher surface area and provide more adsorptive sites for binding H_2 molecules

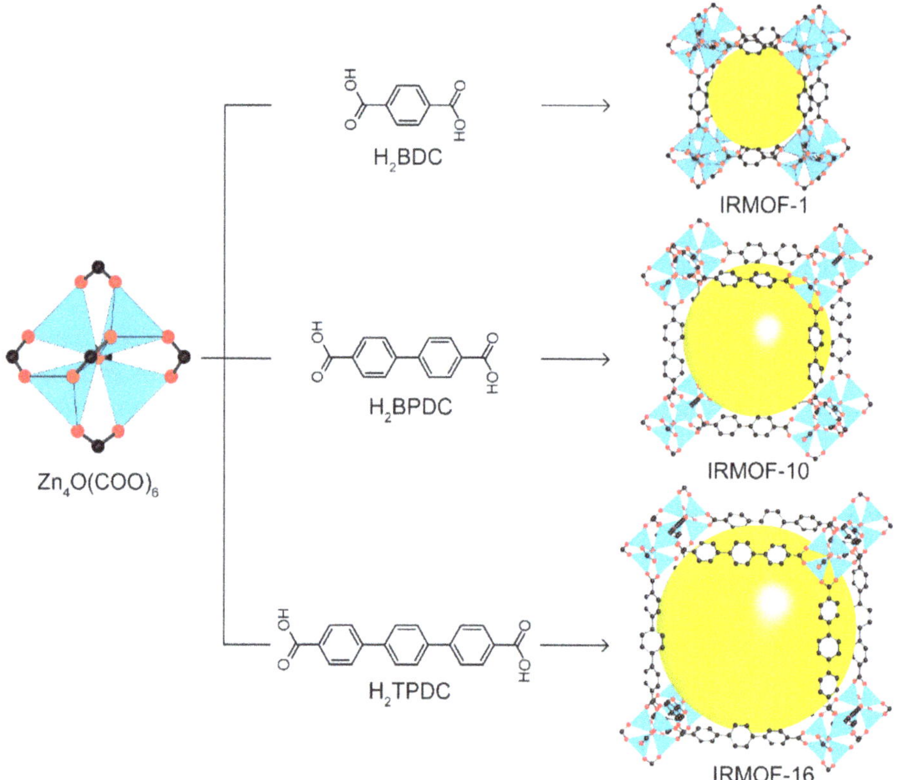

Figure 8.8: Employing di-topic linkers of different lengths for the iso-reticular design of MOFs with different pore sizes. Atom colors: C, black; Zn, light blue; O, red. H atoms are omitted for clarity. Yellow balls provide an indication of the pore sizes and free space [117].

[114]. MOF-210 [32] and DUT-60 [113] are extremely porous and have very high gravimetric adsorption capacities due to their ultrahigh BET areas. As outlined in a recent article, it is noted that ideal H_2 adsorbents must strike a balance between volumetric and gravimetric uptake [20].

To summarize, H_2 uptake behavior depends on how strong the H_2 affinity is to adsorptive sites in the low-pressure range and the uptake capacity appears to solely depend on the magnitude of the surface area and pore volume in the high-pressure range [115]. Apart from being generally correlated with the surface area and pore volume of the framework, the H_2 adsorption capacity also is somewhat dependent on the type of metal ion used to construct the MOF, the chemical and/or electronic environment of the pore surface and the presence of functional groups in the ligand and/or at the vacant coordination sites of the metal ions [116].

8.3.4 H_2 storage at cryogenic temperatures—effect of metal ions

The introduction of metal ions into the pore or additionally to the metal clusters of MOFs may increase the number of binding sites, positively adjust the electrostatic field and/or modify the surface area and pore volume of the MOF leading to tailorable H_2 adsorptive properties (Table 8.4). For example, partially substituting Zn(II) in MOF-5 with Co(II) resulted in two different materials, namely Co_8-MOF-5 and Co_{21}-MOF-5 [118]. Co_{21}-MOF-5 exhibited a small H_2 uptake difference at low pressure, however, a 7.4 % greater H_2 uptake compared to MOF-5 at 77 K and 10 bar. The authors attributed the difference in H_2 adsorption to the incorporation of Co(II) ions into the MOF-5 framework, which broke down the high degree of crystalline order and diminished the robustness and rigidity of the MOF-5 framework [119]. In a separate study, the influence of Zn(II)/Co(II) ratio in MOF-74(Zn) on the material's H_2 adsorption capability was investigated [120]. The results demonstrated an increase in H_2 adsorption capacity for the Co-containing MOF-74 over that of MOF-74(Zn). Specifically, the H_2 uptake capacity values were 26, 43 and 50 % higher than MOF-74(Zn) for Co_{61}-MOF-74, Co_{14}-MOF-74 and Co_{100}-MOF-74, respectively. These differences were attributed to the superior pore textural properties of the Co(II)-doped MOF. Alkali metals (e. g., Li, Na and K) are common doping ions to enhance H_2 uptake capability of MOFs [121, 122]. Previous studies showed an increment of up to 65 % in alkali metal-doped MOFs compared to the pristine MOFs. Additionally, for the same amount of dopant, H_2 uptake at 77 K and 1 bar increased in the order of increasing cation size (i. e., $Li^+ < Na^+ < K^+$).

8.3.5 H_2 storage at cryogenic temperatures—effect of linkers

The utilization of functionalized linkers is a proven strategy for altering H_2 adsorption properties. The type of linkers not only affects the porosity and surface area of MOFs, but also alters the electronic environment of the internal pores to enhance interaction with H_2. For example, investigating the H_2 uptake of two isostructural MOFs of $Cu_2(aobtc)(H_2O)_2{\cdot}3DMA$ (PCN-10, PCN = porous coordination network; aobtc = azobenzene-3,3,5,5-tetracarboxylate) and $Cu_2(sbtc)\text{-}(H_2O)_2{\cdot}3DMA$ (PCN-11, sbtc = trans-stilbene-3, 3,5,5-tetracarboxylate) (Figure 8.9) [123], showed higher H_2 storage capacity for PCN-11. Specifically, the excess H_2 uptakes were 2.34 wt% (18.0 g L^{-1}) and 2.55 wt% (19.1 g L^{-1}) at 77 K and 1 bar for PCN-10 and PCN-11, respectively, which increased to 4.33 wt% (33.2 g L^{-1}) and 5.05 wt% (37.8 g L^{-1}) at 77 K and 20 bar [123]. Since the N=N bonds in PCN-10 have a higher H_2 affinity over C=C double bonds, the opposite tendency is attributed to the better enduring porosity of PCN-11 than PCN-10 after guest removal by thermal activation.

Polarizable linkers favor H_2 uptake through enhanced interactions with H_2 molecules at higher surface coverage [124]. A series of MOFs, IRMOF-1, IRMOF-8, IRMOF-11, IRMOF-18 and MOF-177 were successfully synthesized from linkers bearing polar

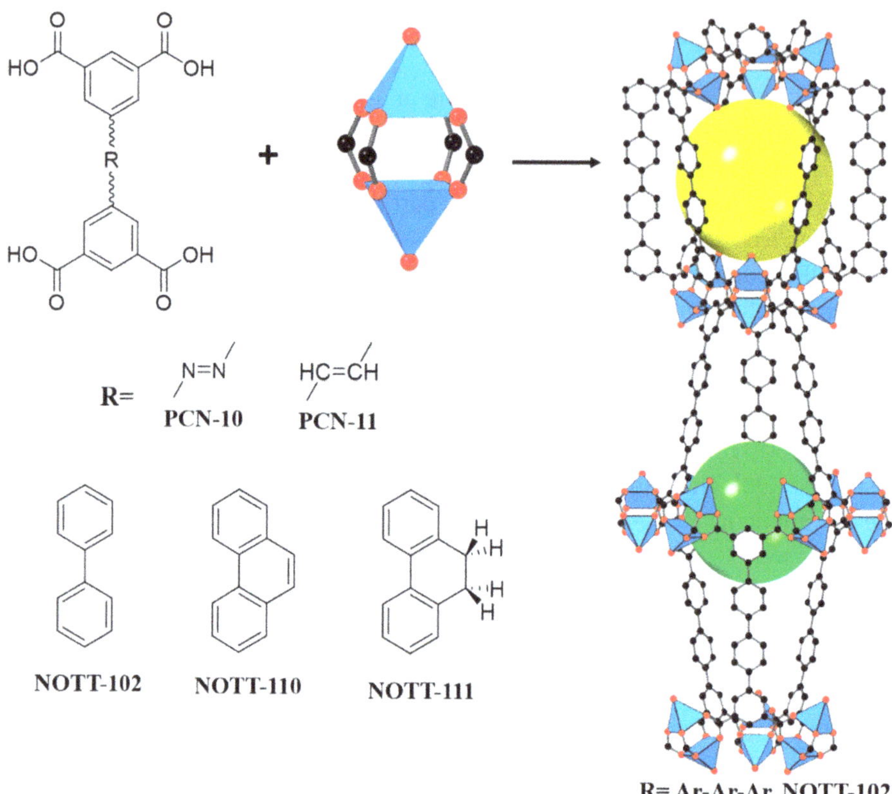

Figure 8.9: **nbo**-based MOFs constructed from different linkers with different functionalities for optimizing H_2 uptake. Atom colors: C, black; Cu, blue polyhedral; O, red. H atoms are omitted for clarity. Yellow and green balls provide an indication of the pore sizes and free space [126, 127].

functionalities [125]. Investigation of the H_2 uptake over these materials showed the influence of the chemical differences of the organic linker. At a pressure of 1 bar and 77 K, IRMOF-11 attained a coverage value per formula unit of 9.3, which is almost double that for IRMOF-18 (4.2) and was found to be proportional to the number of organic units per formula unit. Moreover, IRMOF-11, with the highest H_2 uptake capacity, was the only MOF among the five with a catenated framework. The H_2 storage capacities of IRMOF-1 and IRMOF-18 were quite similar implying that the pendant groups decorating the phenylene spacer do not affect H_2 uptake [125].

Linkers synthesized from aromatic rings can increase the interactions of H_2 molecules with the framework as compared with those synthesized from ethynyl moieties. Additionally, the presence of N atoms in the ring system (e. g., tetrazine rings) can even lead to much stronger interactions because of the formation of an electron-rich conjugated system [128]. For example, post-synthetic modification of amino-factionalized MOFs with a series of anhydrides or isocyanates yield IRMOF-3-AMPh,

IRMOF-3-URPh, UMCM-1-AMPh and DMOF-1-AMPh (UMCM = University of Michigan crystalline material; DMOF = DABCO MO; DABCO = 1,4-diazabicyclo[2.2.2]octane) with different H_2 uptake properties [129]. At 77 K and 1 bar, both IRMOF-3-AMPh and IRMOF-3-URPh revealed enhanced H_2 uptakes (1.73 and 1.54 wt%, respectively) over those obtained for the unmodified IRMOF-3 (1.51 wt%). This resulted from the addition of aromatic moieties in IRMOF-3, which provided a positive effect on H_2 binding affinity to the added phenyl groups. Both gravimetric and volumetric H_2 uptakes of UMCM-1-AMPh (1.54 wt% and 6.61 g L^{-1}) were also augmented when compared to UMCM-1-NH_2 (1.35 wt% and 5.39 g L^{-1}) while DMOF-1-AMPh showed an initial increase of H_2 uptake at a lower pressure relative to DMOF-1-NH_2 (2.08 compared to 1.68 wt%, respectively, at 77 K and 1 bar). This difference in H_2 adsorption behavior depended on the overall porosity of the pristine MOFs (IRMOF-3, UMCM-1-NH_2 and DMOF-1-NH_2, respectively; Figure 8.10) [129].

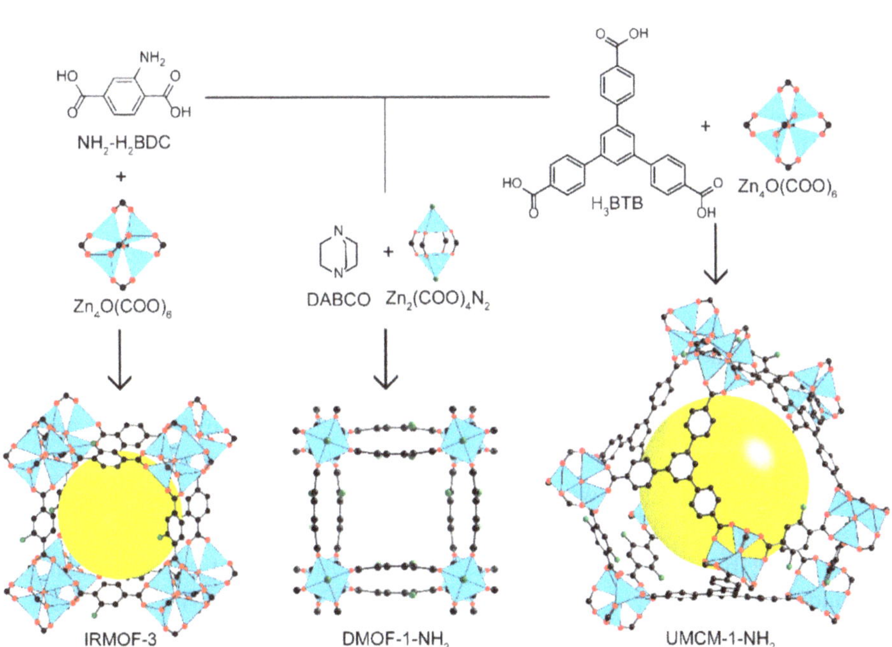

Figure 8.10: MOFs constructed from different linkers and Zn-based SBUs for optimizing H_2 uptake. Atom colors: C, black; Zn, light blue polyhedral; O, red. H atoms are omitted for clarity. Yellow balls provide an indication of the pore sizes and free space [117, 130, 131].

Yang et al. compared the H_2 uptakes of two isostructural MOFs based on the **nbo** topology with composition of $[Cu_2(L)\text{-}(H_2O)_2](DMF)_{7.5}(H_2O)_5$, NOTT-110 (L = (2,7-phenanthrenediyl)-diisophthalate) and NOTT-111 (L = [2,7-(9,10-dihydrophenanthrenediyl)]diisophthalate; NOTT = Nottingham) [126]. These MOFs were isoreticular, but differed only at the 9 and 10 positions of the phenanthrene group (Figure 8.9). The MOFs built from phenanthrene- and hydrophenanthrene-containing linkers resulted in an enhancement of H_2 adsorption over that of their tetraphenyl counterparts (NOTT-102) due to an increase in the affinity between H_2 gas molecules and the phenanthrene and hydrophenanthrene groups. The incorporation of these moieties improved the H_2 total uptake at low pressures (2.64 and 2.56 wt% for desolvated NOTT-110 and NOTT-111, respectively), which were 18 % higher than that of NOTT-102 (2.24 wt%) at 1 bar and 78 K [126].

8.3.6 H_2 storage at ambient conditions

Although there are several examples of MOFs that demonstrate a high H_2 adsorption capacity under cryogenic temperatures, the weak interaction with H_2 leads to a significant decrease of uptake at ambient temperatures. For example, MOF-177 reaches a H_2 uptake of 7.5 wt% at 78 K and 70 bar [73], but this value is reduced to only 0.35 wt% at 40 bar and 298 K [132] and 0.6 wt% at 100 bar and 298 K [48]. This fact has raised the need for materials that have stronger interaction with H_2 at ambient or near ambient temperature.

Based on combined experimental and computational studies, CUSs have been shown to increase the Q_{st} value of H_2 adsorption in a given MOF because H_2 molecules can directly interact with these adsorptive sites. CUSs in MOFs may be constructed by either: (i) employing metal cluster SBUs coordinating solvent molecules, followed by removal of the solvent using thermal treatment or (ii) chelating groups attached to the organic linkers involving covalent functionalization followed by post-synthetic metalation, or attachment of organometallic complexes directly to the aromatic components of the linkers [133]. The ability of MOFs with CUSs at SBUs and organic linkers to bind multiple H_2 molecules at a single binding site has been demonstrated. This MOF strategy was considered one of the most effective means to increase H_2 volumetric capacity in H_2 gas adsorbents [96]. An example is $V_2Cl_{2.8}$(btdd) (H_2btdd = bis(1H-1,2,3-triazolo[4,5-b],[4′,5′-i])-dibenzo[1,4]dioxin), a MOF with square-pyramidal vanadium(II) sites that features exposed vanadium(II) sites capable of back-bonding with weak π acids. This structural feature results in a gravimetric H_2 uptake of 1.64 wt% at 298 K and 100 bar [134]. Additionally, the aforementioned Ni_2(m-dobdc) MOF, known to possess CUSs that can strongly interact with H_2, displayed a usable volumetric capacity between 100 and 5 bar of 11.0 g L^{-1} at 25 °C and 23.0 g L^{-1} with a temperature swing between −75 and 25 °C [106].

In general, adding a CUS reduces the linker contribution to the MOF composition and increases the available space for guest gas adsorption, thereby resulting in increased H_2 gas adsorption [127, 135]. The CUSs have also been confirmed to be the primary and strongest H_2 binding sites in these compounds by a recent neutron diffraction study on Zn_2(dhtp) [136]. While the presence of CUSs culminates in increased H_2

gas uptake capacities as a result of mass reduction, the H_2 gas capacities per volume of MOF samples were also increased (10–20 %) when compared to those lacking CUSs.

8.3.7 Perspective H_2 storage by MOFs

The future of MOF design and synthesis for hydrogen storage must focus on improving volumetric and gravimetric working capacity at ambient conditions to realize the employment of MOFs for onboard H_2 storage. Unlike methane storage, as a result of the weak interaction of H_2 and the adsorbent, employing high-density CUSs that have a high affinity toward H_2 in a given MOF's backbone, is highly desired. Literature has shown design possibility, in which employing N-based linkers, such as H_2btdd, in combination with vanadium may form SBUs with V(II)-based CUSs exhibiting strong binding toward H_2.

Bibliography

[1] Sircar S. Gibbsian surface excess for gas adsorption revisited. Ind Eng Chem Res. 1999;38:3670.
[2] Mason JA, Veenstra M, Long JR. Evaluating metal–organic frameworks for natural gas storage. Chem Sci. 2014;5:32.
[3] Gandara F, Furukawa H, Lee S, Yaghi OM. High methane storage capacity in aluminum metal-organic frameworks. J Am Chem Soc. 2014;136:5271.
[4] Sircar S. Estimation of isosteric heats of adsorption of single gas and multicomponent gas mixtures. Ind Eng Chem Res. 2002;31:1813.
[5] He Y, Zhou W, Qian G, Chen B. Methane storage in metal-organic frameworks. Chem Soc Rev. 2014;43:5657.
[6] Jiang J, Furukawa H, Zhang YB, Yaghi OM. High methane storage working capacity in metal-organic frameworks with acrylate links. J Am Chem Soc. 2016;138:10244.
[7] Yeh S. An empirical analysis on the adoption of alternative fuel vehicles: the case of natural gas vehicles. Energy Policy. 2007;35:5865.
[8] Li H, Eddaoudi M, O'Keeffe M, Yaghi OM. Design and synthesis of an exceptionally stable and highly porous metal-organic framework. Nature. 1999;402:276.
[9] Tu TN, Nguyen MV, Nguyen HL, Yuliarto B, Cordova KE, Demir S. Designing bipyridine-functionalized zirconium metal–organic frameworks as a platform for clean energy and other emerging applications. Coord Chem Rev. 2018;364:33.
[10] Furukawa H, Cordova KE, O'Keeffe M, Yaghi OM. The chemistry and applications of metal-organic frameworks. Science. 2013;341:1230444.
[11] Huang YB, Liang J, Wang XS, Cao R. Multifunctional metal-organic framework catalysts: synergistic catalysis and tandem reactions. Chem Soc Rev. 2017;46:126.
[12] Adil K, Belmabkhout Y, Pillai RS, Cadiau A, Bhatt PM, Assen AH, Maurin G, Eddaoudi M. Gas/vapour separation using ultra-microporous metal-organic frameworks: insights into the structure/separation relationship. Chem Soc Rev. 2017;46:3402.
[13] Konstas K, Osl T, Yang Y, Batten M, Burke N, Hill AJ, Hill MR. Methane storage in metal organic frameworks. J Mater Chem. 2012:22.

[14] Meng X, Wang HN, Song SY, Zhang HJ. Proton-conducting crystalline porous materials. Chem Soc Rev. 2017;46:464.
[15] Nguyen MV, Lo THN, Luu LC, Nguyen HTT, Tu TN. Enhancing proton conductivity in a metal–organic framework at t > 80 °C by an anchoring strategy. J Mater Chem A. 1816;2018:6.
[16] Horcajada P, Chalati T, Serre C, Gillet B, Sebrie C, Baati T, Eubank JF, Heurtaux D, Clayette P, Kreuz C et al. Porous metal-organic-framework nanoscale carriers as a potential platform for drug delivery and imaging. Nat Mater. 2010;9:172.
[17] Horcajada P, Gref R, Baati T, Allan PK, Maurin G, Couvreur P, Ferey G, Morris RE, Serre C. Metal-organic frameworks in biomedicine. Chem Rev. 2012;112:1232.
[18] Peng Y, Krungleviciute V, Eryazici I, Hupp JT, Farha OK, Yildirim T. Methane storage in metal-organic frameworks: current records, surprise findings, and challenges. J Am Chem Soc. 2013;135:11887.
[19] Alezi D, Belmabkhout Y, Suyetin M, Bhatt PM, Weselinski LJ, Solovyeva V, Adil K, Spanopoulos I, Trikalitis PN, Emwas AH et al. MOF crystal chemistry paving the way to gas storage needs: aluminum-based soc-MOF for CH_4, O_2, and CO_2 storage. J Am Chem Soc. 2015;137:13308.
[20] Chen Z, Li P, Anderson R, Wang X, Zhang X, Robison L, Redfern LR, Moribe S, Islamoglu T, Gomez-Gualdron DA et al. Balancing volumetric and gravimetric uptake in highly porous materials for clean energy. Science. 2020;368:297.
[21] Lucena SM, Mileo PG, Silvino PF, Cavalcante CL Jr. Unusual adsorption site behavior in PCN-14 metal-organic framework predicted from Monte Carlo simulation. J Am Chem Soc. 2011;133:19282.
[22] Zhang M, Zhou W, Pham T, Forrest KA, Liu W, He Y, Wu H, Yildirim T, Chen B, Space B et al. Fine tuning of MOF-505 analogues to reduce low-pressure methane uptake and enhance methane working capacity. Angew Chem Int Ed. 2017;56:11426.
[23] He Y, Zhou W, Yildirim T, Chen B. A series of metal–organic frameworks with high methane uptake and an empirical equation for predicting methane storage capacity. Energy Environ Sci. 2013;6.
[24] Wen HM, Li B, Li L, Lin RB, Zhou W, Qian G, Chen B. A metal-organic framework with optimized porosity and functional sites for high gravimetric and volumetric methane storage working capacities. Adv Mater. 2018;30:e1704792.
[25] Wen HM, Shao K, Zhou W, Li B, Chen B. A novel expanded metal-organic framework for balancing volumetric and gravimetric methane storage working capacities. Chem Commun. 2020;56:13117.
[26] Peng Y, Srinivas G, Wilmer CE, Eryazici I, Snurr RQ, Hupp JT, Yildirim T, Farha OK. Simultaneously high gravimetric and volumetric methane uptake characteristics of the metal-organic framework NU-111. Chem Commun. 2013;49:2992.
[27] Li B, Wen HM, Wang H, Wu H, Tyagi M, Yildirim T, Zhou W, Chen B. A porous metal-organic framework with dynamic pyrimidine groups exhibiting record high methane storage working capacity. J Am Chem Soc. 2014;136:6207.
[28] Yan Y, Kolokolov DI, da Silva I, Stepanov AG, Blake AJ, Dailly A, Manuel P, Tang CC, Yang S, Schroder M. Porous metal-organic polyhedral frameworks with optimal molecular dynamics and pore geometry for methane storage. J Am Chem Soc. 2017;139:13349.
[29] Zhang M, Chen C, Wang Q, Fu W, Huang K, Zhou W. A metal–organic framework functionalized with piperazine exhibiting enhanced CH_4 storage. J Mater Chem A. 2017;5:349.
[30] Shao K, Pei J, Wang JX, Yang Y, Cui Y, Zhou W, Yildirim T, Li B, Chen B, Qian G. Tailoring the pore geometry and chemistry in microporous metal-organic frameworks for high methane storage working capacity. Chem Commun. 2019;55:11402.
[31] Tu TN, Nguyen HTD, Tran NT. Tailoring the pore size and shape of the one-dimensional channels in iron-based MOFs for enhancing the methane storage capacity. Inorg Chem Front. 2019;6:2441.

[32] Furukawa H, Ko N, Go YB, Aratani N, Choi SB, Choi E, Yazaydin AO, Snurr RQ, O'Keeffe M, Kim J et al. Ultrahigh porosity in metal-organic frameworks. Science. 2010;329:424.
[33] Lin JM, He CT, Liu Y, Liao PQ, Zhou DD, Zhang JP, Chen XM. A metal-organic framework with a pore size/shape suitable for strong binding and close packing of methane. Angew Chem Int Ed. 2016;55:4674.
[34] Zhang ZH, Fang H, Xue DX, Bai J. Tuning open metal site-free ncb type of metal-organic frameworks for simultaneously high gravimetric and volumetric methane storage working capacities. ACS Appl Mater Interfaces. 2021;13:44956.
[35] Liang CC, Shi ZL, He CT, Tan J, Zhou HD, Zhou HL, Lee Y, Zhang YB. Engineering of pore geometry for ultrahigh capacity methane storage in mesoporous metal-organic frameworks. J Am Chem Soc. 2017;139:13300.
[36] Kundu T, Shah BB, Bolinois L, Zhao D. Functionalization-induced breathing control in metal–organic frameworks for methane storage with high deliverable capacity. Chem Mater. 2019;31:2842.
[37] Mason JA, Oktawiec J, Taylor MK, Hudson MR, Rodriguez J, Bachman JE, Gonzalez MI, Cervellino A, Guagliardi A, Brown CM et al. Methane storage in flexible metal-organic frameworks with intrinsic thermal management. Nature. 2015;527:357.
[38] Kokcam-Demir U, Goldman A, Esrafili L, Gharib M, Morsali A, Weingart O, Janiak C. Coordinatively unsaturated metal sites (open metal sites) in metal-organic frameworks: design and applications. Chem Soc Rev. 2020;49:2751.
[39] Wu H, Zhou W, Yildirim T. High-capacity methane storage in metal-organic frameworks M_2(dhtp): the important role of open metal sites. J Am Chem Soc. 2009;131:4995.
[40] Sillar K, Sauer J. Ab initio prediction of adsorption isotherms for small molecules in metal-organic frameworks: the effect of lateral interactions for methane/CPO-27-Mg. J Am Chem Soc. 2012;134:18354.
[41] Chui SS, Lo SM, Charmant JP, Orpen AG, Williams ID. A chemically functionalizable nanoporous material. Science. 1999;283:1148.
[42] Millward AR, Yaghi OM. Metal-organic frameworks with exceptionally high capacity for storage of carbon dioxide at room temperature. J Am Chem Soc. 2005;127:17998.
[43] Guo Z, Wu H, Srinivas G, Zhou Y, Xiang S, Chen Z, Yang Y, Zhou W, O'Keeffe M, Chen B. A metal-organic framework with optimized open metal sites and pore spaces for high methane storage at room temperature. Angew Chem Int Ed. 2011;50:3178.
[44] Li B, Wen HM, Wang H, Wu H, Yildirim T, Zhou W, Chen B. Porous metal–organic frameworks with Lewis basic nitrogen sites for high-capacity methane storage. Energy Environ Sci. 2015;8:2504.
[45] Hou J, Sapnik AF, Bennett TD. Metal-organic framework gels and monoliths. Chem Sci. 2020;11:310.
[46] Tian T, Zeng Z, Vulpe D, Casco ME, Divitini G, Midgley PA, Silvestre-Albero J, Tan JC, Moghadam PZ, Fairen-Jimenez D. A sol-gel monolithic metal-organic framework with enhanced methane uptake. Nat Mater. 2018;17:174.
[47] Connolly BM, Aragones-Anglada M, Gandara-Loe J, Danaf NA, Lamb DC, Mehta JP, Vulpe D, Wuttke S, Silvestre-Albero J, Moghadam PZ et al. Tuning porosity in macroscopic monolithic metal-organic frameworks for exceptional natural gas storage. Nat Commun. 2019;10:2345.
[48] Singla MK, Nijhawan P, Oberoi AS. Hydrogen fuel and fuel cell technology for cleaner future: a review. Environ Sci Pollut Res Int. 2021;28:15607.
[49] Balat M. Possible methods for hydrogen production. Energy Sources A: Recovery Util Environ Eff. 2008;31:39.
[50] Nassar E, Nassar A. Corrosion behaviour of some conventional stainless steels at different temperatures in the electrolyzing process. Energy Proc. 2016;93:102.

[51] Pareek A, Dom R, Gupta J, Chandran J, Adepu V, Borse PH. Insights into renewable hydrogen energy: recent advances and prospects. Mater Sci Energy Technol. 2020;3:319.
[52] UNFCCC D. Paris Climate Change Conference. 2015.
[53] Aziz M. Liquid hydrogen: a review on liquefaction, storage, transportation, and safety. Energies. 2021;14.
[54] Andersson J, Grönkvist S. Large-scale storage of hydrogen. Int J Hydrog Energy. 2019;44:11901.
[55] Panella B, Hirscher M, Roth S. Hydrogen adsorption in different carbon nanostructures. Carbon. 2005;43:2209.
[56] Ritschel M, Uhlemann M, Gutfleisch O, Leonhardt A, Graff A, Täschner C, Fink J. Hydrogen storage in different carbon nanostructures. Appl Phys Lett. 2002;80:2985.
[57] Samantaray SS, Mangisetti SR, Ramaprabhu S. Investigation of room temperature hydrogen storage in biomass derived activated carbon. J Alloys Compd. 2019;789:800.
[58] Langmi HW, Walton A, Al-Mamouri MM, Johnson SR, Book D, Speight JD, Edwards PP, Gameson I, Anderson PA, Harris IR. Hydrogen adsorption in zeolites A, X, Y and RHO. J Alloys Compd. 2003;356:710.
[59] Langmi HW, Book D, Walton A, Johnson SR, Al-Mamouri MM, Speight JD, Edwards PP, Harris IR, Anderson PA. Hydrogen storage in ion-exchanged zeolites. J Alloys Compd. 2005;404:637.
[60] Zhou L, Zhou Y, Sun Y. Studies on the mechanism and capacity of hydrogen uptake by physisorption-based materials. Int J Hydrog Energy. 2006;31:259.
[61] Zhou L, Zhou Y, Sun Y. Studies on the mechanism and capacity of hydrogen uptake by physisorption-based materials. Int J Hydrog Energy. 2006;31:259.
[62] Xu W, Takahashi K, Matsuo Y, Hattori Y, Kumagai M, Ishiyama S, Kaneko K, Iijima S. Investigation of hydrogen storage capacity of various carbon materials. Int J Hydrog Energy. 2007;32:2504.
[63] Sevilla M, Foulston R, Mokaya R. Superactivated carbide-derived carbons with high hydrogenstorage capacity. Energy Environ Sci. 2010;3:223.
[64] Dillon AC, Jones KM, Bekkedahl TA, Kiang CH, Bethune DS, Heben MJ. Storage of hydrogen in single-walled carbon nanotubes. Nature. 1997;386:377.
[65] Poirier E. Hydrogen adsorption in carbon nanostructures. Int J Hydrog Energy. 2001;26:831.
[66] Poirier E, Chahine R, Bénard P, Cossement D, Lafi L, Mélançon E, Bose TK, Désilets S. Storage of hydrogen on single-walled carbon nanotubes and other carbon structures. Appl Phys A. 2004;78:961.
[67] Chen P, Wu X, Lin J, Tan KL. High H_2 uptake by alkali-doped carbon nanotubes under ambient pressure and moderate temperatures. Science. 1999;285:91.
[68] Reyhani A, Mortazavi SZ, Mirershadi S, Moshfegh AZ, Parvin P, Golikand AN. Hydrogen storage in decorated multiwalled carbon nanotubes by Ca, Co, Fe, Ni, and Pd nanoparticles under ambient conditions. J Phys Chem C. 2011;115:6994.
[69] Ariharan A, Viswanathan B, Nandhakumar V. Nitrogen-incorporated carbon nanotube derived from polystyrene and polypyrrole as hydrogen storage material. Int J Hydrog Energy. 2018;43:5077.
[70] Rosi NL, Eckert J, Eddaoudi M, Vodak DT, Kim J, O'Keeffe M, Yaghi OM. Hydrogen storage in microporous metal-organic frameworks. Science. 2003;300:1127.
[71] Furukawa H, Miller MA, Yaghi OM. Independent verification of the saturation hydrogen uptake in MOF-177 and establishment of a benchmark for hydrogen adsorption in metal–organic frameworks. J Mater Chem. 2007;17.
[72] Li Y, Yang RT. Gas adsorption and storage in metal-organic framework MOF-177. Langmuir. 2007;23:12937.
[73] Farha OK, Yazaydin AO, Eryazici I, Malliakas CD, Hauser BG, Kanatzidis MG, Nguyen ST, Snurr RQ, Hupp JT. De novo synthesis of a metal-organic framework material featuring ultrahigh surface area and gas storage capacities. Nat Chem. 2010;2:944.

[74] Farha OK, Eryazici I, Jeong NC, Hauser BG, Wilmer CE, Sarjeant AA, Snurr RQ, Nguyen ST, Yazaydin AO, Hupp JT. Metal-organic framework materials with ultrahigh surface areas: is the sky the limit? J Am Chem Soc. 2012;134:15016.
[75] Ding L, Yazaydin AO. Hydrogen and methane storage in ultrahigh surface area metal–organic frameworks. Microporous Mesoporous Mater. 2013;182:185.
[76] Ferey G, Latroche M, Serre C, Millange F, Loiseau T, Percheron-Guegan A. Hydrogen adsorption in the nanoporous metal-benzenedicarboxylate M(OH) (O_2C-C_6H_4-CO_2) (M = Al^{3+}, Cr^{3+}), MIL-53. Chem Commun. 2003:2976.
[77] Panella B, Hirscher M, Pütter H, Müller U. Hydrogen adsorption in metal–organic frameworks: Cu-MOFs and Zn-MOFs compared. Adv Funct Mater. 2006;16:520.
[78] Sumida K, Hill MR, Horike S, Dailly A, Long JR. Synthesis and hydrogen storage properties of Be(12) (OH) (12) (1, 3,5-benzenetribenzoate). J Am Chem Soc. 2009;131:15120.
[79] Han SS, Furukawa H, Yaghi OM, Goddard WA. 3rd. Covalent organic frameworks as exceptional hydrogen storage materials. J Am Chem Soc. 2008;130:11580.
[80] Furukawa H, Yaghi OM. Storage of hydrogen, methane, and carbon dioxide in highly porous covalent organic frameworks for clean energy applications. J Am Chem Soc. 2009;131:8875.
[81] Zelenak V, Saldan I. Factors affecting hydrogen adsorption in metal-organic frameworks: a short review. Nanomaterials. 2021;11.
[82] Li H, Wang K, Sun Y, Lollar CT, Li J, Zhou HC. Recent advances in gas storage and separation using metal–organic frameworks. Mater Today. 2018;21:108.
[83] Zou L, Zhou HC. Hydrogen storage in metal-organic frameworks. In: Chen YP, Bashir S, Liu J, editors. Nanostructured Materials for Next-Generation Energy Storage and Conversion. Berlin, Heidelberg: Springer; 2017.
[84] Lee YG, Moon HR, Cheon YE, Suh MP. A comparison of the H_2 sorption capacities of isostructural metal-organic frameworks with and without accessible metal sites: [{Zn_2(abtc) $(dmf)_2$}$_3$] and [{Cu_2(abtc) $(dmf)_2$}$_3$] versus [{Cu_2(abtc)}$_3$]. Angew Chem Int Ed. 2008;47:7741.
[85] Prasad TK, Suh MP. Control of interpenetration and gas-sorption properties of metal-organic frameworks by a simple change in ligand design. Chem-Eur J. 2012;18:8673.
[86] Surble S, Millange F, Serre C, Duren T, Latroche M, Bourrelly S, Llewellyn PL, Ferey G. Synthesis of MIL-102, a chromium carboxylate metal-organic framework, with gas sorption analysis. J Am Chem Soc. 2006;128:14889.
[87] Yuan D, Zhao D, Sun D, Zhou HC. An isoreticular series of metal-organic frameworks with dendritic hexacarboxylate ligands and exceptionally high gas-uptake capacity. Angew Chem Int Ed. 2010;49:5357.
[88] Latroche M, Surble S, Serre C, Mellot-Draznieks C, Llewellyn PL, Lee JH, Chang JS, Jhung SH, Ferey G. Hydrogen storage in the giant-pore metal-organic frameworks MIL-100 and MIL-101. Angew Chem Int Ed. 2006;45:8227.
[89] Wong-Foy AG, Lebel O, Matzger AJ. Porous crystal derived from a tricarboxylate linker with two distinct binding motifs. J Am Chem Soc. 2007;129:15740.
[90] Ibarra IA, Yang S, Lin X, Blake AJ, Rizkallah PJ, Nowell H, Allan DR, Champness NR, Hubberstey P, Schroder M. Highly porous and robust scandium-based metal-organic frameworks for hydrogen storage. Chem Commun. 2011;47:8304.
[91] Sumida K, Her JH, Dincă M, Murray LJ, Schloss JM, Pierce CJ, Thompson BA, FitzGerald SA, Brown CM, Long JR. Neutron scattering and spectroscopic studies of hydrogen adsorption in $Cr_3(BTC)_2$—A metal–organic framework with exposed Cr^{2+} sites. J Phys Chem C. 2011;115:8414.
[92] Panella B, Hirscher M. Hydrogen physisorption in metal-organic porous crystals. Adv Mater. 2005;17:538.
[93] Rowsell JL, Yaghi OM. Effects of functionalization, catenation, and variation of the metal oxide and organic linking units on the low-pressure hydrogen adsorption properties of metal-organic frameworks. J Am Chem Soc. 2006;128:1304.

[94] Koh K, Wong-Foy AG, Matzger AJ. A porous coordination copolymer with over 5000 m^2/g BET surface area. J Am Chem Soc. 2009;131:4184.
[95] Sumida K, Brown CM, Herm ZR, Chavan S, Bordiga S, Long JR. Hydrogen storage properties and neutron scattering studies of Mg_2(dobdc)–a metal-organic framework with open Mg^{2+} adsorption sites. Chem Commun. 2011;47:1157.
[96] Suh MP, Park HJ, Prasad TK, Lim DW. Hydrogen storage in metal-organic frameworks. Chem Rev. 2012;112:782.
[97] Kapelewski MT, Geier SJ, Hudson MR, Stuck D, Mason JA, Nelson JN, Xiao DJ, Hulvey Z, Gilmour E, FitzGerald SA et al. M_2(m-dobdc) (M = Mg, Mn, Fe, Co, Ni) metal-organic frameworks exhibiting increased charge density and enhanced H_2 binding at the open metal sites. J Am Chem Soc. 2014;136:12119.
[98] Kim H, Das S, Kim MG, Dybtsev DN, Kim Y, Kim K. Synthesis of phase-pure interpenetrated MOF-5 and its gas sorption properties. Inorg Chem. 2011;50:3691.
[99] Park HJ, Suh MP. Mixed-ligand metal-organic frameworks with large pores: gas sorption properties and single-crystal-to-single-crystal transformation on guest exchange. Chem-Eur J. 2008;14:8812.
[100] Barman S, Furukawa H, Blacque O, Venkatesan K, Yaghi OM, Berke H. Azulene based metal-organic frameworks for strong adsorption of H_2. Chem Commun. 2010;46:7981.
[101] Sava DF, Kravtsov V, Nouar F, Wojtas L, Eubank JF, Eddaoudi M. Quest for zeolite-like metal-organic frameworks: on pyrimidinecarboxylate bis-chelating bridging ligands. J Am Chem Soc. 2008;130:3768.
[102] Cheon YE, Suh MP. Selective gas adsorption in a microporous metal-organic framework constructed of Co^{II}_4 clusters. Chem Commun. 2009:2296.
[103] Senkovska I, Kaskel S. Solvent-induced pore-size adjustment in the metal-organic framework $[Mg_3(ndc)_3(dmf)_4]$ (ndc = naphthalenedicarboxylate). Eur J Inorg Chem. 2006;2006:4564.
[104] Gygi D, Bloch ED, Mason JA, Hudson MR, Gonzalez MI, Siegelman RL, Darwish TA, Queen WL, Brown CM, Long JR. Hydrogen storage in the expanded pore metal–organic frameworks M_2(dobpdc) (M = Mg, Mn, Fe, Co, Ni, Zn). Chem Mater. 2016;28:1128.
[105] Wahiduzzaman Allmond K, Stone J, Harp S, Mujibur K. Synthesis and electrospraying of nanoscale MOF (metal organic framework) for high-performance CO_2 adsorption membrane. Nanoscale Res Lett. 2017;12:6.
[106] Kapelewski MT, Runcevski T, Tarver JD, Jiang HZH, Hurst KE, Parilla PA, Ayala A, Gennett T, FitzGerald SA, Brown CM et al. Record high hydrogen storage capacity in the metal-organic framework Ni2(m-dobdc) at near-ambient temperatures. Chem Mater. 2018;30.
[107] Dinca M, Dailly A, Liu Y, Brown CM, Neumann DA, Long JR. Hydrogen storage in a microporous metal-organic framework with exposed Mn^{2+} coordination sites. J Am Chem Soc. 2006;128:16876.
[108] Dinca M, Han WS, Liu Y, Dailly A, Brown CM, Long JR. Observation of Cu^{2+}-H_2 interactions in a fully desolvated sodalite-type metal-organic framework. Angew Chem Int Ed. 2007;46:1419.
[109] Balderas-Xicohténcatl R, Schlichtenmayer M, Hirscher M. Volumetric hydrogen storage capacity in metal-organic frameworks. Energy Technol. 2018;6:578.
[110] Boateng E, Chen A. Recent advances in nanomaterial-based solid-state hydrogen storage. Mater Today Adv. 2020;6. https://doi.org/10.1016/j.mtadv.2019.100022.
[111] Du Xm, Wu Ed. Physisorption of hydrogen in A, X and ZSM-5 types of zeolites at moderately high pressures. Chin J Chem Phys. 2006;19:457.
[112] Nijkamp MG, Raaymakers JEMJ, van Dillen AJ, de Jong KP. Hydrogen storage using physisorption – materials demands. Appl Phys A, Mater Sci Process. 2001;72:619.
[113] Honicke IM, Senkovska I, Bon V, Baburin IA, Bonisch N, Raschke S, Evans JD, Kaskel S. Balancing mechanical stability and ultrahigh porosity in crystalline framework materials. Angew Chem Int Ed. 2018;57:13780.

[114] Chae HK, Siberio-Perez DY, Kim J, Go Y, Eddaoudi M, Matzger AJ, O'Keeffe M, Yaghi OM. A route to high surface area, porosity and inclusion of large molecules in crystals. Nature. 2004;427:523.
[115] Lin X, Jia J, Zhao X, Thomas KM, Blake AJ, Walker GS, Champness NR, Hubberstey P, Schroder M. High H2 adsorption by coordination-framework materials. Angew Chem Int Ed. 2006;45:7358.
[116] Langmi HW, Ren J, North B, Mathe M, Bessarabov D. Hydrogen storage in metal-organic frameworks: a review. Electrochim Acta. 2014;128:368.
[117] Eddaoudi M, Kim J, Rosi N, Vodak D, Wachter J, O'Keeffe M, Yaghi OM. Systematic design of pore size and functionality in isoreticular MOFs and their application in methane storage. Science. 2002;295:469.
[118] Botas JA, Calleja G, Sanchez-Sanchez M, Orcajo MG. Cobalt doping of the MOF-5 framework and its effect on gas-adsorption properties. Langmuir. 2010;26:5300.
[119] Yaghi OM, O'Keeffe M, Ockwig NW, Chae HK, Eddaoudi M, Kim J. Reticular synthesis and the design of new materials. Nature. 2003;423:705.
[120] Botas JA, Calleja G, Sánchez-Sánchez M, Orcajo MG. Effect of Zn/Co ratio in MOF-74 type materials containing exposed metal sites on their hydrogen adsorption behaviour and on their band gap energy. Int J Hydrog Energy. 2011;36:10834.
[121] Chu CL, Chen JR, Lee TY. Enhancement of hydrogen adsorption by alkali-metal cation doping of metal-organic framework-5. Int J Hydrog Energy. 2012;37:6721.
[122] Mulfort KL, Hupp JT. Alkali metal cation effects on hydrogen uptake and binding in metal-organic frameworks. Inorg Chem. 2008;47:7936.
[123] Wang XS, Ma S, Rauch K, Simmons JM, Yuan D, Wang X, Yildirim T, Cole WC, López JJ, Meijere Ad et al. Metal–organic frameworks based on double-bond-coupled Di-isophthalate linkers with high hydrogen and methane uptakes. Chem Mater. 2008;20:3145.
[124] Tranchemontagne DJ, Park KS, Furukawa H, Eckert J, Knobler CB, Yaghi OM. Hydrogen storage in new metal–organic frameworks. J Phys Chem C. 2012;116:13143.
[125] Rowsell JL, Millward AR, Park KS, Yaghi OM. Hydrogen sorption in functionalized metal-organic frameworks. J Am Chem Soc. 2004;126:5666.
[126] Yang S, Lin X, Dailly A, Blake AJ, Hubberstey P, Champness NR, Schroder M. Enhancement of H_2 adsorption in coordination framework materials by use of ligand curvature. Chem-Eur J. 2009;15:4829.
[127] Lin X, Telepeni I, Blake AJ, Dailly A, Brown CM, Simmons JM, Zoppi M, Walker GS, Thomas KM, Mays TJ et al. High capacity hydrogen adsorption in Cu(II) tetracarboxylate framework materials: the role of pore size, ligand functionalization, and exposed metal sites. J Am Chem Soc. 2009;131:2159.
[128] Chang Z, Zhang DS, Chen Q, Li RF, Hu TL, Bu XH. Rational construction of 3D. pillared metal-organic frameworks: synthesis, structures, and hydrogen adsorption properties. Inorg Chem. 2011;50:7555.
[129] Wang Z, Tanabe KK, Cohen SM. Tuning hydrogen sorption properties of metal-organic frameworks by postsynthetic covalent modification. Chem-Eur J. 2010;16:212.
[130] Wang Z, Tanabe KK, Cohen SM. Accessing postsynthetic modification in a series of metal-organic frameworks and the influence of framework topology on reactivity. Inorg Chem. 2009;48:296.
[131] Cadman LK, Bristow JK, Stubbs NE, Tiana D, Mahon MF, Walsh A, Burrows AD. Compositional control of pore geometry in multivariate metal-organic frameworks: an experimental and computational study. Dalton Trans. 2016;45:4316.
[132] Saha D, Wei Z, Deng S. Equilibrium, kinetics and enthalpy of hydrogen adsorption in MOF-177. Int J Hydrog Energy. 2008;33:7479.

[133] Kim HK, Yun WS, Kim MB, Kim JY, Bae YS, Lee J, Jeong NC. A chemical route to activation of open metal sites in the copper-based metal-organic framework materials HKUST-1 and Cu-MOF-2. J Am Chem Soc. 2015;137:10009.
[134] Jaramillo DE, Jiang HZH, Evans HA, Chakraborty R, Furukawa H, Brown CM, Head-Gordon M, Long JR. Ambient-temperature hydrogen storage via vanadium(II)-dihydrogen complexation in a metal-organic framework. J Am Chem Soc. 2021;143:6248.
[135] Zhou W, Wu H, Yildirim T. Enhanced H_2 adsorption in isostructural metal-organic frameworks with open metal sites: strong dependence of the binding strength on metal ions. J Am Chem Soc. 2008;130:15268.
[136] Liu Y, Kabbour H, Brown CM, Neumann DA, Ahn CC. Increasing the density of adsorbed hydrogen with coordinatively unsaturated metal centers in metal-organic frameworks. Langmuir. 2008;24:4772.

Patrick Damacet, Karen Hannouche, Rawan Al Natour, and
Mohamad Hmadeh

9 Catalytic transformations in metal-organic framework systems

9.1 Introduction

The field of catalysis has been receiving great attention due to the production of more than 90 % of chemicals worldwide via catalytic processes with a market size that is expected to reach 48 billion United States dollars by 2027 [1, 2]. The first inorganic catalyst developed dates back to 1552 when Valerius Cordus used sulfuric acid to convert ethanol to ethyl ether [3]. This was followed by the development of catalytic systems based on transition metal coordination complexes as well as metal oxides for environmental and biomedical applications between the 17th and 20th centuries [4]. Despite their good performance, these systems were proven to suffer many limitations. For instance, many experience metal leaching into the solution, which in turn creates a major challenge in reusing them in different runs. Moreover, the difficulty of separating and extracting these metal species from the product(s) limits their efficiency as they are mostly nonselective to chiral catalysis and involve complicated synthesis approaches [5–7].

Due to the remarkable increase in the world population and the huge expansion of most industry sectors during the past century, the field of catalysis has played a fundamental role in food processing, petroleum refining and valorization of plastic waste [8–10]. Several key events have been achieved over the course of catalysis development: the Haber–Bosch process in 1909 was employed for the production of ammonia from dinitrogen and dihydrogen using metal species as catalysts, including osmium, iron and uranium. This process is responsible for the production of more than 96 % of the global anhydrous ammonia, which is equivalent to approximately 150 million metric tons yr^{-1} [11]. In addition to the impact that the Haber–Bosch process has had on agriculture and sustainability, another type of catalysts, known as Ziegler–Natta catalysts, has played an enormous role in ethylene polymerization since the 1950s with more than 100 million tons of plastics and rubbers being produced yearly via these catalysts [12]. In total, dozens of scientists have been awarded the Nobel Prize since its inception in 1901 for their work on various catalytic systems. In fact, recently the 2021 Nobel Prize in chemistry was awarded to Benjamin List and David MacMillan for the development of a new type of catalysis primarily known as asymmetric

Patrick Damacet, Karen Hannouche, Rawan Al Natour, Mohamad Hmadeh, Department of Chemistry, Faculty of Arts and Sciences, American University of Beirut, P. O. Box 11-0236 Riad El Solh, Beirut 1107-2020, Lebanon, e-mail: mohamad.hmadeh@aub.edu.lb

https://doi.org/10.1515/9781501524721-009

organocatalysis that is believed by scientists to be a future contributor to the development of new types of drugs and high-tech materials [13]. Today, chemical industries require 42 EJ/year which account for 10 % of the global energy demand and are responsible for the emission of 7 % of global greenhouse gases (GHG). Studying the case of the top 18 large-volume chemicals, which use about 80 % of the yearly energy demand of chemical industries and emit about 75 % of their GHG, the proper improvement of catalytic processes is expected to reduce these intensities by 20–40 % by 2050. In brief, developing new advanced catalysts is a global need and challenge, especially that many extensively studied catalytic systems have reached a stage of incremental improvements. Thus, the exploitation of an emerging field like MOFs is necessary to achieve rapid growth in catalytic processes.

9.2 Catalysis by MOFs: characteristics, advantages and strategies

Metal-organic frameworks (MOFs) possess a wide variety of chemical and physical properties that make them one of the most attractive types of porous, crystalline materials [14]. Known for their unique structural characteristics and versatility, including their high porosities and surface areas, ability to be tailored, modified and functionalized in various manners compared to the other conventional porous materials, MOFs have been widely investigated as heterogeneous catalysts for challenging catalytic reactions and transformations [15]. For instance, due to the exceptional chemical and thermal stability of most MOFs, the opportunity to introduce complementary catalytically active groups into the framework via *de novo* synthesis and post-synthetic modification is feasible in most, if not all, MOFs. Furthermore, their high surface areas and pore volumes enable the integration and uniform dispersion of a high density of catalytically-active elements within the framework, which in turn, lead to an enhancement in the catalytic efficiency of these elements [16]. Moreover, MOFs nanoparticles prohibit the accumulation, aggregation and agglomeration of the active guest molecules by trapping or immobilizing them within their channels [17]. This results in an improvement in the long-term catalytic efficiency. Similarly, MOFs solids can be easily isolated and recovered through centrifugation or filtration while maintaining and preserving a high activity over multiple reaction runs, which is in stark contrast to molecular catalysts that suffer from degradation, decomposition and decreased reactivity after only a few runs [18]. Likewise, in the case where the MOF itself possesses catalytic active sites, the incorporation of active guest species via *in situ* or post-synthetic modification turns the framework into a dual-function heterogeneous catalyst. This results in both the MOF and the guest species to act as active symbiotic centers for various catalytic transformation reactions [19]. Finally, catalytic reaction mechanisms of the interactions between the active sites in the MOF and the molecular reactants can be readily investigated using spectroscopy techniques including transmission electron microscopy coupled with energy dispersive X-ray spectroscopy (TEM-EDC), Fourier-transform infrared spectroscopy (FT-IR), powder X-ray diffraction (PXRD) and X-ray photoelectron spectroscopy (XPS) due to their well-defined atomic structure [20].

Realizing the potential of MOFs as heterogeneous catalysts has been the focus of intensive studies since the discovery of these materials. This can be clearly seen by the swift growth of research on MOF-mediated catalysis with a considerable number of publications being reported for a selected few types of highly stable and porous MOFs (Figure 9.1).

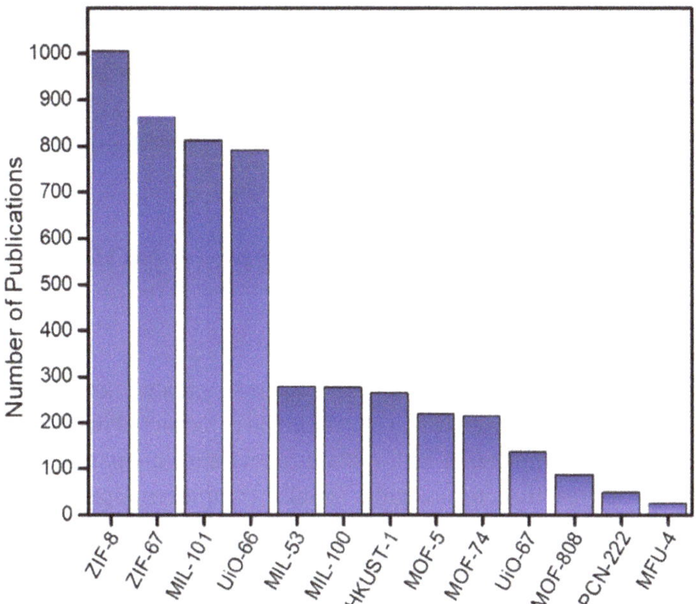

Figure 9.1: Bar graph illustrating the number of publications from Web of Science (as of December 2021) using select types of MOFs as heterogeneous catalysts.

In the following sections, the four main strategies employed to realize MOF catalysis will be thoroughly investigated. As illustrated in Figure 9.2 these include: (i) usage of metal clusters and coordinatively unsaturated metal sites (CUSs) as catalytic sites, (ii) incorporation and usage of functional groups as actives sites on the organic linkers, (iii) loading and encapsulation of active guest species, including metal clusters and enzymes within the pores and (iv) surface of a given MOF.

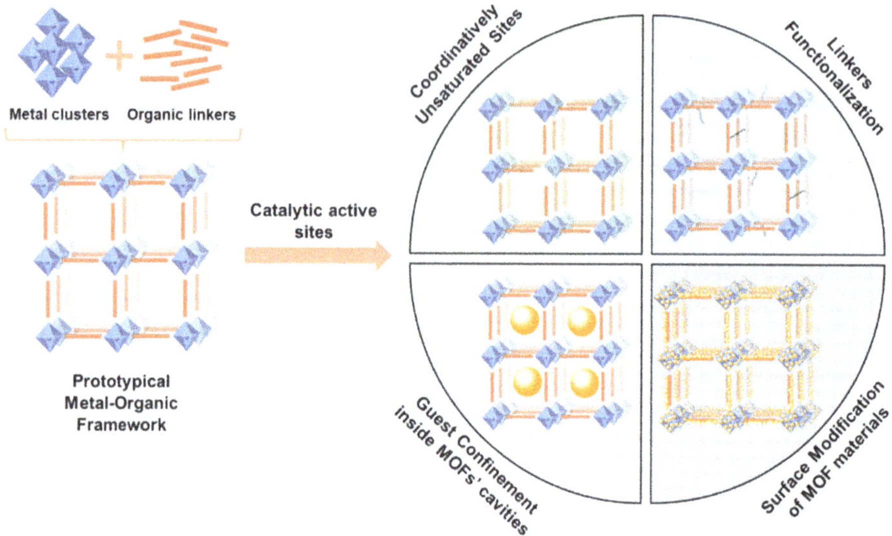

Figure 9.2: Schematic representation of the four main strategies for creating catalytically-active sites in MOFs.

9.3 Metal nodes as active catalytic sites in MOF materials

Metal nodes or metal clusters can form advantageous intrinsic catalytic centers in MOFs and the wide variety of metals from which MOFs can be constructed provides a large catalogue of catalytic possibilities. In fact, metallic active sites are particularly useful due to their analogous reactivity to molecular metallic complexes applied in reported homogeneous catalytic reactions while making use of MOFs structural heterogeneity and characteristics [21]. These metal sites perform their catalytic role in light of their acidic or basic properties. Compared to the extensive work reported for MOFs serving as Lewis catalysts, metal centers acting as Brønsted acids or bases are less explored, yet have recently been targeted [22]. This section aims to discuss in length each catalytic metal category present in MOF SBUs with its performance in selected chemical reactions. Figure 9.3 demonstrates the presence of CUSs in various secondary building units that are used to construct MOFs.

9.3.1 Catalysis by Lewis acid metal nodes

Catalysis by Lewis acid metal nodes in MOFs constitutes the bulk of studied reactions. This type of catalysis is possible when CUSs are present (i. e., sites where the metal node coordination number is lower than the actual number of connected

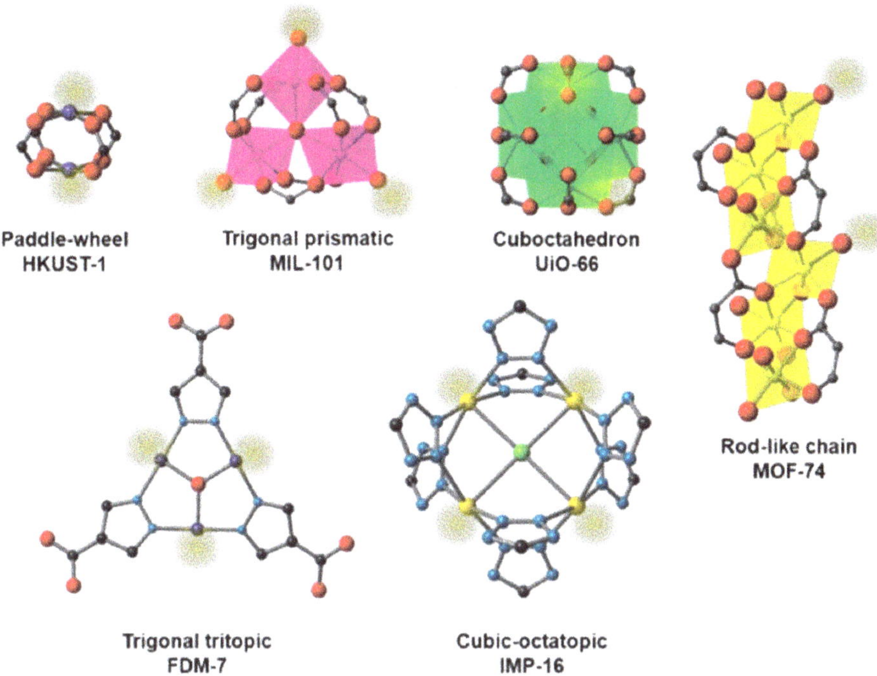

Figure 9.3: Graphical illustration of MOF secondary building units (SBUs) possessing coordinatively unsaturated metal sites as catalytically-active sites. Atom colors: C, gray; O, red; N, cyan and Cl, pink.

molecules) [23]. For example, for a $3d^6$ metal, an octahedral geometry is expected, but a missing ligand would lower the coordination number to five and create one free coordination site [24]. These sites are usually occupied by solvent/modulator molecules originating from the synthesis media and special care must be taken to generate these sites without collapsing the whole framework. It is for this purpose that (1) solvent exchange followed by a thermal treatment, (2) chemical activation or (3) photochemical activation are followed in practice [25]. The coordination state of a metal site can also be altered by adding acidic modulators that compete with the linker in coordinating with metal nodes, thereby creating catalytically-active CUSs as has been reported for UiO-66 (Figure 9.4) [26]. The CUSs are the result of defect creation within the framework that occur via either missing linker or missing cluster defects. After formation, the CUSs are able to accept electron density from the reactive donor molecule and in turn, act as a Lewis acid catalyst in the corresponding reaction [27]. As an aside, the degree of the electron withdrawing or donating characteristic of the linker used distorts the electron density and influences the Lewis catalytic activity of metal centers as well [28].

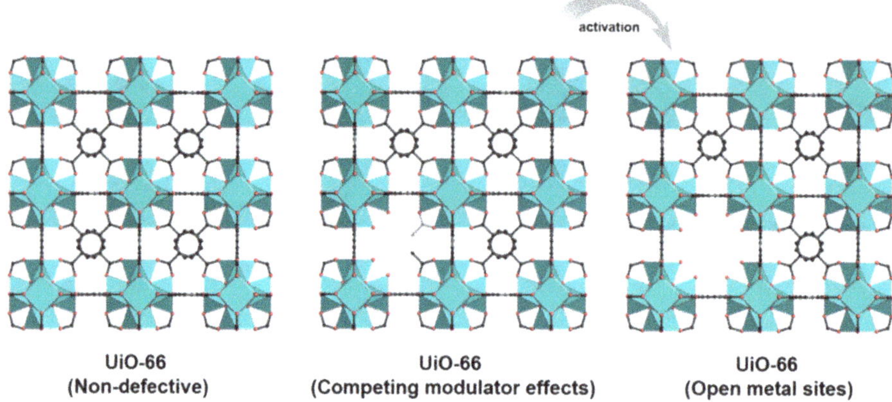

Figure 9.4: Illustration of how modulators compete with linkers in UiO-MOFs to yield CUSs. Atom colors: Zr, blue polyhedra; C, gray and O, red.

Fujita et al. were the first to report a two-dimensional square network material Cd(4,4'-bpy)$_2$(NO$_3$)$_2$ that they showed functioned as a Lewis acid catalyst in the cyanosilylation of aldehydes [29]. Since then, this particular reaction has been used to test and rank the Lewis acidity of MOFs. Numerous active metal nodes/clusters were reported for catalytic activity in MOFs due to their ability to form CUSs. Among them, the Cu$_2$-paddle wheel in HKUST-1 has a CUS within the interior of the pore wall that not only renders it more accessible to substrates, but also enhances its catalytic activity [30]. After activation, the accessible copper sites oriented toward the pore center proved to catalyze the cyanosilylation of benzaldehyde reaching a yield of 57 %. In parallel, the catalytic activity of HKUST-1 was compared to another Cu-based MOF in the acylation of anisole: Cu-MOF-74. This framework adopts a honeycomb-like structure with higher surface area and metal content as compared to the very interconnected paddle-wheel clusters of HKUST-1 [31]. Therefore, when assessed as catalysts for the acylation of anisole, Cu-MOF-74 afforded 98 % conversion in 1 h as compared to 79 % for HKUST-1.

CUSs are selective in their catalytic performance in various ways. The first factor to consider revolves around size constraints, which refer to the bulkiness of the substrate. A representative example is Mn$_3$[(Mn$_4$Cl)$_3$(BTT)$_8$(CH$_3$OH)$_{10}$]$_2$ (BTT = 1,3,5-benzenetristetrazolate) that was shown to selectively catalyze the cyanosilylation of 1-naphtaldehyde (9.7 × 8.4 Å^2 dimensions) in 90 % yield. When 4-phenoxybenzaldehyde or 4-phenylbenzaldehyde were used as the substrates (13.3 × 7.3 Å^2 and 13.1 × 6.7 Å^2, respectively), the cyanosilylation produced the corresponding products with less than 20 % yield [32]. The substitution pattern of the substrate is also related to its bulkiness and influences the performance of a given MOF catalyst. Actually, IRMOF-1 and IRMOF-8 (IRMOF = isoreticular MOF) were shown to selectively catalyze the *tert*-butylation of para-*tert*butylbiphenyl in 96 and 95 % conversion, respectively. This conversion was minimal for the ortho-isomer (3 %) with little to no traces of the dialkylated product; thus outperforming homogeneous metal catalysts in terms of selectivity for this particular reaction [33]. Similar to most common zeolites and metal nanoparticles, the particle size of MOFs plays a role in the selectivity of the chosen catalyst. In fact, in the catalysis of citronellal conversion, Cu-BTC demonstrated an 18 % enhancement in conversion as the particle size was reduced by half. This is primarily due to diffusion limitations observed for bigger crystals citronellal has hindered access to the framework's pores [34].

Some clusters with CUSs were able to show significant catalytic activity upon introducing defects or modified linkers into the framework. For instance, Vermoortele et al. reported the effect of using trifluoroacetic acid (TFA) as a modulator on the catalytic activity of the widely known UiO-66 for the cyclization of citronellal [35]. The modulator creates defects in the framework structure by partially removing/replacing the linkers, thus increasing the number of CUSs on the $Zr_6O_4(OH)_4^{12+}$ cluster. As the concentration of TFA increases from 0 to 20 molar equivalents, the number of CUSs increases in addition to an increase in pore diameter from 0.78 nm to 1–1.1 nm. As a result of this space enlargement around the Zr clusters, catalytic conversion is enhanced from 34 % to 75 % over a reaction time of 10 h. In the same work, the effect of linker functionalization on the Lewis acidic catalytic properties of metal centers was studied. Specifically, an electron-withdrawing nitro group was incorporated onto the linker and was proven to further increase the conversion percentage up to 93 %. A study reported by Blandez et al. further demonstrated the effect of ligand functionalization on the Zr cluster of UiO-66 is far from being solely reliant on electronic effects [28]. UiO-66-X (X = H, NH_2, NO_2, Br, Cl,) was employed as a catalyst for epoxide ring opening and its activity was dependent on multiple parameters: steric hindrance, location, orientation of the functional group within the framework and differences in defect density (Scheme 9.1). The results are reported in Table 9.1 with a general order of reactivity for X as follows: Br > Cl > NH_2 > H > NO_2 – a finding that contradicts the previous article's results. Accordingly, no evident correlation between the type of functionalized ligand and the Lewis metal acidity can be drawn since it is related to a complex interplay between metal, reaction mechanism and reactant properties; however, an optimal structure can be found when taking all factors into account.

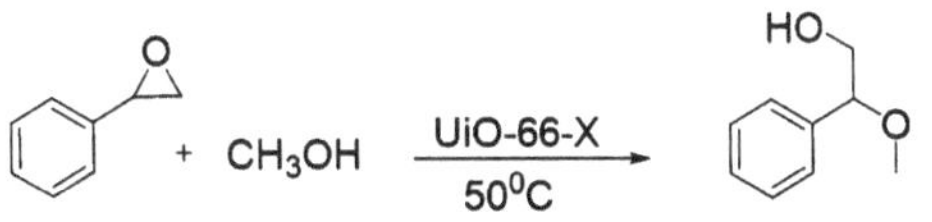

$$\text{Styrene oxide} + CH_3OH \xrightarrow[50^{o}C]{\text{UiO-66-X}} \text{product}$$

Scheme 9.1: Styrene oxide ring opening reaction by methanol and a series of UiO-66 derivatives. Reprinted with permission [28]. Copyright, 2016, Elsevier.

Table 9.1: The initial reaction rates of the styrene oxide ring opening over the various UiO-66-X derivatives. Reprinted with permission [28]. Copyright, 2016, Elsevier.

UiO-66-X	Initial rate (mol s^{-1}) × 10^{-6}
UiO-66-H	9.6
UiO-66-NO_2	20.0
UiO-66-NH_2	50.1
UiO-66-Br	143.1
UiO-66-Cl	128.8

9.3.2 Catalysis by Brønsted acid metal nodes

Although less explored than Lewis acid sites, Brønsted acid MOF chemistry, synthesis and applications have recently been subject to extensive studies [36]. As discussed by Yaghi et al. in a relevant review of the topic, Brønsted acidity can be introduced to MOFs through encapsulation, ligated groups in the SBUs or covalently-bound functional groups on the linker [37]. However, challenges have been identified: (i) a general trend of MOF instability in acidic media, (ii) difficulty in characterizing inhomogeneous Brønsted acidic sites (due to a lack of acid-base equilibrium) and (iii) the progressive loss of acidity of these groups due to their possible participation in the MOF assembly [37, 38].

In this section, we will focus on the metal ligated groups with the other two approaches being discussed later. To start, an approach for achieving Brønsted acidity is based on using metal clusters that have a µ-OH bridging species [39]. For example, the catalytic activity of MIL-53(Al, Ga) in the Friedel–Crafts alkylation of biphenyl with tert-butyl chloride was explored and showed that the Ga-μ_2-OH-Ga species made MIL-53(Ga) (also known as IM-19) mildly and selectively active for this reaction with around 30 % yield for all tested reactants. This is in stark contrast to MIL-53(Al) that did not demonstrate any catalytic activity [40]. This difference was attributed to the tilted arrangement of hydroxyl groups in MIL-53(Ga) leading to a nonzero net dipole that is not observed in MIL-53(Al), in which the dipole moments compensate each other at the molecular level. Accordingly, the acid strength of these embedded hydroxyl groups strongly depends on the type of metal used in the framework. Likewise, the Brønsted acidity of UiO-66(M) (M = Zr, Hf and Ce) was investigated and attributed to the oxophilicity of each metal ion and its impact on the μ_3-OH groups present in the SBUs. UiO-66(Hf), UiO-66(Ce) and UiO-66(Zr) resulted in 94.5 %, 70.1 % and 1.5 % glycerol conversion, respectively, along with a very high turnover frequency (TOF) for UiO-66(Hf) that was 3.6- and 90.7-fold higher than its Ce and Zr counterparts. The results were in alignment with their decreasing acidity and oxophilicity in that same order [41].

Coordinating water or polar solvent molecules to the metal site can also result in Bronsted acidity [38]. In a very recent study, successful control over Bronsted acid sites was achieved by undertaking a hydrothermal instead of a solvothermal synthesis. The resulting MOF, Hf-MOF-808-H_2O, catalyzed the conversion of styrene oxide to near completion in 21 h while the solvothermally-synthesized MOF, Hf-MOF-808-DMF, only achieved 43 % conversion [42]. Once the solid was degassed at high temperature, water molecules no longer coordinated to the metal sites, and as a result, the Brønsted acidity was lost [43]. It is worthwhile to note that the activity of Brønsted acidic metal centers is also subject to defect engineering effects. Liu et al. succeeded in quantizing this effect on the catalytic properties by determining the number of missing linkers per M_6 cluster in UiO-type MOFs via potentiometric acid-base titration. A strong correlation was then established between these estimated numbers and the relative reaction rate as well as the conversion percentage. Multiple UiO-type MOFs were tested as Brønsted acid catalysts for ring opening of styrene oxide with a secondary alcohol. For instance, the relative reaction rate was negligible for UiO-67 without missing linkers with only 4 % conversion after 24 h, while UiO-67 with 1.75 defects per M_6 cluster demonstrated a 0.9 reaction rate relative to Zr-UiO-66 with 34 % conversion [44]. However, it is clear that in most cases, Lewis and Brønsted acid sites can both be present in the same framework and judging whether the former or latter is responsible for the catalysis of a particular reaction can be misleading especially in defective structures. For example, in one of our previous studies, UiO-66 compensated for charge imbalance by coordinating water molecules to vacant CUSs, thereby transforming Lewis into Brønsted acid sites that successfully catalyzed the esterification of butyric acid and butanol [45]. Fully elaborated methods were developed to identify and quantize the Lewis to Brønsted acid ratio, some of which were chemical based consisting of reactions yielding different products based on the type of acidic catalysts used (Figure 9.5). The other methods rely on acid base titrations, gas sorption techniques or spectral analysis, like FT-IR or NMR [37, 46–51].

9.3.3 Catalysis by basic metal centers

Although less commonly reported than acidic MOFs, various frameworks show special catalytic abilities due to intrinsic basic sites at their metal centers. These centers will interfere in a given reaction by receiving a proton from or donating an electron pair to an acidic reactant. Figure 9.6 illustrates the main acidic and basic catalytically active sites that can be generated in a UiO-66 SBU.

The first approach to realize basic active sites focuses on the synthesis of MOFs using alkaline earth metal ions, such as Mg, Ca, Sr and Ba [52]. The ability of these elements to form alkaline earth metal oxides leads to a predictable outcome of basicity as the base strength decreases in the order of BaO > SrO > CaO > MgO [52, 53]. The aldol condensation reaction is one of the most common test reactions for the basicity of a catalyst and it is used to assess the basicity of an Mg-based MOF, $[Mg(Pdc)(H_2O)]_n$ (H_2-Pdc = pyridine-2,5-decarboxylic acid), for several aromatic aldehydes [54]. A sim-

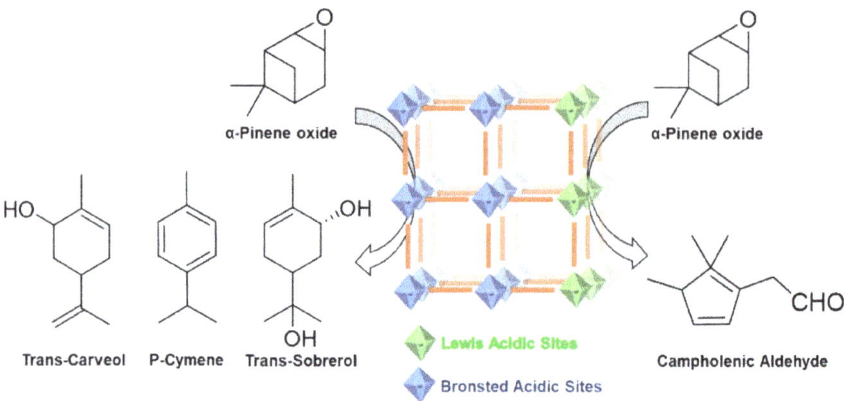

Figure 9.5: Schematic illustration of the isomerization catalytic test reactions of α-Pinene oxide employed for the identification of the type of the acidic active sites in MOFs.

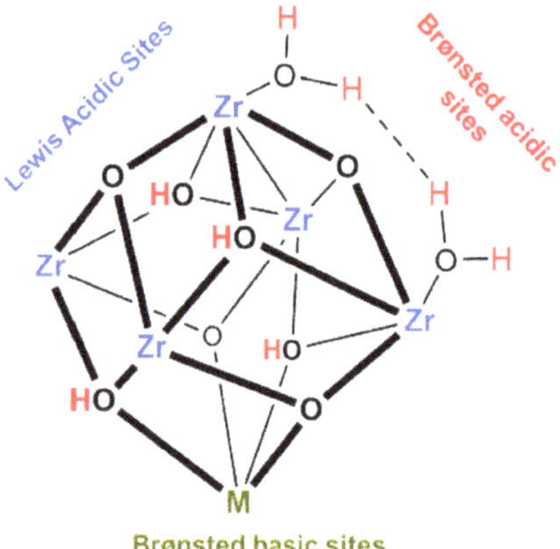

Figure 9.6: Representation of Lewis acidic, Bronsted acidic and basic active sites within a $Zr_6(\mu_3\text{-}O)_4(\mu_3\text{-}OH)_4$ SBU of UiO-66. M is a metal that is introduced via doping or metal-exchange processes and can be Ce(III), Co(II), Ni(II), Mg(II), Sr(II) or Ca(II).

ilar MOF, $[Ba(pdc)]_n$, was used to perform the same set of reactions as in the Mg-MOF case and showed a higher catalytic activity in all performed reactions as highlighted by a yield of 96 % as compared to 68 % for the Mg-MOF in the condensation of acetone with 4-nitrobenzaldehyde and 80 % to 53 % yield in the case of cyclohexanone with 2-nitrobenzaldehyde, respectively [55]. Similarly, Valvekens et al. compared the basic catalytic performance of two MOFs: $M_2(BTC)(NO_3)(DMF)$ (M = Ba or Sr, H_3BTC = 1,3,5-benzenetricarboxylic acid) that was found to be dependent on the activation temperature used. For an optimal activation temperature of 320 °C, the Ba-MOF afforded 99 % conversion for the reaction of benzaldehyde with malononitrile while achieving only 60 % conversion in the case of the Sr-MOF. The Knoevenagel condensation reaction also accounts for this catalytic difference and was performed using reactants of increasing acidity. Interestingly, it proved the correlation between the acidity of the donor molecule and the overall conversion ranging from 6 to 97 % as the pK_a of the reactant decreased from 14.3 to 11.1, a typical feature of basic catalyzed reactions [56]. These results are in agreement with the oxide basicity trend, which stresses the importance of the metal center's role as basic catalytic sites. A second approach is based on the employment of hybrid metal nodes, a strategy that makes use of simultaneous incorporation of two or more metal ions during the crystallization process. Upon heat treatment and due to charge polarization, Lewis basic sites are generated on the hybrid metal nodes. Specifically, the asymmetric W-Cu dimer SBU, with unequal electronegativity, in W-Cu-BTC MOF proved considerably basic for catalyzing CO_2 activation. Two comparative sets were made: the first between Cu-BTC, W-BTC and W-Cu-BTC and the second between Cr-Cu-BTC, Mo-Cu-BTC and W-Cu-BTC based on frontier molecular orbital theory as an attempt to explain the special interaction between CO_2 and the hybrid metal MOF. Results were indeed attributed to the unequal electronegativity in W-Cu dimer as well as the favorable interaction between the lowest unoccupied molecular orbital of CO_2 and the $5d_{z2}$ highest occupied molecular orbital of W-Cu-BTC [57]. Table 9.2 summarizes the different types of catalytic sites present in different MOFs reported with their respective catalytic reactions.

Table 9.2: Catalytic metal sites of MOFs reported with their respective catalytic reactions.

Type of catalytic center	MOF-based catalyst	Catalytic reaction	Ref.
Lewis acidic metal clusters	Cu-BTC MOF	Citronellal conversion	[34]
	Cu-MOF-74	Acylation of anisole	[31]
	HKUST-1	Cyanosilylation of benzaldehyde	[30]
		Acylation of anisole	[31]
	IRMOF-1	Tert-butylation of *p*-tertbutylbiphenyl	[33]
	IRMOF-8		
	$Mn_3[(Mn_4Cl)_3(BTT)_8(CH_3OH)_{10}]_2$	Cyanosilylation of aldehydes	[32]
	UIO-66	Cyclization of citronellal	[35]
	UIO-66-X (X = H, NH_2, NO_2, Br and Cl)	Epoxide ring opening	[35]
Brønsted acidic metal clusters	Hf-MOF-808	Styrene oxide conversion	[42]
	MIL-53(Al, Ga)	Friedel–Crafts alkylation	[40]
	UIO-66	Esterification of butyric acid and butanol	[45]
	UIO-66(M) (M= Zr, Hf and Ce)	Solketal synthesis	[41]
	UIO-67	Ring opening of styrene oxide	[44]
Basic metal clusters	$[Ba(pdc)]_n$ MOF	Aldol condensation	[55]
	$M_2(BTC)(NO_3)(DMF)$ (M = Ba, Sr)	Knoevenagel condensation	[56]
	$[Mg(Pdc)(H2O)]_n$	Aldol condensation	[54]
	W-Cu-BTC MOF	CO_2 activation	[57]

9.4 Catalysis by functionalized linkers

Apart from catalytic metal sites, the wide variety of available organic linkers with different catalytically active functional groups provides design opportunities for many frameworks depending on the type of catalysis being pursued (Figure 9.7). As these functional groups are free from the construction of the framework, they present special affinities to specific substrates in a multitude of reactions [58]. Accordingly, we may classify them under four main categories: (i) basic groups, (ii) acidic groups, (iii) organometallic complexes and (iv) organocatalytic moieties. It is noted that two approaches are generally used in the synthesis of MOFs with functionalized linkers, the first being direct one-pot synthesis where the functionalized linker is among the parent frameworks' building blocks and the second being the exploitation of post-synthetic modification approaches [59]. This section will focus on the first approach and will feature discussions on the variety of readily functionalized linkers and their catalytic performances.

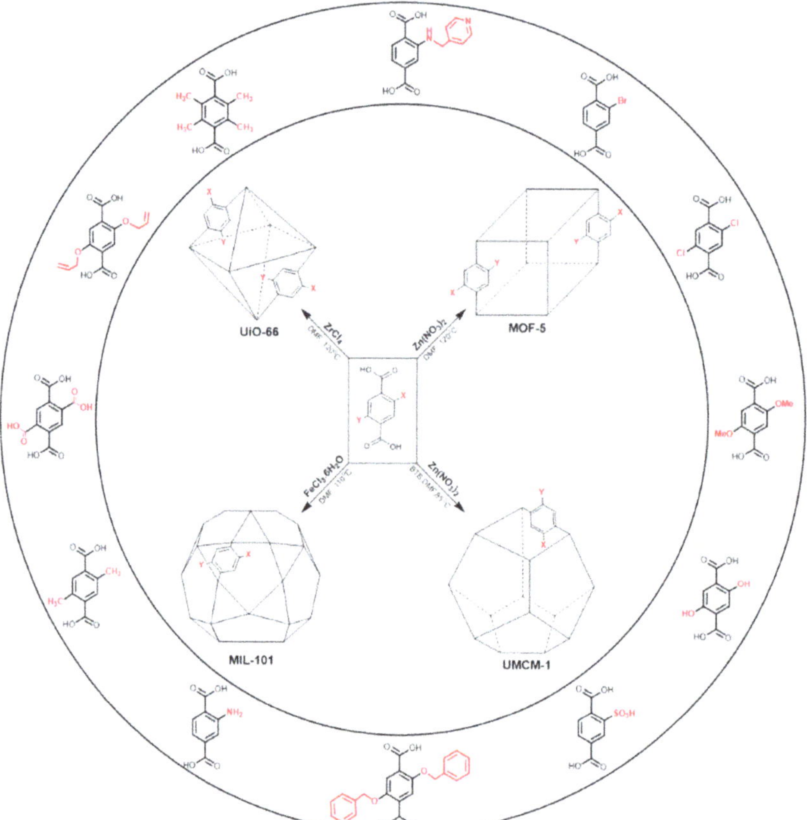

Figure 9.7: Various organic linkers that are employed for the functionalization of UiO-66, MOF-5, MIL-101 and UMCM-1.

9.4.1 Catalysis by basic functional groups

Basic functional groups that render linkers and, subsequently, MOFs suitable for base-driven organic reactions include N-containing ligands, phenolates and thiophenolates. Initially, N-containing ligands were commonly employed as catalytic building blocks in MOFs. Compared to their nonfunctional counterparts, they exhibit pronounced basicity capable of catalyzing many useful reactions, including, but not limited to, Knoevenagel and aldol condensations, Michael additions and transesterification reactions [52]. Among these, NH_2-functionalized MOFs, specifically those containing NH_2-BDC, are the most extensively studied due to their ability to be isoreticular to parent BDC-MOFs, yet with enhancement in basicity. A clear comparison was established between UiO-66 and UiO-66-(NH_2) as catalysts for cross-aldol condensation. The latter showed 67 % conversion, 91 % selectivity and 92 % yield as compared to 30 %, 82 % and 85 % for unfunctionalized UiO-66 using the same reaction parameters [60]. Later, Yang et al. used the same MOFs to catalyze Knoevenagel condensation of aldehydes with 94 % conversion using UiO-66(NH_2) as a catalyst compared to 50 % conversion achieved by UiO-66. The work proposed that the high catalytic activity of UiO-66(NH_2) was not caused by the amino group on its own, but rather revolved around the acid-base bifunctionality of the framework. The presence of unintended Zr(IV) Lewis sites and defects most probably promoted Knoevenagel condensation since UiO-66 with no amino groups showed a significant activity in comparison with the 27 % conversion observed for the blank. To further stress this duality, an organic analogue of UiO-66(NH_2), namely dimethyl 2-amine-1,4-benzenedicarboxylate (DMBDC-NH_2), promoted only 29 % conversion in the same experiment and when mixed with ZrO_2 as a homogeneous catalyst showed 35 % conversion by itself. The results reached 42 % conversion when the mixture was used as a catalyst [61]. Apart from amino-functionalized linkers, the duality effect was highlighted by a recent study on lanthanide-based MOFs having 4,4',4"-nitrilotribenzoic acid (H_3NTB) as a linker. The resulting MOF was expected to catalyze the Knoevenagel condensation of benzaldehyde and malononitrile through its triphenylamine Lewis basic sites, however, the MOFs built from different lanthanide centers (Pr, La, Eu and Tb), but with the same linker displayed distinctive catalytic performance with the best catalyst being Tb-MOF (99 % yield) as a result of the synergistic effect of its metal coordination environment and the linker's basicity [62]. In the same line of work on N-containing linkers, amide linkers show a reduced catalytic potential since they form hydrogen bonds within themselves, which hinders accessibility of substrates. However, $\{[Cd(4\text{-btapa})_2(NO_3)_2]\cdot 6H_2O\cdot 2DMF\}_n$ with 4-btapa = 1,3,5-benzene tricarboxylic acid tris[N-(4-pyridyl)amide] was reported to selectively catalyze the Knoevenagel condensation of malononitrile with 98 % conversion due to the amide groups of its linker. The tritopic linker having pyridyl coordination sites facilitated the role of the three amide groups as guest interaction sites by preventing amide-amide interactions [63].

Structural phenolates create significant basicity in linkers, and as a result, render MOFs catalytically active toward standard base-catalyzed reactions [52]. This property was first demonstrated by Valvekens et al. who worked on a series of isostructural compounds denoted CPO-27-M (M = Mg, Co, Ni, Cu and Zn) (commonly known as MOF-74 for the Zn analogue). The 2,5-dioxidoterephthalate (H_2DOBDC) linker of these MOFs presents two potential protonation sites, the carboxylate and phenolate moieties, with the latter presenting higher proton affinities (946 v. 757 kJ mol^{-1} for carboxylate and phenolate, respectively, in the CPO-27-Ni) and basicity as proven by chemisorption of pyrrole (Scheme 9.2). Phenolate ions were responsible for the catalytic activity of these MOFs in Michael addition and Knoevenagel condensation reactions, with the CPO-27-Ni being the most active with 66 % and 99 % conversion for the two respective reactions. The interplay between phenolate's basicity and the adjacent CUSs are key in catalyzing these reactions [64]. These same series of MOFs were investigated as catalysts in the cyanosilylation of aldehydes and the linkers' phenolates (H_2DOBDC) were substituted by thiophenolates (H_2DSBDC) for comparison. A critical step in the base-driven mechanism of this reaction consists of the interaction of trimethylsilyl cyanide (TMSCN) with the catalytic basic sites to release the cyanide group. Accordingly, interaction energies of M_2DSBDC with TMSCN were measured (–26.5 kcal mol^{-1}) and found to be more negative than those of M_2DOBDC (–10 kcal mol^{-1}), therefore, despite their weaker basicity, thiophenolates have a greater catalytic ability due to their higher polarizability that favor van der Waals interactions [65].

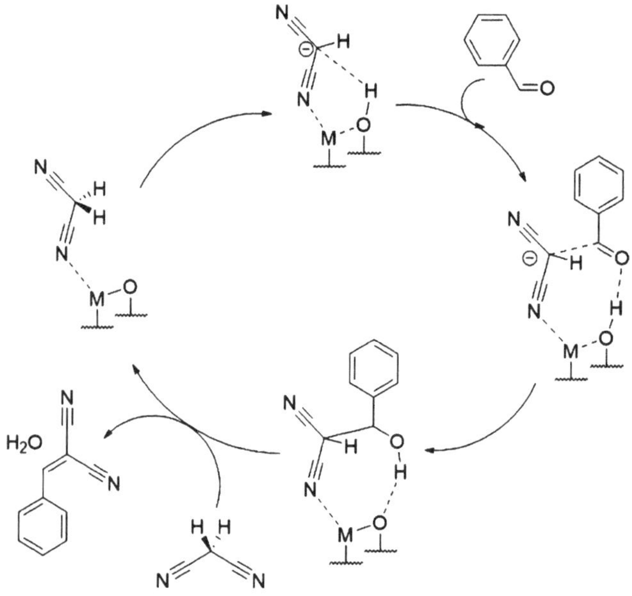

Scheme 9.2: Proposed mechanism for the Knoevenagel condensation reaction of malononitrile and benzaldehyde in CPO-27-M. Reprinted with permission [64]. Copyright, 2014, Elsevier.

9.4.2 Catalysis by Brønsted acidic functional groups

Acidic functional groups incorporated on organic linkers are very useful for introducing Brønsted acidity into MOFs. Among them, carboxyl and hydroxyl groups are of significant importance. In a study by Wang et al., isoreticular structures of UiO-66 were prepared using different functionalized BDC linkers in order to serve as catalysts for cycloaddition reactions. UiO-66-COOH demonstrated the highest conversion of 44 % of 3-methyl-2-butenal compared to the parent, unfunctionalized UiO-66 (20 %), UiO-66-OH (26 %), UiO-66-2OH (30 %) and UiO-2COOH (24 %). The more exposed Brønsted acid active sites in UiO-66-2OH explain its higher activity compared to UiO-66-OH; however, this trend does not apply to the carboxylate-functionalized linkers due to pore size and volume effects in UiO-66-2COOH that hinder the diffusion of substrate molecules to active sites. Interestingly, the researchers performed the same reaction using the corresponding linker of each framework as a catalyst alone. As illustrated in Table 9.3, a different trend was observed in opposition to the MOFs conversion percentages. It was concluded that Brønsted acid sites alone do not account for framework catalytic activity, but rather participate in an interplay between these active sites and pore volumes [66].

Table 9.3: The effect of the Brønsted acid active sites in UiO-66-X linkers on [3 + 3] cycloaddition reactions between 1,3-cyclohexanedione and 3-methyl-2-butenal. Reproduced from ref [66] with permission from the Centre National de la Recherche Scientifique (CNRS) and The Royal Society of Chemistry.

Entry	Catalysts	Conversion percentage (%)
1	UiO-66	19.73
2	UiO-66-OH	26.05
3	UiO-66-2OH	29.60
4	UiO-66-COOH	44.15
5	UiO-66-2COOH	24.20
6	Terephthalic acid	7.81
7	2-Hydroxyterephthalic acid	9.54
8	2,5-Dihydroxyterephthalic acid	7.75
9	1,2,4-Benzenetricarboxylic acid	7.95
10	1,2,4,5-Benzenetetracarboxylic acid	6.37

Sulfonic acid units as functional groups are also favored and extensively studied in Brønsted acidic catalytic reactions by MOFs. As an example, MIL-101-SO_3H realized 15 % conversion of anisole in a Friedel–Crafts acylation reaction as compared to its parent, unfunctionalized MIL-101 counterpart (0.9 % conversion). This activity was further compared to that of UiO-66-SO_3H having a microporous structure with limited availability of acid active sites. The UiO-66-SO_3H produced only a 0.5 % conversion, thereby emphasizing the role of accessibility to sulfonic acid sites [67]. A more pronounced effect for sulfonic acid units was exhibited in esterification reactions of monoacids, diacids, and acid anhydrides when catalyzed by BUT-8(Cr)-SO_3H (Beijing University of Technology) constructed from 4,8-disulfonaphthalene-2,6-dicarboxylatlate organic linker and $Cr_3O(OH)(CO_2)_6$ SBUs. A 91 % yield for the reaction of phthalic anhydride with methanol was obtained in comparison with 58 % for the parent, unfunctionalized BUT-8(Cr) and 75 % for the previously discussed MIL-101-SO_3H. The catalytic performance of BUT-8(Cr)-SO_3H was shown to decrease from a yield of 91 % to trace amounts of product as the alcohol undergoing esterification becomes bulkier [68].

Finally, phosphonate linkers are efficient proton donors when incorporated within MOF linkers though less commonly reported than hydroxyl, carboxyl and sulfonate groups. These linkers are advantageous for having rich coordination possibilities and variable protonation states that are easily controlled by pH change [69]. $Zr(H_4L)$ (where H_8L = tetraphenylsilane tetrakis-4-phosphonic acid) afforded impressive results in the cycloaddition of CO_2 to epoxides: 95 % yield and 238 h^{-1} TOF. These results mostly outperformed previously reported MOF catalysts for this reaction [70, 71]. This MOF benefited from its dense structure that lacked lattice or coordinated solvent molecules, which made its structure ready for catalysis without activation. Its special activity was attributed to free, uncoordinated phosphonate sites in the linker acting as Brønsted acids as well as the Lewis acidity of Zr. Indeed, this work once again emphasized the importance of duality in the catalytic functionality of MOFs [72].

9.4.3 Catalysis by organometallic complexes attached to MOF linkers

Attaching special organometallic complexes on MOF linkers is a fertile space for advances and innovations in the field of catalysis [73]. Reported complexes mainly include metalloporphyrins and *N*-heterocyclic carbene metal complexes, among others [59]. To start, a porphyrin is a biologically-important chemical and its metallo-derived structure is notably useful in biomimetic catalysis [59]. As an example, a metalloporphyrin tetracarboxylic ligand Ir(TCPP)Cl (TCPP = tetrakis(4-carboxyphenyl)porphyrin) was used in the construction of a three-dimensional iridium(III)-porphyrin MOF, termed Ir-PMOF, that was assessed as a catalyst in O-H insertion reactions. In the case of isopropanol reacting with ethyl diazoacetate, the Ir-PMOF catalyst gave an excellent initial rate and a yield of 94 % within 10 mins as compared to 88 % for a homogeneous iridium-porphyrin based catalyst Ir(TPP)(CO)Cl (TPP = tetraphenylporphyrin). This catalytic activity was explained by axial positioned vacant coordination sites of the iridium atoms whose effect was amplified by the framework's structure [74]. As for *N*-heterocyclic carbene (NHC) transition-metal complexes, their catalytic activity arises from functionalizing the imidazole side chain with various substituents [75]. For instance, an imidazole bearing an alcohol donor functional group was used to construct an Ir-NHC metallo-linker that was used as a building block to synthesize an iso-structure of UiO-68. The researchers tackled the catalytic difference between the one-pot synthesis approach where the functionalized linker is readily used to synthesize the MOF and the post-synthetic exchange method (PSE) of incorporating this linker. Interestingly, when used as catalysts for the isomerization of 1-octen-3-ol, the directly synthesized framework gave a significantly high yield of 64 % as compared to 0 % in the absence of a catalyst. However, the PSE version UiO-68 yielded >99 %—a result very similar to that of the homogeneous metallolinker catalyst on its own [76]. Besides these two main classes of organometallic complexes, less common complexes have been utilized and demonstrated to have catalytic activity. Diphosphine pincer linkers offer a considerable deal of electronic and steric diversity, chemical and thermal stability as well as the ability to support a wide catalogue of transition metals. Kassie et al. succeeded in synthesizing three isostructural Zr-MOFs from P^NN^NP-Ru pincer complexes with different ancillary ligands (chloride, CO or phosphines). While these three catalysts showed less than 10 % yield for the hydrosilylation of benzaldehyde, once vacant Ru sites were created in one of the MOFs by deprotonation followed by CO ligand removal, the yield reached 94 %. These observations highlighted the preference of the catalytic reaction occurring on Ru vacant sites rather than Zr-based Lewis acidic sites [77]. Salan-based linkers benefit from a special flexibility and high nitrogen donating ability. In fact, a dipyridyl-functionalized chiral Ti(salan) linker was employed as a building block for a Ti-MOF and was tested for catalysis in the asymmetric oxidation of thioethers to sulfoxides. This chiral linker showed a pronounced enantioselectivity in the reaction when bulky substrates such as 2-naphthylbenzyl sulfide were employed, affording 59 % conversion and 64 % enantiomeric excess. Additionally, the framework exhibited size selective activity given that enantioselectivity was enhanced as the substrate became bulkier reaching a maximum before a very bulky naphthyl dendritic sulfide was not able to access the active sites resulting in less than 2 % conversion [78].

9.4.4 Catalysis by organocatalytic moieties on the MOF linker

First and foremost, organocatalytic moieties refer to metal-free organic compounds that are able to accelerate a chemical reaction by a substoichiometric amount. The development of organocatalysis is mainly due to the availability of optically active organic molecules in nature providing an eco-friendly and cost-effective approach [79]. However, the challenges faced by the homogeneity of these catalysts has propelled interest in immobilizing them on heterogeneous structures like MOFs. Among them, hydrogen-bond-donating organocatalysts (HBD) bearing urea and squaramide moieties are of great interest along with proline chiral organocatalysts [59].

The first example of a MOF-based HBD, NU-601, was reported by Farha et al. in 2012. This urea-based MOF was synthesized by assembling Zn clusters together via a polytopic urea linker along with 4,4'-bipyridine acting as a pillaring agent. HBD catalysts are known as alternatives to Lewis acid activation, thus NU-601 was utilized as a catalyst for a Friedel–Crafts reaction between pyrroles and nitroalkenes. For this reaction, NU-601 showed 90 % conversion in 18 h as compared to 65 % for the homogeneous diphenylurea catalyst. A valid explanation for this difference is that the oligomerization/self-quenching ability of the urea-based homogenous catalyst was overcome once it was embedded in the framework [80]. Later on, the same group succeeded in synthesizing a UiO-67 iso-structure with a squaramide-functionalized 4,4'-biphenyldicarboxylate (bpdc) linker. This MOF outperformed UiO-67 and an urea-based UiO-67 in catalyzing the biorelevant Friedel–Crafts reaction of indole and β-nitrostyrene with 78 % compared to 22 % and 38 % yield, respectively (Figure 9.8) [81].

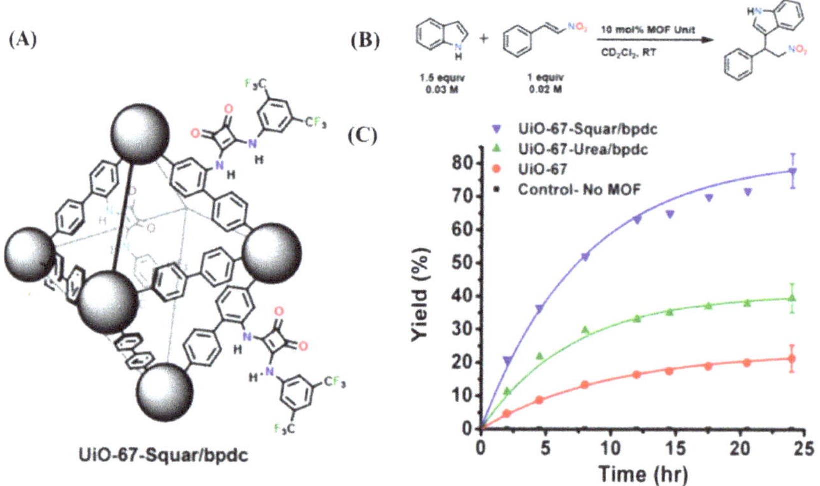

Figure 9.8: (A) Crystal structure of UiO-67-Squar/bpdc. (B) Friedel–Crafts reaction conditions between indole and β-nitrostyrene. (C) The respective yields of the Friedel–Crafts reaction catalyzed by UiO-67-Squar/bpdc, two UiO-67 references, and in the absence of a MOF. Reprinted with permission [81]. Copyright, 2015, American Chemical Society.

Proline chiral organocatalysts can act as enamine or imine catalysts for asymmetric reactions. However, incorporating proline moieties into MOFs faces a major challenge in that its nitrogen atom can be easily protonated during synthesis. This requires a sequence of protecting and deprotecting mechanisms to be carried out in order to maintain its activity [59]. Although this is typically required to be done via post-synthetic modification, Kutzscher et al. reported an *in situ* de-protection of Pro-Boc units [Pro: Proline, Boc: N-(tert-butyloxycarbonyl)] to form H_2bpdc-NHProBoc and H_2tpdc-NHProBoc chiral linkers for constructing UiO-67-NHPro and UiO-68-NHPro, respectively [82]. Once the frameworks were formed, the hard Lewis-acidity of Zr(IV) was responsible for the *in situ* deprotection of the linkers, a feature that is not possible using the low acidity of Zn(II) in DUT-32-NHProBoc (DUT = Dresden University of Technology) and an iso-reticular IRMOF-Pro-Boc, for example [83, 84]. Interestingly, these MOFs not only produced high yields in the aldol addition of 4-nitrobenzaldehyde and cyclohexanone, but also exhibited a reversed diastereomeric selectivity favoring *syn* over *anti* adducts in stark contrast to homogeneous catalysts and previously reported MOFs. This diastereoselectivity was attributed to the high concentration of proline units (at least six proline linkers surrounding each pore) of which several groups may participate in the transition state [85]. Table 9.4 summarizes the different types of linkers used as catalytically-active species in MOFs with their respective catalytic reactions.

Table 9.4: List of all reported MOFs with different catalytically active functional groups attached to linkers.

Type of catalytic center	Type of linker	MOF-catalyst	Catalytic reaction	Ref.
Basic linkers	N-containing ligands	$[Cd(4\text{-btapa})_2(NO_3)_2]\cdot 6H_2O\cdot 2DMF_n$	Knoevenagel condensation	[63]
		Lanthanide MOFs	Knoevenagel condensation	[62]
		UiO-66-(NH_2)	Cross aldol condensation	[60]
		UiO-66-(NH_2)	Knoevenagel condensation of aldehydes	[61]
	Phenolates	CPO-27-M (M = Mg, Co, Ni, Cu and Zn)	Cyanosilylation of aldehydes	[65]
			Knoevenagel condensation Michael addition	[56]
	Thiophenolates	CPO-27-M (M = Mg, Co, Ni, Cu and Zn)	Cyanosilylation of aldehydes	[65]
Brønsted acidic linkers	Carboxylates and hydroxylates	UiO-66-COOH UiO-66-2COOH UiO-66-OH UiO-66-2OH	Cycloaddition reactions	[66]
	Sulfonic acid groups	BUT-8(Cr)-SO_3H MOF	Esterification reactions	[68]
		MIL-101-SO_3H UiO-66-SO_3H	Friedel Crafts acylation	[67]
	Phosphonates	$Zr(H_4L)$	Cycloaddition of CO2 to epoxides	[69]
Linker-organometallic complexes	Metalloporphyrins	Ir(TCPP)Cl	O-H insertion reactions	[74]
	N-heterocyclic carbene transition-metal complexes	Ir-NHC isostructure of UIO-68	Isomerisation of 1-octen-3-ol	[76]
	Diphosphine pincer ligands	P^NN^NP-Ru pincer Zr-MOFs	Hydrosilylation of benzaldehyde	[77]
	Salan based ligands	Ti-MOF	Asymmetric oxidation of thioethers	[78]
Linker-organocatalytic moieties	Hydrogen-bond-donating organocatalysts	NU-601	Friedel–Crafts reactions	[80]
		UiO-67	Friedel–Crafts reactions	[81]
	Proline chiral organocatalysts	UiO-67-NHPro	Aldol addition	[85]

9.5 Post-synthetic modification for introducing catalytic active sites in MOFs

Despite the roles played by CUSs and the functionalities present on the organic linkers as catalytically active sites, the possibility of introducing suitable, compatible and complex functional groups and nanoparticles to the framework during its assembly remains limited by its often stringent synthesis procedures [86]. Post-synthetic modification (PSM) in MOFs has recently emerged as a promising and powerful strategy to enhance the catalytic activity of guest molecules encapsulated within MOFs that act as scaffolds. This approach is widely utilized to increase structural stability, reproducibility and activity of the parent frameworks while maintaining their topology, porosity, and crystal structure [14, 87]. Various post-synthetic strategies have been employed in order to modify MOF structures and create new active sites for catalytic applications including metal-based PSM, ligand-based PSM and guest-based PSM (Figure 9.9).

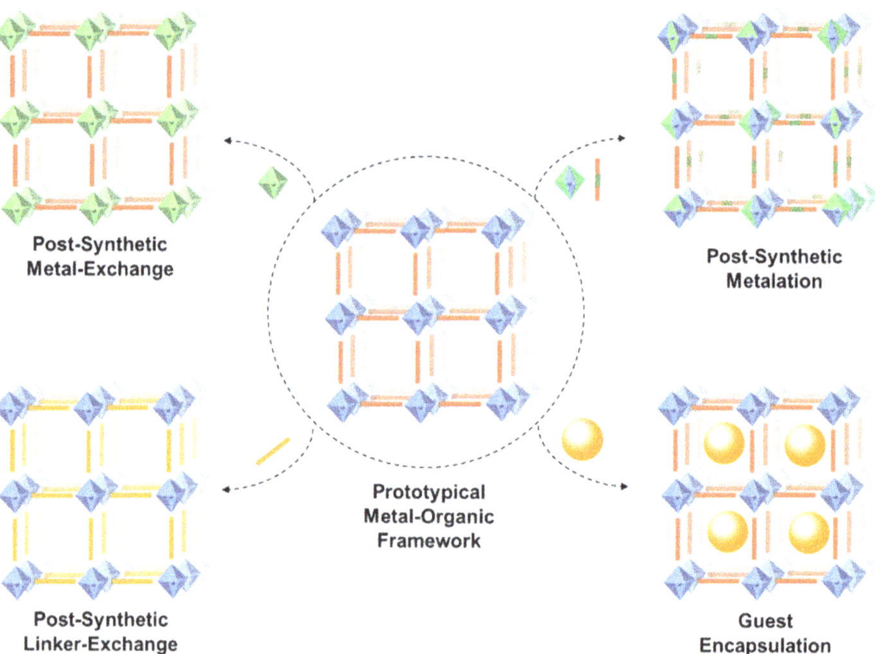

Figure 9.9: Introduction of catalytic active sites into MOF structures via post-synthetic modification (PSM).

9.5.1 Post-synthetic metal exchange (PSME) in MOFs

PSME allows for the complete substitution of metal centers in MOFs by other metallic species having higher catalytic activity [88]. It results in the generation of frameworks with new metal clusters and specific topologies that could not have been formed in case of a direct hydro or solvothermal synthesis. One advantage of this approach is that the number of metal cations found in each cluster remains the same after the metal-exchange process in contrast to the post-synthetic metalation (PSM) approach, which results in the addition rather than the exchange of the framework's metal ions [89]. Although the first PSME in robust MOFs dates back to 2007 with the replacement of Mn(II) guest cations in $Mn_3[(Mn_4Cl)_3(BTT)_8(CH_3OH)_{10}]_2$ with Li(I), Fe(II), Co(II), Cu(I), Cu(II), Ni(II) and Zn(II) cations, it was not until 2015 that the usability of PSME for a MOF-based catalyst was achieved by Dincă et al. [90, 91]. Specifically, Dinca et al. successfully reported the complete substitution of Zn(II) ions in MOF-5 with Fe(II) resulting in Fe(II)-MOF-5 having an identical topology and crystallinity to the parent structure. As illustrated in Figure 9.10, the newly formed MOF was used as a catalyst in the disproportionation of nitric oxide (NO) into nitrous oxide (N_2O).

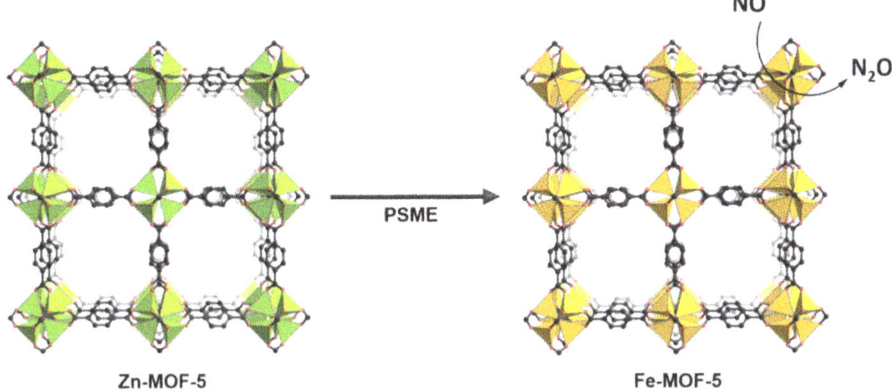

Figure 9.10: Illustration depicting PSME of MOF-5 and its transformation into Fe-MOF-5 with new catalytic properties. Atom colors: Zn, green polyhedra; Fe, yellow polyhedra; C, grey and O, red.

Despite the fact that most of the iron-based catalytic systems used in the reductive disproportionation of nitric oxide require the presence of bimetallic Fe-Fe metal centers for the generation of N–N bonds, this was the first example where the N–N coupling step, necessary for the iron nitrosyl intermediate, as well as the nitrous compound formation required a single iron atom [92]. The formation and decomposition of the intermediate, in addition to the formation of the product, were constantly monitored via diffuse reflectance infrared Fourier transform spectroscopy (DRIFTS) and residual gas analyzer (RGA-MS). Then Dincă and Volkmer followed this by reporting for the first time the synthesis and post-synthetic modification of a MOF-based heterogeneous catalyst, namely MFU-4l ($Zn_5Cl_4(BTDD)_3$, H_2BTDD = bis(1H-1,2,3-triazolo[4,5-b],[4',5'-i]) dibenzo[1, 4] dioxin) [93]. This work demonstrated that by PSME of the zinc metal centers with different heterometal cations including Cu(I), Cu(II), Ni(II), Ti(III), Cr(III) and Co(II) species, obtained frameworks could be utilized for a variety of catalytic applications [94–96]. Subsequently, Comito et al. reported the post-synthetic modification of four novel MFU-4l frameworks by exchanging the zinc metal clusters with Cr(II), Cr(III), Ti(III) and Ti(IV) ions, respectively. These framework-based catalysts were tested in the polymerization of ethylene with the newly substituted metal nodes playing the role of the catalytic active sites. At an ethylene pressure of 40 bar, and in the presence of methyl-aluminoxane, the four modified MOFs showed a high effectiveness in catalytically polymerizing ethylene resulting in TOF values ranging between 1,300 and 15,000 h^{-1} [94]. To further affirm the role of the metal catalytic active sites in Cr(II)-MFU-4l, Cr(III)-MFU-4l, Ti(III)-MFU-4l and Ti(IV)-MFU-4l, the same experiment was repeated using the standard Zn(II)-MFU-4l where no ethylene polymerization was detected. The catalytic activity of the framework was attributed to the two exchanged metal cations. In another study, Metzger et al. successfully exchanged Zn(II) ions in MFU-4l with Ni(II) without altering the topology nor the porosity of the framework. Ni(II)-MFU-4l was employed in ethylene dimerization reaction where it displayed an excellent catalytic activity accompanied by a TOF of 41,500 h^{-1} [95]. A similar catalytic protocol was applied to olefin polymerization where the importance of the added catalytic sites was also assessed by performing the same experiment under the same conditions using the parent Zn-based MFU-4l. Unsurprisingly, no activity was detected since no 1-butene was generated in the case of the parent MFU-4l. It is important to mention that the PSME process took place by soaking the parent framework comprised of Zn(II) in a large excess of different metal chloride salts in DMF at different temperatures followed by washing with DMF and acetone [96, 97].

9.5.2 Post-synthetic metalation in MOFs

Unlike the PSME approach that involves the substitution of metal atoms already existing within the cluster, post-synthetic metalation adds new organometallic moieties onto the presynthesized cluster and within the pore cavities turning the framework into an active, recyclable and robust catalyst [98–100]. This is carried out by using organic linkers possessing extra secondary functional groups that remain free and uncoordinated during MOF synthesis. The main purpose of these groups is to complex extraneous metal species. Important examples include thioethers, hydroxides, thiols, porphyrins and 2,2'-bypiridine building units. Post-synthetic metalation is mostly employed when pre-synthetic metalation is not desired nor applicable due to structural instability of the MOF.

Different strategies have been employed for the impregnation of catalytically active metal and organometallic species in MOFs including atomic layer deposition and solvothermal deposition [101]. A prominent example of grafting NbO_x species on the Zr-based MOF, NU-1000, for cyclohexene epoxidation was developed by Farha and Hupp [102]. In the study, they presented different reactivity data for the epoxidation reaction including the initial rate, TOF, total yield and selectivity using different synthesized heterogeneous catalysts as well as control materials. In addition to the higher Nb(V) loading that was detected in NU-1000 compared to the control catalysts, Nb/ZrO_2 and Nb/SiO_2, a higher oxygenate yield was achieved by both Nb-SIM(H)-NU-1000 and Nb-AIM(H)-NU-1000 post-synthetically modified via solution-phase grafting in a MOF (SIM) and vapor-phase atomic layer deposition (ALD) in a MOF (AIM) technique, respectively. This was explained by the lengthy organic linkers that NU-1000 possesses, which eliminates surface diffusion of the NbO_x species and allows their complete distribution within the cluster of the MOF. This, in turn, enhances the catalytic activity of the Nb(V) materials [103]. In another research study conducted in 2020 by Cui et al., two novel Zr-based chiral MOFs (CMOFs) with *flu* topology were post-synthetically modified and employed as heterogeneous catalysts for the hydrogenation of α-dehydroamino acid esters (Figure 9.11) [104]. The two CMOFs, denoted as 1^{flu} and 2^{flu}, were first synthesized via solvothermal reactions using enantiopure 1,1'-biphenol-derived tetra-carboxylate linkers followed by a post-synthetic modification of the free dihydroxy functional groups of the linkers with tris(dimethylamino)phosphine $P(NMe_2)_3$ and iridium [Ir] complexes. These post-synthetic modifications yielded the two Zr-based CMOFs catalysts, 1^{flu}-P-Ir and 2^{flu}-P-Ir. To assess the role of the MOFs in ameliorating the catalytic activities of the iridium complexes, Me_4 L^1-P-Ir (Me_4 L^1 = dimethyl 2',2''-dihydroxy-5',5''-bis(4-(methoxycarbonyl)phenyl)-4',6''-dimethyl-[1,1':3',1'':3'',1'''-quaterphenyl]-4,4'''-dicarboxylate) catalyst was also examined in asymmetric hydrogenation reactions of different α-dehydroamino acid esters. In particular, when using (*Z*)-methyl-α-acetamidocinnamate as a substrate, a conversion of 96 and 97 % was achieved for 1^{flu}-P-Ir and 2^{flu}-P-Ir, respectively, compared to only 62 % in the case of Me_4 L^1-P-Ir. Furthermore, with an Ir loading of 0.02 mol%, the 1^{flu}-P-Ir catalyst afforded a 62 % conversion compared to 8 % for Me_4 L^1-P-Ir. This demonstrated the importance of the framework in ameliorating the activity of the iridium complex during the asymmetric hydrogenation process where it successfully prevented the unnecessary exchanges of linkers.

9.5.3 Post-synthetic ligand exchange (PSLE) in MOFs

PSLE has developed into a well-used strategy for functionalizing MOFs when the functional groups on the organic linkers can be presynthesized to include new active sites that otherwise cannot be incorporated via *de novo* synthesis [105]. Zhang et al. reported the incorporation of a free nucleophilic *N*-heterocyclic carbene into UiO-68-(NH_2) via PSLE for the selective hydrosilylation of CO_2 [106]. The exchange process

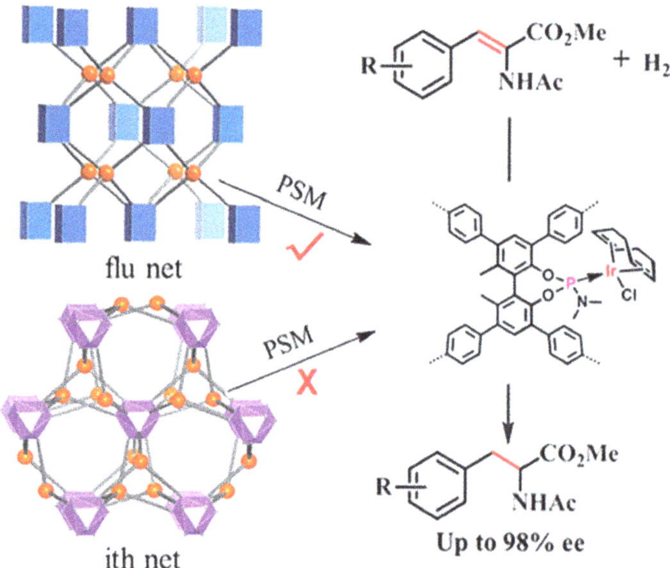

Figure 9.11: Incorporation of an iridium-phosphorus complex catalyst into a *flu*-type cluster-based Zr-MOF via post-synthetic metalation to use for the hydrogenation of α-dehydroamino acid esters. Reprinted with permission [104]. Copyright, 2020, American Chemical Society.

took place by incubating UiO-68-(NH_2) particles in a mixture of DMF/H_2O containing the NHC complex, namely 1,3-bis(4-carboxybenzyl)-1H-imidazol-3-ium bromide, at room temperature. The NHC complex proved to have a high catalytic activity in the CO_2 reduction reaction when attached to the organic linker giving a 99 % methanol yield at ambient temperature as compared to only 5 % yield using the unmodified, parent UiO-68-NH_2. In a similar study, PSLE was performed using UiO-67-CHO to add ethylenediamine and was examined in the Knoevenagel condensation of benzaldehyde with malononitrile [107]. The PSLE process was carried out via a tandem Schiff base condensation where solid UiO-67-CHO was dispersed in a DMF solution containing an excess amount of ethylenediamine. Both IR and NMR techniques were employed to confirm imine formation on the linkers in the framework. This was followed by the reduction of the imino groups with $NaBH_4$. The addition of the catalytic active diamine groups post-synthetically to the MOF structure resulted in an 81 % conversion to 2-benzylidenemalononitrile in 30 min, in contrast to only 8 % conversion achieved by the parent UiO-67-CHO. Fu et al. applied PSLE strategies to two water stable Zr-MOFs, UMCM-309 (UMCM = University of Michigan crystalline material) and MOF-808, which were then examined in desulfurization reactions [108]. In particular, formate ions were substituted with methanol by soaking the synthesized MOFs in different MeOH solutions at room temperature followed by conventional washing and activation processes. The replacement of the coordinated formate ions resulted in the formation of free methoxy groups along with vacant Zr(IV) sites. Indeed, an improvement in the catalytic activity was achieved in these MOF catalysts, where the conversion of benzothiophene at 60 °C increased from 18 % to 49 % for UMCM-309 and from 60 % to ca. 90 % when using MOF-808.

9.5.4 Integration of guest molecules into MOFs

The encapsulation of different active guest molecules, including metal nanoparticles, metal clusters, polyoxometalates, quantum dots and enzymes within the preexisting pores of MOFs is considered an exciting, straightforward approach for establishing MOF-based heterogeneous catalysts with enhanced activity and stability [109–111]. In particular, the porous nature of MOFs allows for facile distribution of guest molecules within the channels and on the surface of the framework, which in turn, prevents the aggregation and fusion of these molecules and enhances their catalytic properties and activity [112, 113]. Currently, post-synthetic incorporation of guest molecules into MOF structures occurs rationally via two primary approaches: (i) "ship in a bottle" and (ii) the "aperture opening" pathways.

9.5.4.1 The "ship in a bottle" approach

This strategy is recognized as one of the most famous approaches to post-synthetically functionalize MOFs. By definition, it involves the encapsulation of small catalytically-active guest molecules into the preformed channels of a MOF followed by a more advanced treatment of reduction or thermal decomposition [114]. Li et al. recently reported the synthesis of a series of Pd@UiO-66 catalysts via a novel impregnation reaction method for ethanol upgrading [115]. The concentration effect on different loadings of Pd nanoparticles in the catalytic performance of the frameworks was studied and it was found that the presence of Pd nanoparticles within the framework enhanced the catalytic performance and selectivity of the catalyst toward *n*-butanol. As an illustrative example, 2 wt% Pd@UiO-66 provided a 50 % conversion of ethanol with a selectivity of 50.1 % toward *n*-butanol compared to 2.6 % ethanol conversion and 0.3 % selectivity toward *n*-butanol for UiO-66. The catalytic activity and selectivity enhancement upon impregnation was attributed to the close synergy and dispersion of the Pd nanoparticles within the various channels of UiO-66. Likewise, Jiang et al. successfully introduced bimetallic CuNi nanoparticles into the giant cavities of mesoporous Cr-MIL-101 by a facile double solvent method [116]. The catalytic behavior of the newly established CuNi@MIL-101 catalyst was examined for the hydrogenation of nitroarene compounds into anilines by coupling with NH_3BH_3 dehydrogenation. The cascade reactions that took place at room temperature resulted in aniline yields of more than 99 % for the various nitroarenes used bearing methyl, amino, ketone and nitrile reducing groups. The authors clearly asserted that the role of the CuNi NPs encapsulated in the cavities of MIL-101 were as catalytically active species that enhanced catalytic activity and recyclability compared to Cu/Ni graphene catalysts and Cr-MIL-101 [117].

Few studies have been reported thus far on the post-synthetic encapsulation of biomacromolecules within the channels of various MOFs [118, 119]. In a recent study developed by Farha et al., organophosphate acid anhydrolase, a nerve agent hydrolyzing enzyme known as OPAA, was successfully encapsulated within the channels of the water stable NU-1003 for diisopropyl fluorophosphate (DFP) hydrolysis [120]. The enzyme's catalytic activity was examined after the immobilization process by investigating the degradation process of DFP in the cases of free OPAA, NU-1003 and OPAA@NU-1003 with different crystal sizes. While NU-1003 did not display any catalytic activity in DFP degradation, a diisopropyl fluorophosphate conversion of 100 % was achieved in the first 2 minutes of the reaction using OPAA@NU-1003-300nm, which was three times faster than the conversion obtained using free OPAA. These findings point to the role of MOFs in enhancing the catalytic activity of guest molecules by eliminating their aggregation in solution media and enhancing their kinetic activity.

9.5.4.2 The aperture opening approach

This relatively new approach was developed by the Byers and Tsung research groups [121]. It focuses on the encapsulation of guest molecules having a larger volume than the aperture size of a given MOF material (Figure 9.12).

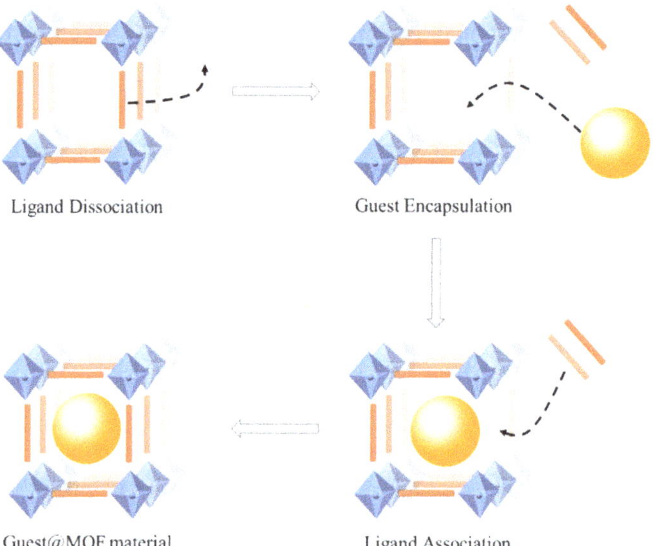

Figure 9.12: Schematic representation of the aperture opening approach used to encapsulate guest molecules in MOFs.

Li et al. reported the encapsulation of a ruthenium complex, (tBuPNP)Ru(CO)HCl (tBuPNP = 2,6-bis((di-tert-butylphosphino)methyl)pyridine), into presynthesized UiO-66 via the aperture opening approach [122]. The resulting catalyst, [Ru]@UiO-66, was investigated for CO_2 hydrogenation in a DMF/1,8-diazabicyclo(5.4.0)undec-7-ene mixture over five cycles. An enhancement in the stability, recyclability and resistance to poisoning of [Ru]@UiO-66 was detected when compared to the homogeneous Ru catalyst [123]. These results affirmed the role of the MOF in boosting the catalytic activity and properties of the active species while preserving their structure within the framework.

9.6 Catalysis through MOF composites

Desirable species, such as metal nanoparticles, polyoxometalates, enzymes, silica, polymers and a second MOF, can be encapsulated/coated on the surface of MOFs (Figure 9.13). Although most of these substances are more commonly introduced into MOF cavities, having them on the surface can accommodate bulky groups that are far larger than the pore sizes and/or act as shelters that improve stability, strength and/or catalytic performance [124]. MOF structures and their properties are able to resolve many of the limitations encountered when these substances are used as free catalysts by exploiting existing chemical synergies [125].

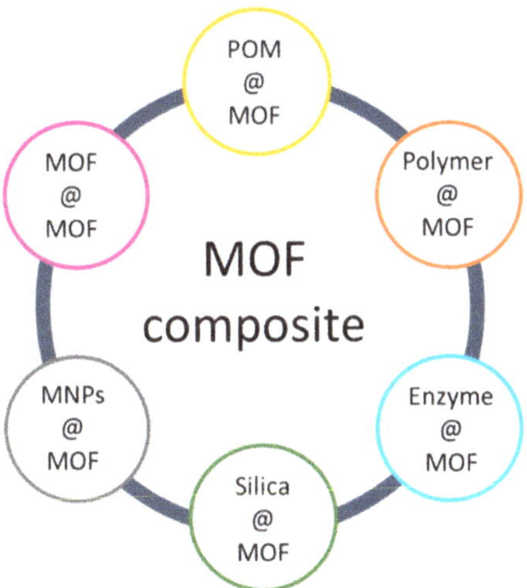

Figure 9.13: Various types of MOF composites used for catalytic applications.

9.6.1 MOF-metal nanoparticle composites

Compared to their bulky counterparts, metal nanoparticles show greater heterogeneous catalytic ability due to their rich active surface atoms, high surface to volume ratio and favorable electronic structures; however, these tend to aggregate and fuse easily [126]. Grafting nanoparticles (NPs) on the surface of MOFs allows them to perform their catalytic activity while the framework acts as a stabilizer and a regulator of the NP's electron environment [124]. For instance, an Au-Pd bimetallic alloy was dispersed on the surface of MIL-101 via a simple colloidal method and the composite was tested as a heterogeneous catalyst for the oxidation of cyclohexane. Results show 40 % conversion and an 80 % selectivity, both of which exceed the performance of pure metals and an Au and Pd physical mixture, with a proposed mechanism illustrated in Figure 9.14 [127]. Another approach witnesses the synthesis of a three-layered MOF/MNPs/MOF composite in a sandwich-like structure. Typically, pre-synthesized nanoparticles are assembled on the surface of a MOF and a different MOF with the same lattice structure as the core MOF will constitute the outer shell. These composites are advantageous since the MOF shell covering the NP middle layer enhances stability and controls the size selectivity of the catalysis [128]. Liu et al. exploited this strategy to synthesize Pt/MIL-100(Fe)@MIL-100(Fe) and Pt/NH_2-MIL-101(Al)@NH_2-MIL-101(Al) composites and used them as catalysts for the hydrogenation of cinnamaldehyde. The materials increased the selectivity to 96 %, which was considerably higher than the 55 % selectivity achieved by uncoated Pt/MOFs [129].

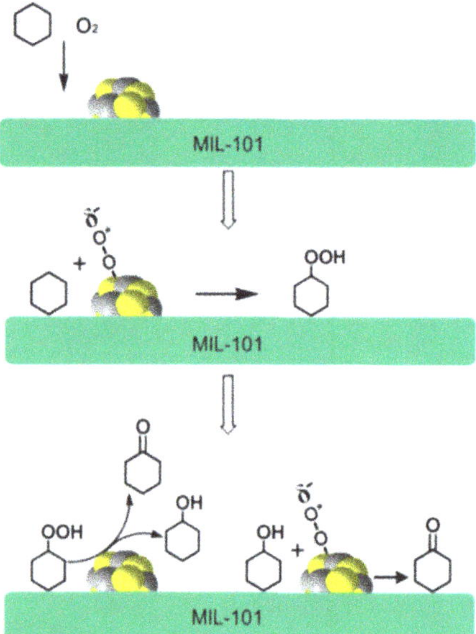

Figure 9.14: Proposed catalytic mechanism for the cyclohexane aerobic oxidation over Au-Pd/Mil-101 composite. Reprinted with permission [127]. Copyright, 2013, American Chemical Society.

9.6.2 MOF-POM composites

Polyoxometalates (POMs) are discrete molecular metal oxide anions that have unique catalytic activity due to their ability to not only play the role of Brønsted acids, but also Brønsted bases and Lewis acids [130]. Since POMs are usually soluble in both water and organic solvents, their low surface area and instability under reaction conditions necessitate their immobilization in order to generate and optimize their use in heterogeneous catalysis [131]. The use of MOFs for this aim has risen dramatically. Most reported examples focus on embedding POMs into MOF cavities; however, when MOF apertures are smaller than a given POM size, anchoring POMs on the surface of MOFs is possible. For example, when ZIF-8 surface supported a Keplerate-type POM, known as Mo_{132} nanoballs, it was employed as a catalyst for the oxidative desulfurization of dibenzothiophene. The composite demonstrated a 92 % conversion compared to 55 % for pristine, parent ZIF-8 under the same reaction conditions. This improvement was mechanistically attributed to the Mo atoms, which enhanced the formation of metal peroxides resulting from nucleophilic attack of the oxidant [132]. Granadeiro et al. prepared a sandwich-type composite with POMs forming the middle layer. The $[Eu(PW_{11}O_{39})_2]^{11-}$ POM was supported on both NH_2-MIL-53(Al) and MIL-101(Cr) where two composites were obtained: POM/MIL(Al) and POM/MIL(Cr). Both were evaluated as catalysts in the oxidative desulfurization of a multicomponent diesel model. POM/MIL(Al) achieved the highest performance with 99.9 % desulfurization compared to 82 % for POM/MIL(Cr), both of which exceeded the catalytic performance of the homogeneous POM (73.9 %) [133].

9.6.3 MOF-enzyme composites

Enzyme's catalytic performance is limited by their structural sensitivity to temperature, pH and organic solvents, as well as the inevitable contamination of products resulting from enzyme-catalyzed reactions. If they are to be practically used, this requires them to be immobilized on different matrices, notably MOFs [134]. When it comes to surface immobilization of enzymes on MOFs, it is noteworthy that the structure, orientation and conformational mobility of the enzyme may be modified due to the interactions between the protein and the MOF support. Even though these interactions can stabilize the nonactive structure and/or promote the spread of the enzyme, they may also diminish its functionality, which makes the design of the MOF-enzyme composite very critical [134]. Within this scope, bacillus subtilis lipase (BSL2) was grafted on the surface of HKUST-1 and the resulting composite successfully catalyzed the esterification of lauric acid and benzyl alcohol with an initial rate 17 times greater than that obtained when BSL2 was used on its own. Although improvement was observed for all reaction conditions, temperature and pH showed optimal values of 30 °C and 8.0, respectively [135]. A more recent comprehensive study by Zhou et al. studied the immobilization of cellulase on four different MOFs, ZIF-8, UiO-66-NH_2, MIL-100-Fe and PCN-250, due to their physicochemical robustness and water stability. All selected MOFs presented high adsorption capacity with the highest being 176.16 mg g^{-1} for ZIF-8. The ZIF-8-cellulase composite outperformed all counterparts with a yield 92.92 % for the hydrolysis of bagasse compared to 50 % for the free enzyme [136].

9.6.4 MOF-silica composites

Silica, also called silicon dioxide, is suitable for performing many nanoscale functions of which catalysis is of considerable importance. Combining MOFs and silica may lead to one of two types of composites, SiO_2@MOFs and MOFs@SiO_2, with both realizing enhanced properties. In this section, focus will be placed on MOFs@SiO_2 since it involves silica as a shell coating of a MOF surface or MOFs growing on a silica support [124]. When coating MIL-101(Cr) with a very thin hydrophobic layer of silica, the stability of the composite was improved and the diffusion of reactants and products through the catalytic MOF was facilitated (Figure 9.15). In fact, while silica alone does not exhibit any catalytic behavior in the oxidation of indene, MIL-101Cr@mSiO_2 was proven catalytic with a TOF of 95.2 mmol g^{-1} h^{-1}, which was 1.24 times higher than that of the parent MIL-101(Cr) (76.8 mmol g^{-1} h^{-1}). The effect of the silica shell may be counterintuitive since it should slow the diffusion rate of molecules and partially restrict access to active sites. However, a deeper investigation proved that indene, a hydrophobic reactant, favorably diffuses through rough silica shells that are more hydrophobic than MIL-101(Cr), thus enhancing substrate-enzyme interactions [137]. Concerning the second approach, Cu-BTC MOF was grown on dendrimer-like porous silica nanoparticles (DPSNs) forming DPSNs@Cu-BTC nanocomposites. This composite yielded >99 % conversion in the epoxidation of cyclooctene as compared to 87 % for Cu-BTC even though the latter is thought to be responsible for the catalytic activity while DPSNs are just carriers. A plausible explanation is that Cu-BTC's size was reduced via the confined growth in the channels of DPSNs and it was evenly distributed allowing for better exposure of active sites [138].

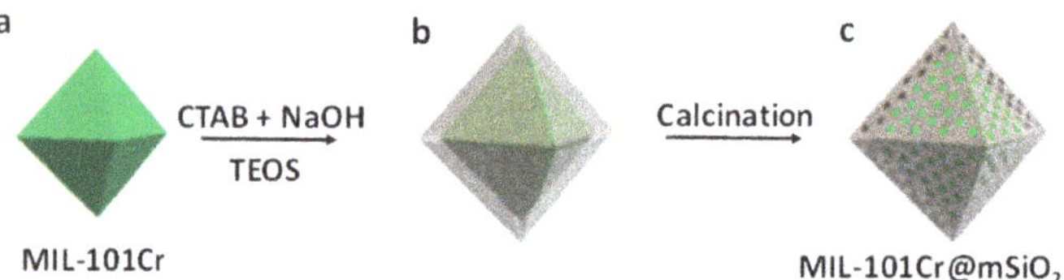

Figure 9.15: Schematic illustration for the direct coating of MIL-101(Cr) with a hydrophobic mesoporous silica shell. Reprinted with permission [137]. Copyright, 2018, American Chemical Society.

9.6.5 MOF-polymer composites

Polymers consisting of macromolecules with multiple repeating subunits display unique properties, such as softness and thermal and chemical stability. Coating MOFs with hydrophobic polymers is a viable method for improving the moisture/water stability of MOFs [139]. As a proof of concept, Yu et al. coated the surface of multiple MOFs (MOF-5, HKUST-1 and ZnBT) with hydrophobic polydimethysiloxane (PDMS). While coated and uncoated, HKUST-1 showed very similar yields for the cyanosilylation of benzaldehyde and trimethylsilylcyanide (48.2 compared to 50.1 %, respectively), where the difference is noticed upon saturated water vapor treatment. While HKUST-1 showed a sharp decrease to 19.6 % yield after treatment, the PDMS coating prevented water molecules from reaching the Cu(II) sites, and thus allowed coated HKUST-1 to retain its activity [140]. In a study conducted by Zhao et al., UiO-66 was successfully coated with a functional nanocomposite shell containing Pd nanoparticles and polypyrrole (PPy), a conducting polymer. The resulting UiO-66/Pd@PPy nanocomposite displayed catalytic activity in the reduction of *p*-nitrophenol (4-NP) by $NaBH_4$ with an apparent reaction constant, $k_{app} = 2.86 \times 10^{-2}\,s^{-1}$ which is higher than the no catalytic activity for UiO-66 alone and $8.87 \times 10^{-3}\,s^{-1}$ for Pd/PPy. This improvement is attributed to electron transfer from PPy to Pd NPs creating an electron-rich region suitable for the adsorption of both reactants on the Pd site. UiO-66 particles facilitated the adsorption of 4-NP and the dispersion of Pd NPs, further proving the power of hosting different active species by MOFs for catalytic purposes [141].

9.6.6 MOF-MOF composites

MOF@MOF structures can be achieved by synthesizing a core MOF and enveloping it with another MOF shell to access properties that are not provided by conventional frameworks [142]. Gong et al. successfully synthesized a UiO-66@UiO-67-BPY-Ag composite by growing UiO-67-BPY on UiO-66 followed by post-synthetic metalation of Ag. As a catalytic site for CO_2 carboxylation of phenylacetylene, Ag yielded 96 % product when loaded into the MOF@MOF composite as compared to 75 % in the case of UiO-67-BPY-Ag. Interestingly, a lower Ag content was needed to achieve high yields in the core-shell structure as compared to the single MOF catalyst [143]. Similarly, a MoO_x@UiO-66@MoO_x@UiO-bpy bilayer composite was synthesized having MoO_x as active catalytic sites for selective oxidation of cyclopentene with H_2O_2. Introducing such species into a bilayer MOF-MOF composite suppressed the leaching of MoO_x, improved the transport of reactants and products and ensured good trapping of intermediates, all of which led to a superior activity as compared to catalysis carried out by MoO_x@UiO-66 and commercial MoO_3 [144].

9.7 Multilinker and/or -metal multivariate MOFs as catalytic systems

Catalytic properties can be enhanced through the introduction of numerous interchangeable linkers having similar structure, but different functionalities, or of various metal ions within the same secondary building units within one crystalline network (Figure 9.16) [145]. These multivariate (MTV) MOFs possess many advantages over the parent MOFs having the same topology. In particular, various catalytic active sites can be incorporated within the same framework resulting in better efficiency, performance and stereoselectivity over monocatalytic frameworks in tandem reactions.

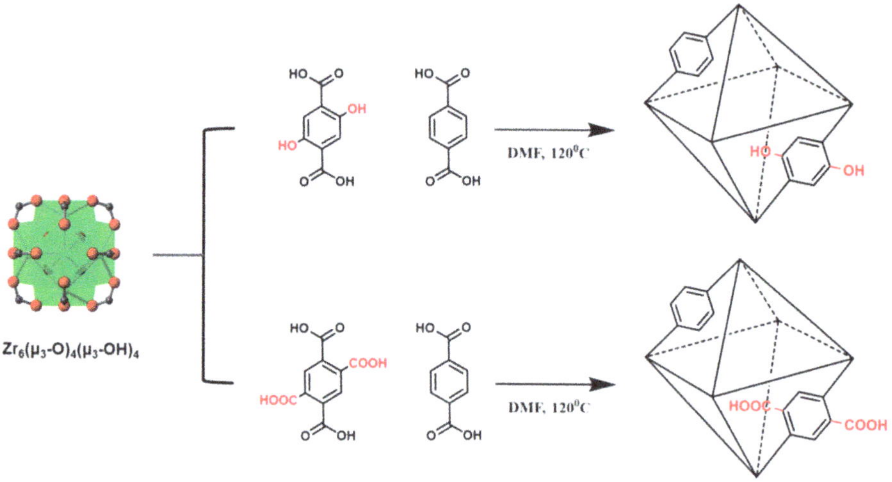

Figure 9.16: Schematic representation of the synthesis procedure of two MTV-UiO-66 materials. Atom colors: Zr, green polyhedra; C, grey and O, red.

Ahn et al. successfully synthesized a bifunctional Brønsted acid-base heterogeneous catalyst, named MIL-101-NH_2-SO_3H, via a solvothermal route using two different terephthalate linkers, H_2BDC-NO_2 and H_2BDC-SO_3Na, followed by a reduction process of the NO_2 group on the linkers to NH_2 [146]. The resulting framework achieved a remarkable selectivity of 99.3 % and conversion of 100 % toward *trans*-1-nitro-2-phenylethylene formation from benzaldehyde dimethyl acetal in a one-pot tandem de-acetalization-nitroaldol reaction. This performance was explained by the presence of both Brønsted acid and base catalytic sites on the linkers that circumvented the weak Lewis acidity of the CUSs found in MIL-101. In a similar fashion, Ahmad et al. reported the synthesis of two MTV-UiO-66 systems, termed MTV-UiO-66$(COOH)_2$ synthesized from BDC and pyromellitic acid linkers and MTV-UiO-66$(OH)_2$ constructed from BDC and 2,5-dihydroxyterephthalic acid (DHTA) linkers. Both of these MTV-MOFs were employed as heterogeneous catalysts in the esterification reaction of butyric acid and butanol and showed outstanding and improved catalytic activity over UiO-66, UiO-66$(OH)_2$ and UiO-66$(COOH)_2$ due to the catalytic active sites present on the added linkers in addition to the high surface areas and pore accessibility they maintained [147]. In particular, a conversion rate of 92.2 % to butyl butyrate was achieved in 24 hours using MTV-UiO-66$(OH)_2$ that incorporated 52 % of DHTA linker compared to only 75.1 % and 80.5 % conversion rate for UiO-66 and UiO-66$(OH)_2$, respectively. The mechanism is described as a dual acid-base activation and it begins with the protonation of the hydroxyl group on the organic linker by 1-butanol, followed by the donation of a pair of electrons from the oxygen atom of the alcohol to the carbon atom of the acid's carboxyl group. Then the π electrons of the latter attack the Lewis acid active site of Zr(IV). A water molecule is later released upon the regeneration of the C=O leading to the formation of the MOF-ester intermediate, followed by the release of the ester. In a similar manner, Wright et al. reported the synthesis of a mixed-metal MIL-100(Sc_{60}, Fe_{40}) for the one-pot tandem deacetalization Friedel–Crafts addition-oxidation of methylketol and 1-ethoxy-2,2,2-trifluoroethanol in presence of *tert*-butyl hydroperoxide [148]. The presence of both iron and scandium metal cations within the framework resulted in a catalytic conversion of 96 % compared to only 10 % for MIL-100(Sc) and 60 % for MIL-100(Fe). This proved the synergistic effect of the Lewis acid catalytic sites, namely Fe(III) and Sc(III) in addition to the oxidation sites in Fe(III) generated via the mixed-metal MOF.

9.8 MOFs-based photocatalysis

Inspired by natural photosynthesis, photocatalysis involves the use of a catalyst, typically a semiconductor, which is able to absorb light, most desirably the clean, safe and inexhaustible solar energy, to change the rate of useful chemical reactions such as H_2 production, organic molecules photodegradation and CO_2 photoreduction into solar fuels [149]. Ever since the pioneering discovery of water photolysis on a TiO_2 electrode in 1972, many homogeneous and heterogeneous photocatalysts were reported, among the latest being MOFs [150]. Potential photocatalytically active centers in MOFs include metal nodes that may act as isolated semiconductor quantum dots, organic linkers regarded as light absorption antenna and/or guest species incorporated into the framework by post-synthetic modification [21]. Compared to classical semiconductors, MOFs are advantageous due to their excellent optical tunability that is expressed by the rich catalogue of building blocks that could be selected *a priori*, their localized electrons limiting the electron-hole recombination rate and the possibility of optically favored active site engineering [151, 152]. Numerous engineering strategies have been developed and remain promising for future improvements as summarized in Figure 9.17 [153]. Upon exposure to light with sufficient energy, MOFs exhibit a photo-

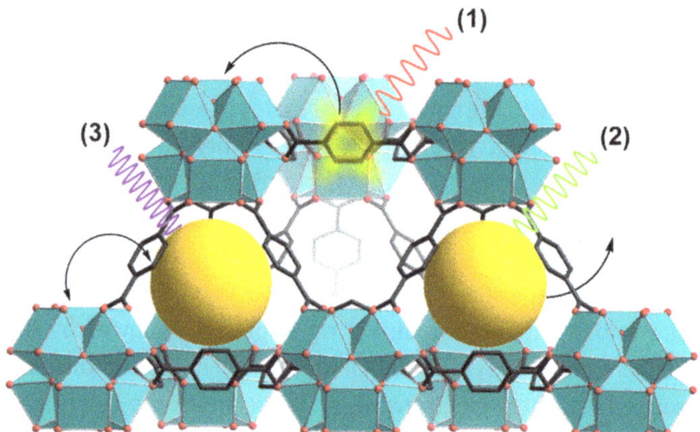

Figure 9.17: The three major strategies employed in introducing photocatalytic active sites in MOFs. (1) Usage of electron-rich organic ligands that exhibit optical absorption in the visible region and result in a charge transfer to the inorganic clusters of the MOF. (2) Encapsulation of guest species that absorb light strongly and act as photocatalysts. (3) Encapsulation of guest species that result in a charge transfer between them and the MOF structure. Atom colors: Zr, blue polyhedra; C, grey; O, red. Yellow balls provide an indication of the open space.

excitation of electrons from the highest occupied molecular orbitals (HOMO), where holes are subsequently generated to the lowest unoccupied molecular orbitals (LUMO) of active centers (i. e., metal nodes, organic linkers and/or hosted guests). Based on the moieties used, LMCT (ligand-to-metal charge transfer), MLCT (metal-to-ligand charge transfer) or MMCT (metal-to-metal charge transfer) can occur, thus promoting oxidation reactions that reduce the generated holes while the excited electrons become capable of reducing desired species [154].

MOF-5 was one of the earliest MOFs to be reported for photocatalytic activity [155]. Although after this report, the number of documented photocatalytic MOFs increased drastically, the scope of this section will be limited to selected remarkable MOFs showing distinctive performance in versatile reactions.

To start, CO_2 conversion into value-added products, notably solar fuels, is of great environmental importance [156]. Within this scope, a recent report proved that AUBM-4 achieved one of the highest conversions of CO_2 to formate (366 $\mu mol\, g^{-1}\, h^{-1}$) compared to its counterparts. This new Zr MOF was synthesized by incorporating the photoactive bis(4'-(4-carboxyphenyl)-terpyridine)ruthenium(II) (Ru(cptpy)2) linker into the framework's backbone having a simple ZrO_8 cluster forming the highly stable and efficient one dimensional-structure (Figure 9.18). Density functional theory (DFT) calculations suggest that photoexcitation promotes the fast Ru-to-cptpy MLCT followed by a reduction of metal centers by a sacrificial agent, while CO_2 is being reduced by the $cptpy^-$ radical anion. This study was among few to verify the source of the obtained product using labeled $^{13}CO_2$ since carbon contamination on the surface and pores of MOFs may lead to false positive results [157].

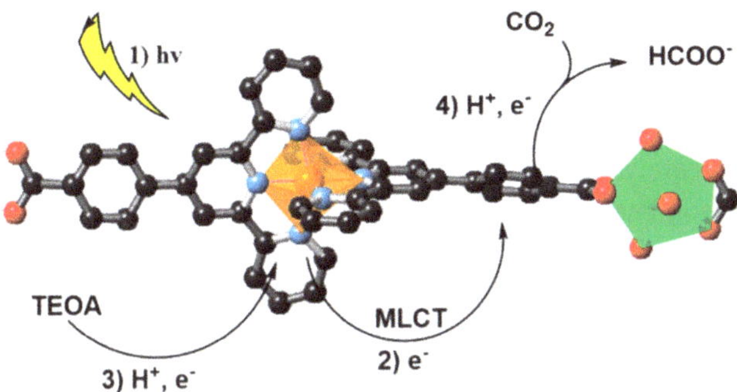

Figure 9.18: Proposed mechanism for the photoreduction of CO_2 into formic acid over a Zr-based MOF incorporating a photoactive bis(4'-(4-carboxyphenyl)terpyridine) ruthenium (II) linker. Atom colors: Zr, green; Ru, orange; N, blue; C, black, O, red. Reprinted with permission [157]. Copyright, 2019, American Chemical Society.

A second major reaction to consider is H_2 production, being a promising candidate for renewable energy. H_2 can be produced from formic acid dehydrogenation; however, this reaction is also competitive with possible dehydration that generates toxic CO [158]. An interesting study reported the use of Au@Pd/UiO-66($Zr_{85}Ti_{15}$) for selective photocatalysis of formic acid to H_2. The framework's design made use of Pd, being the most practical heterogeneous metal catalyst for this reaction, refined its disadvantages by forming an Au@Pd core-shell structure, and introduced the obtained nanoparticles into a Ti-doped framework, having amino-functionalized linkers. This composite emphasizes the power of multicomponent-driven synergistic effects in MOFs. A plausible mechanism starts by the absorption of visible light by the amino-functionalized linker followed by a LMCT reaching and activating the Pd nanoparticles. Meanwhile, Au strongly absorbs visible light due to its localized surface plasmon resonance, which enables it to transfer hot electrons to neighboring Pd, thus further increasing the electron density on activated Pd that leads to the dehydrogenation of formic acid with a 42000 $mL\,h^{-1}\,g^{-1}$ production rate and a TOF of 200 h^{-1} [159]. Another interesting approach involves the use of bifunctional MOFs having photocatalytic capability along with Lewis acid/base chemistry. In 2021, Li et al. documented the activity of a bifunctional CdS-MOF material in cascade reactions, including first a photocatalytic oxidation followed by acid- or base-driven reactions. CdS/NH_2-MIL-125 profits from loaded CdS nanoparticles for the photocatalysis of benzyl alcohol to benzaldehyde while the basicity of the amino-functionalized linker accounts for the Knoevenagel reaction of benzaldehyde and malononitrile yielding 97 % conversion and 93 % selectivity. Similarly, in CdS@MIL-101, CdS photocatalytic activity was coupled with the Lewis acidity of MIL-101 for the aldimine condensation of the photocatalytically generated benzaldehyde with aniline to give *N*-benzylideneaniline [160]. Such novel reports provide deep insights into the wide range of engineering possibilities and synergistic roles of building blocks and hosted substances in MOFs.

9.9 Conclusion and future directions

Notwithstanding the various technical difficulties and challenges faced in the synthesis of new heterogeneous catalysts used for organic transformations, the utilization of MOFs as heterogeneous solid catalysts has witnessed an explosion in interest and a marked upsurge in recent decades owing to their structural features, properties and characteristics. Such catalytic transformations can be achieved by tuning any of the three main constituents of MOFs: metal clusters, organic linkers and void spaces. In particular, the high porosities and surface areas that most MOFs possess allow for an easy incorporation of catalytically-active species, including metal nanoparticles, enzymes and metal complexes, into the framework pores and/or onto their external surfaces. Additionally, intrinsic catalytic activity can be achieved in MOFs via the creation of CUSs through missing-linker and missing-cluster defects and the modification of organic linkers resulting in the generation of acidic and basic catalytic active sites. These active sites are mostly produced during MOFs' synthesis or via different post-synthetic treatments. Additionally, MTV-MOFs are a promising platform for multiple active site cooperation, which is useful for optimizing heterogeneity and reactivity.

Nevertheless, this field is still in need of many improvements to overcome the limitations hindering the extended use of MOFs in catalysis. To start, large scale production of various MOF structures with high yields for catalytic processes remains challenging. For instance, some organic linkers are not commercially available, and their synthesis method is complicated and requires multiple steps resulting in a high production cost and a soaring energy consuming process. Hence, it is crucial to expand the catalytic trials to cover additional types of MOFs and reactions apart from the most conventionally reported and to investigate MOFs with smaller pore sizes, for which confinement effects are considerable. Second, a well-tailored catalytic correlative progressive design should be subjected to stability assessment since many of the main types of MOFs show poor structural and mechanical stability. In particular, it is important that the MOF catalyst maintains its activity, stability and selectivity during a catalytic process under real reaction conditions to be considered effective. This can help in solving their regeneration process issues where the removal of the product(s) and intermediate(s) commonly occur at high temperatures. Moreover, the multifunctionality and active sites synergy provided by MOFs should inspire future structural designs to accommodate reaction mechanisms involving multiple intrinsic, encapsulated and/or grafted catalytic moieties at the same time. This specific feature is key for the development of advantageous tandem reactions employing a single MOF as a catalyst for the whole sequence. Adding to that, experimental failures should be subjected to mechanistic studies in order to identify the pertinent problem, understand secondary interactions within the framework and gain some control over the microenvironment of the reaction media. In parallel, spectroscopic methods combined with DFT should be used to characterize MOFs even in their working state in order to gain evidence about the changes occurring during the reaction as well as during potential catalyst deactivation. There exists extensive room for innovation and incremental progress within MOF catalysis, which will hopefully lead to the further flourishment of this domain.

Bibliography

[1] Global Catalyst (Chemical Compounds, Zeolites, Metals) Market Size, Share & Trends Analysis Report 2020–2027 – ResearchAndMarkets.com, in Business Wire. Business Wire, Inc.; 2021.
[2] Xia Y et al. Introduction: advanced materials and methods for catalysis and electrocatalysis by transition metals. Chem Rev. 2021;121:563–6.
[3] Wisniak J. The history of catalysis. From the beginning to Nobel prizes. Educ Quimica. 2010;21:60–9.
[4] Moulijn JA, van Leeuwen PWNM, van Santen RA. Chapter 1 history of catalysis. In: Studies in surface science and catalysis. Elsevier; 1993. p. 3–21.
[5] Zhou QL. Transition-metal catalysis and organocatalysis: where can progress be expected? Angew Chem, Int Ed. 2016;55:5352–3.
[6] Ali ME et al. Heterogeneous metal catalysts for oxidation reactions. J Nanomater. 2014;2014:192038.
[7] Yang W et al. Metallic ion leaching from heterogeneous catalysts: an overlooked effect in the study of catalytic ozonation processes. Environ Sci Water Res Technol. 2017;3:1143–51.
[8] Li N et al. Conversion of plastic waste into fuels: a critical review. J Hazard Mater. 2022;424:127460.
[9] Vollmer I et al. Plastic waste conversion over a refinery waste catalyst. Angew Chem, Int Ed. 2021;60:16101–8.
[10] Ebikade EO et al. A review of thermal and thermocatalytic valorization of food waste. Green Chem. 2021;23:2806–33.
[11] Smith C, Hill AK, Torrente-Murciano L. Current and future role of Haber-Bosch ammonia in a carbon-free energy landscape. Energy Environ Sci. 2020;13:331–44.
[12] Kumawat J, Gupta VK. Fundamental aspects of heterogeneous Ziegler-Natta olefin polymerization catalysis: an experimental and computational overview. Polym Chem. 2020;11:6107–28.
[13] Digest CI. Pioneers of asymmetric organocatalysis win 2021 chemistry Nobel prize. Mumbai: Chemical Industry Digest; 2021.
[14] Cui Y et al. Metal–organic frameworks as platforms for functional materials. Acc Chem Res. 2016;49:483–93.
[15] Gangu KK, Jonnalagadda SB. A review on metal–organic frameworks as congenial heterogeneous catalysts for potential organic transformations. Front Chem 2021;9.
[16] Wu CD, Zhao M. Incorporation of molecular catalysts in metal–organic frameworks for highly efficient heterogeneous catalysis. Adv Mater. 2017;29:1605446.
[17] Meledina M et al. Ru catalyst encapsulated into the pores of MIL-101 MOF: direct visualization by TEM. Materials. 2021;14:4531.
[18] Veisi H et al. Pd immobilization biguanidine modified Zr-UiO-66 MOF as a reusable heterogeneous catalyst in Suzuki-Miyaura coupling. Sci Rep. 2021;11:21883.
[19] Ma D et al. Bifunctional MOF heterogeneous catalysts based on the synergy of dual functional sites for efficient conversion of CO2 under mild and co-catalyst free conditions. J Mater Chem A. 2015;3:23136–42.

[20] Nuri A et al. Synthesis and characterization of palladium supported amino functionalized magnetic-MOF-MIL-101 as an efficient and recoverable catalyst for Mizoroki-Heck cross-coupling. Catal Lett. 2020;150:2617–29.

[21] Dhakshinamoorthy A, Li Z, Garcia H. Catalysis and photocatalysis by metal organic frameworks. Chem Soc Rev. 2018;47:8134–72.

[22] Yang D, Gates BC. Catalysis by metal organic frameworks: perspective and suggestions for future research. ACS Catal. 2019;9:1779–98.

[23] Geng L et al. Redox property switching in MOFs with open metal sites for improved catalytic hydrogenation performance. J Alloys Compd. 2021;888:161494.

[24] Kökçam-Demir Ü et al. Coordinatively unsaturated metal sites (open metal sites) in metal–organic frameworks: design and applications. Chem Soc Rev. 2020;49:2751–98.

[25] Abednatanzi S et al. Metal- and covalent organic frameworks as catalyst for organic transformation: comparative overview and future perspectives. Coord Chem Rev. 2022;451:214259.

[26] Wang J et al. Engineering effective structural defects of metal–organic frameworks to enhance their catalytic performances. J Mater Chem A. 2020;8:4464–72.

[27] Hall JN, Bollini P. Structure, characterization, and catalytic properties of open-metal sites in metal organic frameworks. React Chem Eng. 2019;4:207–22.

[28] Blandez JF et al. Influence of functionalization of terephthalate linker on the catalytic activity of UiO-66 for epoxide ring opening. J Mol Catal A, Chem. 2016;425:332–9.

[29] Fujita M et al. Preparation, clathration ability, and catalysis of a two-dimensional square network material composed of Cadmium(II) and 4, 4'-bipyridine. J Am Chem Soc. 1994;116:1151–2.

[30] Schlichte K, Kratzke T, Kaskel S. Improved synthesis thermal stability and catalytic properties of the metal–organic framework compound Cu3(BTC)2. Microporous Mesoporous Mater. 2004;73:81–8.

[31] Calleja G et al. Copper-based MOF-74 material as effective acid catalyst in Friedel-Crafts acylation of anisole. Catal Today. 2014;227:130–7.

[32] Horike S et al. Size-selective Lewis acid catalysis in a microporous metal–organic framework with exposed Mn2+ coordination sites. J Am Chem Soc. 2008;130:5854–5.

[33] Ravon U et al. MOFs as acid catalysts with shape selectivity properties. New J Chem. 2008;32:937–40.

[34] Vandichel M et al. Insight in the activity and diastereoselectivity of various Lewis acid catalysts for the citronellal cyclization. J Catal. 2013;305:118–29.

[35] Vermoortele F et al. Synthesis modulation as a tool to increase the catalytic activity of metal–organic frameworks: the unique case of UiO-66(Zr). J Am Chem Soc. 2013;135:11465–8.

[36] Hu Z et al. Direct synthesis of hierarchically porous metal–organic frameworks with high stability and strong Brønsted acidity: the decisive role of hafnium in efficient and selective fructose dehydration. Chem Mater. 2016;28:2659–67.

[37] Jiang J, Yaghi OM. Brønsted acidity in metal–organic frameworks. Chem Rev. 2015;115(14):6966–97.

[38] Gong W et al. Metal–organic frameworks as solid Brønsted acid catalysts for advanced organic transformations. Coord Chem Rev. 2020;420:213400.

[39] Ravon U et al. Engineering of coordination polymers for shape selective alkylation of large aromatics and the role of defects. Microporous Mesoporous Mater. 2010;129:319–29.

[40] Ravon U et al. Investigation of acid centers in MIL-53(Al,Ga) for Brønsted-type catalysis: in situ FTIR and ab initio molecular modeling. ChemCatChem. 2010;2:1235–8.

[41] Bakuru VR et al. Exploring the Brønsted acidity of UiO-66 (Zr, Ce, Hf) metal–organic frameworks for efficient solketal synthesis from glycerol acetalization. Dalton Trans. 2019;48:843–7.

[42] Rojas-Buzo S et al. Tailoring Lewis/Brønsted acid properties of MOF nodes via hydrothermal and solvothermal synthesis: simple approach with exceptional catalytic implications. Chem Sci. 2021;12:10106–15.
[43] Trickett CA et al. Identification of the strong Brønsted acid site in a metal–organic framework solid acid catalyst. Nat Chem. 2019;11:170–6.
[44] Liu Y et al. Probing the correlations between the defects in metal–organic frameworks and their catalytic activity by an epoxide ring-opening reaction. ChemComm. 2016;52:7806–9.
[45] Jrad A et al. Structural engineering of Zr-based metal–organic framework catalysts for optimized biofuel additives production. Chem Eng Sci. 2020;382:122793.
[46] Jiang J et al. Superacidity in sulfated metal–organic framework-808. J Am Chem Soc. 2014;136:12844–7.
[47] Valekar AH et al. Catalytic transfer hydrogenation of furfural to furfuryl alcohol under mild conditions over Zr-MOFs: exploring the role of metal node coordination and modification. ACS Catal. 2020;10:3720–32.
[48] Lin Foo M et al. Ligand-based solid solution approach to stabilisation of sulphonic acid groups in porous coordination polymer Zr6O4(OH)4(BDC)6 (UiO-66). Dalton Trans. 2012;41:13791–4.
[49] Van Humbeck JF et al. Ammonia capture in porous organic polymers densely functionalized with Brønsted acid groups. J Am Chem Soc. 2014;136:2432–40.
[50] Zaera F. New advances in the use of infrared absorption spectroscopy for the characterization of heterogeneous catalytic reactions. Chem Soc Rev. 2014;43:7624–63.
[51] Lieder C et al. Adsorbate effect on AlO4(OH)2 centers in the metal–organic framework MIL-53 investigated by solid-state NMR spectroscopy. J Phys Chem C. 2010;114:16596–602.
[52] Zhu L et al. Metal–organic frameworks for heterogeneous basic catalysis. Chem Rev. 2017;117:8129–76.
[53] Leo P et al. A double basic Sr-amino containing MOF as a highly stable heterogeneous catalyst. Dalton Trans. 2019;48:11556–64.
[54] Saha D et al. Porous magnesium carboxylate framework: synthesis, X-ray crystal structure, gas adsorption property and heterogeneous catalytic aldol condensation reaction. Dalton Trans. 2012;41:7399–408.
[55] Saha D et al. Heterogeneous catalysis over a barium carboxylate framework compound: synthesis, X-ray crystal structure and aldol condensation reaction. Polyhedron. 2012;43:63–70.
[56] Valvekens P et al. Base catalytic activity of alkaline Earth MOFs: a (micro)spectroscopic study of active site formation by the controlled transformation of structural anions. Chem Sci. 2014;5:4517–24.
[57] Zhang Q et al. Catalyzed activation of CO2 by a Lewis-base site in W-Cu-BTC hybrid metal organic frameworks. Chem Sci. 2012;3:2708–15.
[58] Kang YS et al. Metal–organic frameworks with catalytic centers: from synthesis to catalytic application. Coord Chem Rev. 2019;378:262–80.
[59] Zhang Y, Yang X, Zhou HC. Synthesis of MOFs for heterogeneous catalysis via linker design. Polyhedron. 2018;154:189–201.
[60] Vermoortele F et al. An amino-modified Zr-terephthalate metal–organic framework as an acid-base catalyst for cross-aldol condensation. ChemComm. 2011;47:1521–3.
[61] Yang Y et al. Amino-functionalized Zr(IV) metal–organic framework as bifunctional acid-base catalyst for Knoevenagel condensation. J Mol Catal A, Chem. 2014;390:198–205.
[62] Yao QX et al. A series of microporous and robust Ln-MOFs showing luminescent properties and catalytic performances towards Knoevenagel reactions. Dalton Trans. 2021;50:17785–91.
[63] Hasegawa S et al. Three-dimensional porous coordination polymer functionalized with amide groups based on tridentate ligand: selective sorption and catalysis. J Am Chem Soc. 2007;129:2607–14.

[64] Valvekens P et al. Metal-dioxidoterephthalate MOFs of the MOF-74 type: microporous basic catalysts with well-defined active sites. J Catal. 2014;317:1–10.
[65] de Oliveira A et al. Acidic and basic sites of M2DEBDC (M = Mg or Mn and E = O or S) acting as catalysts for cyanosilylation of aldehydes. Polyhedron. 2018;154:98–107.
[66] Wang P et al. Preparation of MOF catalysts and simultaneously modulated metal nodes and ligands via a one-pot method for optimizing cycloaddition reactions. New J Chem. 2020;44:9611–5.
[67] Leo P et al. Catalytic activity and stability of sulfonic-functionalized UiO-66 and MIL-101 materials in friedel-crafts acylation reaction. Catal Today. 2022;390–391:258–64.
[68] Dou Y et al. Highly efficient catalytic esterification in an -SO3H-functionalized Cr(III)-MOF. Ind Eng Chem Res. 2018;57:8388–95.
[69] Gagnon KJ, Perry HP, Clearfield A. Conventional and unconventional metal–organic frameworks based on phosphonate ligands: MOFs and UMOFs. Chem Rev. 2012;112:1034–54.
[70] Zhou Z et al. Metal–organic polymers containing discrete single-walled nanotube as a heterogeneous catalyst for the cycloaddition of carbon dioxide to epoxides. J Am Chem Soc. 2015;137:15066–9.
[71] Ren Y et al. A chiral mixed metal–organic framework based on a Ni(saldpen) metalloligand: synthesis, characterization and catalytic performances. Dalton Trans. 2013;42:9930–7.
[72] Gao CY et al. An ultrastable zirconium-phosphonate framework as bifunctional catalyst for highly active CO2 chemical transformation. ChemComm. 2017;53:1293–6.
[73] Dincă M, Gabbaï FP, Long JR. Organometallic chemistry within metal–organic frameworks. Organometallics. 2019;38:3389–91.
[74] Cui H et al. A stable and porous iridium(iii)-porphyrin metal–organic framework: synthesis, structure and catalysis. CrystEngComm. 2016;18:2203–9.
[75] Ibáñez S, Poyatos M, Peris E. N-heterocyclic carbenes: a door open to supramolecular organometallic chemistry. Acc Chem Res. 2020;53:1401–13.
[76] Carson F et al. Effect of the functionalisation route on a Zr-MOF with an Ir-NHC complex for catalysis. ChemComm. 2015;51:10864–7.
[77] Kassie AA et al. Synthesis and reactivity of Zr MOFs assembled from PNNNP-Ru pincer complexes. Organometallics. 2019;38:3419–28.
[78] Xuan W et al. A chiral porous metallosalan-organic framework containing titanium-oxo clusters for enantioselective catalytic sulfoxidation. Chem Sci. 2013;4:3154–9.
[79] Ferré M et al. Recyclable organocatalysts based on hybrid silicas. Green Chem. 2016;18:881–922.
[80] Roberts JM et al. Urea metal–organic frameworks as effective and size-selective hydrogen-bond catalysts. J Am Chem Soc. 2012;134:3334–7.
[81] McGuirk CM et al. Turning on catalysis: incorporation of a hydrogen-bond-donating squaramide Moiety into a Zr metal–organic framework. J Am Chem Soc. 2015;137:919–25.
[82] Dong XW et al. Heterogenization of homogeneous chiral polymers in metal–organic frameworks with enhanced catalytic performance for asymmetric catalysis. Green Chem. 2018;20:4085–93.
[83] Kutzscher C et al. Proline functionalization of the mesoporous metal–organic framework DUT-32. Inorg Chem. 2015;54:1003–9.
[84] Lun DJ, Waterhouse GIN, Telfer SG. A general thermolabile protecting group strategy for organocatalytic metal–organic frameworks. J Am Chem Soc. 2011;133:5806–9.
[85] Kutzscher C et al. Proline functionalized UiO-67 and UiO-68 type metal–organic frameworks showing reversed diastereoselectivity in aldol addition reactions. Chem Mater. 2016;28:2573–80.
[86] Cohen SM. Postsynthetic methods for the functionalization of metal–organic frameworks. Chem Rev. 2012;112:970–1000.

[87] Tanabe KK, Cohen SM. Engineering a metal–organic framework catalyst by using postsynthetic modification. Angew Chem, Int Ed. 2009;48:7424–7.
[88] Lollar CT et al. Interior decoration of stable metal–organic frameworks. Langmuir. 2018;34:13795–807.
[89] Evans JD, Sumby CJ, Doonan CJ. Post-synthetic metalation of metal–organic frameworks. Chem Soc Rev. 2014;43:5933–51.
[90] Dincă M, Long JR. High-enthalpy hydrogen adsorption in cation-exchanged variants of the microporous metal–organic framework Mn3[(Mn4Cl)3(BTT)8(CH3OH)10]2. J Am Chem Soc. 2007;129:11172–6.
[91] Brozek CK et al. NO disproportionation at a mononuclear site-isolated Fe2+ center in Fe2+-MOF-5. J Am Chem Soc. 2015;137:7495–501.
[92] Collman JP et al. A functional nitric oxide reductase model. Proc Natl Acad Sci. 2008;105:15660.
[93] Denysenko D et al. Elucidating gating effects for hydrogen sorption in MFU-4-type triazolate-based metal–organic frameworks featuring different pore sizes. Eur J Chem. 2011;17:1837–48.
[94] Comito RJ et al. Single-site heterogeneous catalysts for olefin polymerization enabled by cation exchange in a metal–organic framework. J Am Chem Soc. 2016;138:10232–7.
[95] Metzger ED et al. Selective dimerization of ethylene to 1-butene with a porous catalyst. ACS Cent Sci. 2016;2:148–53.
[96] Denysenko D et al. Scorpionate-type coordination in MFU-4l metal–organic frameworks: small-molecule binding and activation upon the thermally activated formation of open metal sites. Angew Chem, Int Ed. 2014;53:5832–6.
[97] Denysenko D et al. Postsynthetic metal and ligand exchange in MFU-4l: a screening approach toward functional metal–organic frameworks comprising single-site active centers. Eur J Chem. 2015;21:8188–99.
[98] Kalaj M, Cohen SM. Postsynthetic modification: an enabling technology for the advancement of metal–organic frameworks. ACS Cent Sci. 2020;6:1046–57.
[99] Mandal S et al. Post-synthetic modification of metal–organic frameworks toward applications. Adv Funct Mater. 2021;31:2006291.
[100] Chen F et al. Metal–organic frameworks as versatile platforms for organometallic chemistry. Inorganics. 2021;9(4):27.
[101] Li Z et al. Metal–organic framework supported cobalt catalysts for the oxidative dehydrogenation of propane at low temperature. ACS Cent Sci. 2017;3:31–8.
[102] Ahn S et al. Stable metal–organic framework-supported niobium catalysts. Inorg Chem. 2016;55:11954–61.
[103] Wang TC et al. Scalable synthesis and post-modification of a mesoporous metal–organic framework called NU-1000. Nature. 2016;11:149–62.
[104] Jiang H et al. Topology-based functionalization of robust chiral Zr-based metal–organic frameworks for catalytic enantioselective hydrogenation. J Am Chem Soc. 2020;142:9642–52.
[105] Bien C, Cai EZ, Wade CR. Using postsynthetic X-type ligand exchange to enhance CO2 adsorption in metal–organic frameworks with Kuratowski-type building units. Inorg Chem. 2021;60:11784–94.
[106] Zhang X et al. In situ generation of an N-heterocyclic carbene functionalized metal–organic framework by postsynthetic ligand exchange: efficient and selective hydrosilylation of CO2. Angew Chem, Int Ed. 2019;58:2844–9.
[107] Xi FG et al. Aldehyde-tagged zirconium metal–organic frameworks: a versatile platform for postsynthetic modification. Inorg Chem. 2016;55:4701–3.
[108] Fu G, Bueken B, De Vos D. Zr-metal–organic framework catalysts for oxidative desulfurization and their improvement by postsynthetic ligand exchange. Small Methods. 2018;2:1800203.

[109] Rösler C et al. Encapsulation of bimetallic metal nanoparticles into robust zirconium-based metal–organic frameworks: evaluation of the catalytic potential for size-selective hydrogenation. Eur J Chem. 2017;23:3583–94.
[110] Liu J et al. Encapsulation of Au55 clusters within surface-supported metal–organic frameworks for catalytic reduction of 4-Nitrophenol. ACS Appl Nano Mater. 2021;4:522–8.
[111] Kumar S et al. Incorporation of homogeneous organometallic catalysts into metal–organic frameworks for advanced heterogenization: a review. Catal Sci Technol. 2021;11:5734–71.
[112] Chen L, Luque R, Li Y. Encapsulation of metal nanostructures into metal–organic frameworks. Dalton Trans. 2018;47:3663–8.
[113] Hardian R et al. Tuning the properties of MOF-808 via defect engineering and metal nanoparticle encapsulation. Eur J Chem. 2021;27:6804–14.
[114] Wang T et al. Rational approach to guest confinement inside MOF cavities for low-temperature catalysis. Nat Commun. 2019;10:1340.
[115] Jiang D et al. Multifunctional Pd@UiO-66 catalysts for continuous catalytic upgrading of ethanol to n-butanol. ACS Catal. 2018;8:11973–8.
[116] Zhou YH et al. Low-cost CuNi@MIL-101 as an excellent catalyst toward cascade reaction: integration of ammonia borane dehydrogenation with nitroarene hydrogenation. ChemComm. 2017;53:12361–4.
[117] Yu C et al. CuNi nanoparticles assembled on graphene for catalytic methanolysis of ammonia borane and hydrogenation of nitro/nitrile compounds. Chem Mater. 2017;29:1413–8.
[118] Liang S et al. Metal–organic frameworks as novel matrices for efficient enzyme immobilization: an update review. Coord Chem Rev. 2020;406:213149.
[119] Sha F et al. Stabilization of an enzyme cytochrome C in a metal–organic framework against denaturing organic solvents. iScience. 2021;24:102641.
[120] Li P et al. Nanosizing a metal–organic framework enzyme carrier for accelerating nerve agent hydrolysis. ACS Nano. 2016;10:9174–82.
[121] Morabito JV et al. Molecular encapsulation beyond the aperture size limit through dissociative linker exchange in metal–organic framework crystals. J Am Chem Soc. 2014;136(36):12540–3.
[122] Li Z et al. Aperture-opening encapsulation of a transition metal catalyst in a metal–organic framework for CO2 hydrogenation. J Am Chem Soc. 2018;140:8082–5.
[123] Filonenko GA et al. Highly efficient reversible hydrogenation of carbon dioxide to formates using a ruthenium PNP-pincer catalyst. ChemCatChem. 2014;6:1526–30.
[124] Chen L, Xu Q. Metal–organic framework composites for catalysis. Matter. 2019;1:57–89.
[125] Liu KG et al. Metal–organic framework composites as green/sustainable catalysts. Coord Chem Rev. 2021;436:213827.
[126] Gao C, Lyu F, Yin Y. Encapsulated metal nanoparticles for catalysis. Chem Rev. 2021;121:834–81.
[127] Long J et al. Selective oxidation of saturated hydrocarbons using au-pd alloy nanoparticles supported on metal–organic frameworks. ACS Catal. 2013;3(4):647–54.
[128] Yang Q, Xu Q, Jiang HL. Metal–organic frameworks meet metal nanoparticles: synergistic effect for enhanced catalysis. Chem Soc Rev. 2017;46:4774–808.
[129] Liu H et al. Nanocomposites of platinum/metal–organic frameworks coated with metal–organic frameworks with remarkably enhanced chemoselectivity for cinnamaldehyde hydrogenation. ChemCatChem. 2016;8:946–51.
[130] Samaniyan M et al. Heterogeneous catalysis by polyoxometalates in metal–organic frameworks. ACS Catal. 2019;9:10174–91.
[131] Buru CT, Farha OK. Strategies for incorporating catalytically active polyoxometalates in metal–organic frameworks for organic transformations. ACS Appl Mater Interfaces. 2020;12:5345–60.

[132] Ghahramaninezhad M, Pakdel F, Niknam Shahrak M. Boosting oxidative desulfurization of model fuel by POM-grafting ZIF-8 as a novel and efficient catalyst. Polyhedron. 2019;170:364–72.
[133] Granadeiro CM et al. Influence of a porous MOF support on the catalytic performance of Eu-polyoxometalate based materials: desulfurization of a model diesel. Catal Sci Technol. 2016;6:1515–22.
[134] Liang W et al. Metal–organic framework-based enzyme biocomposites. Chem Rev. 2021;121:1077–129.
[135] Cao Y et al. Immobilization of Bacillus subtilis lipase on a Cu-BTC based hierarchically porous metal–organic framework material: a biocatalyst for esterification. Dalton Trans. 2016;45:6998–7003.
[136] Zhou M et al. Development of an immobilized cellulase system based on metal–organic frameworks for improving ionic liquid tolerance and in situ saccharification of bagasse. ACS Sustain Chem Eng. 2019;7:19185–93.
[137] Ying J et al. Nanocoating of hydrophobic mesoporous silica around MIL-101Cr for enhanced catalytic activity and stability. Inorg Chem. 2018;57:899–902.
[138] Zhou Z et al. Growth of Cu-BTC MOFs on dendrimer-like porous silica nanospheres for the catalytic aerobic epoxidation of olefins. New J Chem. 2020;44:14350–7.
[139] He S et al. A generalizable method for the construction of MOF@polymer functional composites through surface-initiated atom transfer radical polymerization. Chem Sci. 2019;10:1816–22.
[140] Zhang W et al. A facile and general coating approach to moisture/water-resistant metal–organic frameworks with intact porosity. J Am Chem Soc. 2014;136:16978–81.
[141] Zhao Y et al. Controlled synthesis of metal–organic frameworks coated with noble metal nanoparticles and conducting polymer for enhanced catalysis. J Colloid Interface Sci. 2019;537:262–8.
[142] Dai S, Tissot A, Serre C. Recent progresses in metal–organic frameworks based core-shell composites. Adv Energy Mater. 2021;12(4):2100061.
[143] Gong Y et al. Core-shell metal–organic frameworks and metal functionalization to access highest efficiency in catalytic carboxylation. J Catal. 2019;371:106–15.
[144] Niu Q et al. Bilayer MOF@MOF and MoOx species functionalization to access prominent stability and selectivity in cascade-selective biphase catalysis. Mol Catal. 2021;513:111818.
[145] Helal A et al. Multivariate metal–organic frameworks. Natl Sci Rev. 2017;4:296–8.
[146] Lee YR, Chung YM, Ahn WS. A new site-isolated acid-base bifunctional metal–organic framework for one-pot tandem reaction. RSC Adv. 2014;4:23064–7.
[147] Jrad A et al. Efficient biofuel production by MTV-UiO-66 based catalysts. Chem Eng Sci. 2021;410:128237.
[148] Mitchell L et al. Mixed-metal MIL-100(Sc,M) (M=Al, Cr, Fe) for Lewis acid catalysis and tandem C–C bond formation and alcohol oxidation. Eur J Chem. 2014;20:17185–97.
[149] Qian Y, Zhang F, Pang H. A review of MOFs and their composites-based photocatalysts: synthesis and applications. Adv Funct Mater. 2021;31:2104231.
[150] Fujishima A, Honda K. Electrochemical photolysis of water at a semiconductor electrode. Nature. 1972;238:37–8.
[151] Grau-Crespo R et al. Modelling a linker mix-and-match approach for controlling the optical excitation gaps and band alignment of zeolitic imidazolate frameworks. Angew Chem, Int Ed. 2016;55:16012–6.
[152] Nasalevich MA et al. Metal–organic frameworks as heterogeneous photocatalysts: advantages and challenges. CrystEngComm. 2014;16:4919–26.
[153] Wang Q et al. Recent advances in MOF-based photocatalysis: environmental remediation under visible light. Inorg Chem Front. 2020;7:300–39.

[154] Al Natour R, AlSabeh G, Hmadeh M. Chapter 14 – Metal–organic framework photocatalysts for carbon dioxide reduction. In: Nguyen V-H, Vo D-VN, Nanda S, editors. Nanostructured photocatalysts. 2021. p. 389–420.

[155] Alvaro M et al. Semiconductor behavior of a Metal-Organic Framework (MOF). Eur J Chem. 2007;13:5106–12.

[156] Mazari SA et al. An overview of catalytic conversion of CO2 into fuels and chemicals using metal organic frameworks. Process Saf Environ Prot. 2021;149:67–92.

[157] Elcheikh Mahmoud M et al. Metal–organic framework photocatalyst incorporating Bis(4'-(4-carboxyphenyl)-terpyridine)ruthenium(II) for visible-light-driven carbon dioxide reduction. J Am Chem Soc. 2019;141:7115–21.

[158] Yoo JS et al. Effect of boron modifications of palladium catalysts for the production of hydrogen from formic acid. ACS Catal. 2015;5:6579–86.

[159] Wen M et al. Plasmonic Au@Pd nanoparticles supported on a basic metal–organic framework: synergic boosting of H2 production from formic acid. ACS Energy Lett. 2017;2:1–7.

[160] Li SR et al. Photocatalytic cascade reactions and dye degradation over CdS-metal–organic framework hybrids. RSC Adv. 2021;11:35326–30.

Rana R. Haikal, Worood A. El-Mehalmey, Hicham Idriss, and Mohamed H. Alkordi

10 MOFs for energy conversion and storage through water electrolysis reactions

10.1 Introduction

The tremendous growth of global population as well as industrial revolution place a significant demand on energy sources, which are now primarily dominated by fossil fuels. Fossil fuels are produced from the decomposition of buried, carbon-based organisms that perished millions of years ago. They produce carbon-rich deposits that are extracted and burned for energy use purposes. Fossil fuels, including coal, gas and oil are nonrenewable energy sources that supply three-quarters of the total energy demand, where the transportation sector is heavily dependent on petroleum (90 %), while residential heating on gas (42 %) and industrial sector on natural gas (40 %) as shown in Figure 10.1.

According to the U. S. Energy Information Administration, it is expected that global energy consumption will rise by 28 % from 2015 to 2040. Nuclear power has been devoted totally to electricity production in power plants, since its usage requires the most controlled technology to avoid radiation exposure and to manage the radioactive waste generated from the fission process. Burning fossil fuels have negative footprint on the environment, as they emit greenhouse gases that contribute to global warming, and more importantly, pose serious threat on human health [1–3]. Therefore, there is a noticeable movement toward the utilization of renewable energy sources.

Renewable energy, often known as clean energy, is derived from natural sources which are regenerative, and inexhaustible such as wind, geothermal, hydropower, tidal and solar energy. The largest source of renewable electricity in the United States is hydropower, which uses water to generate electricity by rotating turbine blades in a generator. The water used in hydropower is often fast-moving water in a large river or quickly descending water from a high point. Despite the growing share of renewable energy sources in the U. S. (summing up to ~12 % of the total energy production) [1, 2], there remains a large room for increasing its share especially for the electricity supply, a sector that consumes 37 % of the natural gas and 90 % of the coal consumed within the U. S. Several solutions are being actively developed to resolve key challenges encountered for reliable, uninterrupted and cost-effective energy supply from renewable

Rana R. Haikal, Worood A. El-Mehalmey, Mohamed H. Alkordi, Center for Materials Science, Zewail City of Science and Technology, 6th of October, Giza 12578, Egypt, e-mail: malkordi@zewailcity.edu.eg
Hicham Idriss, Institute of Functional Interfaces (IFG), Karlsruhe Institute of Technology (KIT), Hermann-von-Helmholtz-Platz, 176344 Eggenstein-Leopoldshafen, Germany

https://doi.org/10.1515/9781501524721-010

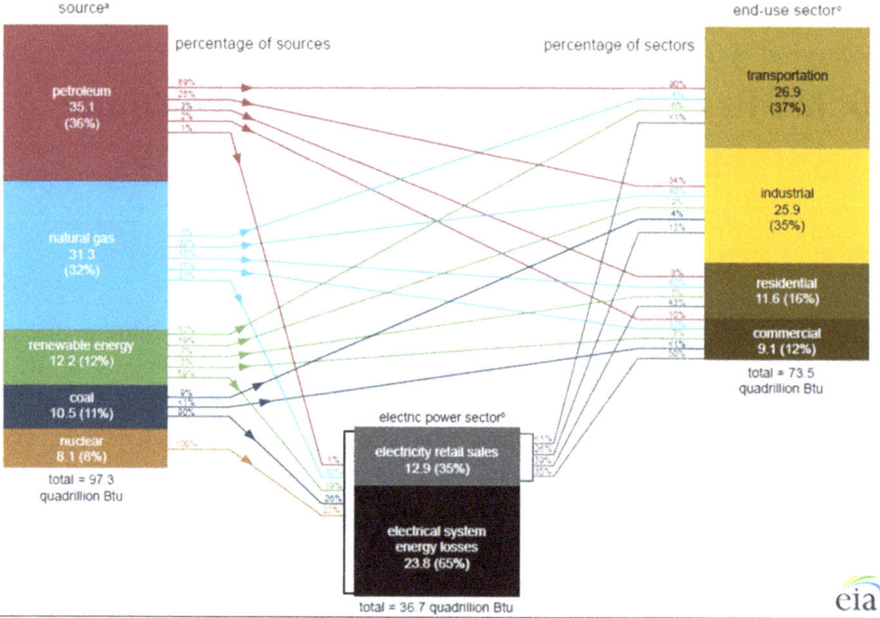

Figure 10.1: U. S. energy consumption by source and sector, 2021 in quadrillion British thermal units (Btu). Reproduced with permission from [1].

resources [4]. Critical to tap into those energy sources is the ability to convert mechanical, thermal or photo energy into electrical energy that can be easily transported from the production sites to consumption sites, Figure 10.2. This initial challenge is being actively addressed by several technologies, including windmills, hydro- and thermal-power plants, solar cells (photovoltaic cells) and solar concentrators.

Transporting electricity is a mature technology that has been developed since the early days of constructing coal power plants in the 19th century and is still witnessing notable developments. A key challenge to face nowadays for reliable dependence on renewable energy resources is energy storage. As most renewable energy resources are intermittent, perhaps except for the hydropower from dams, a reliable uninterrupted supply of electrical energy from such resources requires power management systems with considerable energy storage capacity. Toward this goal, a number of promising energy storage technologies are actively being developed. Any installation or technology, typically subject to independent control, that makes it feasible to store energy produced in a power system and use it in the power system, as needed, can be referred to as energy storage in a power system [5]. The energy storage process can be divided into three stages: (i) charging, (ii) storage and (iii) and discharging. Energy storage must have the proper rated power and energy capacity in order to maintain a balance between power and energy in the power system under each of these three regimes. Due to their unique qualities, many energy storage technologies coexist in a variety of

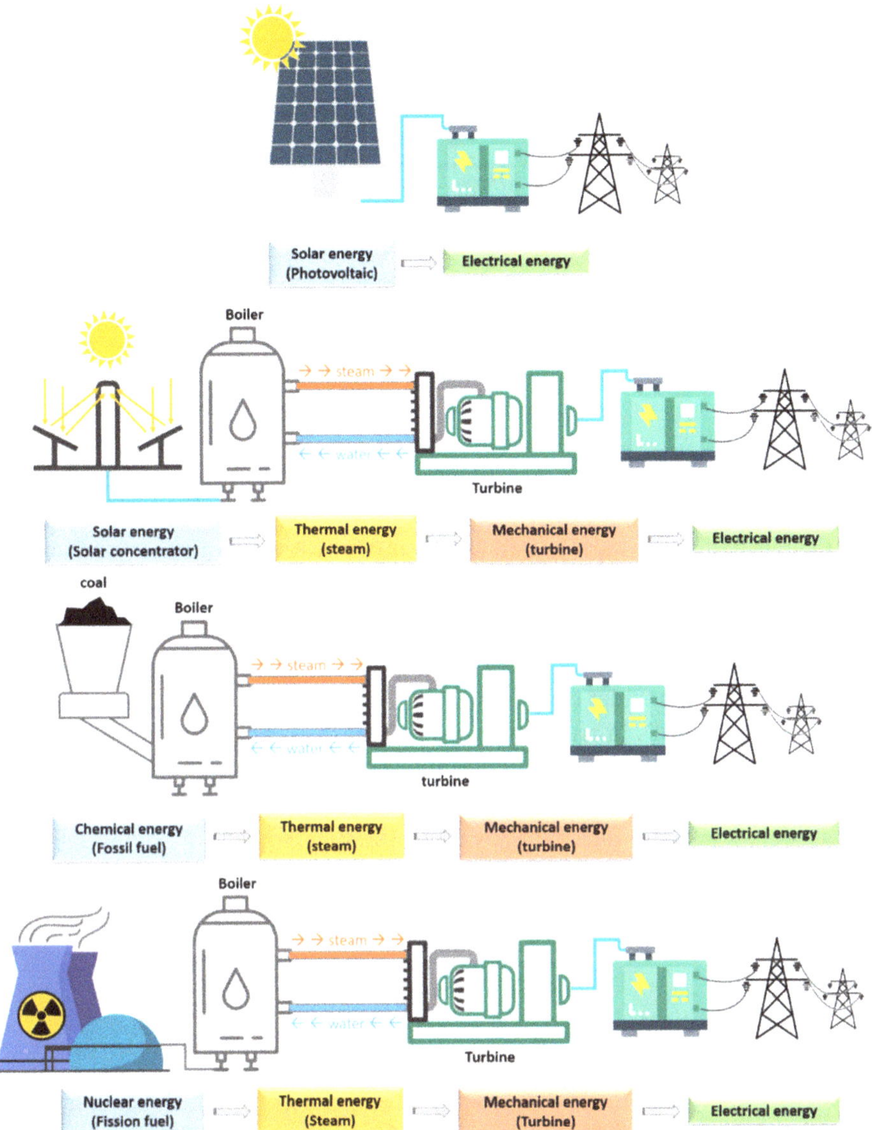

Figure 10.2: Technologies used for energy conversion into electrical power.

applications. Such technologies extend from the range of mechanical storage in centripetal flywheel systems and water pumping, to thermal storage using molten metal and molten salt, to chemical storage in batteries and electrolyzers (Figure 10.3) [6–8].

In the context of recent attempts to achieve efficient storage of electrical energy generated from renewable sources, this chapter will review the recent progress in utilizing porous hybrid solids, MOFs, as catalysts in electrochemical energy conversion.

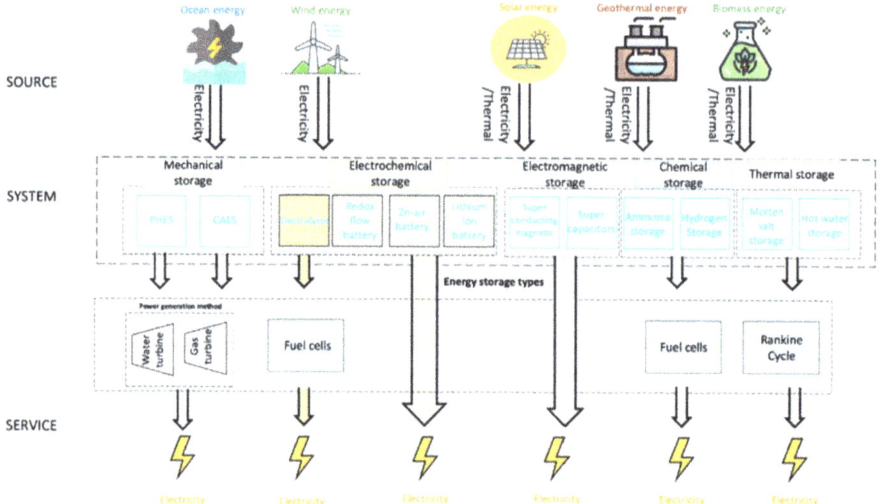

Figure 10.3: Different technologies for electrical energy storage. Reproduced with permission from [8].

We will attempt first to (i) give a brief introduction of the key electrochemical energy conversion reactions under consideration, followed by (ii) an account on the intrinsic properties of MOFs in an attempt to highlight their potential into this field. We then (iii) highlight recent reports on the utilization of MOFs, their composites or MOF-derived material in electrochemical energy conversion reactions, and finally (iv) give an overview of the most relevant aspects of MOFs that requires further investigations in order to derive further their successful implementation into energy conversion systems.

10.2 Water splitting reactions

While nuclear fuel has the highest energy density (~800 million MJ kg^{-1}), hydrogen is considered a carbon-free energy source with significantly higher energy density (140 MJ kg^{-1}) compared to liquid fossil fuels like gasoline (46 MJ kg^{-1}). The primary method utilized globally to produce hydrogen is steam methane reforming (SMR), which is one of the most energy intensive processes in the chemical industry [9]. It releases, stoichiometrically, 5.5 kg carbon dioxide (CO_2) per kg of hydrogen (H_2) as shown in equation (10.1). This process results in the production of one billion tons of CO_2 annually, which is incomparable to the ~36 billion tons produced using fossil fuels, but still contributes to the overall CO_2 emissions and subsequently to global warming [10, 11]. Therefore, efforts are being made to replace the present processes for H_2 production using renewable resources. However, it is quite challenging, as all chemicals (not fuel) consumed worldwide once burnt contribute about 3 % in CO_2 global emission. Nevertheless, the environmental significance of using H_2 appears once it is utilized as a fuel for transport, heat and other energy applications, as these sectors produce over 70 % of the CO_2 emission worldwide. Thus, developing a clean pathway that is not dependent on carbon sources or high-temperature processes is crucial for utilizing hydrogen.

$$CH_4 + 2H_2O \rightarrow CO_2 + 4H_2 \tag{10.1}$$

Water electrolysis is an alternative solution for this issue where it is integrated with renewable energy sources such as photovoltaic (PV) cells to produce a highly stable output of H_2 (this was first discovered in 1789 using gold electrodes) [12]. Electrolysis may become competitive when electricity generated from renewables drops to US$ 0.02 per kilowatt hour or below and with an overall solar to hydrogen (STH) efficiency of 30 % [13]. While the last few years witnessed promising electricity cost reduction, as looked at in more detail by Esposito [14], this low-cost electricity is available for a small fraction of the day. Subsequently, electrolyzers are required to be connected to a grid integrated by other renewable energy sources to work for 24 h. Although the process of making hydrogen from water using PV electrolysis is mature, there is a considerable knowledge gap in scaling it up [15]. At present, the largest electrolyzer in the world operates at 10 MW (single-stack alkaline-water electrolysis system at the Fukushima Energy Research Field (FH2R) in Japan that produces 2.5 metric tons day^1. To put things into context, a typical ammonia plant needs 250 metric tons day^{-1}.

Since the electrical conductivity of pure water is very low, therefore, both an electrolyte (an acid or a base or a salt), as well as electrocatalysts are used to conduct water electrolysis at appreciable rate. In an alkaline environment, the overall process occurs as highlighted in equation (10.4), where the water molecules are reduced at the cathode, producing molecular hydrogen, equation (10.2), while the oxidation reaction takes place at the anode as hydroxyl ions release their electrons to the anode during water electrolysis forming molecular oxygen, equation (10.3).

At the cathode:

$$4H_2O_{(l)} + 4e^- \rightarrow 2H_{2(g)} + 4OH^- \quad (E^\circ_{red} = -0.83\text{ V with respect to SHE}) \tag{10.2}$$

At the anode:

$$4OH^- \rightarrow O_{2(g)} + 2H_2O_{(l)} + 4e^- \quad (E^\circ_{ox} = -0.40\text{ V with respect to SHE}) \tag{10.3}$$

Net reaction:

$$2H_2O_{(l)} \rightarrow 2H_{2(g)} + O_{2(g)} \tag{10.4}$$

On the other hand, in an acidic environment, the processes occur as shown in the following equations;

At the cathode:

$$2H^+ + 2e^- \rightarrow H_{2(g)} \quad (E^\circ = +0.0\text{ V with respect to SHE}) \tag{10.5}$$

At the anode:

$$H_2O_{(l)} \rightarrow \frac{1}{2}O_{2(g)} + 2H^+ + 2e^- \quad (E^\circ = +1.23\text{ V with respect to SHE}) \tag{10.6}$$

Net reaction:

$$H_2O_{(l)} \rightarrow H_{2(g)} + \frac{1}{2}O_{2(g)} \tag{10.7}$$

The Gibbs free energy change for the overall reaction is given by the expression:

$$\Delta G^\circ = -nFE^\circ \tag{10.8}$$

ΔG° is the standard Gibbs free energy, n is the number of electrons, F is Faraday's constant (96,485 s A mol^{-1}) and E° is the standard cell potential. At 298 K and 1 mol L^{-1}, 1 bar, the electrochemical cell voltage E° is –1.23 V (1 V = J A^{-1} s^{-1}), and $n = 2$ which corresponds to a Gibbs free energy change of +237 kJ mol^{-1} H_2; a thermodynamically uphill reaction. Figure 10.4 illustrates the overall catalytic cycles for the two types of electrolyzers.

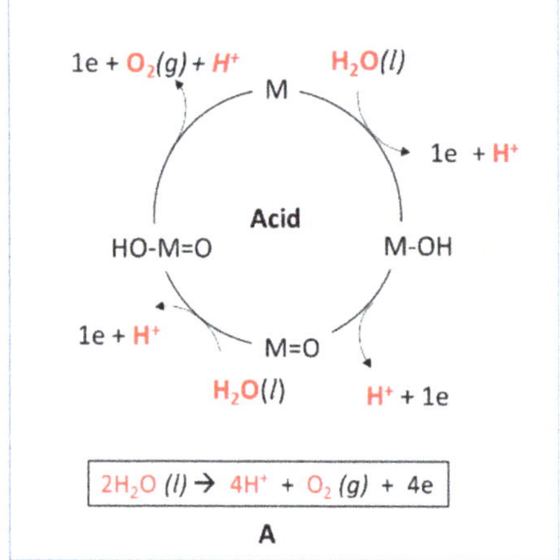

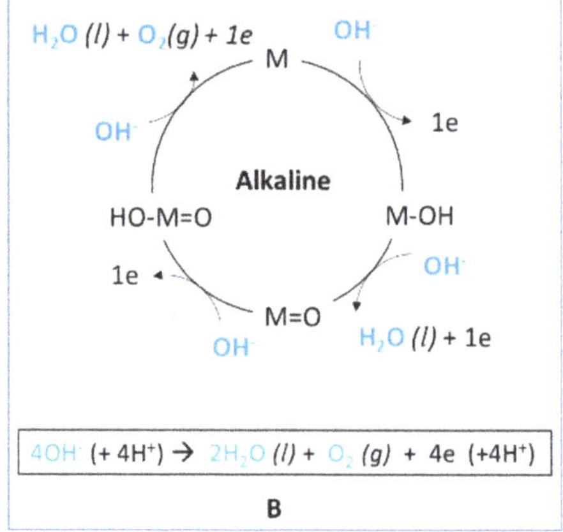

Figure 10.4: A simplified reaction cycle for water splitting to molecular hydrogen and oxygen on the catalytic surfaces in acidic (A) and alkaline (B) media.

According to the pH of the water in the electrolyzer, two types of electrolyzers are commonly utilized, where alkaline electrolyzers (operating with electrolyte of 20–30 % KOH) are utilized in the large-scale hydrogen production, while lately those operating in neutral/slightly acidic environment are increasingly being implemented. The conventional alkaline water electrolyzer typically uses a diaphragm-type separator. The presence of the separator requires that the distance between the anode and cathode electrodes is about 2–3 mm, but it is an essential component of the electrolyzer to prevent gas crossover, but in effect this configuration limits the operating current density to a max of 400 mA cm^{-2}. In order to increase the current density, alkaline stable anion exchange membranes technology are pursued [16]. Typical electrocatalysts are based on Ni, Co and Fe hydroxides for both electrodes with an overall overpotential close to 0.5 V.

The polymer electrolyte membrane (PEM) provides higher proton conductivity, and can operate at high pressure [17]. To use PEM the two catalysts (for OER and HER) layers are coated on both sides of a Nafion-type membrane [18], often termed zero-gap electrocatalysts to maximize electron transfer rates. Catalysts used in commercial PEM water electrolyzer, usually consist of Pt/C (HER), and IrO_2 or RuO_2 for (OER) (Figure 10.5).

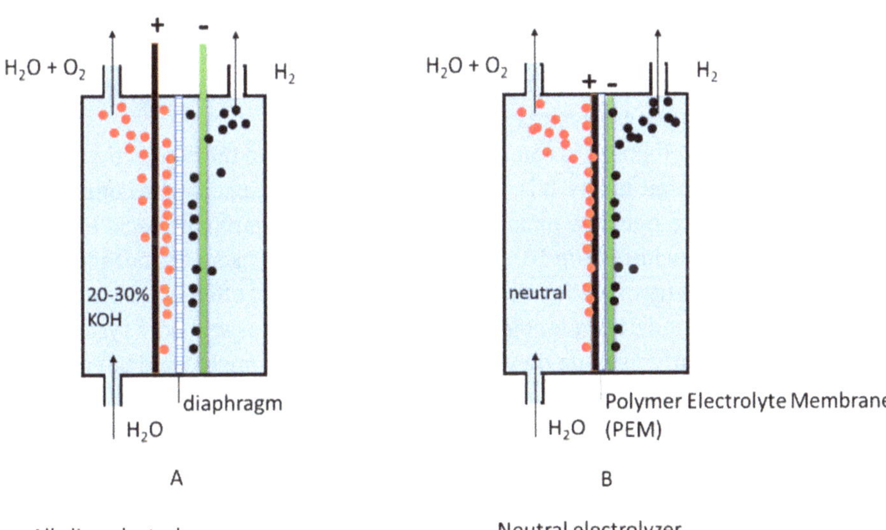

Alkaline electrolyzer
Anode (+): Ni-Co-based catalysts
Cathode (-): Ni-Mo-based catalysts

Neutral electrolyzer
Anode (+): Ir/Ru oxide-based catalysts
Cathode (-): Pt-based catalysts

Figure 10.5: A schematic description of alkaline and neutral (PEM-based) water electrolyzers.

10.3 Intrinsic properties of metal-organic frameworks

10.3.1 Hybrid conditions

Metal-organic frameworks (MOFs) are constructed through formation of coordination bonds between metal ions or metal ions clusters and organic linkers containing heteroatoms capable of sustaining coordination interaction(s) to transition metals. The organic linkers commonly used are those based on the aromatic benzene ring, functionalized at least by two coordinating groups, to be able to bridge at least two metal ions (clusters) and form a coordination network. The literature contains a large number of MOFs constructed from ligands through coordination interactions with aromatic nitrogen, carboxylate or phenoxide linkers (ligands) [19]. Common to the organic linkers used is presence of Lewis base group capable to coordinate transition metal ions (or clusters). The 3D structure of MOFs is supported by the polytopic nature of the organic linkers as well as the polytopic nature of the transition metal ions(clusters) with more or less symmetric coordination sphere. This hybrid organic/inorganic composition of MOFs is rather interesting from catalytic point of view, as it offers opportunities in catalysis pertinent to the ability of MOFs to selectively bind specific substrate mediated by the organic component in the framework. Additionally, the transition metal ions (clusters) are known catalysts for many chemical transformations. Therefore, this hybrid composition of MOFs brings about the benefits of catalytic activity and substrate selectivity into one framework material.

MOFs underlying topologies can be classified according to the connectivity of the linkers and the inorganic nodes. It is common to use organic linkers with connectivity of 2, 3 or 4, best represented by terepthalic acid, benzenetricarboxylic acid and benzenetetracarboxylic acid (Figure 10.6) [20]. Increasing the linker size but maintaining the overall geometry of the functional groups results most often in an isoreticular MOF (sharing same underlying topology) with expanded cages and pore system (Figure 10.7a) [21]. Similarly, keeping the dimension of the ligand unchanged but utilizing functionalized version of it also commonly led to isolation of isoreticular MOFs but expressing the additional functionality in their cages and pore system (Figure 10.7b) [22].

10.3.2 Permanent porosity

In order to attain open structures with permanent porosity, i. e., maintained porosity after removal of solvent or guest molecules, MOFs are frequently constructed form rigid organic multitopic linkers and cationic transition metal ions. Carboxylate based organic linkers (as terephthalate, isopthalate, benzene tricarboxylate and benzenetribenzoate) are the most reported in permanently porous MOFs [23]. This can be

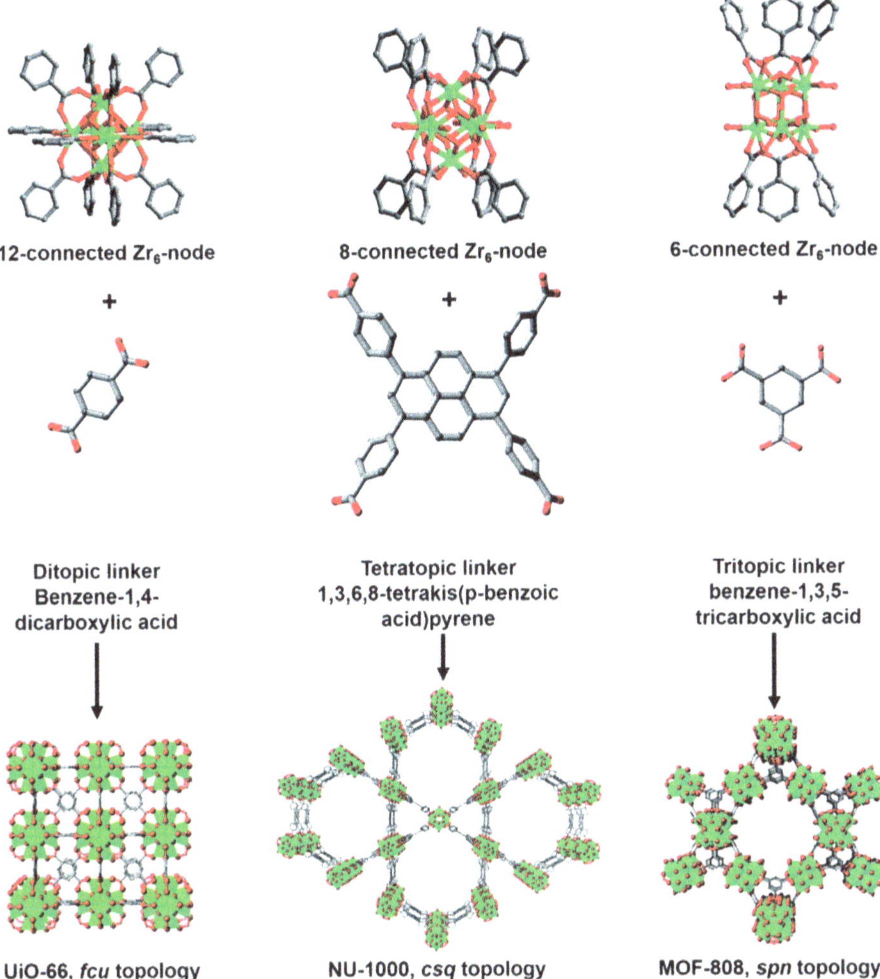

Figure 10.6: The diverse topologies obtained by having organic linkers with different connectivity. Reproduced with permission from [20].

ascribed to two main factors, namely: (i) the rigidity of the linkers preventing structure collapse upon removal of the guest molecules [24] and (ii) the anionic nature of the linkers enhances structural stability due to electrostatic interactions with the metal ion(s) [25]. Moreover, the aromatic linkers can be functionalized with a wide number of functional groups but still can produce the same overall topology as the pristine MOF [26]. This is a very attractive attribute of MOFs as the pore chemistry can be altered without having to change the pore dimensions, allowing chemists to explore more chemical space within a frame of reference, to be able to reach some meaningful and transferrable conclusions. Permanent porosity can be assessed di-

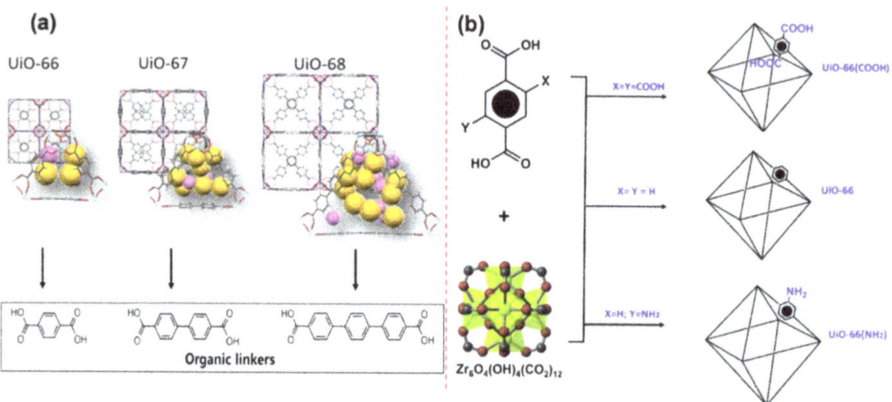

Figure 10.7: (a) The MOF cages and pore system expansion upon increasing the length of the organic linker. Reproduced with permission from [21]. (b) The introduction of different functionalities into the MOF cages. Reproduced with permission from [22].

rectly by measuring the surface area of the MOFs using N_2, He or CO_2 adsorption isotherms. It is commonly observed that the specific surface area of MOFs decreases upon loading with the catalyst due to increasing the weight/density of the framework. Although it is beneficial to reassess the porosity of the MOF after conducting an electrocatalytic reaction, in most reports this is not a commonly conducted characterization understandably due to the requirement of appreciable mass of the solid for meaningful measurement of its surface area, and the intrinsically small amount of material commonly utilized in electrochemical catalytic reactions. One solution to this would be the use of large area electrodes, like carbon cloth, to be able to retrieve usable mass of the catalyst for proper characterization of its surface area.

10.3.3 Post-synthetic modification

The organic linkers can be substituted with a wide range of functional groups in non-binding sites, allowing isolation of isostructural MOFs (those sharing same underlying topology and similar cage/pore dimensions) yet with additional functionality lining the pores [27]. This approach is unique to the MOFs chemistry and allows for anchorage of catalytic species after isolation of the MOF (post synthetic modification) [26, 28]. The ability to generate a catalyst-loaded and a pristine MOF facilitates conducting control experiments that can demonstrate clearly the effect of the catalyst and to contrast its performance to the catalyst-free sample.

An additional benefit of MOFs is that they allow for matrix isolation of the catalytic species, retarding or preventing altogether aggregation and subsequent deactivation of the loaded catalysts. Moreover, the porosity of the MOFs allows, at least in principle, accessibility of the catalytically active centers by the reactants from solution [29].

However, as solid heterogeneous catalysts, MOFs demonstrate diffusion-controlled kinetics [30], which is to be expected when compared to homogeneous catalysts counterparts.

10.3.4 Structural stability in aqueous environment

Certain families of MOFs appear to have pronounced water stability, where such stability is largely defined by retained crystallinity after exposure to aqueous environment for relatively long period of time in the range of at least few hours (Figure 10.8) [19]. Common to those MOFs is the presence of strong oxygen-metal bonds, dominated by carboxylate-high valent transition metal ions complexes.

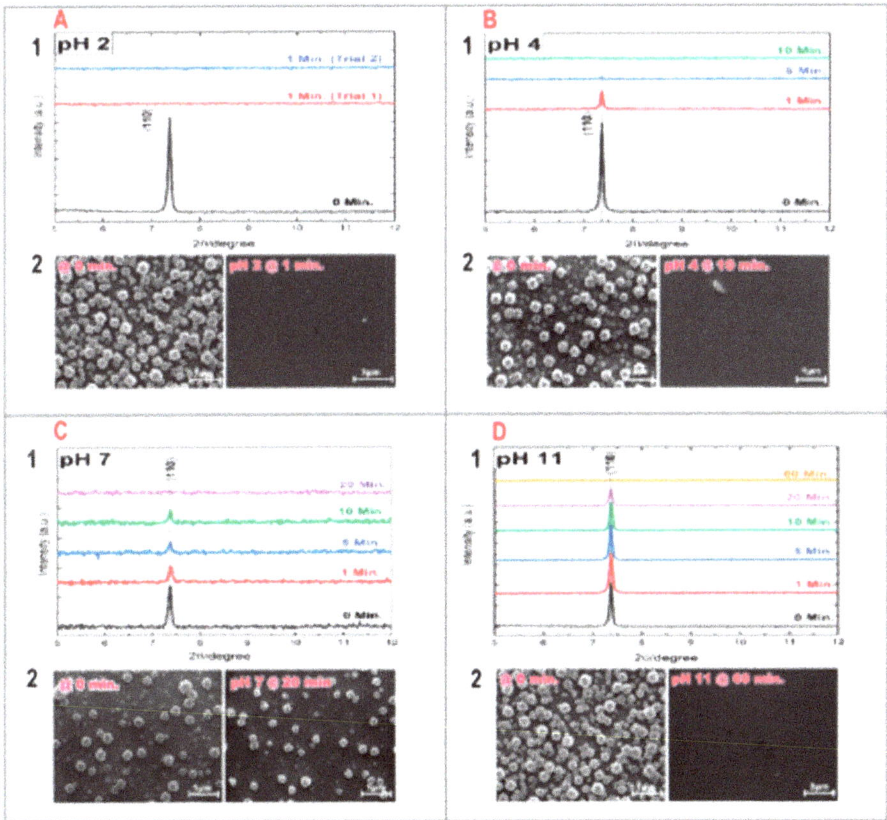

Figure 10.8: XRD patterns and SEM images for surface-supported MOFs (SURMOFs) where ZIF-8 (top panel) showed maintained crystallinity only at basic conditions, while UiO-66-NH2 (lower panel) demonstrated structural stability at neutral and acidic conditions. [19].

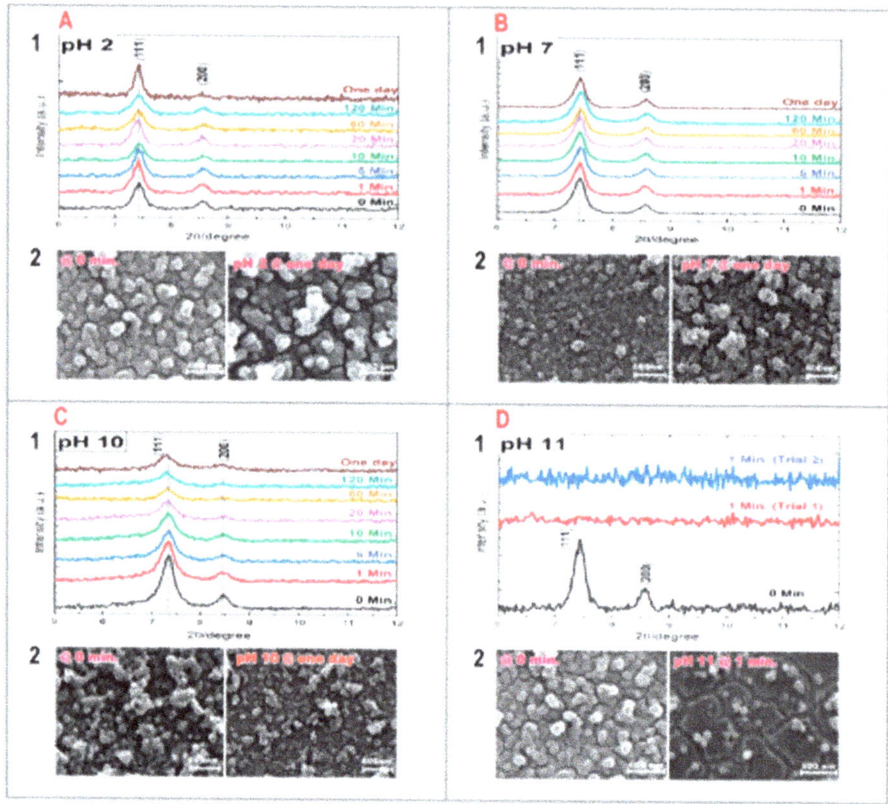

Figure 10.8: (continued)

However, the true underlying principles for water stability of such MOFs should not be simply ascribed to strong interactions between high-valence cationic metal ions (e. g., Zr^{IV}, Ti^{IV}, Cr^{III}, Fe^{III} and Al^{III}) with the carboxylate groups from the polytopic organic linkers ligands. This is commonly referred to in current literature as hard-acid hard-base interaction (HAHB), which is a rather oversimplification of the complex cascade of events taking place when MOFs with high valence metal ions is in contact with water molecules as pointed out recently [31]. A critical factor that was identified by Alkordi and coworkers [31], and pointed out to be a major player in noticed stability of certain MOFs, is the formation of metal hydrates (M-OH_2), and the consequent auto-deprotonation process depending on the pKa of the M-coordinated water molecules and the pH of the medium (Figure 10.9). In this recent study, a comprehensive reaction pathway that justifies observed trends in stability of several MOFs was postulated taking into consideration (i) rapid formation of metal hydrates upon contacting the MOFs with aqueous environment, where the pKa values were shown to depend strongly on the nature of the metal ion, (ii) protonation stages of bridging oxide/hydroxide ions found in MOFs based on metal-oxide clusters (e. g., UiO-66, MIL-100 family) and (iii)

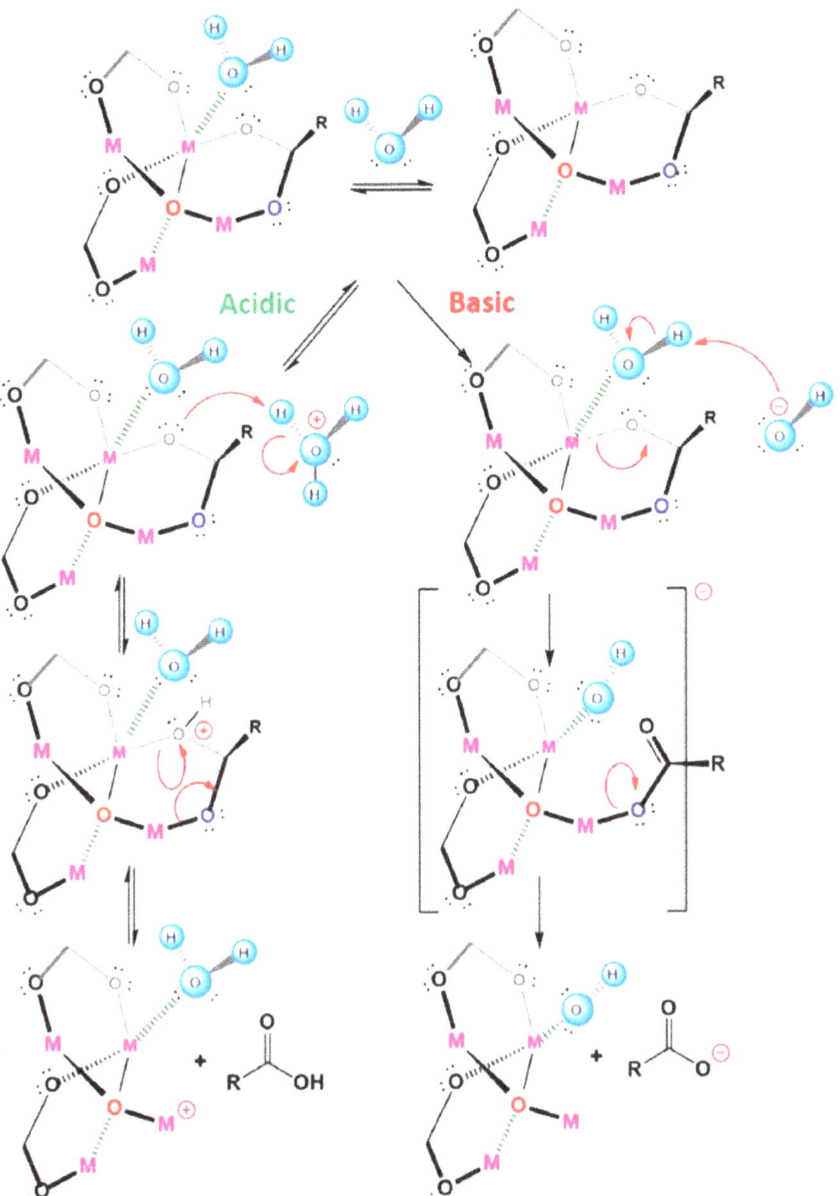

Figure 10.9: Proposed degradation pathways under acidic and basic conditions for MOFs based on metal–carboxylate clusters (a hypothetical metal-carboxylate cluster is used here as an illustrative example, with which the basic principles shown in this model can be readily transferrable to other types of metal–carboxylate MOFs). Metal-ion hydration is assumed to be the initial stage in the process, followed by either ligand protonation (in acidic medium) or deprotonation of coordinated water molecule (neutron to basic medium). The strong interactions between metal ions and hydroxyl ions are assumed to cause irreversible damage to the nodes of a particular MOF. Reproduced with permission from [31].

competitive binding of hydroxyl ions (from water auto-deprotonation) and carboxylates (ligands) to the metal ions in the nodes.

10.4 MOFs as electrocatalysts

10.4.1 MOFs as catalysts for hydrogen evolution reactions

Pt and Pt group metals are considered the benchmark catalysts for electrochemical hydrogen evolution reaction (HER) owing to their low over potential, superior activity and excellent stability. Yet, their scarcity and high cost limit their applicability towards commercial hydrogen production. Therefore, the search for non-noble metals as well as nonmetal-based catalysts that are cost effective with high electrocatalytic performance is exceedingly necessary. MOFs with their hybrid inorganic–organic composition exhibit unique properties that render them ideal candidates as HER electrocatalysts. These include permanent porosity, high surface areas and tailorable structures and functionalities. However, their poor electrical conductivity and chemical stability in acidic and alkaline media impose challenges for the utilization of these coordination polymers in electrochemical catalysis. Nevertheless, numerous efforts have been dedicated in attempt to alleviate these shortcomings [32, 33]. In general, strategies to utilize MOFs as HER electrocatalysts can be classified into three main approaches: (i) pristine MOFs, (ii) MOFs as porous scaffolds/supports and (iii) MOFs as sacrificial precursors.

10.4.1.1 Pristine MOFs

In order for MOFs to serve as active electrochemical catalysts, the judicial selection of the organic linkers and metal nodes is crucial. The inclusion of redox-active metal clusters and/or redox-active organic ligands was found to yield highly active MOF catalysts. Gu *et al.* have utilized a Fe_3 cluster as the metal building block and a carboxylic acid as the organic ligand to generate a stable MOF [34]. The hard Lewis acid ion (Fe^{III}) and the carboxylate linker were specifically chosen due to their renowned strong electrostatic interaction, thereby leading to structural stability of the MOF in aqueous as well as acidic/basic media [31]. Replacing an iron atom by another metal to form the bimetallic cluster, Fe_2M, where M = Ni, Co, Zn or Mn, further enhanced the electrochemical performance of the respective MOFs, where Fe_2Zn-MOF displayed a low overpotential (η_{10} = 221 mV) and small Tafel slope (174 mV dec^{-1}) in 0.1 M KOH. On the other hand, Roy and coworkers employed cobaloxime hydrogen evolving complex (HEC) as redox-active linkers connected via zirconium-oxo clusters to fabricate a novel 3D MOF, UU-100 (Figure 10.10) [35]. Thin films of UU-100 grown atop conductive glassy

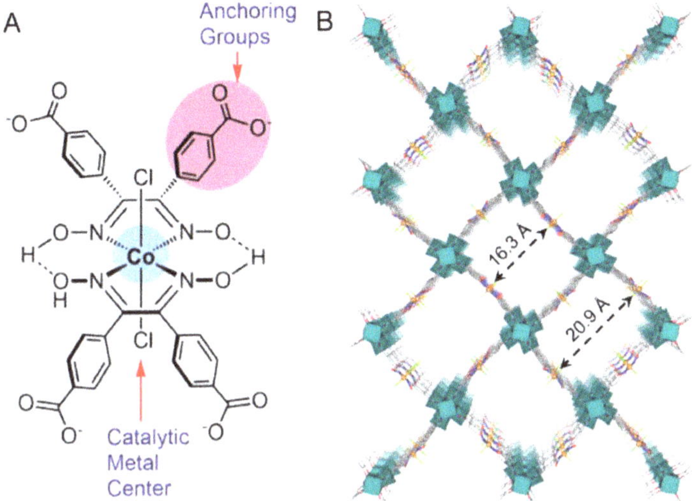

Figure 10.10: (A) Structure of the cobaloxime linker in UU-100(Co), and (B) structural model of UU-100(Co) MOF viewed along [001]. Reproduced with permission from [35].

carbon exhibited superior electrocatalytic activity towards HER with excellent stability for over 18 h. Furthermore, conductive MOFs consisting of planar metal nodes and 2D π-conjugated systems have shown promise towards electrocatalysis [36]. Cheng *et al.* reported the deposition of ultrathin nanosheets of conductive Cu-MOF atop iron hydro(oxy)oxide [$Fe(OH)_x$] nanoboxes as solid support via a simple template-based solvothermal reaction followed by redox-etching [37]. The abundance of highly active coordinatively unsaturated Cu^I-O_2 centers lead to the exceptional HER catalytic activity with an overpotential of 112 mV at 10 mV cm^{-2} and Tafel slope of 76 mv dec^{-1} in 1 M KOH.

While synthesis of novel conductive MOFs is still challenging, alternative approaches have been adopted to enhance the conductivity, thereby electrocatalytic activity, of non-conductive MOFs. Loading of single atom catalysts (SACs) into MOFs can modify their local electronic structures, thus enhancing their electrochemical catalytic performance while maintaining their structural integrity. Sun and co-workers successfully incorporated atomically dispersed Ru into Ni-benzenedicarboxylate (Ni-BDC) MOF [38]. An outstanding HER performance was displayed by the optimized $NiRu_{0.13}$-BDC, compared to pristine Ni-BDC, with the lowest overpotential of 36 mV at 10 mA cm^{-2} and lowest Tafel slope of 32 mV dec^{-1} in 1 M phosphate buffer saline (PBS) solution. Such superior performance behavior can be attributed to modulation of H_2O and H* adsorption strength due to tuning of the Ni electronic structure by the newly introduced Ru single atom.

10.4.1.2 MOFs as porous scaffolds/supports

The high surface area of MOFs has directed their utilization as porous scaffolds or solid supports for catalytic moieties [39]. These loaded sorbates can then also impart electrical conductivity to MOFs, thus benefiting from the synergistic relationship between the dual components of the resultant composite. For example, Nie *et al.* employed a Zn-based MOF to encapsulate Pd NPs, where the Pd/MOF composite showed a low onset overpotential of 105 mV with excellent stability in acidic media [40]. Aside from metal NPs, Ni-containing polyoxometalates (POMs) were also embedded within MIL-100(Fe) and UiO-67, which resulted in enhancement of HER electrocatalytic activity of POMs after composite formation [41]. The electrostatic interaction between the POMs and MOFs was believed to alter the local microenvironment, thereby causing slight peak shifting, relative to bare POMs, as well as hindering formation of reduced metal oxide layers. Furthermore, Dai and co-workers reported a simple one-pot solvothermal methodology to anchor molybdenum polysulfide (MoS_x) onto amine-functionalized Zr-carboxylate MOF (UiO-66-NH_2), which demonstrated a low overpotential (η_{10} = 200 mV) and Tafel slope (59 mV dec^{-1}) [42]. The improved HER performance of the attained composite was ascribed to the increased surface area and accessible active sites, reduced charge-transfer impedance, and enhanced proton conductivity.

10.4.1.3 MOFs as sacrificial precursors

While the utilization of MOFs as synthesized or as porous supports for electrocatalytic hydrogen production is a promising route to fabricate non-Pt catalysts, yet the poor electrical conductivity and stability of such systems remain challenging. Therefore, post-synthetic pyrolysis or carbonization to prepare MOF-derived non-noble, metal-based catalysts remains one of the most widely adopted strategies [43]. Co-based ZIF-67 has been used to generate carbon-supported cobalt oxide (Co_3O_4/C), which exhibited a decent electrocatalytic activity towards HER in alkaline medium (182.7 mV overpotential and 68.37 mV dec^{-1} Tafel slope) [44]. Further *in situ* selenization via carbonization in presence of Se powder yielded cobalt diselenide ($CoSe_2$) NPs atop nitrogen-doped graphitic carbon, which demonstrated superior HER electrocatalytic performance with 150 mV onset potential and 42 mV dec^{-1} Tafel slope in 0.5 M H_2SO_4 [45]. Moreover, a 1D cobalt-based MOF nanowires produced cobalt phosphide nanocrystals confined within N-doped carbon nanowires (CoP/NCNWs) after carbonization and subsequent low-temperature phosphorization processes [46]. The HER catalyst, CoP/NCNWs displayed an excellent electrocatalytic performance in both acidic and alkaline media, which was attributed to the synergy between the catalytically active CoP NPs and the shielding effect of N-doped carbon frameworks resulting in a highly active and stable hybrid catalyst. Additionally, Ni-based MOFs have also been exploited as precursors to achieve porous HER catalysts. He et al. have demonstrated that phosphatization of MOF-74-Ni resulted in carbon-supported nickel phosphide (Ni_2P/C) microrods (Figure 10.11) [47]. The porous morphology and good electrical conductivity of the attained composite brought about the improved HER electrocatalytic activity with a low onset overpotential (98 mV) and exceptional stability in acidic media.

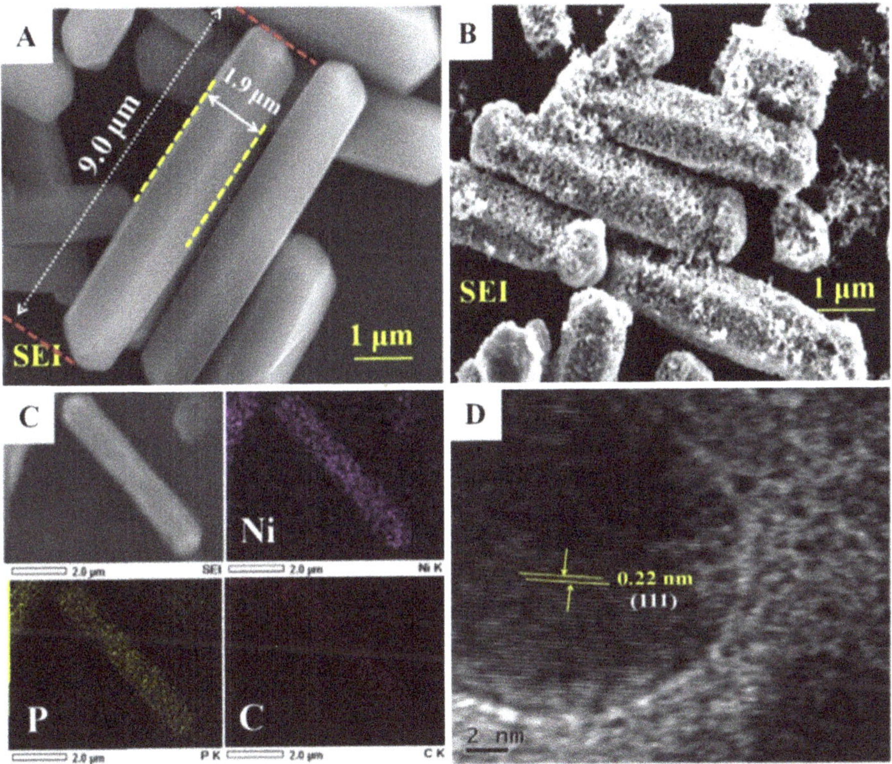

Figure 10.11: SEM images of MOF-74-Ni (A) and Ni_2P/C (B); Elemental mappings (C) of Ni, P and C and HRTEM image (D) of Ni_2P/C. Reproduced with permission from [47].

10.4.2 MOFs as electrocatalysts for the OER reactions

Compared to the HER, the oxygen-evolution reaction (OER) suffers from sluggish kinetics via the multiple electron transfer process. Currently, noble-metal-based composites, such as Pt, RuO_2 and IrO_2 are considered to be the benchmark catalysts for OER due to their outstanding performance. However, their high price and scarcity hinder their large-scale application. Hence, exploring alternative electro-catalysts with low cost, high activity, long stability and earth abundance is of extreme urgency. As transition metal oxides/hydroxides (especially those of Ni, Co and Fe) are known for their excellent electrocatalytic activity toward the alkaline OER reactions, it became evident that extending their chemistry to scaffold materials like MOFs or MOF-based composites can be associated with notable performance enhancement [48].

10.4.2.1 Pristine MOFs

Indeed, the OER activity of the MIL-53 MOF based on trinuclear Fe^{III} cluster was investigated by Li et al. As it is well documented that catalytic activity of metal hydroxides can be greatly affected by mixed metal hydroxide formation, the MIL-53 presented the opportunity to investigate the effect of varying the mole ratio of Fe in the clusters due to partial substitution with Co and/or Ni ions. In their report, Li et al. demonstrated that the best performance was attained for the catalyst with the composition $Fe/Ni_{2.4}/Co_{0.4}$, recording an over potential of 236 mV to derive a current density of 20 mA cm^{-2} with a Tafel slope of 52.2 mV dec^{-1} [49]. In this report, the bimetallic MOFs Fe/Ni_x-MIL-53/C ($x = 1.6, 2.0, 2.4$) exhibited low over potentials of 258, 258 and 244 mV at the current density of 10 mA cm^{-2}, respectively. Those values compare favorably with the commercial Ir/C showing an over potential of 310 mV to derive the same current density. The produced MOFs also show lower Tafel slopes of 37.8, 45.5 and 48.7 mV dec^{-1} again lower than that of Ir/C (53.8 mV dec^{-1}). This report demonstrated the direct use of MOFs in OER instead of a MOF-derived material, which is remarkable due to the harsh caustic conditions used for OER, a truly challenging environment for many MOFs.

It was demonstrated that the trinuclear trigonal-prismatic metal-carboxylate nodes, as those present in the MIL-53, have sufficient structural stability in basic media as pointed out recently by Alkordi et al. [31]. In their work, the authors suggested that the cationic $M_3(CO_2)_6$ clusters, M representing a trivalent transition metal cation like Cr^{III} and Fe^{III}, can easily exchange the halide counterion present in the structure by hydroxide ions from solution. Furthermore, the M-bound water of hydration were shown to be more acidic due to coordination to electropositive transition metal ions. It was also suggested that the multiple M-carboxylate bonds holding this cluster together provided kinetic inertness and, therefore, the MOF can be water stable for extended period (Figure 10.12).

A recent report by Zhu and coworkers highlighted a chemically stable and catalytically optimized mixed metal FeNi-thiophenediarboxylate MOF (Figure 10.13) [50]. This MOF presented the opportunity to fine tune its catalytic activity by tuning the Fe:Ni ratio in the MOF, reaching optimal catalytic activity at equimolar ratio of Ni to Fe. In this report, the maintained crystallinity of the MOF was demonstrated through X-ray diffraction measurements that clearly demonstrated maintained structure under the testing conditions. Additionally, in this report the authors built the MOF directly on the electrode made of Ni foam, to circumvent the limited electronic conductivity of the MOF by forming a thin layer of the MOF anchored on conductive support.

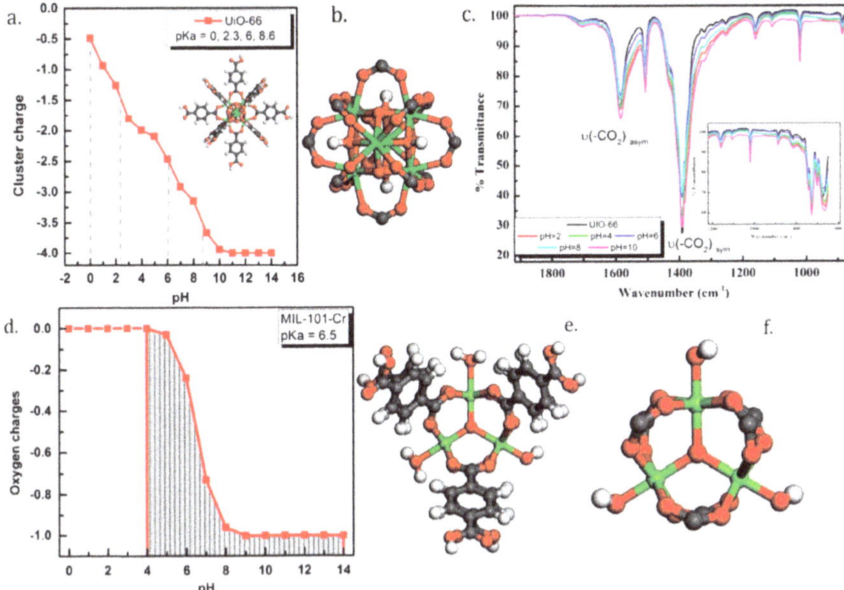

Figure 10.12: (a) Calculated cluster charge and the corresponding pKa values (dotted lines) for the four bridging hydroxyl ions in the $Zr_6(m_3\text{-}O)_4(m_3\text{-}OH)_4(CO_2)_{12}$ clusters in UiO-66 MOFs, (b) Enlarged view of the geometry-optimized Zr cluster, (c) FTIR spectra for UiO-66 crystalline powder acquired after brief exposure to aqueous solution with variable pH values for 5 min at RT, (d) Calculated charge as a function of pH on the Cr-coordinated water molecules and the corresponding pKa value. © Geometry-optimized hydrated Cr carboxylate cluster, (e) Geometry-optimized cluster coordinated by hydroxyl ions from deprotonated water molecules, showing the immediate environment around the Cr^{III} ions [31].

The best performing MOF demonstrated low overpotential reported as 239 and 308 mV to derive current densities of 50 and 200 mA cm^2, respectively. Additionally, the MOF demonstrated high stability over 43 days of continuous electrolysis (measured at fixed current density of 100 mA cm^{-2}).

Most recently, a novel class of metal-hydroxide organic frameworks (MHOFs) made by transforming layered hydroxides into two-dimensional sheets cross-linked using aromatic carboxylate linkers demonstrated exceptional activity toward OER reaction with maintained structural integrity after prolonger continuous electrolysis [51]. Key to the high stability of such a class of compounds was attributed to either higher kinetic barrier or decreased thermodynamic driving forces due to enhanced π-π interactions between the organic linkers in a series of compounds with varying organic linkers. Performance optimization of the Ni-MHOF nanosheets was obtained through insertion of electron-withdrawing metal ions (e. g., La^{III}, Y^{III}, Al^{III}, Sc^{III} and Fe^{III}) in the range of 12.5 to 17.9 atom%, resulting in an optimized catalyst demonstrating high mass activity (80 A g^{-1}), high TOF (0.30 O_2 s^{-1}) at an overpotential of 0.3 V, as recorded in 0.1 M KOH for 20 h.

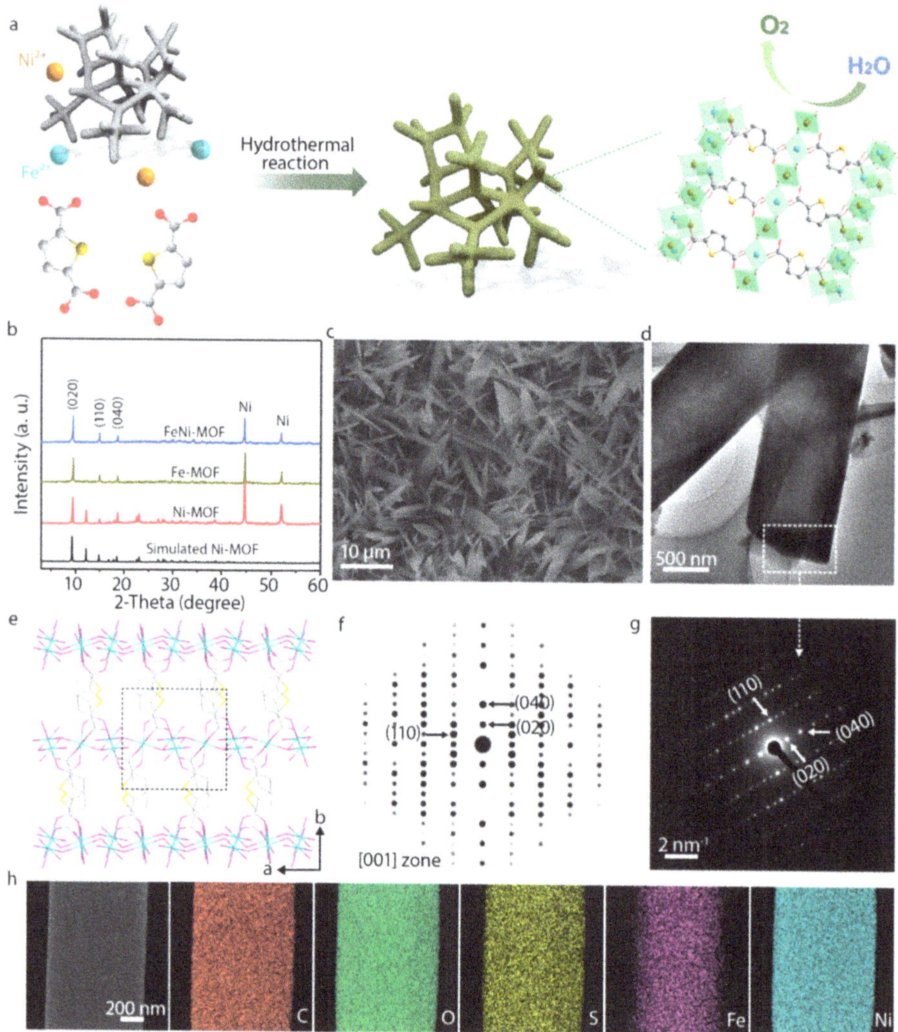

Figure 10.13: Design strategy, morphology and structural characterizations of FeNi-MOF nanoarrays. (a) Schematic illustration of the synthesis of FeNi-MOF nanoarrays via a hydrothermal method, (b) XRD patterns of FeNi-MOFs, Fe-MOFs and Ni-MOFs and simulated data from the reported crystal structure, (c) SEM image of FeNi-MOF nanoarrays grown on NF, (d) TEM image of FeNi-MOF nanobelts, (e) Crystal structure and (f) Simulated SAED pattern of FeNi-MOF projected along the [001] zone axis, (g) Experimental SAED pattern of FeNi-MOF obtained from the region in (d), (h) High-angle annular dark-field scanning TEM (HAADF-STEM) and the corresponding EDS elemental mapping images of the C, O, S, Fe and Ni elements for FeNi-MOF nanobelts.

10.4.2.2 MOFs as precursors for metal-oxides/hydroxides

Due to their relative chemical instability in highly caustic conditions usually used in OER reactions, MOFs can be utilized as precursors to deposit electrochemically active catalytic centers of metal oxides/hydroxides. An apparent advantage that this approach provides is better control over the composition of the end deposited catalyst, as compared to simple wet synthesis of similar metal oxide/hydroxide species. Additionally, as the metal itching and subsequent deposition as insoluble oxide/hydroxide is derived from a porous precursor, this approach opens more doors in accessing more open structures of the resulting species. Moreover, starting from a MOF made of one type of transition metal and doped with another metal ion have resulted in deposition of highly active mixed hydroxide phases. An early example on the use of MOFs as precursors to deposit catalytically active $Ni(OH)_2$ was demonstrated through the use of Ni-loaded Zr-carboxylate (UiO-66-NH_2) MOF built on graphene nanoplatelets [52]. When the Ni-loaded MOF (UiO-66-NH_2-Ni@G) was deposited on the electrode and subjected to anodic potential immediately after immersion in the 1 M KOH solution, deposition of the active $Ni(OH)_2$-$Zr(OH)_4$@G was derived through electrophoretic deposition. The basic etching of the MOF resulted in formation of mixed hydroxides, which in the highly alkaline solution will acquire a negative surface charge due to adsorption of hydroxyl ions. This then will cause electrophoretic deposition of the mixed hydroxides on the conductive graphene sheets. This catalyst demonstrated a remarkable activity toward the OER with an overpotential at current density of 10 mA cm^{-2} (η_{10} = 0.38 V vs. RHE) and demonstrated a long-time stability of 10 hours under continuous electrolysis. The significance of including the electronically conductive graphene sheets into the MOF-graphene composite was evidenced by the inactivity of the control compound UiO-66-NH2-Ni toward OER (Figure 10.14).

A recent report also demonstrated the use of MOFs as precursors to yield catalytically-active Co-based OER centers through *in situ* etching of the MOF and deposition of the active metal hydroxide phase [53]. The authors demonstrated that two different types of Co-MOFs resulted in attainment of two different $Co(OH)_2$ phases, namely the ZIF-67 and CoBDC (the former using 2-methylimidazole and the later using 1,4-benzenedicarboxylic acid as linkers) resulted in formation and deposition of α- and β-$Co(OH)_2$ phase, respectively. It was also reported that the deposited α-phase is the active OER catalyst, directly linking the nature of the starting MOF to the structure and activity of the end electrocatalyst. The authors described that electrochemical activation was more readily attained for the ZIF-67 due to the presence of the Co-N coordination bonds as compared to Co-O bonds found in the CoBDC MOF.

The significant role of conductive carbon and the sacrificial role of MOFs to deposit OER-active metal hydroxides was again recently demonstrated in composites of Ni-MOFs based on the 1,3,5-benzenetricarboxylic acid (BTC) and conductive carbon [54].

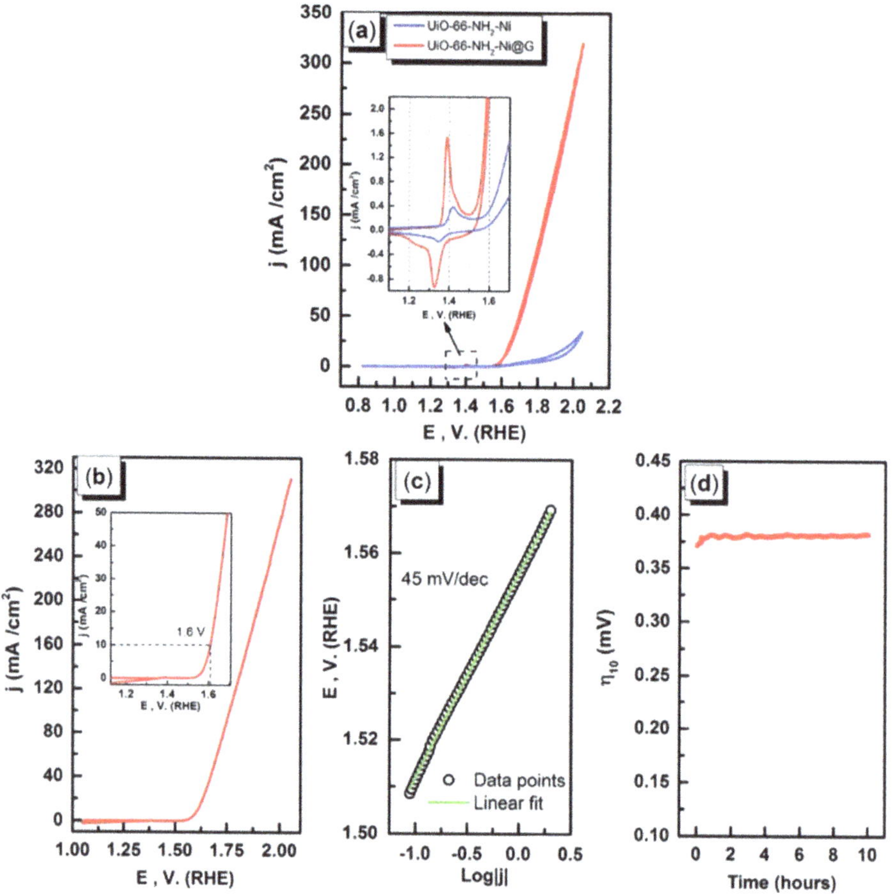

Figure 10.14: (a) CV scans, with a magnified part of the CVs showing the $Ni^{II/III}$ redox couple in the inset, of UiO-66-NH_2-Ni (blue trace), and UiO-66-NH_2-Ni@G (red trace), demonstrating OER activity only for the latter, (b) LSV scan of UiO-66-NH_2-Ni@G with a magnified image in the inset, (c) the Tafel plot with a linear fit for UiO-66-NH_2-Ni@G and (d) the corresponding chronopotentiometry showing stability for 10 hours under continuous electrolysis. All measurements were recorded in 1 M KOH with a scan rate of 100 mV s^{-1} for CVs and 10 mV s^{-1} for LSV. Reproduced with permission from [52].

10.5 Conclusion and outlook

The rich chemistry of MOFs presents ample room for structural modifications and functionalization that can have direct impact on performance in electrocatalysis. While the investigation of MOFs as electrocatalysts for water splitting is in its early beginnings, promising performance is very evident reviewing recent reports on OER and HER reactions catalyzed by MOFs or MOF-derived material. Two major challenges stand out in the way of applicability of MOFs toward energy storage applications through water splitting, namely structural stability and electronic conductivity. While the poor electronic conductivity of MOFs, in general, hindered their applications in electrochemical processes, different strategies to enhance their electronic conductivity appeared in the recent literature but were not reviewed here. Maintaining structural integrity of the MOF or the MOF-derived catalyst is essential for its durability, a factor that is highlighted herein with examples of certain MOFs demonstrated exceptional structural stability under sufficiently acidic or basic conditions. Overall, MOFs represent a unique platform material to immobilize OER and HER catalytic centers on electrodes, with the added advantage of porosity, structural tunability and post-synthetic modification ability. Such attributes of microporous solids are expected to open more doors in heterogeneous water splitting applications, given proper engineering of the material was implemented to ensure structural stability, catalyst accessibility and good electronic conductivity, simultaneously.

Bibliography

[1] Administration, U. S.E.I.U. S. Energy Consumption by Source and Sector, 2020 Quadrillion British Thermal Units 2020.
[2] Moffitt AH. Am J Orthod Dentofac Orthop. 2022;161:276.
[3] Gray HB. Powering the planet with solar fuel. Nat Chem. 2009;1:7.
[4] Kousksou T, Bruel P, Jamil A, El Rhafiki T, Zeraouli Y. Energy storage: applications and challenges. Sol Energy Mater Sol Cells. 2014;120:59–80.
[5] Nair NKC, Battery GN. Energy storage systems: assessment for small-scale renewable energy integration. Energy Build. 2010;42:2124–30.
[6] Ibrahim H, Ilinca A, Perron J. Energy storage systems-characteristics and comparisons. Renew Sustain Energy Rev. 2008;12:1221–50.
[7] Baker J. New technology and possible advances in energy storage. Energy Policy. 2008;36:4368–73.
[8] AlShafi M, Thermodynamic BY. Performance comparison of various energy storage systems from source-to-electricity for renewable energy resources. Energy. 2021;219:119626.
[9] Worrel E, Phylipsen D, Einstein D, Martin N. Energy Use and Energy Intensity of the U. S. Chemical Industry. Lawrence Berkeley Natl Lab 2000;LBNL-44314:34.
[10] Blank TK, Molly P. Hydrogen's Decarbonization Impact for Industry. 2020;13.
[11] Sun P, Elgowainy A. Updates of Hydrogen Production from SMR Process in GREET. 2019.
[12] De Levie R. The electrolysis of water. J Electroanal Chem. 1999;476:92–3.
[13] Alsayegh SO, Varjian R, Alsalik Y, Katsiev K, Isimjan TT, Methanol IH. Production using ultrahigh concentrated solar cells: hybrid electrolysis and CO_2 capture. ACS Energy Lett. 2020;5:540–4.
[14] Membraneless EDV. Electrolyzers for low-cost hydrogen production in a renewable energy future. Joule. 2017;1:651–8.
[15] Hydrogen Council. Path to Hydrogen Competitiveness: a Cost Perspective. 2020.
[16] Liu Z, Sajjad SD, Gao Y, Yang H, Kaczur JJ, Masel RI. The effect of membrane on an alkaline water electrolyzer. Int J Hydrog Energy. 2017;42:29661–5.
[17] Li X, Zhao L, Yu J, Liu X, Zhang X, Liu H, Zhou W. Water Splitting: from electrode to green energy system. Nano-Micro Lett. 2020;12:131.
[18] Xu W, Scott K. The effects of ionomer content on PEM water electrolyser membrane electrode assembly performance. Int J Hydrog Energy. 2010;35:12029–37.

[19] Hashem T, Valadez Sanchez EP, Bogdanova E, Ugodchikova A, Mohamed A, Schwotzer M, Alkordi MH, Wöll C. Stability of monolithic mof thin films in acidic and alkaline aqueous media. Membranes (Basel). 2021;11:1–12.
[20] Bobbitt NS, Mendonca ML, Howarth AJ, Islamoglu T, Hupp JT, Farha OK, Snurr RQ. Metal–organic frameworks for the removal of toxic industrial chemicals and chemical warfare agents. Chem Soc Rev. 2017;46:3357–85.
[21] Al-Jadir TM, Siperstein FR. The influence of the pore size in metal–organic frameworks in adsorption and separation of hydrogen sulphide: a molecular simulation study. Microporous Mesoporous Mater. 2018;271:160–8.
[22] Jrad A, Abu Tarboush BJ, Hmadeh M, Ahmad M. Tuning acidity in zirconium-based metal organic frameworks catalysts for enhanced production of butyl butyrate. Appl Catal A, Gen. 2019;570:31–41.
[23] Zhang X, Chen Z, Liu X, Hanna SL, Wang X, Taheri-Ledari R, Maleki A, Li P, Farha OK. A historical overview of the activation and porosity of metal–organic frameworks. Chem Soc Rev. 2020;49:7406–27.
[24] Lv XL, Yuan S, Xie LH, Darke HF, Chen Y, He T, Dong C, Wang B, Zhang YZ, Li JR et al. Ligand rigidification for enhancing the stability of metal–organic frameworks. J Am Chem Soc. 2019;141:10283–93.
[25] Yuan S, Feng L, Wang K, Pang J, Bosch M, Lollar C, Sun Y, Qin J, Yang X, Zhang P et al. Stable metal–organic frameworks: design, synthesis, and applications. Adv Mater. 2018;30:1704303.
[26] Islamoglu T, Goswami S, Li Z, Howarth AJ, Farha OK, Hupp JT. Postsynthetic tuning of metal–organic frameworks for targeted applications. Acc Chem Res. 2017;50:805–13.
[27] Deria P, Mondloch JE, Karagiaridi O, Bury W, Hupp JT, Farha OK. Beyond post-synthesis modification: evolution of metal–organic frameworks via building block replacement. Chem Soc Rev. 2014;43:5896–912.
[28] Lee J, Farha OK, Roberts J, Scheidt KA, Nguyen ST, Hupp JT. Metal–organic framework materials as catalysts. Chem Soc Rev. 2009;38:1450–9.
[29] Young RJ, Huxley MT, Pardo E, Champness NR, Sumby CJ, Doonan CJ. Isolating reactive metal-based species in metal–organic frameworks – viable strategies and opportunities. Chem Sci. 2020;11:4031–50.
[30] Greifenstein R, Ballweg T, Hashem T, Gottwald E, Achauer D, Kirschhöfer F, Nusser M, Brenner-Weiß G, Sedghamiz E, Wenzel W et al. MOF-hosted enzymes for continuous flow catalysis in aqueous and organic solvents. Angew Chem, Int Ed. 2022;61:e202117144.
[31] Safy MEA, Amin M, Haikal RR, Elshazly B, Wang J, Wang Y, Wöll C, Alkordi MH. Probing the water stability limits and degradation pathways of metal–organic frameworks. Chem Eur J. 2020;26:7109–17.
[32] Zaman N, Noor T, Iqbal N. Recent advances in the metal–organic framework-based electrocatalysts for the hydrogen evolution reaction in water splitting: a review. RSC Adv. 2021;11:21904–25.
[33] Han W, Li M, Ma Y, Yang J. Cobalt-based metal–organic frameworks and their derivatives for hydrogen evolution reaction. Front Chem. 2020;8:1–18.
[34] Gu M, Wang SC, Chen C, Xiong D, Yi FY. Iron-based metal–organic framework system as an efficient bifunctional electrocatalyst for oxygen evolution and hydrogen evolution reactions. Inorg Chem. 2020;59:6078–86.
[35] Roy S, Huang Z, Bhunia A, Castner A, Gupta AK, Zou X, Ott S. Electrocatalytic hydrogen evolution from a cobaloxime-based metal–organic framework thin film. J Am Chem Soc. 2019;141:15942–50.
[36] Budnikova YH. Recent advances in metal–organic frameworks for electrocatalytic hydrogen evolution and overall water splitting reactions. Dalton Trans. 2020;49:12483–502.

[37] Cheng W, Zhang H, Luan D, Wen X, Lou D. Exposing unsaturated Cu^{I}-O_2 sites in nanoscale Cu-MOF for efficient electrocatalytic hydrogen evolution. Sci Adv. 2021;7:eabg2580.
[38] Sun Y, Xue Z, Liu Q, Jia Y, Li Y, Liu K, Lin Y, Liu M, Li G, Su CY. Modulating electronic structure of metal–organic frameworks by introducing atomically dispersed ru for efficient hydrogen evolution. Nat Commun. 2021;12:1–8.
[39] Dhakshinamoorthy A, Garcia H. Catalysis by metal nanoparticles embedded on metal–organic frameworks. Chem Soc Rev. 2012;41:5262–84.
[40] Nie M, Sun H, Lei D, Kang S, Liao J, Guo P, Xue Z, Novel XF. Pd/MOF electrocatalyst for hydrogen evolution reaction. Mater Chem Phys. 2020;254:123481.
[41] Shah WA, Ibrahim S, Abbas S, Naureen L, Batool M, Imran M, Nadeem MA. Nickel containing polyoxometalates incorporated in two different metal–organic frameworks for hydrogen evolution reaction. J Env Chem Eng. 2021;9:106004.
[42] Dai X, Liu M, Li Z, Jin A, Ma Y, Huang X, Sun H, Wang H, Molybdenum ZX. Polysulfide anchored on porous Zr-metal organic framework to enhance the performance of hydrogen evolution reaction. J Phys Chem C. 2016;120:12539–48.
[43] Wen X, Recent GJ. Progress on MOF-derived electrocatalysts for hydrogen evolution reaction. Appl Mater Today. 2019;16:146–68.
[44] Gothandapani K, Grace AN, Venugopal V. Mesoporous carbon-supported Co_3O_4 derived from ZIF-67 metal organic framework (MOF) for hydrogen evolution reaction in acidic and alkaline medium. Int J Energy Res. 2022;46:3384–95.
[45] Lin J, He J, Qi F, Zheng B, Wang X, Yu B, Zhou K, Zhang W, Li Y, Chen Y. In-situ selenization of co-based metal–organic frameworks as a highly efficient electrocatalyst for hydrogen evolution reaction. Electrochim Acta. 2017;247:258–64.
[46] Yu Y, Qiu X, Zhang X, Wu Z, Wang H, Wang W, Wang Z, Tan H, Peng Z, Guo X et al. Metal–organic frameworks derived bundled N-doped carbon nanowires confined cobalt phosphide nanocrystals as a robust electrocatalyst for hydrogen production. Electrochim Acta. 2019;299:423–9.
[47] He S, He S, Bo X, Wang Q, Zhan F, Wang Q, Porous ZC. Ni_2P/C microrods derived from microwave-prepared MOF-74-Ni and its electrocatalysis for hydrogen evolution reaction. Mater Lett. 2018;231:94–7.
[48] Stern LA, Enhanced HX. Oxygen evolution activity by NiO_x and $Ni(OH)_2$ nanoparticles. Faraday Discuss. 2014;176:363–79.
[49] Li FL, Shao Q, Huang X, Lang JP. Nanoscale trimetallic metal–organic frameworks enable efficient oxygen evolution electrocatalysis. Angew Chem, Int Ed. 2018;57:1888–92.
[50] Wang CP, Feng Y, Sun H, Wang Y, Yin J, Yao Z, Bu XH, Self-Optimized ZJ. Metal–organic framework electrocatalysts with structural stability and high current tolerance for water oxidation. ACS Catal. 2021;11:7132–43.
[51] Yuan S, Peng J, Cai B, Huang Z, Garcia-Esparza AT, Sokaras D, Zhang Y, Giordano L, Akkiraju K, Zhu YG et al. Tunable metal hydroxide–organic frameworks for catalysing oxygen evolution. Nat Mater. 2022;21:673–80.
[52] Hassan MH, Soliman AB, Elmehelmey WA, Abugable AA, Karakalos SG, Elbahri M, Hassanien A, Alkordi MH. A Ni-loaded, metal–organic framework-graphene composite as a precursor for in situ electrochemical deposition of a highly active and durable water oxidation nanocatalyst. Chem Commun. 2019;55:31–4.
[53] Cai X, Peng F, Luo X, Ye X, Zhou J, Lang X, Shi M. Understanding the evolution of cobalt-based metal–organic frameworks in electrocatalysis for the oxygen evolution reaction. ChemSusChem. 2021;14:3163–73.
[54] Sondermann L, Jiang W, Shviro M, Spieß A, Woschko D, Rademacher L, Nickel-Based JC. Metal–organic frameworks as electrocatalysts for the oxygen evolution reaction (OER). Molecules. 2022;27:1–16.

Victor Quezada-Novoa, Christopher Copeman,
P. Rafael Donnarumma, and Ashlee J. Howarth

11 Metal-organic frameworks for clean water generation: from purification to harvesting

11.1 Introduction

Water is essential for sustainable human development, the existence of ecosystems and the eradication of poverty and hunger. It has been estimated that 4 billion people face severe water scarcity for at least one month each year and over half a billion people face water scarcity throughout the entire year [1, 2]. This issue is expected to worsen by 2050, when it is predicted that nearly 6 billion people will face clean water scarcity issues [3]. Furthermore, the problem will be aggravated by other challenges such as COVID-19, global conflicts and climate change [4]. Although our planet does not lack water, 97 % of it is contained in oceans and much of the remaining 3 % is inaccessible in the form of glaciers (50 %) and groundwater (49 %) with less than 1 % accessible in rivers and lakes [5]. Water vapor in the atmosphere accounts for 13 quadrillion liters, an amount that is equivalent to ca. 0.001 % of the water on earth. The water demand for agriculture (3.1 quadrillion liters), industry (0.8 quadrillion liters), municipal and domestic use (0.6 quadrillion liters), increases by approximately 2 % each year, while the accessible, sustainable and reliable water supply from groundwater and surface water (4.2 quadrillion liters) is not sufficient for this demand. As such, there is expected to be a 40 % deficit (2.8 quadrillion liters) in relation to the total water demand by 2030 (6.9 quadrillion liters) [1]. Therefore, promising strategies to cope with water scarcity involve the development of high-performance materials for atmospheric water harvesting, seawater desalination and improved wastewater purification for industrial and municipal sources. It is notable that these water-related research directions simultaneously contribute to several of the sustainable development goals (SDGs) set by the United Nations (UN) for 2030, such as zero hunger (SDG 2), good health and well-being (SDG 3), clean water and sanitation (SDG 6) and sustainable cities and communities (SDG 11) [6].

Towards these goals, it is essential to design multifunctional materials for the creation and improvement of clean water generation technologies. The structural and chemical versatility of metal-organic frameworks (MOFs) make this diverse class of materials competitive against the multiple challenges related to clean water generation. Section 11.2 reviews current progress on the study of MOFs for water purifica-

Victor Quezada-Novoa, Christopher Copeman, P. Rafael Donnarumma, Ashlee J. Howarth, Department of Chemistry and Biochemistry, and Centre for NanoScience Research, Concordia University, 7141 Sherbrooke St W., Montréal, QC, Canada, e-mail: ashlee.howarth@concordia.ca

https://doi.org/10.1515/9781501524721-011

tion by adsorption and degradation of organic or inorganic pollutants, microfiltration and oil-water separation. Section 11.3 discusses the study of MOFs for applications in desalination, which involves saltwater evaporation-adsorption-based systems and membranes. Section 11.4 addresses the potential of MOFs for atmospheric water harvesting, plus insights into structure-property relationships that give rise to water adsorption.

11.2 Water purification

11.2.1 Adsorption of pollutants

Due to their high porosity, versatile composition and aqueous-phase stability, MOFs are adept for use in adsorption applications for the removal of contaminants or pollutants in water. The nature of the target adsorbates is as diverse as the MOFs themselves and can be categorized into two main groups: organic and inorganic pollutants.

11.2.1.1 Adsorption of organic contaminants

As industrial development moves forward and the global production of goods increases in volume and diversity, new emerging contaminants are appearing more regularly [7–9]. The effects of these emerging contaminants are often unaccounted for and under studied. This is not only true in regards to pharmaceutical and personal care products (PPCPs) [8–10], but also with other contaminants as well, like artificial sweeteners [11] or herbicides [12]. If left unchecked, some of these organic contaminants can have significant adverse effects, such as antibiotics contributing to the apparition of antibiotic-resistant super strains of bacteria, or herbicides acting as endocrine disrupting compounds [7–12]. Traditionally, carbon-based materials (e. g., activated carbon, graphene and carbon nanotubes) have been studied and used for the adsorptive removal of these chemicals [13, 14], but these materials have experienced adsorptive performance limitations for the removal of all organic contaminants, particularly those that are hydrophilic, and present costly regeneration processes together with low volumetric uptakes of contaminants [15]. More recently, MOFs have been studied for the adsorptive removal of organic chemicals, exhibiting promise based on their advantages over carbon-based materials including high surface area and porosity, chemically tunable adsorption sites and high capacity [16–48].

The two most common approaches used to design MOFs for the adsorption of organic contaminants involve utilizing (i) functional groups that promote noncovalent interactions, or (ii) coordinatively unsaturated metal sites (CUSs) for coordination. Showcasing the former approach, Seo et al. focused their attention on the highly

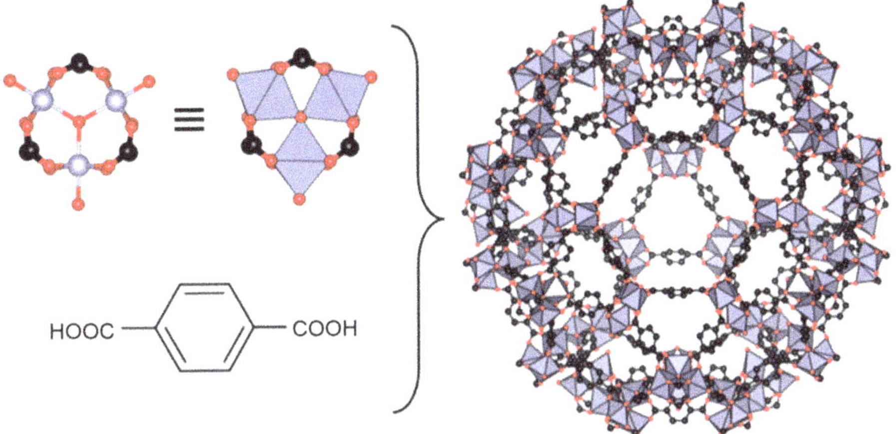

Figure 11.1: Structure of Cr-MIL-101 highlighting the trinuclear Cr(III)-cluster node and 1,4-benzenedicarboxylic acid linker. Atom colors: Cr, light purple polyhedra; C, black and O, red.

porous, Cr-based MOF MIL-101 [49, 50]. Cr-MIL-101 is comprised of trinuclear Cr(III)-cluster nodes bridged by 1,4-benzenedicarboxylate (BDC) linkers (Figure 11.1). Each node contains two potential CUSs that are capped by two $-H_2O$ ligands and an additional CUS capped by –OH or a halide. Cr-MIL-101 can be post-synthetically functionalized by grafting ligands to these CUS or by performing chemistry on the BDC linker core to append diverse functional groups. Seo et al. applied both of these post-synthetic approaches to facilitate the capture of hazardous materials through hydrogen bonding, electrostatic and acid-base interactions in Cr-MIL-101. In one example, the adsorption and removal of the PPCPs naproxen, ibuprofen and oxybenzone was achieved by functionalizing the CUS of Cr-MIL-101 with –OH and $-(OH)_2$ using ethanolamine and diethanolamine, and by post-synthetically adding $-NO_2$ and $-NH_2$ groups to the BDC linker core (Figure 11.2) [49]. It was shown in this work that the quantity adsorbed (q) per mass unit of adsorbent was directly related to the presence of functional groups capable of acting like H-donors, giving rise to the following trend for q: MIL-101-OH

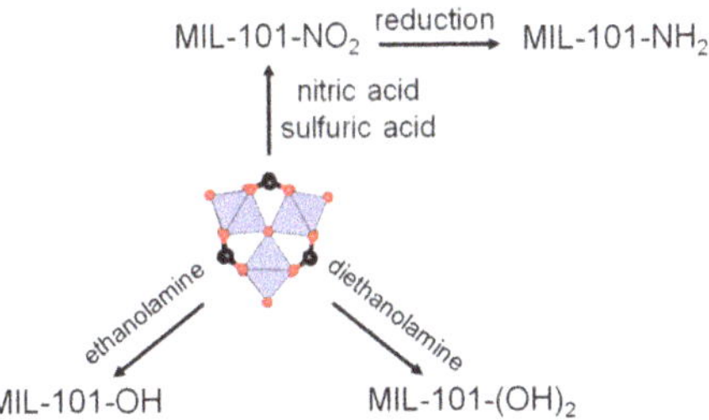

Figure 11.2: Post synthetic modification of Cr-MIL-101 to give Cr-MIL-101 analogues with $-NO_2$, $-NH_2$, –OH and $-(OH)_2$ functional groups. Atom colors: Cr, light purple polyhedra; C, black and O, red.

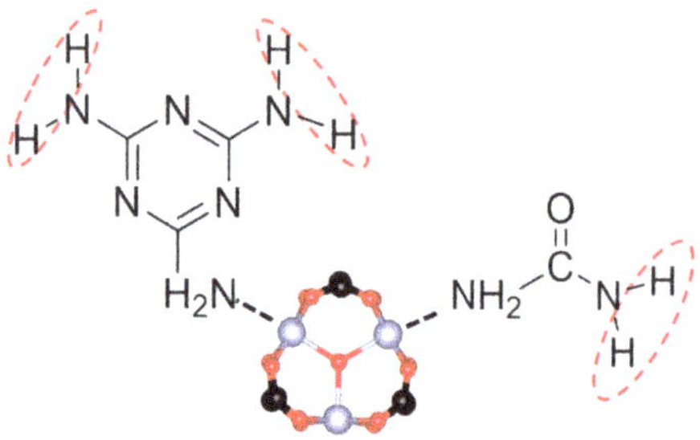

Figure 11.3: Grafting of melamine and urea ligands on the CUSs present in Cr-MIL-101. Atom colors: Cr, light purple; C, black and O, red.

> MIL-101-NH_2 > MIL-101-$(OH)_2$ > MIL-101 > activated carbon (AC) > MIL-101-NO_2 with values ranging from ~170 to 50 mg g^{-1}. The authors deduced then that the higher performance of a material was directly related to its capacity to participate in H-bonding.

In addition to the better performance over conventional AC, it was reported that both the functionalized and pristine MIL-101 show recyclability over four cycles with only a slight decrease in adsorption capacity. In a separate work, Seo et al. decorated MIL-101 through grafting of urea and melamine (Figure 11.3) onto the CUS to study the adsorptive removal of three artificial sweeteners (saccharine, acesulfame and cyclamate) [50]. Similar to the previous work, pristine MIL-101 had higher values of q per mass unit of adsorbent than AC, but the two decorated MIL-101 analogues outperformed pristine MIL-101 as follows: urea-MIL-101 > melamine-MIL-101 > MIL-101 > AC > MIL-101-NO_2 with values ranging from 86 to 19 mg g^{-1}. Again, the enhanced uptake observed in urea- and melamine-grafted MIL-101 was explained through the potential for H-bonding between $-NH_2$ groups on the MOF and R–OH or C=O groups in the artificial sweeteners.

Zr-based MOFs have been extensively studied for water purification because of their structural diversity (born in the ability of Zr to form hexanuclear Zr(IV)-clusters with varying connectivity) [51] and high stability (owing to the strength of the Zr(IV)–O bond) [52]. NU-1000 is one example that has been studied for adsorptive removal of organic pollutants in water [53, 54] due to its architecture that gives rise to a substantially large surface area (~ 2200 m^2 g^{-1}), accessible metal sites and π-conjugated linkers and an overall superb water stability. NU-1000 is comprised of hexanuclear Zr(IV)-cluster nodes bridged by 1,3,6,8-(p-benzoate)pyrene linkers (Figure 11.4a). More specifically, the Zr_6-nodes are 8-connected, meaning that there are four terminal –OH and $-OH_2$ ligands available for substitution chemistry.

Akpinar et al. studied the use of NU-1000 for the removal of atrazine, an endocrine disruptor emerging from agricultural use as a herbicide [53]. When compared to other Zr_6-based MOFs, NU-1000 was found to remove 95 % of atrazine from 10 mL of a 10 ppm solution, compared to 2 % for UiO-66 (Figure 11.4b) and UiO-66-NH_2 and up to 84 % for NU-901 (Figure 11.4c). The ability of NU-1000 to remove atrazine lies in the

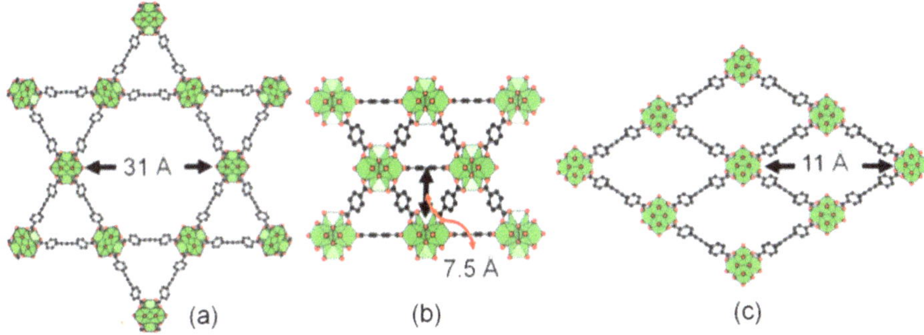

Figure 11.4: Structure of the NU-1000 (a), UiO-66 (b) and NU-901 (c) highlighting the pore aperture size of each of these MOFs. Atom colors: Zr, green polyhedra; C, black and O, red.

collusion between the structure of its linker, a pyrene-based tetra-carboxylate capable of participating in π–π interactions with the analyte, and the large pores (~30 Å in diameter). In addition, NU-1000 is found to reach its saturation uptake for atrazine in <1 min, as compared to 6 h or more for other adsorbents tested, with recyclability over three cycles and high atrazine uptake (>90 %) even in the presence of interferents (i. e., $CaCl_2$, Na_2SO_4, $NaNO_3$, NaCl). In a different study, Li et al. used NU-1000 (Figure 11.4a) to remove per- and polyfluoroalkyl substances (PFASs) from water [54]. PFASs are used ubiquitously in a wide array of products from non-stick cookware to cosmetics [55]. In this study, NU-1000 was exposed to nine different PFASs and its adsorption capacity was compared to other MOFs (MOF-74, UiO-66 and ZIF-8), revealing that NU-1000 had the highest adsorption capacity for these contaminants with values ranging from 201 to 620 mg g^{-1}. Again, NU-1000 showed fast adsorption kinetics by reaching its saturation uptake in <1 min and outperforming conventional adsorbents such as AC and ion-exchange resins in terms of both kinetics and equilibrium capacity. In this case, the mechanism for the adsorption of the pollutants is related to a combination of hydrophobic interactions between the PFAS and the pyrene-based linker as well as coordination to the Zr_6-node through sulfonate and carboxylate groups on the PFASs. Furthermore, the recyclability of NU-1000 and subsequent uptake of PFASs was determined to be outstanding after 5 cycles. Even more interestingly, NU-1000 was tested against a complex matrix taken from contaminated groundwater at U. S. Air Force bases and proved to be efficient in the removal of 75–98 % of the PFASs present showing that interference of other salts or molecules is minimal.

In another example involving linker interactions with organic contaminants, Wang et al. synthesized and studied two Zr-MOFs, BUT-12/NU-1200 and BUT-13, for the simultaneous detection and removal of antibiotics and organic explosives. BUT-12/NU-1200 and BUT-13 are isostructural MOFs comprised of 8-connected hexanuclear Zr(IV)-cluster nodes bridged by tritopic linkers (Figure 11.5) [56]. It was found that both BUT-12/NU-1200 and BUT-13 are able to adsorb antibiotics, such as nitrofurazone and nitrofurantoin, and explosives, such as 2,4,6-trinitrophenol and 4-nitrophenol, with

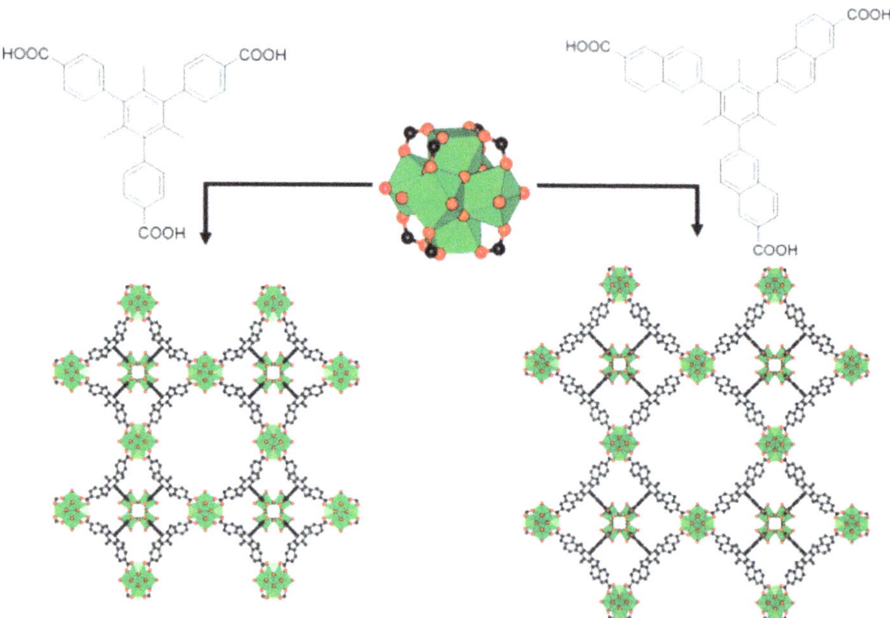

Figure 11.5: Structures of BUT-12/NU-1200 and BUT-13 highlighting the hexanuclear Zr(IV)-cluster node and tritopic linkers (5'-(4-carboxyphenyl)-2',4',6'-trimethyl-[1,1':3',1"-terphenyl]-4,4"-dicarboxylic acid and 6,6',6"-(2,4,6-trimethylbenzene-1,3,5-triyl)tris(2-naphthoic acid)). Atom colors: Zr, green polyhedra; C, black and O, red.

high uptake and good recyclability indicators. The mechanism of adsorption was attributed to the suitable pore size (21 and 28 Å for BUT-12/NU-1200 and BUT-13, respectively) and hydrophobic-hydrophobic interactions between the structural organic linkers in the MOF with the organic contaminants.

Rojas et al. targeted the removal of atenolol, a PPCP used to treat high blood pressure that is poorly absorbed by the body and ends up being excreted into the water system [57]. In their study, eight different MOFs were screened for atenolol uptake, including UiO-66, UiO-66-NH_2 (Figure 11.4b), MIL-163, MIL-127, MIL-53, MIL-53-$(OH)_2$, Ni_8BDP_6 and $Ni_8(BDP\text{-}NH_2)_6$, where BDP = 1,4-bis(1*H*-pyrazol-4-yl)benzene. It was found that Ni_8BDP_6 demonstrated the highest adsorption capacity, removing 95 % of atenolol in only 1 hour when exposed to 0.5 molar equivalents of the contaminant. The uptake is attributed to the larger pore aperture sizes of Ni_8BDP_6 (12 Å) compared to the other MOFs (<10 Å) as well as hydrophobic-hydrophobic interactions between the BDP linker and atenolol. Interestingly, when missing linker defects were introduced into the MOF to give KOH@Ni_8BDP_6, the removal was enhanced from 95 to 99 % with very limited degradation (<1 %). This highlights the fact that increasing the aperture size of a MOF by introducing defects can be used as a strategy to improve contaminant uptake, though a fine balance is required to not disrupt the important chemical functionalities required for adsorption. Beyond lab scale testing, Rojas et al. tested KOH@Ni_8BDP_6 in river water matrices under continuous flow experiments ($34\,L\,m^{-2}\,h^{-1}$) using a MOF-pellet packed column and did not observe breakthrough of atenolol until day 12. This study demonstrated the potential of this MOF for practical applications in water treatment.

11.2.1.2 Adsorption of inorganic pollutants

Inorganic species typically targeted for adsorption are heavy metals such as chromium, cadmium, mercury and lead, as well as other potentially dangerous elements like arsenic and selenium [58]. Water containing these elements may be acutely toxic at high enough concentrations or may be harmful with continuous exposure or through bioaccumulation, which may cause cancer, neurotoxicity or developmental problems, among other issues [59]. These elements may be naturally occurring, but are more often released during industrial activities, such as mining, manufacturing and the combustion of fossil fuels [60]. While the adsorptive properties of MOFs are typically used to decontaminate water containing harmful inorganic species, MOFs have also been explored for harvesting and concentrating valuable resources from seawater, such as gold or uranium [61, 62].

One approach that is used to design MOF adsorbents for water treatment involves creating cationic or anionic frameworks for ion-exchange treatment. Zhang et al. post-synthetically modified the pyridyl sites of the 2,2'-bipyridine-5,5'-dicarboxylate linker of MOF-867 with methyl groups to give the cationic MOF, ZJU-101 [63]. ZJU-101 is comprised of hexanuclear Zr(IV)-clusters bridged by ditopic linkers and is isostructural to UiO-67 (Figure 11.6). MOF-867 was stirred in trifluoromethanesulfonate for 48 hours, followed by washing with $CHCl_3$ and 0.1 M $NaNO_3$, resulting in 70 % conversion to the methylated ZJU-101. The linker modification was confirmed by the appearance of N^+-CH_3 groups and NO_3^- counterions via proton nuclear magnetic resonance (^1H-NMR) and infrared (IR) spectroscopy. As a result of the cationic nature of ZJU-101, an increase in the uptake of $Cr_2O_7^{2-}$ was observed with a removal capacity of $53\,mg\,g^{-1}$ for MOF-867 and $245\,mg\,g^{-1}$ for ZJU-101. The authors also demonstrated the selectivity of the MOF towards $Cr_2O_7^{2-}$ adsorption showing an 81 % removal efficiency even when exposed to a 12-fold molar excess of a competing anion solution consisting of halides F^-, Cl^-, Br^-, I^- and oxyanions NO_3^- and SO_4^{2-}. Adsorption kinetics were also studied, demonstrating very rapid uptake of $Cr_2O_7^{2-}$ with over 87 % removal within 5 minutes and 50 % adsorption in less than 1 minute.

In another example taking advantage of ion-exchange, Desai et al. designed a highly water stable cationic nickel(II) based MOF $[\{Ni_2(tris(4\text{-}(1H\text{-}imidazol\text{-}1\text{-}yl)phenyl)amine)_3(SO_4)(H_2O)_3\}\cdot SO_4]_n$ incorporating both coordinated and uncoordinated sulfate anions (Figure 11.7) [64]. The sulfate anions were shown to be present in the MOF pores as observed by single crystal X-ray diffraction (SCXRD), which allowed

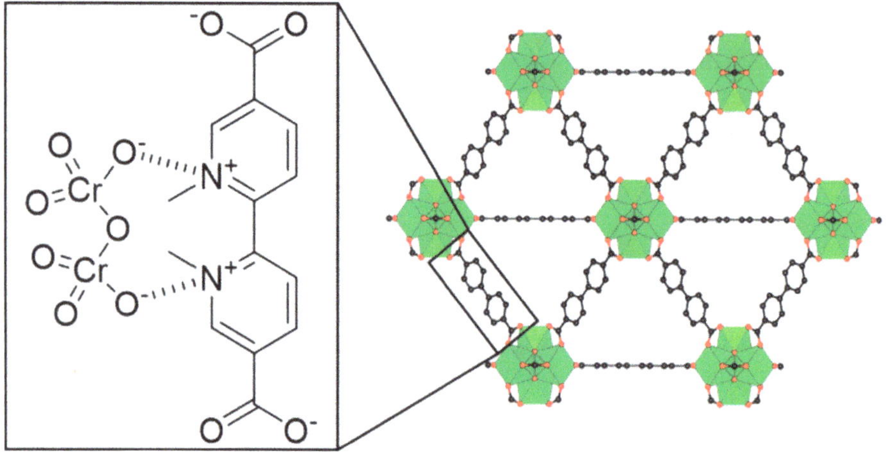

Figure 11.6: Structure of ZJU-101 highlighting the electrostatic interaction between the organic linker of the MOF with $Cr_2O_7^{2-}$. Atom colors: Zr, green polyhedra; C, black and O, red.

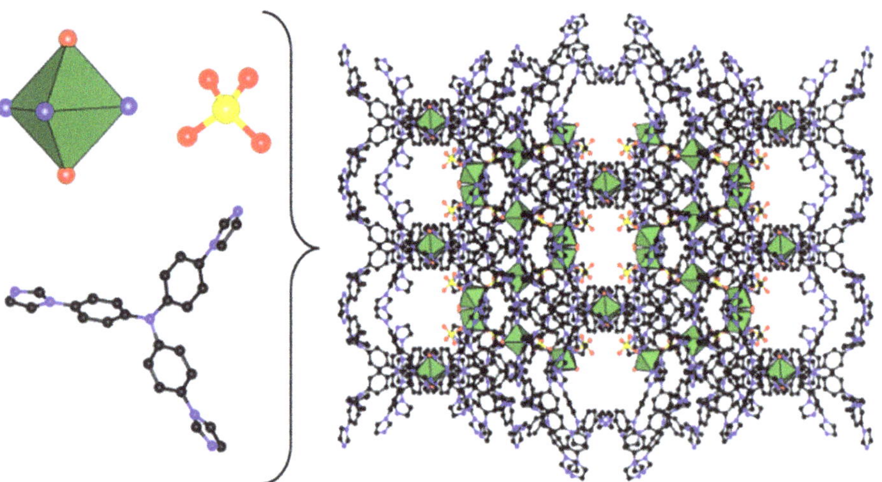

Figure 11.7: Structure of [{Ni_2(tris(4-(1H-imidazol-1-yl)phenyl)amine)$_3$(SO_4)(H_2O)$_3$}·SO_4]$_n$ highlighting the Ni(II)-node, tris(4-(1H-imidazol-1-yl)phenyl)amine linker and the presence of coordinated SO_4^{2-}. Atom colors: Ni, green polyhedra; C, black; N, blue; S, yellow and O, red.

for the MOF to be investigated for anion exchange applications. Anion exchange was found to occur most readily with oxyanions that share similar geometry to the departing tetrahedral SO_4^{2-}, including $Cr_2O_7^{2-}$, where one-half of the dichromate anion exhibits a tetrahedral geometry. Ion exchange experiments showed a maximum gravimetric $Cr_2O_7^{2-}$ uptake of 166 mg g^{-1} over 7 days. The authors also showed the potential for multicycle regeneration, where up to 94 mg g^{-1} of adsorbed $Cr_2O_7^{2-}$ could be removed by exposing the MOF to a 45 μM Na_2SO_4 solution. Furthermore, the regenerated MOF reaches the same maximum uptake in a second adsorption cycle.

Further work was performed with this cationic Ni_2-MOF by Sharma et al. utilizing the exchangeable sulfate counterion and selectivity for tetrahedral analytes to remove the oxyanions of selenium and arsenic, SeO_4^{2-}, and $HAsO_4^{2-}$ [65]. The MOF proved effective in the removal of these analytes with adsorption capacities of 100 mg g^{-1} and 84 mg g^{-1}, respectively. As in the previous study, additional experiments confirmed adsorption performance in the presence of competing anions including NO_3^-, HCO_3^-, CO_3^{2-}, SO_4^{2-} and Cl^-. The mode of anion capture was further studied by SCXRD, in which the SeO_4^{2-} ions were found to be located in sites previously occupied by SO_4^{2-} ions. Additionally, the authors observed that $HAsO_4^{2-}$ is further stabilized by H-bonding with Ni-bound water molecules. In a follow-up study, Sharma et al. developed a new MOF, iMOF-3C [{Ni(tris(4-(1H-imidazol-1-yl)phenyl)amine)$_2$}·(SO_4) (Figure 11.8), which exhibits a high affinity and uptake capacity for selenium and arsenic oxyanions SeO_3^{2-}, SeO_4^{2-}, and $HAsO_4^{2-}$, with maximum adsorption capacities of 140, 72 and 75 mg g^{-1}, respectively [66]. The authors performed computational studies to probe the mechanism of adsorption showing that the hydrogen atoms on the imidazole linker form H-bonds with the oxygen atoms in the oxyanions, which helps to stabilize the oxyanion. Furthermore, the size and shape of the oxyanions was confirmed to play a role in the strength of the interaction with the MOF.

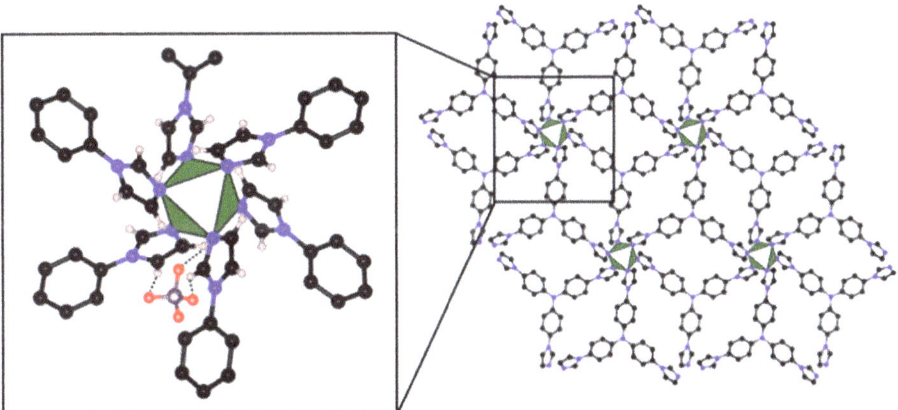

Figure 11.8: Structure of iMOF-3C highlighting the Ni(II)-node, tris(4-(1H-imidazol-1-yl)phenyl)amine linker, and the presence of the AsO_4^{2-}. Atom colors: Ni, green polyhedra; C, black; N, blue; As, purple and O, red.

Toward utilizing linker functional groups for the capture of contaminants, Rudd et al. designed a series of luminescent dual-linker MOFs by combining 1,1,2,2-tetrakis-(4-(pyridine-4-yl)phenyl)ethane (tppe) with [9-oxo-9H-fluorene-2,7-dicarboxylate] (ofdc), [9H-fluorene-2,7-dicarboxylate] (fdc) or [dibenzo[b,d]thiophene-3,7-dicarboxylate 5,5-dioxide] (dbtdcO_2) and Zn_2-nodes to produce LMOF-261, LMOF-262 and LMOF-

263, respectively (Figure 11.9) [67]. The authors proposed that the sulfone group, present in the dbtdcO_2 linker of LMOF-263, acts as a soft Lewis base leading to preferential adsorption of softer metals like Hg(II) compared to hard metal ions like Ca(II) and Mg(II). LMOF-263 exhibits a high Hg(II) uptake capacity of 380 mg g^{-1} with demonstrated uptake of 99.1 % from a 10 ppm solution in under 30 minutes. In addition, the incorporation of the fluorescent type linker results in a MOF with adsorption and sensing capabilities as the inclusion of heavy metals Hg(II) or Pb(II) quenches the blue fluorescence with a detection limit of 3.3 and 19.7 ppb, respectively. The selectivity towards soft heavy metals Hg(II) and Pb(II) compared to hard light metals results in a selective detection ratio for Hg(II)/Ca(II) of 209.5 and Hg(III)/Mg(II) of 167.4.

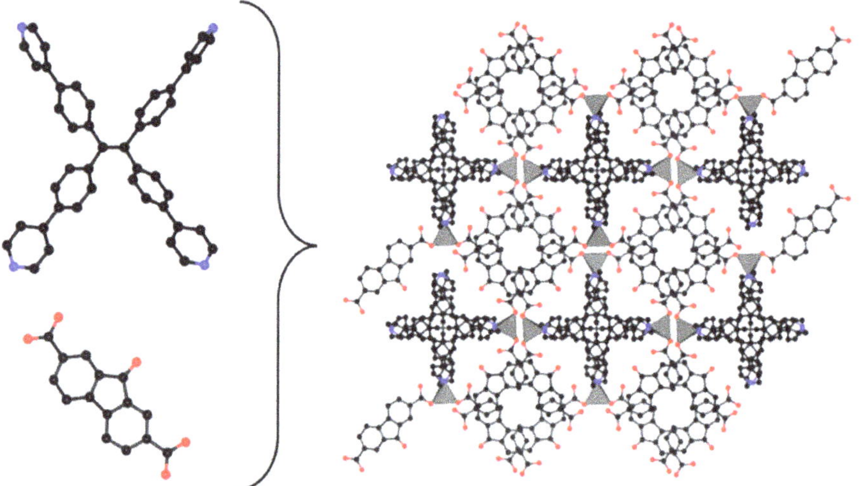

Figure 11.9: 4-Fold interpenetrated structure of LMOF-261 highlighting the 1,1,2,2-tetrakis(4-(pyridine-4-yl)phenyl)ethane (tppe) and [9-oxo-9H-fluorene-2,7-dicarboxylate] (ofdc) linkers. Atom colors: Zn, grey polyhedra; C, black; N, blue and O, red.

In another example taking advantage of linker-based adsorption, Carboni et al. targeted the adsorption of uranyl ions using a series of Zr_6-MOFs that feature amino-*p*,*p*'-terphenyldicarboxylate (UiO-68-NH_2), diethoxyphosphorylurea (UiO-68-P(O)$(OEt)_2$) and dihydroxyphosphorylurea (UiO-68-P(O)$(OH)_2$) [68]. The UiO-68 analogues contain pores that allow the passage of molecules up to 10 Å in size, a necessity for large, hydrated actinide ions. UO_2^{2+} was found to interact with the phosphorylurea groups in a monodentate fashion with each UO_2^{2+} coordinating to two adjacent linkers in the MOF pore. Furthermore, UiO-68-NH_2, showed no affinity towards UO_2^{2+}. The authors showed that protonation of the phosphorylurea groups improves adsorption, with UiO-68-P(O)$(OEt)_2$ performing better than UiO-68-P(O)$(OH)_2$ in simulated seawater at pH 2.5 with maximum adsorption capacities of 217 and 188 mg g^{-1}, respectively.

Open and accessible CUSs in MOFs can also be used to facilitate the adsorption of inorganic contaminants. Howarth et al. screened a variety of Zr_6-based MOFs for the adsorption of selenium oxyanions, SeO_3^{2-} and SeO_4^{2-} [69]. MOFs comprised of hexanuclear Zr(IV)-clusters were evaluated, including NU-1000 (Figure 11.4a), with an 8-connected node and tetratopic pyrene-based linker alongside UiO-66 (Figure 11.4b) and UiO-67, with 12-connected nodes constructed using BDC and biphenyl-4,4'-dicarboxylate (BPDC), respectively. While the $-NH_2$ functionalization of BDC in UiO-66 led to a higher affinity towards SeO_3^{2-} than the parent MOF, the uptake kinetics and capacity of NU-1000 could not be matched due to its much larger 30 Å pore aperture and higher number of accessible metal node binding sites (CUSs). Upon removal of the node-bound structure-directing benzoate ligands, NU-1000 exhibits eight accessible zirconium CUSs per node, occupied only by substitutionally labile $-OH_2$ and $-OH$ ligands. Using differential pair distribution function analysis, a $\eta_2\mu_2$ bonding mode was found where SeO_3^{2-} and SeO_4^{2-} bridge two zirconium atoms in the cluster (Figure 11.10). The maximum adsorption capacity of NU-1000 was determined to be 95 and 85 $mg\,g^{-1}$ for SeO_3^{2-} and SeO_4^{2-}, respectively. To test the applicability of the MOF towards industrial application, the authors performed uptake experiments at concentrations of 1000 ppb, 40 °C and a pH of 6, mimicking standard conditions for flue gas desulfurization. NU-1000 displayed fast kinetics, removing 98 % of SeO_3^{2-} and SeO_4^{2-} in 5 minutes and reducing the concentration to levels well below those considered safe for consumption. In related work, Rangwani et al. demonstrated that NU-1000 could be applied towards the adsorption of $Sb(OH)_6^-$ by the same mechanism resulting in a maximum uptake capacity of 260 $mg\,g^{-1}$ [70]. Howarth et al. also employed NU-1000 as an adsorbent for SO_4^{2-} in the same fashion resulting in a maximum uptake of 56 $mg\,g^{-1}$ in less than 1 minute [71].

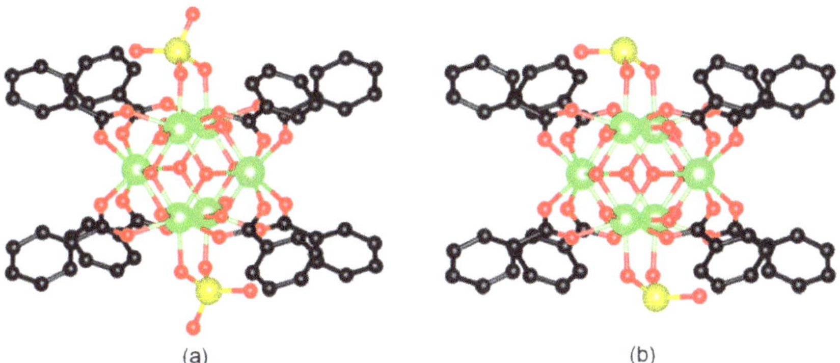

Figure 11.10: Hexanuclear Zr(IV)-cluster node of NU-1000 with coordinated SeO_4^{2-} (a) and SeO_3^{2-} (b), respectively. Atom colors: Zr, green; C, black; Se, yellow and O, red.

Another strategy for designing adsorbent MOF-based materials involves integration with other materials (i. e., polymers) that have a known affinity for a specific contaminant to enhance porosity and accessibility of adsorption sites. Sun et al. targeted the adsorption of the heavy metals, Pb(II) and Hg(II), by combining an amine and catechol-rich polymer, polydopamine (PDA), with MIL-100 [72]. MIL-100 is comprised of trinuclear Fe(III)-cluster nodes bridged by 1,3,5-benzenetricarboxylate (BTC) linkers giving rise to large cages of 25 and 29 Å in diameter (Figure 11.11). PDA is suitable for the capture of heavy metal ions due to its abundance of –OH and $-NR_3$ groups that can coordinate to metals. In this study, the authors demonstrate the utility of combining a metal-scavenging, hydrophilic polymer with a highly porous MOF as a solid-state support for water remediation. The MOF-polymer composite was synthesized by taking advantage of the redox-active Fe(III) CUSs in the MOF that are oxidative and promote the polymerization of dopamine within the MOF pores. The Fe(III) also acts as a binding site for –O and –N atoms of polydopamine, resulting in materials synthesized with loadings of 19, 28, 38 and 42 mass% PDA, exhibiting surface areas of 1135, 760, 490 and 165 $m^2 g^{-1}$, respectively. The MOF-polymer composite showed remarkable stability with no leaching of either the metal or polymer after 2 months of exposure to an environmental water sample from the Rhone River. After screening a variety of heavy metals (e. g., As(III), Cd(II), Cr(VI), Hg(II) and Pb(II)), the MOF-polymer composite was found to be most effective in the removal of Hg(II) and Pb(II) ions. Environmental water samples with up to 1000 ppb of Pb(II) were reduced by 99.7 % in under 1 minute, which were down to levels safe for consumption. Environmental water samples with 860 ppb of Hg(II) were reduced by ~99 % within 1 minute. The extremely fast kinetics and high removal efficiencies, even in the presence of natural environmental interferences, is promising for the real-world application of MOF-polymer composites. Building on this knowledge, Sun et al. extended the idea to synthesize redox-active polymer decorated MOFs to target other metals, like Au(III) [61]. The researchers used the same MOF, MIL-100, but functionalized the material using poly(para-phenylenediamine (PpPDA). This MIL-100-PpPDA composite was synthesized by introducing the monomer into the MOF pore, where nodal Fe(III) oxidizes the $-NH_2$ and –OH groups to =NH or =O leading to polymerization. The authors showed the resulting redox-active MOF-polymer composite to be highly effective at reducing Au(III) ions to Au(0) and subsequently capturing Au(0) from environmental samples and electronics waste. The authors soaked 10 mg of the MOF-polymer composite in 10 L of 0.8 ppm Au(III) for 3 weeks resulting in over 8 mg of extracted Au. After extraction, the material was calcined at 900 °C and treated with HCl leaving only 23.9 karat gold particles remaining. To prove applicability in the emerging field of electronics waste recycling, the authors dissolved metals from computer processing units in aqueous solutions of *N*-bromosuccinimide and pyridine, which created a sample with 1470 ppm Cu(II), 95 ppm Ni(II) and 7.3 ppm Au(III). Within 30 minutes, 95 % of the Au(III) in this solution was removed representing a 662-fold selectivity for gold over the competing ions and demonstrating promise for the material in extracting gold from electronics waste.

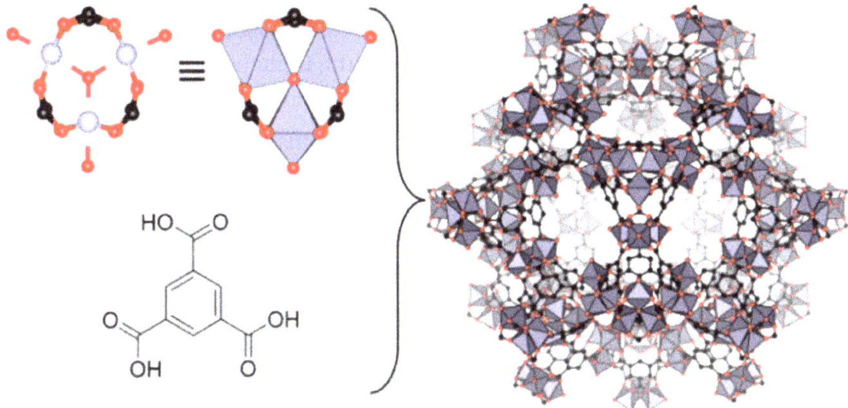

Figure 11.11: Structure of MIL-100 highlight the trinuclear Fe(III)-cluster nodes and 1,3,5-benzenetricarboxylic acid (BTC) linker. Atom colors: Fe, purple polyhedra; C, black and O, red.

In addition to polymers, MOF-nanoparticle composites can also be produced to take advantage of the distinct adsorption properties of both materials. Boix et al. developed a MOF-nanoparticle microbead formed by continuous flow spray drying using a thiolated derivative of UiO-66 (Figure 11.4b) with CeO_2 nanoparticles [73]. By combining these two materials that both have an affinity toward different analytes, the authors endeavored to create a single material with broad-based adsorption potential. UiO-66-$(SH)_2$ exhibits strong adsorption of As(III), As(V), Hg(II), Pb(II), Cr(III) and Cd(II) while the CeO_2 nanoparticles display affinity toward As(V), Pb(II), Cr(IV) and Cd(II). Furthermore, by encapsulating and dispersing the nanoparticles throughout the MOF structure, the exposed surface area of the CeO_2 was increased while preventing aggregation. In a solution at pH = 5 with 100 ppb concentration of each metal, the CeO_2@UiO-66-$(SH)_2$ beads with 3.3 % CeO_2 loading were able to remove 99 % Pb(II), 99 % Cu(II), 98 % Hg(II), 93 % Cr(III and VI), 87 % Cd(II) and 56 % As(III and V) in 3 hours. To demonstrate the practical applicability of the material, a continuous flow column was built and after three cycles through the column, the concentration of each metal ion was reduced to below 1 ppb from 100 ppb. The microbeads were also evaluated using a simulated environmental river system water mimicking heavily polluted waterways; however, the higher natural pH of this water reduced the effectiveness of the material toward some metals such as As(V) and Cd(II). Finally, to facilitate the recovery of the microbeads post adsorption, the group explored the simultaneous incorporation of magnetic Fe_3O_4 nanoparticles, which allowed for facile isolation of the microbeads from solution by magnetic attraction.

Another important aspect to consider when studying MOFs for water treatment applications is how to formulate or support the MOF crystallites for final application. Yang et al. prepared a series of MOF-polymer bead composites using poly(acrylic acid) (PAA), sodium alginate and Ca(II) ions [74]. The MOF-polymer beads were formed by dropping an aqueous solution of alginate-containing dispersed MOF into a solution of PAA and Ca(II) causing rapid gelation and the formation of a bead. The authors demonstrated that the addition of alginate to PAA and Ca(II) causes cross-linking between both polymers around Ca(II), coagulating to form a bead and encapsulating the MOF. The broad applicability of the process was demonstrated by encapsulating a variety of MOFs including MIL-127-Fe, MIL-100 (Figure 11.11), HKUST-1 (Figure 11.12a), Cu-TDPAT, Ni-pyrazolate MOF, MIL-101(Cr) (Figure 11.1), ZIF-8 (Fig. 11.12b), ZIF-67, UiO-66 (Figure 11.4b), NH_2-MIL-53(Al), $Eu_2(BDC)_3$ and $Tb_2(BDC)_3$. In addition, the MIL-100/PDA composite previously discussed was encapsulated in the polymer bead with a MOF loading of 92.9 wt%. Only a small reduction in surface area of the composite was observed after encapsulation in the beads from 997 to 915 $m^2\,g^{-1}$, which is ascribed to the high 250,000 M_W PAA, which does not occlude the MOF pores. The effect of encapsulating the MIL-100/PDA composite within a polymer bead was evaluated, and the beads were found to remove 90 % of Pb(II) from water in 1 hour compared to 99 % in under 2 minutes for the composite alone. In order to test the performance in environmental settings, the authors prepared a column containing 1 g of MIL-100/PDA beads and used it to purify over 10 L of river water contaminated with 600 ppb of Pb(II) reducing the concentration to below 15 ppb.

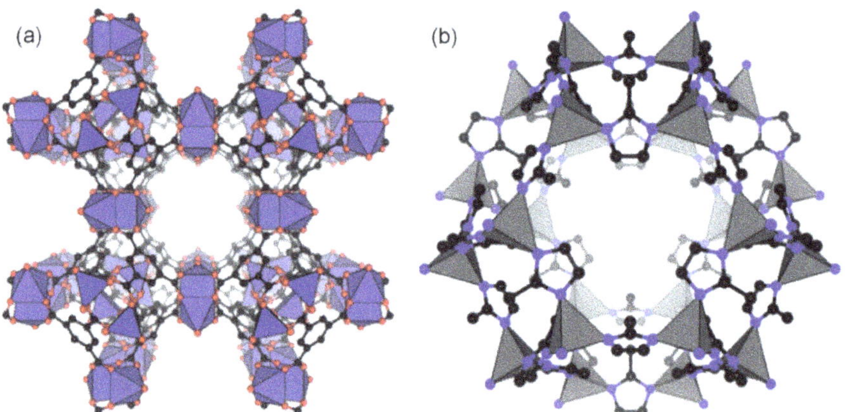

Figure 11.12: Structure of HKUST-1 (a) ZIF-8 and (b). Atom colors: Zn, grey polyhedra; Cu, blue polyhedra; C, black; N, blue and O, red.

11.2.1.3 Dual adsorption of organic and inorganic pollutants

While the tuning of MOFs allows for targeting specific analytes, some materials are also explored and employed to adsorb a wide variety of contaminants that can be classified as both organic and inorganic. By combining the design features discussed above, hybrid systems can be developed that are capable of simultaneously adsorbing heavy metals and/or oxyanions and organic contaminants. Furthermore, these MOFs can be anchored on a support or integrated into a polymer. Examples of this include the adsorptive removal of Cr(VI) and methyl orange [75, 76] or Hg(II) and nonpolar residues such as oils [77].

11.2.2 Catalytic degradation

11.2.2.1 Degradation of organics

MOFs have shown promise toward the decontamination of polluted water through the catalytic breakdown of pollutants to more innocuous products. Recently, Rojas et al. studied the performance of UiO-66-X (X = H, NH_2 and OH) and the effect of crystallite size on the photocatalytic degradation of sulfamethazine (SMT), a representative antibiotic used in livestock [78]. Microsized UiO-66-H (Figure 11.4b) was able to degrade 77 % of SMT in 24 hours using UV-vis irradiation in tap water. The MOF remained stable through the reaction maintaining its performance over 4 cycles. To evaluate the effect of crystallite size and light harvesting capability, nanoUiO-66-X (X = H, NH_2, and OH) was synthesized. Although no adsorption of SMT was observed in UiO-66-H, both UiO-66-OH and $-NH_2$ show some SMT adsorption, which was attributed to hydrophilic and hydrogen bonding interactions. Regardless of the differences in adsorption, nanoUiO-66-H, $-NH_2$ and –OH, were found to photodegrade 100 %, 90 % and 95 % of SMT in 4 h, though nanoUiO-66-NH_2 and –OH suffer from amorphization during the process. These results demonstrate that modification of the absorption spectrum of the MOF by linker functionalization does not improve SMT photodegradation performance. Degradation products were identified using ultrahigh performance liquid chromatography coupled to mass spectrometry (UHPLC-MS) as the *N*-(4,6-dimethylpyrimidin-2-yl)benzene-1,4-diamine and 2-amino-4,6-dimetoxypyrimidine, products are classified as irritants, but not health hazards like SMT. The exact mechanism of the photodegradation was not studied, but the products suggest the presence of reactive oxygen species, such as hydroxyl radicals. A detailed review on the generation of reactive oxygen species for pollutant degradation using MOFs was recently published by our group [79].

MOFs can also act as catalyst supports for the degradation of pollutants. In a recent example, Castillo-Blas et al. used solvothermal incorporation in MOFs (SIM) to incorporate binuclear Fe-oxo clusters of Fe(II) in MOF-808. Fe-MOF-808 can capture and degrade bisphenol A (BPA) through a Fenton process using H_2O_2 [80]. Synchrotron X-ray techniques, such as total X-ray scattering and X-ray absorption fine structure analysis combined with DFT modeling, performed on Fe-MOF-808 demonstrated the existence of small Fe(II)-oxo clusters at low Fe-loadings (0.5 Fe/node) and larger Fe(III)-oxo clusters at higher Fe-loadings (1.2 Fe/node). Furthermore, the Fe-MOF-808 sample with 0.5 Fe/node and Fe(II) clusters was able to degrade 40 % of BPA in 1 hour, a value much higher than that found for Fe-MOF-808 with 1.2 Fe/node loading and Fe(III). This is consistent with Fe(II) being more active for Fenton catalysis [79].

11.2.2.2 Degradation of inorganics

Although most examples of catalytic pollutant degradation involve organic contaminants, the degradation of inorganic contaminants is also possible. In a representative example, Zheng et al. reported the use of UiO-67-bpydc (bpydc = 2,2'-bipyridine-5,5'-dicarboxylate), in which a portion of the linkers were functionalized with $[Ru(bpy)_2]^{2+}$ (bpy = 2,2'-bipyridine) to give $Ru(bpy)_3$-like moieties throughout the UiO-67-bpydc backbone [81]. The remaining unfunctionalized bpy linkers in the MOF were post-synthetically modified by N-methylation to yield the cationic Ru-UiO-67-dmbpy (Figure 11.13). Ru-UiO-67-dmbpy was found to absorb visible light up to 780 nm compared to only 420 nm for the pristine UiO-67-bpy. As such, Ru-UiO-67-dmbpy was applied for the photoreduction of toxic Cr(VI) using visible light, where an electron-hole pair is generated and the photoexcited electron is used for the reduction. The MOF was found to reduce 80 % of Cr(VI) with a rate constant of (k_1) of 0.011 min^{-1}, which was faster than that of pristine UiO-67-bpy (0.003 min^{-1}) and nonmethylated Ru-UiO-67-bpy (0.007 min^{-1}). The Cr(VI) photoreduction activity of Ru-UiO-67-dmbpy under visible light irradiation is attributed to the photosensitizer incorporation leading to improved light harvesting, and the cationic nature of the framework that enhances interactions with $Cr_2O_7^{2-}$. Furthermore, when benzyl alcohol was used as a sacrificial electron donor to regenerate the ground state $Ru(bpy)_3$ complex, the MOF achieves 100 % photoreduction within 30 min at a rate of 13.3 mg g^{-1} min^{-1} for $Cr_2O_7^{2-}$ concentrations lower than 100 ppm. Additionally, Ru-UiO-67-dmbpy remains stable for at least 7 successive runs.

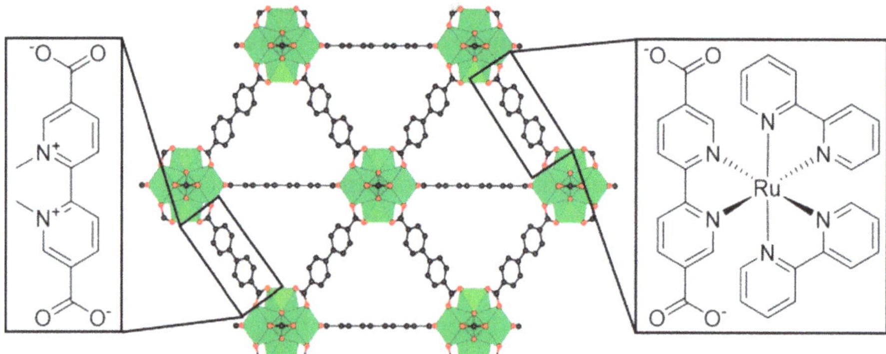

Figure 11.13: Overall structure of Ru-UiO-67-dmpy highlighting the methylated 2,2'-bipyridine-5,5'-dicarboxylate linkers and post-synthetically appended $Ru(bpy)_3$. Atom colors: Zr, green polyhedra; C, black and O, red.

11.2.3 Microfiltration and ultrafiltration

MOFs have been studied in-depth for various applications related to gas and liquid phase separations. As a subset of this work, MOFs have been proposed as components of membranes where the incorporation of MOF particles can introduce channels of known width into a membrane that can be used to control separation processes. For example, to exclude or retain molecules too large to pass, while allowing smaller particles such as water to pass through, potentially leading to a material with both high retention and selectivity, without sacrificing total flux.

In one example, Zhang et al. developed a MOF-membrane composite targeting the separation of the common organic dye methylene blue from water [82]. The group sought to overcome issues associated with MOF crystallite aggregation at high membrane loadings by performing an *in situ* self-assembly of ZIF-8 (Figure 11.12b) and poly(sodium 4-styrenesulfonate) (PSS) onto a polyacrylonitrile (PAN) substrate. By exposing a base layer of hydrolyzed PAN to a solution of Zn(II), the metal was evenly dispersed before being exposed to a mixture of Hmim linker (2-methylimidazole) and PSS. This resulted in the coordination of both Hmim and PSS sulfonate groups to the Zn(II) ions leading to a well-integrated MOF-based membrane where the ZIF-8 crystallite size could be modified by adjusting the concentration of MOF precursors. The membrane was found to reject methylene blue while allowing water to pass through with separation increasing as the number of MOF-membrane layers increases. The optimal number of deposited MOF/PSS layers is two, above which a significant drop in membrane flux is observed. Furthermore, the coordination of Zn(II) to both Hmim and sulfonate groups of PSS causes an increase in hydrophilicity compared to the typically hydrophobic MOF. This is confirmed by a decrease in water contact angle from 91.8 to 58.9°. This increased hydrophilicity allows the two-layer MOF-membrane to exhibit a flux of $265\,L\,m^{-2}\,h^{-1}\,MPa^{-1}$, while also maintaining efficient separation capability and retaining over 98 % of the methylene blue dye even after 350 minutes of continuous filtration. Moreover, the membrane can be reused multiple times simply by rinsing it with water.

The hydrophobic nature of some MOFs results in the aggregation of particles during membrane fabrication, leading to poor dispersion, decreased membrane stability and reduced performance. Sun et al. explored a different approach for incorporating MOFs into membranes by applying a hydrophilic polymer coating around the MOF nanocrystals prior to membrane fabrication [83]. To accomplish this, a zwitterionic polymer, poly(sulfobetaine methacrylate) (PSBMA) was post-synthetically tethered to UiO-66-NH_2. The $-NH_2$ groups on the MOF react with α-bromoisobutyryl bromide (BiBB) to form a UiO-66-BiBB intermediate that is decorated with PSBMA polymer strands via atom transfer radical polymerization to form coated UiO-66-PSBMA. The authors showed that by covering the MOF surface in hydrophilic polymer groups, the nanoparticles exhibited much higher dispersibility, which improves the exchange rate between the solvent (*N*-methyl-2-pyrrolidone) and antisolvent (H_2O) during membrane formation. A polysulfone MOF-membrane composite was prepared using UiO-66-PSBMA, which resulted in a flux of over 600 L m^{-2} h^{-1} compared to 240 L m^{-2} h^{-1} for a polysulfone membrane without any MOF and 294 L m^{-2} h^{-1} for a polysulfone membrane with UiO-66-NH_2. The antifouling capability of the membrane was also improved as tested by studying bovine serum albumin (BSA) filtration, where the hydrophilic nature of the MOF helps to prevent BSA adsorption/deposition within the membrane while still filtering the protein from water.

11.2.4 Oil and water separation

An important type of application based on the adsorption of organic molecules by MOFs is that of oil and water separation. Oil spills can lead to long-term environment issues having a negative impact on ecosystems. In addition, oil spills are very challenging and costly to remediate. Various factors like response time, location and type of oil, among others, can impact the effectiveness of treatments. One possible solution is to develop and study MOFs that have both a high stability against aqueous environments and a high affinity and selectivity for the adsorption of hydrocarbon chemicals or for the adsorption of water [84–98].

One way of achieving water stability and hydrocarbon affinity involves tuning the hydrophobicity of MOFs via the chemical structure of the organic linker. Yang et al. studied the incorporation of perfluorinated organic linkers to synthesize fluorous MOFs (FMOFs) with hydrophobic character [99]. Using Ag(I) as the metal node and 3,5-bis(trifluoromethyl)-1,2,4-triazolate as the linker, FMOF-1 (Figure 11.14a) and FMOF-2 (Figure 11.14b) were synthesized under solvothermal conditions. Water adsorption in FMOF-1, as tested through SCXRD, IR and water adsorption isotherms, demonstrates that no water molecules can be incorporated in the MOF channels even after being soaked in distilled water for several days. On the other hand, the material shows exceptional capacity and high affinity for C_6-C_8 hydrocarbons of oil components such as *n*-hexane (190 kg/m^3), cyclohexane (300 kg/m^3), benzene (290 kg/m^3), toluene (270 kg/m^3) and *p*-xylene (265 kg/m^3). The ability to exclude water while adsorbing large amounts of hydrocarbons is attributed to the hydrophobicity engendered by the fluorinated linkers. Regardless, the use of Ag(I) as the metal component makes the commercial use and applications of these MOFs less attractive. Mukherjee et al. applied the same concept by using a different perfluorinated linker (4,4'-{[3,5-bis(trifluoromethyl)phenyl]azanediyl}dibenzoate) with dinuclear Cu(II)-cluster nodes to form UHMOF-100 [100]. UHMOF-100, as an FMOF, is ultrahydrophobic demonstrating no water uptake and a high affinity for C_6-C_8 hydrocarbons such as benzene, ethyl benzene, toluene and *p*-xylene. In this work, a proof-of-concept experiment was performed using a UHMOF-100 membrane fabricated with polydimethylsiloxane on a polypropylene support, from which clear separation is visually observed in an oil(hexadecane)/water mixture.

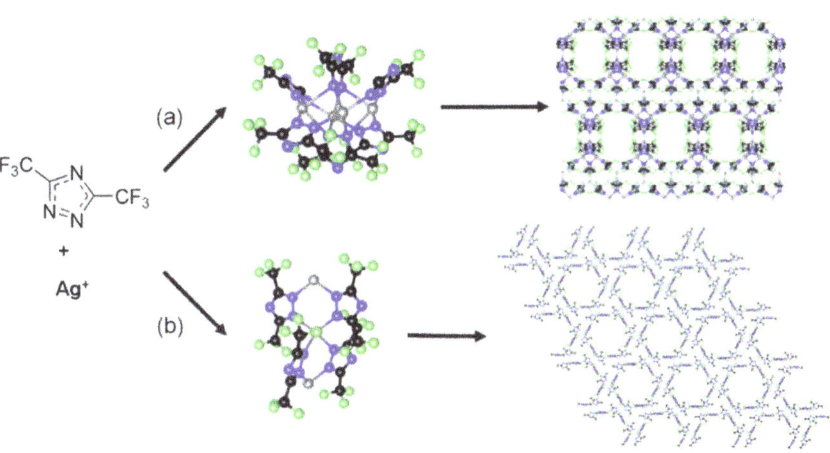

Figure 11.14: Structure of FMOF-1 (a) and FMOF-2 (b), highlighting the tetranuclear and metal-chain Ag(I) nodes, respectively. Atom colors: Ag, grey; C, black; N, blue and F, light green.

In a different approach involving post-synthetic modification, Eom et al. synthesized an extended linker version of Mg-MOF-74 using 4,4'-dioxidobiphenyl-3,3'-dicarboxylate and anchored primary amines with different alkyl chains length (n = 3–16) to the CUSs in the MOF inorganic node [101]. It was observed through contact angle measurements that the material became hydrophobic with an alkyl chain length of $n > 7$ and a proof-of-concept experiment showed visual separation of water from hexadecane using the MOF embedded in gauze as a membrane.

Another strategy for oil and water separation using MOFs is to design and synthesize hydrophilic MOFs. Through defect engineering (i. e., the generation of missing linkers and clusters), Huang et al. tuned Zr-UiO-66 (Figure 11.4b) to generate d-UiO-66, which demonstrated super hydrophilicity in air and superoleophobicity when under water [102]. Zr-UiO-66, comprised of hexanuclear Zr(IV)-cluster nodes bridged by BDC linkers, is ideal to study the effects of missing linker and cluster defects due to its high connectivity (i. e., 12 linkers per node) that can be reduced through the use of modulators without major effects on framework stability. It was observed that UiO-66 analogues with more defects gave rise to lower water contact angles and higher water uptake, proving that the hydrophilicity of UiO-66 can be adjusted by simply modifying the modulator ratio (i. e., acetic acid) during synthesis. In a proof-of-concept experiment, d-UiO-66 was coated onto stainless steel mesh and the material was shown to visibly separate an oil (n-hexane) and water mixture.

11.3 Desalination

Given that 97 % of earth's water is salt water found in oceans, the development of desalination technologies is important for coping with global water scarcity [103]. Desalination can be accomplished by reverse osmosis, thermal treatment, electrochemical methods or adsorption [104]. Among these approaches, adsorption-based desalination is promising from the standpoint of energy usage and cost [105]. Adsorption-based desalination systems typically consist of an evaporator, condenser and adsorbent, where seawater is evaporated and the water vapor is adsorbed by the adsorbent material. The adsorbent is then heated to release water vapor, which is then condensed to give freshwater. As such, many of the materials properties that are relevant for water harvesting, as discussed in Section 11.4 below, are also relevant for desalination applications.

In an example highlighting the performance of MOFs compared to traditional silica gel adsorbents, Elsayed et al. evaluated the use of CPO-27(Ni)/MOF-74(Ni), aluminium fumarate, and MIL-101(Cr) for water desalination [105]. These three MOFs were chosen due to their high capacity for water adsorption of 0.47, 0.53 and 1.47 $g\,g^{-1}$, respectively. A Simulink model was used to explore the adsorption-based desalination performance of the MOFs under different operating conditions and, in nearly all cases, the MOFs outperformed traditional silica gel-based adsorbents with specific daily water production in the range of 2.6–6.3 $m^3\,ton^{-1}\,d^{-1}$.

Another approach for water desalination using MOFs involves developing membranes that exclude hydrated ions and only allow the passage of pure water. Xiao et al. prepared thin-film nanocomposite membranes by incorporating UiO-66 (Figure 11.4b) and UiO-66-NH_2 nanoparticles into polyamide during the polymerization process at different loadings and using different addition methods [106]. The membranes exhibit higher water permeability and salt rejection compared to pristine thin-film composites without MOF including >98 % rejection of Na_2SO_4, 90–98 % rejection of $MgSO_4$, and 20–30 % rejection of NaCl. In addition, the amino functional group of UiO-66-NH_2 was found to reduce agglomeration of the MOF nanoparticles giving rise to a membrane with higher water flux than that containing UiO-66. Using a different synthetic strategy to obtain UiO-66-membranes, Li et al. grew the MOF on a mullite substrate coated with TiO_2, to give a uniform 1 μm thick layer of MOF. The membrane demonstrates 99.9 % rejection of salt, good antifouling properties and a high flux of 37.4 $L\,m^{-2}\,h^{-1}$ [107]. Such high salt rejection is attributed to the near defect-free membrane allowing for size exclusion dictated by the pore window of UiO-66 (0.6 nm) since hydrated ions such as Na^+ and Cl^- have effective diameters of ~0.716 and ~0.664 nm, respectively [108].

11.4 Water adsorption in MOFs

In order to develop a suitable porous material for water adsorption-related applications, including direct harvesting from the atmosphere, five main criteria are considered and targeted: (i) high hydrolytic stability upon adsorption and desorption of water, (ii) large porosity (pore volume) for high water uptake, (iii) steep water uptake at the desired relative humidity (RH) with a step-shaped (Type V) isotherm, (iv) reversible adsorption-desorption to achieve energy efficient cycling and (v) highly reproducible cycling performance requiring only mild regeneration conditions. To develop all these parameters in a single material is a difficult task; however, the chemical and structural tunability of MOFs offers the flexibility required to optimize all these parameters at once. Early transition metal carboxylates [109, 110] and metal azolates [111, 112] that are stable to water are logical choices as building blocks. In general, water stability is dictated by metal-linker bond strength, where high-valent metal ions (hard acids), like Al(III), Cr(III) and Zr(IV), produce stable MOFs in combination with relatively hard carboxylate-based organic linkers. Also, the combination of soft Lewis bases, such as N donor ligands (i. e., 1,2,4-triazole), and soft Lewis acids (i. e., Cu(II), Zn(II), Cd(II)) yield a similar outcome [113–115]. Furthermore, functionalization of organic linkers with hydrophobic moieties that repel water from the inorganic nodes can help to improve water stability [116, 117].

Porosity and pore volume are also important factors that dictate water uptake in MOFs. To avoid capillary condensation that can lead to irreversible sorption behavior, the pore diameter of a MOF adsorbent must be below the critical diameter (D_C) of the working fluid. Below the D_C, adsorption involves reversible continuous pore filling whereas above the D_C, hysteretic capillary condensation occurs [118]. For water, D_C is 20.76 Å at 298 K meaning that an adsorbent with a pore diameter of ~20 Å is ideal for maximizing the potential for reversible water adsorption. For uptake at low water concentrations (<30 % RH), pore hydrophilicity must also be sufficient to allow for water nucleation and pore filling to occur, giving rise to steep uptake and the potential for high working capacity. For most applications, realistic day/night temperature and RH fluctuation conditions need to be considered, such as in the Atacama [119], the Sonoran [120] and the Arabian [121] deserts, for example, with daytime values of 318 K and 5 % RH to nighttime values of 298 K and 35 % RH.

Theoretical simulations have been performed to gain insight into the mechanism of water adsorption in MOFs and how different structural and chemical features affect this process. Models have been developed and used to accurately simulate water-framework and water-water interactions affording theoretical water sorption isotherms that match well with experimental data [122, 123]. These computational studies have provided information regarding: (i) the relationship between increasing hydrophobicity of a MOF and water uptake occurring at higher relative pressures [123], (ii) the cause of hysteresis in water adsorption isotherms for hydrophobic materials being related to the stepwise filling of neighboring MOF cavities and the existence of metastable states [124], (iii) hydrogen-bonding interactions between water molecules and with extra framework counterions being a driving force for water uptake [125] and (iv) factors affecting thermal conductivity of a MOF, a property required for coping with the energy-intensive adsorption and desorption cycles [126].

Atmospheric water harvesting

In an early example of MOFs studied for water adsorption applications, the linkers imidazole-4,5-dicarboxylate (IDC^{3-}) and piperazine (prz) were used to coordinate Zn(II) ions to construct a framework with 1D channels (Figure 11.15). The flexible MOF, $[Zn_6(IDC)_4(OH)_2(Hprz)_2]_n$, selectively adsorbs water over organic solvents (i. e., ethanol, acetone, tetrahydrofuran, benzene, toluene and xylene) and is reusable over multiple adsorption–desorption cycles. The hydrophilicity of the MOF is attributed to the composition of the pores that are built mainly from μ_2-OH^-, $Hprz^+$ and μ_3-IDC^{3-} groups [127].

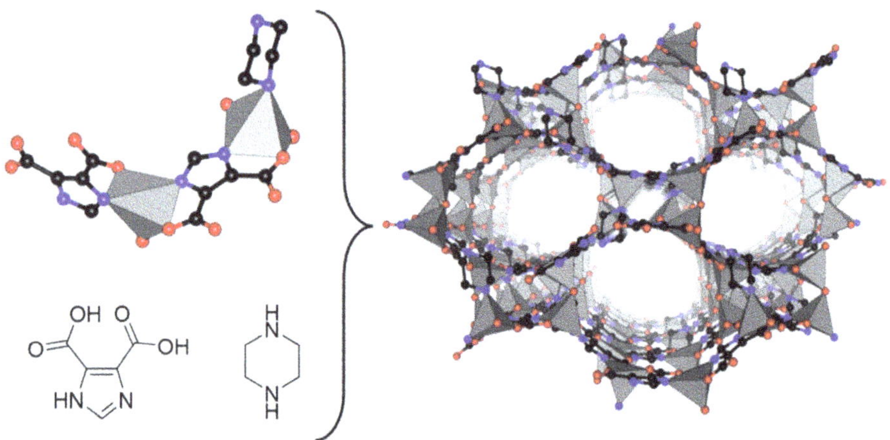

Figure 11.15: Structure of $[Zn_6(IDC)_4(OH)_2(Hprz)_2]_n$ highlighting the Zn(II)-nodes, imidazole-4,5-dicarboxylate (IDC^{3-}) linkers, and piperazine (prz) capping ligands. Atom colors: Zn, grey polyhedra; C, black; N, blue and O, red.

Later, MIL-101 (Figure 11.1), MIL-100(Fe) (Figure 11.11), HKUST-1 (Figure 11.12a), ZIF-8 (Figure 11.12b) and DUT-4 were studied using water and N_2 sorption analysis to compare differences between specific surface area, pore size and pore volume obtained using each adsorbate and to gain insight into the hydrophobicity and moisture stability of the materials [109]. HKUST-1 and DUT-4 were found to be unstable in water when submerged for 24 hours at 323 K, whereas MIL-101, MIL-100(Fe) and ZIF-8 were stable. Although HKUST-1 has long-term water stability issues, the MOF shows the highest affinity for water of the series due to the presence of hydrophilic pores of 9 Å in diameter and the coordination of water molecules to the open Cu(II) sites on the MOF node. ZIF-8, on the other hand, is more stable in water but shows hydrophobic behavior with the onset of water adsorption occurring at 80 % RH rendering the material unsuitable for water adsorption applications. MIL-101 and MIL-100(Fe) are both hydrolytically stable and good candidates for water adsorption applications. MIL-101 is comprised of Cr_3-clusters interconnected by terephthalate linkers, assembling mesoporous cages of 29 and 34 Å in diameter. The onset of water adsorption in MIL-101 occurs at 40 % RH, with only one step in the isotherm suggesting that both pores have similar hydrophilicity. MIL-100(Fe) also has two mesoporous cages, but of 25 and 29 Å in diameter, and shows two steps in the water sorption isotherm at 30 % and 40 % RH. This two-step adsorption process indicates the consecutive filling of the 25 and 29 Å mesoporous cages. The isosteric heats of adsorption (q_{st}) for water in MIL-100(Fe), MIL-101 and ZIF-8 are 48.83, 45.13 and 44.68 kJ mol^{-1}, respectively. These energies are close to the molar enthalpy of evaporation for water (40.69 kJ mol^{-1}) corresponding to the energy required to form hydrogen bonds, which is consistent with hydrogen bonded water clusters forming within the MOF pores.

In 2010, Akiyama et al. reported three MIL-100(Cr) analogues (Figure 11.11) with different halide ligands capping the Cr_3-cluster node [128]. MIL-100 was synthesized using hydrofluoric acid, hydrochloric acid and sulfuric acid to give fluoride, chloride and sulfate capped analogues with the chemical formula $Cr_3XO[C_6H_3(CO_2)_3]_2$ (X = F, Cl, SO_4). The MIL-100 analogues show a water uptake capacity over 0.6 g g^{-1} at less than 60 % of RH. Furthermore, the fluoride containing MIL-100 maintained its capacity after two thousand cycles. The identity of the capping ligand was found to play an important role in controlling the RH at which the adsorption process is triggered (i. e., the onset of steep water adsorption). For example, the sulfate analogue demonstrated steep uptake at lower RH values than the fluoride and chloride analogues, which is attributed to the hydration energies (ΔG_h) of Cl^-, F^- and SO_4^{2-}, being −347, −472 and −1090 kJ mol^{-1}, respectively [129]. This trend was also observed in the Q_{st} with the largest value for the sulfate analogue (47.9 kJ mol^{-1}) and the smallest for the chloride one (47.6 kJ mol^{-1}). In a related study, Akiyama et al. performed a systematic study on the effect of different linker functional groups on the water sorption behavior in MIL-101 [110]. Water adsorption in MIL-101, MIL-101-NO_2, MIL-101-NH_2 and MIL-101-SO_3H (Figure 11.1) was studied at 298 K, showing capacities between 0.8–1.2 g g^{-1}. MIL-101 demonstrates the highest water uptake of 1.2 g g^{-1} at <60 % RH, consistent with the results reported previously by Küsgens et al. and discussed above [109]. The water adsorption isotherm of MIL-101 shows one step corresponding to chemisorption on the open metal sites (~40 % RH) and adsorption of water molecules in the smaller cages (29 Å) and then a second, more subtle step due to adsorption in the larger cages of 34 Å. In MIL-101-NH_2 and MIL-101-SO_3H, steep water uptake is found at lower RH values compared to MIL-101 due to the increased hydrophilicity of the materials, whereas MIL-101-NO_2 showed almost identical adsorption performance as MIL-101 because of the lower hydrophilicity of the NO_2 group.

In an example where water adsorption leads to reversible and irreversible phase changes, O'Nolan et al. synthesized a 2D square grid (**sql**) composed of ditopic N-containing linkers connected to Cu(II) or Ni(II) ion nodes. The 2D sheets are then pillared by hexafluorosilicate (SIFSIX) anions giving a 3D **pcu** topology [130]. The MOFs are named SIFSIX-1-Cu $[Cu(SIFSIX)(4,4'\text{-bipyridyl})_2]_n$ (Figure 11.16a), SIFSIX-2-Cu-i

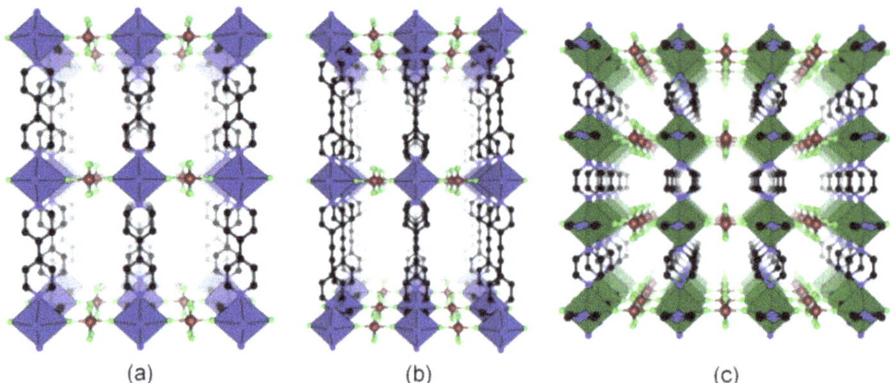

Figure 11.16: Structure of SIFSIX-1-Cu (a) SIFSIX-2-Cu-i, (b) SIFSIX-3-Ni and (c). Interpenetration is omitted for clarity. Atom colors: Zr, dark blue polyhedra; Ni, green polyhedra; Si, brown; C, black; N, blue and F, light green.

[Cu(SIFSIX)(1,2-bis(4-pyridyl)acetylene)$_2$]$_n$ (Figure 11.16b), SIFSIX-3-Ni [Ni(SIFSIX)-(pyrazine)$_2$]$_n$ (Figure 11.16c) and SIFSIX-14-Cu-i [Cu(SIFSIX)(1,2-bis(4-pyridy)diazene)$_2$]$_n$ where i means the structure is interpenetrated. Dynamic water vapor sorption experiments revealed that SIFSIX-1-Cu, SIFSIX-3-Ni and SIFSIX-14-Cu-i show water uptake at low RH followed by a negative water adsorption step in the range of 40–50 % RH attributed to a phase transformation from the porous 3D **pcu** net to the nonporous 2D **sql** and **sql-c*** net. The phase transformation is irreversible for SIFSIX-1-Cu and SIFSIX-14-Cu-i, but SIFSIX-3-Ni can be regenerated upon heating to 353 K under vacuum. Furthermore, SIFSIX-3-Ni can be recycled 10 times without impact on its water adsorption behavior.

Taking advantage of the robust hydrolytic stability of Zr_6-MOFs, Furukawa et al. reported the water adsorption properties of 20 MOFs, including 10 Zr(IV)-based materials. UiO-66, DUT-67, PIZOF-2, MOF-801-P (powder), MOF-801-SC (single crystals), –802, –805, –806, –808, –812 and –841 are all comprised of $Zr_6O_4(OH)_4(-CO_2)_n{}^{(12-n)+}$ secondary building units (n = 6, 8, 10 or 12) with di-, tri- or tetratopic carboxylate linkers [121]. Water adsorption measurements reveal that MOF-801-P (Figure 11.17a) has the highest water uptake at low RH (10 %) whereas MOF-841 (Figure 11.17b) outperforms the other Zr-MOFs at 30 % RH. Both materials demonstrate hydrolytic stability, high and consistent capacity after five adsorption/desorption cycles and can be regenerated at room temperature. MOF-801, reported previously by Wißmann et al. [131] exhibits an **fcu** topology with fumarate linkers featuring two crystallographically independent tetrahedral cavities with sizes of 5.6 and 4.8 Å in diameter and an octahedral cavity with a diameter of 7.4 Å. MOF-801-P shows steep water uptake in the range of 5 to 10 % RH with a maximum water uptake of 450 $cm^3\,g^{-1}$ (0.36 $g\,g^{-1}$) at 90 % RH. MOF-841 has an **flu** topology and is obtained by the combination of an 8-connected Zr_6-cluster node and tetrahedral (MTB = 4,4',4'',4'''-methanetetrayltetrabenzoate) link-

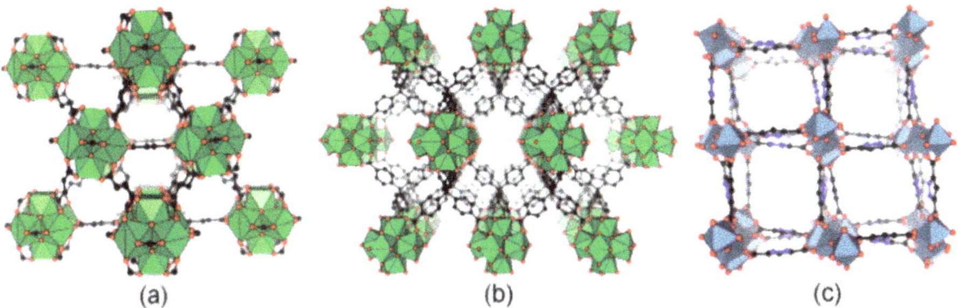

Figure 11.17: Structure of MOF-801 (a) MOF-841, (b) MOF-303 (c). Atom colors: Zr, green polyhedra; Al, blue polyhedra; C, black; N, blue; and O, red.

ers featuring a cage of approximately 11.6 Å in diameter. Interestingly, these MOFs showed higher water uptake than those comprised of hydroxyl functionalized linkers suggesting that increasing hydrophilicity in the pores may not be as critical for water adsorption as the overall network structure, pore size and pore shape. SCXRD and powder neutron diffraction studies reveal the position of the water adsorption sites in MOF-801 highlighting the role of μ_3-OH ligands on the Zr_6-nodes at low RH and the importance of water-water hydrogen bonding interactions for pore filling at higher RH. Neutron diffraction performed on MOF-801-P exposed to D_2O at 30 % RH confirms similar adsorption behavior as that found by SCXRD. A recent study by Katz et al. also highlights the importance of hydrophilic Zr_6-clusters for water uptake showing adsorption of one water molecule per Zr center in a series of Zr_6-MOFs with **fcu** topology [132]. Furthermore, they found that extending the length of the ditopic linker increases pore hydrophobicity, a result that is also consistent with MOF-801 having high water capacity owing to the short fumarate linkers. A few years after the initial study, Kim et al. incorporated MOF-801 in a device for the adsorption of atmospheric water at ambient conditions using only sunlight for water desorption and regeneration [133]. The system allows for the collection of 2.8 L of water per kilogram of MOF per day with RH ranging from 65 % to 20 % with the latter being a value consistent with dry desert conditions. Later, MOF-801 was integrated in an improved device to achieve complete water saturation during the night (20–40 % RH) and to concentrate the sunlight (i. e., heat) reaching the device during the day (10–20 % RH). The latter leads to an ~5× increase in thermal efficiency and enables complete regeneration of MOF-801 [134]. The second- generation device uses an air-cooled condenser to increase the effective RH experienced by the MOF at night allowing for MOF-801 to take up high volumes of water below ~20 % RH and giving rise to the collection of 0.25 L of water per kg of MOF per day under these extremely dry conditions. Later in the same year, a prototype device was tested in the Arizona desert using 1.2 kg of MOF-801 producing 0.1 L of water per kg of MOF per day using an air-cooled condenser and ambient sunlight as a source of energy [135]. Additionally, a newly designed aluminium-based MOF, MOF-303, was found to collect over 0.2 L of water per kg of MOF per day using the same device. MOF-303 is comprised of Al-chain nodes bridged by 1*H*-pyrazole-3,5-dicarboxylate (HPDC) linkers giving rise to the **xhh** topology and exhibiting hydrophilic 1D pores of 6 Å in size and with pore volume of 0.54 $cm^3 g^{-1}$ (Figure 11.17c). MOF-303 demonstrates a Type IV water sorption isotherm with an adsorption onset at 15 % RH reaching a saturation of ~0.25 $g g^{-1}$ at 30 % RH. In a related study, Hanikel et al. demonstrated that MOF-303 outperforms other commercial sorbents (i. e., zeolite 13X, SAPO-34, Basolite A520 [Al-fumarate]), owing in part to the rapid (~3 minute) adsorption–desorption cycle observed for this MOF under mild temperature swings (303 K adsorption and 358 K desorption) [136]. When integrated into a next generation solar powered device similar to those described previously but designed for multiple adsorption–desorption cycles per day, MOF-303 generates 1.3 L of water per kg of MOF per day in an indoor arid environment (32 % RH, 300 K, 9 cycles) and 0.7 L per kg of MOF per day in the Mojave Desert (10 % RH, 300 K) showing an improvement from previous prototype devices. This study highlights that adsorption–desorption kinetics and the ability to perform fast cycles with minimal energy input has similar importance to high uptake at low RH.

In an example highlighting the importance of network structure, pore connectivity and pore volume, Towsif et al. reported Cr-**soc**-MOF-1, which is hydrolytically stable and capable of capturing twice its weight in adsorbed water [137]. Cr-**soc**-MOF-1 was obtained through the single-crystal to single-crystal transmetalation of Fe-**soc**-MOF-1 resulting in near-complete exchange (~98 %) with chromium. Cr-**soc**-MOF-1 is comprised of rectangular tetratopic linkers ($TCPT^{4-}$ = 3,3",5,5"-tetrakis(4-carboxyphenyl)-*p*-terphenyl) bridging trinuclear Cr(III)-cluster nodes giving the **soc** topology with 1D cubic-shaped channels and cages of ~17 Å in size (Figure 11.18a). Cr-**soc**-MOF-1 features a surface area of 4550 $m^2 g^{-1}$ with a Type V water sorption isotherm showing water uptake of 1.95 $g g^{-1}$ at 75 % RH and 298 K. Reduction of the RH to 25 % at 298 K is enough to fully desorb the water from Cr-s**oc**-MOF-1, maintaining its performance over more than 100 adsorption–desorption cycles. The water adsorption performance of Cr-soc-MOF-1 is attributed to its hydrolytic stability, high micropore volume (2.1 $cm^3 g^{-1}$), and the presence of cubic channels/cages that allow for water nucleation at pore vertices and edges followed by the formation of large water clusters.

Another approach for controlling water uptake behavior involves tuning the hydrophilic and hydrophobic properties of MOF channels. Wade et al. reported novel hydrophobic dipyrazole ligands containing naphthalenediimide cores (H_2NDI) functionalized as H_2NDI-H, H_2NDI-NHEt and H_2NDI-SEt, which were used to obtain a series of isoreticular MOFs named Zn(NDI-X) [138]. Zn(NDI-X) features infinite chains of Zn(II) ions bridged by NDI-pyrazolate linkers giving ~16 Å channels and BET surface areas of 1460, 1240 and 890 $m^2 g^{-1}$ for Zn(NDI-H), Zn(NDI-NHEt) and Zn(NDI-SEt), respectively. These materials show Type V water adsorption isotherms with the onset of water adsorption occurring in the 40–50 % RH range, which are consistent with the hydrophobic pore surface governed by the linkers [139]. Post-synthetic oxidation of Zn(NDI-SEt) using dimethyldioxorane was performed to generate ethyl sulfoxide and

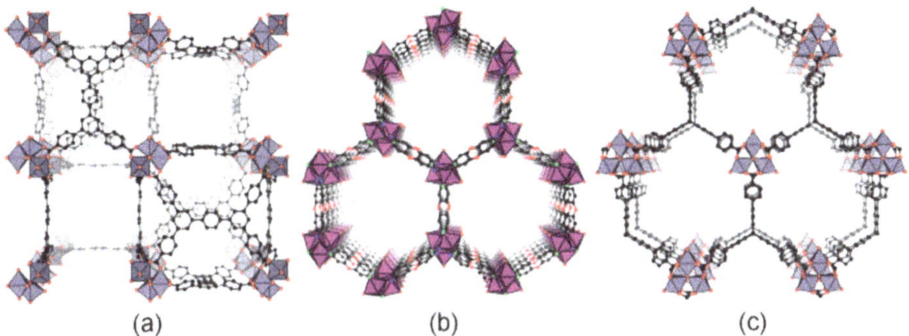

Figure 11.18: Structure of Cr-**soc**-MOF-1 (a) Mn_2Cl_2(BTDD), (b) NU-1500-Cr and (c). Atom colors: Cr, purple polyhedra; Mn, purple; C, black; N, blue; and O, red.

ethyl sulfone analogues producing more hydrophilic channels. Indeed, Zn(NDI-SOEt) and Zn(NDI-SO_2Et) analogues demonstrated water adsorption at an earlier onset in the 20–30 % and 30–40 % RH range, respectively. This is consistent with sulfoxides exhibiting greater hydrophilic character than sulfones [140]. In another example, Rieth et al., reported the water adsorption capabilities of the mesoporous MOF M_2Cl_2(BTDD) (M = Mn, Co, Ni and BTDD = bis(1*H*-1,2,3-triazolo[4,5-*b*],[4',5'-*i*])dibenzo[1,4]dioxin)) that features metal chain nodes and channels of 22 Å matching the D_C for water adsorption (Figure 11.18b) [141]. All MOFs demonstrate reversible Type IV water sorption isotherms with the onset of steep water uptake occurring at approximately 28 % RH. Mn_2Cl_2(BTDD) collapses upon water adsorption around 28 % RH, but Co_2Cl_2(BTDD) and Ni_2Cl_2(BTDD) remain crystalline with BET surface areas after water adsorption of 1910 and 1760 $m^2\,g^{-1}$, respectively. Co_2Cl_2(BTDD) shows superior water uptake compared to the Ni(II) or Mn(II) analogues reaching 0.9 $g\,g^{-1}$ water uptake below 30 % RH. Furthermore, Co_2Cl_2(BTDD) reveals an initial deliverable capacity of 0.85 $g\,g^{-1}$ under simulated desert daytime–nighttime conditions and declining by only 0.05 $g\,g^{-1}$ over 6 consecutive cycles. Following this study, Rieth et al. explored the mechanism of water uptake in Co_2Cl_2BTDD [141] as a function of RH by diffuse reflectance infrared Fourier transform spectroscopy and many-body molecular dynamics simulations [142]. The authors demonstrate that water binds initially to the hydrophilic open metal sites at low RH and subsequently forms 1D chains of hydrogen bonded water molecules that bridge neighbouring Co(II) sites. With increasing RH (up to 30 %), the 1D chains of water act as nucleation sites for pore filling leading to the formation of concentric cylindrical water layers within the 22 Å pores of the MOF. These mechanistic insights can help with the design of novel porous materials for water harvesting, suggesting that materials with pores on the order of the D_C of water and lined with open metal sites are an asset. Later, Rieth et al. demonstrated the effect of tuning the bridging halide ligand in Ni_2Cl_2BTDD on water adsorption [143]. In this MOF, the Cl^- can be easily exchanged by F^-, or Br^- post-synthetically. The authors hypothesized that changing the bridging halide should modulate the hydrogen bonding interactions between water and the MOF chain node, and thus, change the RH at which the onset of steep water uptake occurs. The ultimate effect, however, was a decrease in unit cell parameters in the order of F > Cl > Br, which ultimately leads to more pore confinement and the onset of steep water uptake at lower RH for Ni_2Br_2BTDD. Pore filling occurs in Ni_2Br_2BTDD at 24 % RH compared to 32 % RH for Ni_2F_2BTDD and Ni_2Cl_2BTDD, with a high-water uptake capacity for the Br-analogue below 25 % RH of 0.64 g g^{-1} maintaining its porosity and crystallinity upon water cycling for at least 400 cycles.

In 2019, Chen et al. reported the use of a rigid hexatopic triptycene linker (H_6PET) to obtain a series of 6-c noncatenated M(III)-**acs**-MOFs, NU-1500-M (M = Fe(III), Cr(III) or Sc(III)). NU-1500-M is comprised of trinuclear metal clusters featuring hexagonal channels of 14 Å (Figure 11.18c) and BET surface areas ranging from 3580 to 4280 m^2 g^{-1} [144]. Transmetalation was used to obtain NU-1500-Cr from NU-1500-Fe giving rise to a MOF with impressive hydrolytic stability. In fact, NU-1500-Cr can be activated directly from water to give a BET surface area of 3580 m^2 g^{-1}, which is comparable to that obtained when the MOF is activated from acetone. NU-1500-Cr features high water uptake of 1.09 g g^{-1} at 90 % RH and 298 K and recyclability over 20 cycles. In addition, the uptake of 1.02 g g^{-1} of water in NU-1500-Cr at 60 % RH outperforms that of Co_2Cl_2(BTDD) [141]. Cr-**soc**-MOF-1 [137], Ni-MOF-74-TPP [145] and MCM-41 [121]. Later, the research group evaluated the hydrolytic stability and water adsorption performance of three Zr-based MOFs NU-901 (**scu**) (Figure 11.19a), NU-1000 (**csq**) (Figure 11.19b), and the novel material NU-950 (**sqc**) (Figure 11.19c), highlighting the relationship between topology and hydrolytic stability [146]. Furthermore, a method for enhancing the water stability of Zr-based MOFs was proposed, which involves using hydrophobic trifluoroacetic acid (TFA) as a capping ligand on the hexanuclear Zr(IV)-cluster node. All 3 MOFs were comprised of 8-connected Zr_6-clusters bridged by tetratopic 1,3,6,8-tetrakis(*p*-benzoic acid)pyrene (H_4TBAPy) linkers, but with different pore size and shape (Figure 11.19). NU-1000 and NU-950, with and without TFA capping ligands, are more stable than NU-901 owing to the nature of their topologies. NU-1000 with **csq** topology has interconnected triangular and hexagonal pores, which has been shown to increase water stability, whereas the **sqc** topology has alternating linker orientations that help to impart stability [147, 148]. On the other hand, NU-901 only has one type of pore and linker orientation making it more flexible and less water stable. NU-1000-TFA demonstrates reversible water adsorption with an uptake of 1.32 g g^{-1} over two cycles, superior to the parent NU-1000, which shows a more than 65 % decrease in water uptake during the second adsorption cycle (1.42 to 0.50 g g^{-1} in cycles 1 and 2, respectively). In addition to the importance of topology, this work highlights that the presence of TFA capping ligands can prevent direct hydrolysis of the Zr_6 cluster.

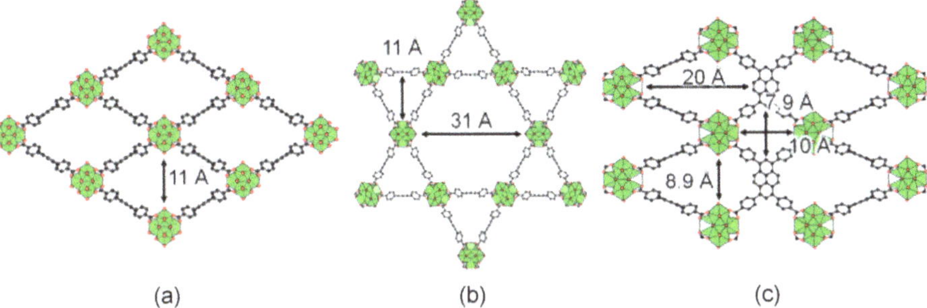

Figure 11.19: Structure of NU-901 (a) NU-1000 (b) and NU-950 (c) highlighting pore aperture dimensions. Atom colors: Zr, green polyhedra; C, black; and O, red.

11.5 Conclusion

MOFs have shown great promise for the future of clean water generation technologies, from adsorption and degradation of water pollutants to desalination of seawater and atmospheric water harvesting. To address global water scarcity issues going forward, several solutions will need to be pursued and applied. The porosity and high degree of structural and functional tunability of MOFs allows for these materials to be designed and studied for the wide-ranging solutions required. By continuing to probe and understand the ideal structural properties and interactions that lead to analyte and water adsorption or the mechanism of catalysis (i. e., increasing CUSs or linker-based generation of ROS), or the best membrane performance, MOF design and discovery will continue to improve and move forward toward use in these water-related applications. When it comes to applicability, the exploration of ferromagnetic MOFs or composites of MOFs with magnetic nanoparticles could help with ease of extraction of spent MOF from large volumes of water once a pollutant is adsorbed or catalytically degraded. In addition, the removal of microplastics, biotoxins and chemical warfare agents from wastewater and groundwater are topics that require further research due to their relevance in public health and national defense. Moreover, there is a need for more studies on pollutant degradation using MOFs in water with different levels of hardness, addressing adsorbent selectivity or catalyst poisoning and extensive characterization of degradation products to rule out the formation of more toxic compounds. On the other hand, the selectivity for water over environmental pollutants, in atmospheric water capture systems, needs to be addressed to expand this technology for large scale clean water generation. Finally, as more MOF-based products are commercialized [149, 150], the cost of MOF building blocks and large-scale production will decrease to enhance the library of materials that can be deployed for real-world applications.

Bibliography

[1] Charting Our Water Future. 2009 (accessed Jan 30, 2022, at https://www.mckinsey.com/business-functions/sustainability/our-insights/charting-our-water-future).
[2] Mekonnen MM, Hoekstra AY. Four billion people facing severe water scarcity. Sci Adv. 2016;2:1–6.
[3] Boretti A, Rosa L. Reassessing the projections of the world water development report. npj Clean Water. 2019;2:1–6.
[4] UN World Water Development Report 2021 (accessed Jan 30, 2022, at https://www.unwater.org/publications/un-world-water-development-report-2021/).
[5] How Much Water is There on Earth? | U. S. Geological Survey (accessed Mar 14, 2022, at https://www.usgs.gov/special-topics/water-science-school/science/how-much-water-there-Earth).
[6] THE 17 GOALS | Sustainable Development (accessed Jan 30, 2022, at https://sdgs.un.org/goals).
[7] Richardson SD, Kimura SY. Water analysis: emerging contaminants and current issues. Anal Chem. 2016;88:546–82.
[8] Kümmerer K. Antibiotics in the aquatic environment – a review – Part I. Chemosphere. 2009;75:417–34.
[9] Sui Q, Cao X, Lu S, Zhao W, Qiu Z, Yu G. Occurrence, sources and fate of pharmaceuticals and personal care products in the groundwater: a review. Emerg Contam. 2015;1:14–24.
[10] Blair BD, Crago JP, Hedman CJ, Klaper RD. Pharmaceuticals and personal care products found in the Great Lakes above concentrations of environmental concern. Chemosphere. 2013;93:2116–23.
[11] Lange FT, Scheurer M, Brauch H-J. Artificial sweeteners—a recently recognized class of emerging environmental contaminants: a review. Anal Bioanal Chem. 2012;403:2503–18.
[12] de Souza RM, Seibert D, Quesada HB, de Jesus Bassetti F, Fagundes-Klen MR, Bergamasco R. Occurrence, impacts and general aspects of pesticides in surface water: a review. Process Saf Environ Prot. 2020;135:22–37.
[13] Jung C, Son A, Her N, Zoh K-D, Cho J, Yoon Y. Removal of endocrine disrupting compounds, pharmaceuticals, and personal care products in water using carbon nanotubes: a review. J Ind Eng Chem. 2015;27:1–11.
[14] Carmalin SA, Lima EC, Allaudeen N, Rajan S. Application of graphene based materials for adsorption of pharmaceutical traces from water and wastewater- a review. Desalin Water Treat. 2016;57:27573–86.
[15] Xiang Y, Xu Z, Wei Y, Zhou Y, Yang X, Yang Y, Yang J, Zhang J, Luo L, Zhou Z. Carbon-based materials as adsorbent for antibiotics removal: mechanisms and influencing factors. J Environ Manag. 2019;237:128–38.
[16] Chen Y, Wang B, Wang X, Xie L-H, Li J, Xie Y, Li J-R. A copper(II)-paddlewheel metal–organic framework with exceptional hydrolytic stability and selective adsorption and detection ability of aniline in water. ACS Appl Mater Interfaces. 2017;9:27027–35.
[17] Negro C, Pérez-Cejuela H M, Simó-Alfonso EF, Herrero-Martínez JM, Bruno R, Armentano D, Ferrando-Soria J, Pardo E. H. Efficient removal of neonicotinoid insecticides by thioether-based (multivariate) metal–organic frameworks. ACS Appl Mater Interfaces. 2021;13:28424–32.
[18] Peng J, Li Y, Sun X, Huang C, Jin J, Wang J, Chen J. Controlled manipulation of metal–organic framework layers to nanometer precision inside large mesochannels of ordered mesoporous silica for enhanced removal of bisphenol a from water. ACS Appl Mater Interfaces. 2019;11:4328–37.

[19] Ren Z, Luo J, Wan Y. Enzyme-like metal–organic frameworks in polymeric membranes for efficient removal of aflatoxin B1. ACS Appl Mater Interfaces. 2019;11:30542–50.
[20] Vellingiri K, Deng Y-X, Kim K-H, Jiang J-J, Kim T, Shang J, Ahn W-S, Kukkar D, Boukhvalov DW. Amine-functionalized metal–organic frameworks and covalent organic polymers as potential sorbents for removal of formaldehyde in aqueous phase: experimental versus theoretical study. ACS Appl Mater Interfaces. 2019;11:1426–39.
[21] Zhao C, Du Y, Zhang J, Mi Y, Su H, Fei T, Li S, Pang S. Highly efficient separation of anionic organic pollutants from water via construction of functional cationic metal–organic frameworks and mechanistic study. ACS Appl Mater Interfaces. 2020;12:22835–44.
[22] Xu J, Li K, Zhang S, Chen Y, Ning L. Removal of endocrine-disrupting chemicals from environment using a robust platform based on metal–organic framework nanoparticles. ACS Appl Nano Mater. 2020;3:3646–51.
[23] Yuan N, Gong X-R, Han B-H. Hydrophobic fluorous metal–organic framework nanoadsorbent for removal of hazardous wastes from water. ACS Appl Nano Mater. 2021;4:1576–85.
[24] Tabatabaii M, Khajeh M, Oveisi AR, Erkartal M, Sen U. Poly(lauryl methacrylate)-grafted amino-functionalized zirconium-terephthalate metal–organic framework: efficient adsorbent for extraction of polycyclic aromatic hydrocarbons from water samples. ACS Omega. 2020;5:12202–9.
[25] Souza BE, Möslein AF, Titov K, Taylor JD, Rudić S, Tan J-C. Green reconstruction of MIL-100 (Fe) in water for high crystallinity and enhanced guest encapsulation. ACS Sustain Chem Eng. 2020;8:8247–55.
[26] Borjigin T, Sun F, Zhang J, Cai K, Ren H, Zhu G. A microporous metal–organic framework with high stability for GC separation of alcohols from water. Chem Commun. 2012;48:7613–5.
[27] Peng Y, Yao R, Yang W. A poly(amidoamine) nanoparticle cross-linked two-dimensional metal–organic framework nanosheet membrane for water purification. Chem Commun. 2019;55:3935–8.
[28] Haghighat GA, Sadeghi S, Saghi MH, Ghadiri SK, Anastopoulos I, Giannakoudakis DA, Colmenares JC, Shams M. Zeolitic imidazolate frameworks (ZIFs) of various morphologies against eriochrome black-T (EBT): optimizing the key physicochemical features by process modeling. Colloids Surf A, Physicochem Eng Asp. 2020;606:125391.
[29] Wu M-K, Yi F-Y, Fang Y, Xiao X-W, Wang S-C, Pan L-Q, Zhu S-R, Tao K, Han L. An ultrastable metal–organic framework with open coordinated sites realizing selective separation toward cationic dyes in aqueous solution. Cryst Growth Des. 2017;17:5458–64.
[30] Ye X, Liu D. Metal–organic framework UiO-68 and its derivatives with sufficiently good properties and performance show promising prospects in potential industrial applications. Cryst Growth Des. 2021;21:4780–804.
[31] Chen Z, Wang X, Noh H, Ayoub G, Peterson GW, Buru CT, Islamoglu T, Farha OK. Scalable, room temperature, and water-based synthesis of functionalized zirconium-based metal–organic frameworks for toxic chemical removal. CrystEngComm. 2019;21:2409–15.
[32] Siddique A. IM, Rawat P, Singh RN, Shahid M, Trivedi S, Gautam A, Zeeshan M. A new Zn(II) MOF assembled from metal–organic cubes (MOCs) as a highly efficient adsorbent for cationic dyes. CrystEngComm. 2021;23:2316–25.
[33] Li J-J, Wang C-C, Fu H, Cui J-R, Xu P, Guo J, Li J-R. High-performance adsorption and separation of anionic dyes in water using a chemically stable graphene-like metal–organic framework. Dalton Trans. 2017;46:10197–201.
[34] Zhao S, Li S, Zhao Z, Su Y, Long Y, Zheng Z, Cui D, Liu Y, Wang C, Zhang X, Zhang Z. Microwave-assisted hydrothermal assembly of 2D copper-porphyrin metal–organic frameworks for the removal of dyes and antibiotics from water. Environ Sci Pollut Res. 2020;27:39186–97.

[35] Akpinar I, Yazaydin AO. Rapid and efficient removal of carbamazepine from water by UiO-67. Ind Eng Chem Res. 2017;56:15122–30.

[36] Ma H-F, Liu Q-Y, Wang Y-L, Yin S-G. A water-stable anionic metal–organic framework constructed from columnar zinc-adeninate units for highly selective light hydrocarbon separation and efficient separation of organic dyes. Inorg Chem. 2017;56:2919–25.

[37] Lv J, Chen Q, Liu J-H, Yang H-S, Wang P, Yu J, Xie Y, Wu Y-F, Li J-R. Effective removal of clenbuterol and ractopamine from water with a stable Al(III)-based metal–organic framework. Inorg Chem. 2021;60:1814–22.

[38] Liu C, Sun Z-C, Pei W-Y, Yang J, Xu H-L, Zhang J-P, Ma J-F. A porous metal–organic framework as an electrochemical sensing platform for highly selective adsorption and detection of bisphenols. Inorg Chem. 2021;60:12049–58.

[39] Saleh HAM, Mantasha I, Qasem KMA, Shahid M, Akhtar MN, AlDamen MA, Ahmad M. A two dimensional Co(II) metal–organic framework with bey topology for excellent dye adsorption and separation: exploring kinetics and mechanism of adsorption. Inorg Chim Acta. 2020;512:119900.

[40] Imanipoor J, Mohammadi M, Dinari M, Ehsani MR. Adsorption and desorption of amoxicillin antibiotic from water matrices using an effective and recyclable MIL-53(Al) metal–organic framework adsorbent. J Chem Eng Data. 2021;66:389–403.

[41] DeChellis DM, Ngule CM, Genna DT. Removal of hydrocarbon contaminants from water with perfluorocarboxylated UiO-6X derivatives. J Mater Chem A. 2020;8:5848–52.

[42] Huang Z, Lee HK. Micro-solid-phase extraction of organochlorine pesticides using porous metal–organic framework MIL-101 as sorbent. J Chromatogr A. 2015;1401:9–16.

[43] Alamgir TK, Wang B, Liu J-H, Ullah R, Feng F, Yu J, Chen S, Li J-R. Effective adsorption of metronidazole antibiotic from water with a stable Zr(IV)-MOFs: insights from DFT, kinetics and thermodynamics studies. J Environ Chem Eng. 2020;8:103642.

[44] Karami A, Sabouni R, Ghommem M. Experimental investigation of competitive co-adsorption of naproxen and diclofenac from water by an aluminum-based metal–organic framework. J Mol Liq. 2020;305:112808.

[45] Guo X, Kang C, Huang H, Chang Y, Zhong C. Exploration of functional MOFs for efficient removal of fluoroquinolone antibiotics from water. Microporous Mesoporous Mater. 2019;286:84–91.

[46] Zango ZU, Jumbri K, Sambudi NS, Bakar NHHA, Abdullah NAF, Basheer C, Saad B. Removal of anthracene in water by MIL-88(Fe), NH2-MIL-88(Fe), and mixed-MIL-88(Fe) metal–organic frameworks. RSC Adv. 2019;9:41490–501.

[47] Wu M, Guo X, Zhao F, Zeng B. A poly(ethylenglycol) functionalized ZIF-8 membrane prepared by coordination-based post-synthetic strategy for the enhanced adsorption of phenolic endocrine disruptors from water. Sci Rep. 2017;7:8912.

[48] Zhou Y, Yang Q, Zhang D, Gan N, Li Q, Cuan J. Detection and removal of antibiotic tetracycline in water with a highly stable luminescent MOF. Sens Actuators B, Chem. 2018;262:137–43.

[49] Seo PW, Bhadra BN, Ahmed I, Khan NA, Jhung SH. Adsorptive removal of pharmaceuticals and personal care products from water with functionalized metal–organic frameworks: remarkable adsorbents with hydrogen-bonding abilities. Sci Rep. 2016;6:34462.

[50] Seo PW, Khan NA, Hasan Z, Jhung SH. Adsorptive removal of artificial sweeteners from water using metal–organic frameworks functionalized with urea or melamine. ACS Appl Mater Interfaces. 2016;8:29799–807.

[51] Chen Y, Zhang X, Mian MR, Son FA, Zhang K, Cao R, Chen Z, Lee S-J, Idrees KB, Goetjen TA, Lyu J, Li P, Xia Q, Li Z, Hupp JT, Islamoglu T, Napolitano A, Peterson GW, Farha OK. Structural diversity of zirconium metal–organic frameworks and effect on adsorption of toxic chemicals. J Am Chem Soc. 2020;142:21428–38.

[52] Howarth AJ, Liu Y, Li P, Li Z, Wang TC, Hupp JT, Farha OK. Chemical, thermal and mechanical stabilities of metal–organic frameworks. Nat Rev Mater. 2016;1:1–15.
[53] Akpinar I, Drout RJ, Islamoglu T, Kato S, Lyu J, Farha OK. Exploiting $\pi-\pi$ interactions to design an efficient sorbent for atrazine removal from water. ACS Appl Mater Interfaces. 2019;11:6097–103.
[54] Li R, Alomari S, Stanton R, Wasson MC, Islamoglu T, Farha OK, Holsen TM, Thagard SM, Trivedi DJ, Wriedt M. Efficient removal of per- and polyfluoroalkyl substances from water with zirconium-based metal–organic frameworks. Chem Mater. 2021;33:3276–85.
[55] Kucharzyk KH, Darlington R, Benotti M, Deeb R, Hawley E. Novel treatment technologies for PFAS compounds: a critical review. J Environ Manag. 2017;204:757–64.
[56] Wang B, Lv X-L, Feng D, Xie L-H, Zhang J, Li M, Xie Y, Li J-R, Zhou H-C. Highly stable Zr(IV)-based metal–organic frameworks for the detection and removal of antibiotics and organic explosives in water. J Am Chem Soc. 2016;138:6204–16.
[57] Rojas S, Navarro JAR, Horcajada P. Metal–organic frameworks for the removal of the emerging contaminant atenolol under real conditions. Dalton Trans. 2021;50:2493–500.
[58] Feng M, Zhang P, Zhou H-C, Sharma VK. Water-stable metal–organic frameworks for aqueous removal of heavy metals and radionuclides: a review. Chemosphere. 2018;209:783–800.
[59] Tchounwou PB, Yedjou CG, Patlolla AK, Sutton DJ. Molecular, clinical and environmental toxicology: volume 3: environmental toxicology. In: Luch A, editor. Experientia supplementum. Basel: Springer; 2012. p. 133–64.
[60] Bradl H. Heavy metals in the environment: origin, interaction and remediation. Elsevier; 2005.
[61] Sun DT, Gasilova N, Yang S, Oveisi E, Queen WL. Rapid, selective extraction of trace amounts of gold from complex water mixtures with a metal–organic framework (MOF)/polymer composite. J Am Chem Soc. 2018;140:16697–703.
[62] Carboni M, Abney CW, Liu S, Lin W. Highly porous and stable metal–organic frameworks for uranium extraction. Chem Sci. 2013;4:2396–402.
[63] Zhang Q, Yu J, Cai J, Zhang L, Cui Y, Yang Y, Chen B, Qian G. A porous Zr-cluster-based cationic metal–organic framework for highly efficient Cr2O72– removal from water. Chem Commun. 2015;51:14732–4.
[64] Desai AV, Manna B, Karmakar A, Sahu A, Ghosh SK. A water-stable cationic metal–organic framework as a dual adsorbent of oxoanion pollutants. Angew Chem, Int Ed. 2016;55:7811–5.
[65] Sharma S, Desai AV, Joarder B, Ghosh SK. A water-stable ionic MOF for the selective capture of toxic oxoanions of SeVI and AsV and crystallographic insight into the ion-exchange mechanism. Angew Chem, Int Ed. 2020;59:7788–92.
[66] Sharma S, Let S, Desai AV, Dutta S, Karuppasamy G, Shirolkar MM, Babarao R, Ghosh SK. Rapidselective capture of toxic oxo-anions of Se(IV), Se(VI) and As(V) from water by an ionic metal–organic framework (iMOF). J Mater Chem A. 2021;9:6499–507.
[67] Rudd ND, Wang H, Fuentes-Fernandez EMA, Teat SJ, Chen F, Hall G, Chabal YJ, Li J. Highly efficient luminescent metal–organic framework for the simultaneous detection and removal of heavy metals from water. ACS Appl Mater Interfaces. 2016;8:30294–303.
[68] Carboni M, Abney CW, Liu S, Lin W. Highly porous and stable metal–organic frameworks for uranium extraction. Chem Sci. 2013;4:2396–402.
[69] Howarth AJ, Katz MJ, Wang TC, Platero-Prats AE, Chapman KW, Hupp JT, Farha OK. High efficiency adsorption and removal of selenate and selenite from water using metal–organic frameworks. J Am Chem Soc. 2015;137:7488–94.
[70] Rangwani S, Howarth AJ, DeStefano MR, Malliakas CD, Platero-Prats AE, Chapman KW, Farha OK. Adsorptive removal of Sb(V) from water using a mesoporous Zr-based metal–organic framework. Polyhedron. 2018;151:338–43.
[71] Howarth AJ, Wang TC, Al-Juaid SS, Aziz SG, Hupp JT, Farha OK. Efficient extraction of sulfate from water using a Zr-metal–organic framework. Dalton Trans. 2015;45:93–7.

[72] Sun DT, Peng L, Reeder WS, Moosavi SM, Tiana D, Britt DK, Oveisi E, Queen WL. Rapid, selective heavy metal removal from water by a metal–organic framework/polydopamine composite. ACS Cent Sci. 2018;4:349–56.
[73] Boix G, Troyano J, Garzón-Tovar L, Camur C, Bermejo N, Yazdi A, Piella J, Bastus NG, Puntes VF, Imaz I, Maspoch D. MOF-beads containing inorganic nanoparticles for the simultaneous removal of multiple heavy metals from water. ACS Appl Mater Interfaces. 2020;12:10554–62.
[74] Yang S, Peng L, Syzgantseva OA, Trukhina O, Kochetygov I, Justin A, Sun DT, Abedini H, Syzgantseva MA, Oveisi E, Lu G, Queen WL. Preparation of highly porous metal–organic framework beads for metal extraction from liquid streams. J Am Chem Soc. 2020;142:13415–25.
[75] Guo M, Guo H, Liu S, Sun Y, Guo X. A microporous cationic metal–organic framework for the efficient removal of dichromate and the selective adsorption of dyes from water. RSC Adv. 2017;7:51021–6.
[76] Hashem T, Ibrahim AH, Wöll C, Alkordi MH. Grafting zirconium-based metal–organic framework UiO-66-NH2 nanoparticles on cellulose fibers for the removal of Cr(VI) ions and methyl orange from water. ACS Appl Nano Mater. 2019;2:5804–8.
[77] Shi M, Lin D, Huang R, Qi W, Su R, He Z. Construction of a mercapto-functionalized Zr-MOF/melamine sponge composite for the efficient removal of oils and heavy metal ions from water. Ind Eng Chem Res. 2020;59:13220–7.
[78] Rojas S, Torres A, Dato V, Salles F, Ávila D, García-González J, Horcajada P. Towards improving the capacity of UiO-66 for antibiotic elimination from contaminated water. Faraday Discuss. 2021;231:356–70.
[79] Bicalho HA, Quezada-Novoa V, Howarth AJ. Metal–organic frameworks for the generation of reactive oxygen species. Chem Phys Rev. 2021;2:041301.
[80] Castillo-Blas C, Romero-Muñiz I, Mavrandonakis A, Simonelli L, Platero-Prats AE. Unravelling the local structure of catalytic Fe-oxo clusters stabilized on the MOF-808 metal organic-framework. Chem Commun. 2020;56:15615–8.
[81] Zheng H-Q, He X-H, Zeng Y-N, Qiu W-H, Chen J, Cao G-J, Lin R-G, Lin Z-J, Chen B. Boosting the photoreduction activity of Cr(VI) in metal–organic frameworks by photosensitiser incorporation and framework ionization. J Mater Chem A. 2020;8:17219–28.
[82] Zhang R, Ji S, Wang N, Wang L, Zhang G, Li J-R. Coordination-driven in situ self-assembly strategy for the preparation of metal–organic framework hybrid membranes. Angew Chem, Int Ed. 2014;53:9775–9.
[83] Sun H, Tang B, Wu P. Development of hybrid ultrafiltration membranes with improved water separation properties using modified superhydrophilic metal–organic framework nanoparticles. ACS Appl Mater Interfaces. 2017;9:21473–84.
[84] Sun T, Hao S, Fan R, Qin M, Chen W, Wang P, Yang Y. Hydrophobicity-adjustable MOF constructs superhydrophobic MOF-rGO aerogel for efficient oil–water separation. ACS Appl Mater Interfaces. 2020;12:56435–44.
[85] Wang M, Zhang Z, Wang Y, Zhao X, Men X, Yang M. Ultrafast fabrication of metal–organic framework-functionalized superwetting membrane for multichannel oil/water separation and floating oil collection. ACS Appl Mater Interfaces. 2020;12:25512–20.
[86] Wang R, Zhao X, Jia N, Cheng L, Liu L, Gao C. Superwetting oil/water separation membrane constructed from in situ assembled metal–phenolic networks and metal–organic frameworks. ACS Appl Mater Interfaces. 2020;12:10000–8.
[87] Lin K-YA, Yang H, Petit C, Hsu F-K. Removing oil droplets from water using a copper-based metal organic frameworks. Chem Eng J. 2014;249:293–301.
[88] Lei Z, Deng Y, Wang C. Multiphase surface growth of hydrophobic ZIF-8 on melamine sponge for excellent oil/water separation and effective catalysis in a knoevenagel reaction. J Mater Chem A. 2018;6:3258–63.

[89] Zhang M, Xin X, Xiao Z, Wang R, Zhang L, Sun D. A multi-aromatic hydrocarbon unit induced hydrophobic metal–organic framework for efficient C2/C1 hydrocarbon and oil/water separation. J Mater Chem A. 2017;5:1168–75.
[90] Pakdel E, Wang J, Varley R, Wang X. Recycled carbon fiber nonwoven functionalized with fluorine-free superhydrophobic PDMS/ZIF-8 coating for efficient oil-water separation. J Environ Chem Eng. 2021;9:106329.
[91] Cai Y, Chen D, Li N, Xu Q, Li H, He J, Lu J. Nanofibrous metal–organic framework composite membrane for selective efficient oil/water emulsion separation. J Membr Sci. 2017;543:10–7.
[92] Cao J, Su Y, Liu Y, Guan J, He M, Zhang R, Jiang Z. Self-assembled MOF membranes with underwater superoleophobicity for oil/water separation. J Membr Sci. 2018;566:268–77.
[93] Zhu M, Liu Y, Chen M, Sadrzadeh M, Xu Z, Gan D, Huang Z, Ma L, Yang B, Zhou Y. Robust superhydrophilic and underwater superoleophobic membrane optimized by Cu doping modified metal–organic frameworks for oil-water separation and water purification. J Membr Sci. 2021;640:119755.
[94] Cai Y, Chen D, Li N, Xu Q, Li H, He J, Lu J. Superhydrophobic metal–organic framework membrane with self-repairing for high-efficiency oil/water emulsion separation. ACS Sustain Chem Eng. 2019;7:2709–17.
[95] Li W, Shi J, Zhao Y, Huo Q, Sun Y, Wu Y, Tian Y, Superhydrophobic JZ. Metal–organic framework nanocoating induced by metal-phenolic networks for oily water treatment. ACS Sustain Chem Eng. 2020;8:1831–9.
[96] Yogapriya R, Kasibhatta KRD. Hydrophobic-superoleophilic fluorinated graphene nanosheet composites with metal–organic framework HKUST-1 for oil–water separation. ACS Appl Nano Mater. 2020;3:5816–25.
[97] Shi M, Huang R, Qi W, Su R, He Z. Synthesis of superhydrophobic and high stable Zr-MOFs for oil-water separation. Colloids Surf A, Physicochem Eng Asp. 2020;602:125102.
[98] Xue J, Xu M, Gao J, Zong Y, Wang M, Ma S. Multifunctional porphyrinic Zr-MOF composite membrane for high-performance oil-in-water separation and organic dye adsorption/photocatalysis. Colloids Surf A, Physicochem Eng Asp. 2021;628:127288.
[99] Yang C, Kaipa U, Mather QZ, Wang X, Nesterov V, Venero AF, Fluorous OMA. Metal–organic frameworks with superior adsorption and hydrophobic properties toward oil spill cleanup and hydrocarbon storage. J Am Chem Soc. 2011;133:18094–7.
[100] Mukherjee S, Kansara AM, Saha D, Gonnade R, Mullangi D, Manna B, Desai AV, Thorat SH, Singh PS, Mukherjee A, Ghosh SK. An ultrahydrophobic fluorous metal–organic framework derived recyclable composite as a promising platform to tackle marine oil spills. Chem Eur J. 2016;22:10937–43.
[101] Eom S, Kang DW, Kang M, Choe JH, Kim H, Kim DW, Hong CS. Fine-tuning of wettability in a single metal–organic framework via postcoordination modification and its reduced graphene oxide aerogel for oil–water separation. Chem Sci. 2019;10:2663–9.
[102] Huang Y, Jiao Y, Chen T, Gong Y, Wang S, Liu Y, Sholl DS, Walton KS. Tuning the wettability of metal–organic frameworks via defect engineering for efficient oil/water separation. ACS Appl Mater Interfaces. 2020;12:34413–22.
[103] US Department of Commerce NO and A. A., Where is all of the Earth's water? (accessed Mar 17, 2022, at https://oceanservice.noaa.gov/facts/wherewater.html).
[104] Pattarachai S, Su X, Jeyong Y, Doron A, Volker P. Charge-transfer materials for electrochemical water desalination, ion separation and the recovery of elements. Nat Rev Mater. 2020;5:517–38.
[105] Elsayed E, AL-Dadah R, Mahmoud S, Anderson Paul A, Elsayed A, Youssef PG. CPO-27(Ni), aluminium fumarate and MIL-101(Cr) MOF materials for adsorption water desalination. Desalination. 2017;406:25–36.

[106] Xiao F, Hu X, Chen Y, Zhang Y. Porous Zr-based metal–organic frameworks (Zr-MOFs)-incorporated thin-film nanocomposite membrane toward enhanced desalination performance. ACS Appl Mater Interfaces. 2019;11:47390–403.
[107] Li H, Fu M, Wang S-Q, Zheng X, Zhao M, Yang F, Tang CY, Stable DY. Zr-based metal–organic framework nanoporous membrane for efficient desalination of hypersaline water. Environ Sci Technol. 2021;55:14917–27.
[108] Nightingale ER. Phenomenological theory of ion solvation. Effective radii of hydrated ions. J Phys Chem. 1959;63:1381–7.
[109] Küsgens P, Rose M, Senkovska I, Fröde H, Henschel A, Siegle S, Kaskel S. Characterization of metal–organic frameworks by water adsorption. Microporous Mesoporous Mater. 2009;120:325–30.
[110] Akiyama G, Matsuda R, Sato H, Hori A, Takata M, Kitagawa S. Effect of functional groups in MIL-101 on water sorption behavior. Microporous Mesoporous Mater. 2012;157:89–93.
[111] Choi HJ, Dincă M, Dailly A, Long JR. Hydrogen storage in water-stable metal–organic frameworks incorporating 1, 3- and 1, 4-benzenedipyrazolate. Energy Environ Sci. 2010;3:117–23.
[112] Colombo V, Galli S, Choi HJ, Han GD, Maspero A, Palmisano G, Masciocchi N, Long JR. High thermal and chemical stability in pyrazolate-bridged metal–organic frameworks with exposed metal sites. Chem Sci. 2011;2:1311–9.
[113] Demessence A, D'Alessandro DM, Foo ML, Long JR. Strong CO2 binding in a water-stable, triazolate-bridged metal–organic framework functionalized with ethylenediamine. J Am Chem Soc. 2009;131:8784–6.
[114] Zhang J-P, Zhu A-X, Lin R-B, Qi X-L, Chen X-M. Pore surface tailored SOD-type metal–organic zeolites. Adv Mater. 2011;23:1268–71.
[115] He H, Zhu Q-Q, Li C-P, Du M. Design of a highly-stable pillar-layer zinc(II) porous framework for rapid, reversible, and multi-responsive luminescent sensor in water. Cryst Growth Des. 2019;19:694–703.
[116] Jasuja H, Huang Y, Walton KS. Adjusting the stability of metal–organic frameworks under humid conditions by ligand functionalization. Langmuir. 2012;28:16874–80.
[117] Liu B, Vikrant K, Kim K-H, Kumar V, Kailasa SK. Critical role of water stability in metal–organic frameworks and advanced modification strategies for the extension of their applicability. Environ Sci Nano. 2020;7:1319–47.
[118] Canivet J, Fateeva A, Guo Y, Coasne B, Farrusseng D. Water adsorption in MOFs: fundamentals and applications. Chem Soc Rev. 2014;43:5594–617.
[119] Cáceres L, Gómez-Silva B, Garró X, Rodríguez V, Monardes V, McKay CP. Relative humidity patterns and fog water precipitation in the Atacama Desert and biological implications. J Geophys Res. 2007;112:1–11.
[120] Unland HE, Houser PR, Shuttleworth WJ, Yang Z-L. Surface flux measurement and modeling at a semi-arid Sonoran Desert site. Agric For Meteorol. 1996;82:119–53.
[121] Furukawa H, Gándara F, Zhang Y-B, Jiang J, Queen WL, Hudson MR, Yaghi OM. Water adsorption in porous metal–organic frameworks and related materials. J Am Chem Soc. 2014;136:4369–81.
[122] Zang J, Nair S, Sholl DS. Prediction of water adsorption in copper-based metal–organic frameworks using force fields derived from dispersion-corrected DFT calculations. J Phys Chem C. 2013;117:7519–25.
[123] Paranthaman S, Coudert F-X, Fuchs AH. Water adsorption in hydrophobic MOF channels. Phys Chem Chem Phys. 2010;12:8124–30.
[124] Zhang H, Snurr RQ. Computational study of water adsorption in the hydrophobic metal–organic framework ZIF-8: adsorption mechanism and acceleration of the simulations. J Phys Chem C. 2017;121:24000–10.

[125] Skarmoutsos I, Eddaoudi M, Maurin G. Highly efficient rare-Earth-based metal–organic frameworks for water adsorption: a molecular modeling approach. J Phys Chem C. 2019;123:26989–99.
[126] Lamaire A, Wieme J, Hoffman AEJ, Speybroeck VV. Atomistic insight in the flexibility and heat transport properties of the stimuli-responsive metal–organic framework MIL-53(Al) for water-adsorption applications using molecular simulations. Faraday Discuss. 2021;225:301–23.
[127] Gu J-Z, Lu W-G, Jiang L, Zhou H-C, Lu T-B. 3D porous metal–organic framework exhibiting selective adsorption of water over organic solvents. Inorg Chem. 2007;46:5835–7.
[128] Akiyama G, Matsuda R, Kitagawa S. Highly porous and stable coordination polymers as water sorption materials. Chem Lett. 2010;39:360–1.
[129] Marcus Y. In: Ion properties. New York: Marcel Dekker; 1997. p. 117–35.
[130] O'Nolan D, Kumar A, Zaworotko MJ. Water vapor sorption in hybrid pillared square grid materials. J Am Chem Soc. 2017;139:8508–13.
[131] Wißmann G, Schaate A, Lilienthal S, Bremer I, Schneider AM, Behrens P. Modulated synthesis of Zr-fumarate MOF. Microporous Mesoporous Mater. 2012;152:64–70.
[132] Lawrence MC, Katz MJ. Analysis of the water adsorption isotherms in UiO-based metal–organic frameworks. J Phys Chem C. 2022;126:1107–14.
[133] Kim H, Yang S, Rao SR, Narayanan S, Kapustin EA, Furukawa H, Umans AS, Yaghi OM, Wang EN. Water harvesting from air with metal–organic frameworks powered by natural sunlight. Science. 2017;356:430–4.
[134] Kim H, Rao SR, Kapustin EA, Zhao L, Yang S, Yaghi OM, Wang EN. Adsorption-based atmospheric water harvesting device for arid climates. Nat Commun. 2018;9:1191.
[135] Fathieh F, Kalmutzki MJ, Kapustin EA, Waller PJ, Yang J, Yaghi OM. Practical water production from desert air. Sci Adv. 2018;4:1–9.
[136] Hanikel N, Prévot MS, Fathieh F, Kapustin EA, Lyu H, Wang H, Diercks NJ, Glover TG, Yaghi OM. Rapid cycling and exceptional yield in a metal–organic framework water harvester. ACS Cent Sci. 2019;5:1699–706.
[137] Towsif Abtab SM, Alezi D, Bhatt PM, Shkurenko A, Belmabkhout Y, Aggarwal H, Weseliński ŁJ, Alsadun N, Samin U, Hedhili MN, Reticular EM. Chemistry in action: a hydrolytically stable MOF capturing twice its weight in adsorbed water. Chem. 2018;4:94–105.
[138] Wade CR, Corrales-Sanchez T, Narayan TC, Dincă M. Postsynthetic tuning of hydrophilicity in pyrazolate MOFs to modulate water adsorption properties. Energy Environ Sci. 2013;6:2172–7.
[139] Leo AJ. Methods in enzymology. In: Molecular design and modeling: concepts and applications part A: proteins, peptides, and enzymes. vol. 202. Academic Press; 1991. p. 544–91.
[140] Caron G, Gaillard P, Carrupt P-A, Testa B. Lipophilicity behavior of model and medicinal compounds containing a suilfide, sulfoxide, or sulfone moiety. Helv Chim Acta. 1997;80:449–62.
[141] Rieth AJ, Yang S, Wang EN, Dincă M. Record atmospheric fresh water capture and heat transfer with a material operating at the water uptake reversibility limit. ACS Cent Sci. 2017;3:668–72.
[142] Rieth AJ, Hunter KM, Dincă M, Paesani F. Hydrogen bonding structure of confined water templated by a metal–organic framework with open metal sites. Nat Commun. 2019;10:4771.
[143] Rieth AJ, Wright AM, Skorupskii G, Mancuso JL, Hendon CH, Record-Setting DM. Sorbents for reversible water uptake by systematic anion exchanges in metal–organic frameworks. J Am Chem Soc. 2019;141:13858–66.
[144] Chen Z, Li P, Zhang X, Li P, Wasson MC, Islamoglu T, Stoddart JF, Farha OK. Reticular access to highly porous acs-MOFs with rigid trigonal prismatic linkers for water sorption. J Am Chem Soc. 2019;141:2900–5.

[145] Zheng J, Vemuri RS, Estevez L, Koech PK, Varga T, Camaioni DM, Blake TA, McGrail BP, Motkuri RK. Pore-engineered metal–organic frameworks with excellent adsorption of water and fluorocarbon refrigerant for cooling applications. J Am Chem Soc. 2017;139:10601–4.
[146] Yang L, Idrees KB, Chen Z, Knapp J, Chen Y, Wang X, Cao R, Zhang X, Xing H, Islamoglu T, Farha OK. Nanoporous water-stable Zr-based metal–organic frameworks for water adsorption. ACS Appl Nano Mater. 2021;4:4346–50.
[147] Jiang H-L, Feng D, Wang K, Gu Z-Y, Wei Z, Chen Y-P, Zhou H-C. An exceptionally stable, porphyrinic Zr metal–organic framework exhibiting pH-dependent fluorescence. J Am Chem Soc. 2013;135:13934–8.
[148] Chen Z, Jiang H, Li M, O'Keeffe M, Eddaoudi M. Reticular chemistry 3.2: typical minimal edge-transitive derived and related nets for the design and synthesis of metal–organic frameworks. Chem Rev. 2020;120:8039–65.
[149] Faust T. MOFs move to market. Nat Chem. 2016;8:990–1.
[150] Frameworks for commercial success. Nat Chem. 2016;8:987–987.

Muhammad Alif Mohammad Latif,
Mostafa Yousefzadeh Borzehandani, and
Mohd Basyaruddin Abdul Rahman

12 Insights into host-drug interactions in metal-organic frameworks

12.1 Introduction

There exist many challenges with optimizing structural properties of nanomaterials. For example, carbon nanotubes are well-known drug carriers, although controlling their pore size to properly encapsulate the drug molecules remains a challenge [1]. Large pores and open gates in mesoporous silica materials have made it possible for receiving guest molecules but they too are restricted due to low hydrothermal stability [2]. Zeolites may have long-term chemical and biological stability, but their capability has been limited because of their small pore size and poor structural diversity [3]. Therefore, it is advantageous to gain molecular insights into such materials as they play a critical role in designing, promoting and optimizing their use for practical applications.

Over the past decade, nanosized reticular materials have become the choice of immobilization studies due to the variety of pore size and shapes, framework structures and extraordinary properties exhibited by metal-organic frameworks (MOFs), zeolitic imidazolate frameworks (ZIFs) and covalent–organic frameworks (COFs). Reticular chemistry is the chemistry of linking molecular building blocks by strong bonds [4]. MOFs are constructed rationally using organic linker and inorganic metal node molecular building blocks via coordination bonds that give rise to porous two- and three-dimensional architectures [5]. MOFs are employed for a broad range of purposes due to their high internal surface areas, extensive porosity, high degree of crystallinity and thermal and chemical stability [6].

Muhammad Alif Mohammad Latif, Integrated Chemical BioPhysics Research, Faculty of Science, Universiti Putra Malaysia, 43400 UPM Serdang, Selangor, Malaysia; and Centre of Foundation Studies for Agricultural Science, Universiti Putra Malaysia, 43400 UPM Serdang, Selangor, Malaysia; and Foundry of Reticular Materials for Sustainability (FORMS), Institute of Nanoscience and Nanotechnology, Universiti Putra Malaysia, 43400 UPM Serdang, Selangor, Malaysia, e-mail: aliflatif@upm.edu.my
Mostafa Yousefzadeh Borzehandani, Integrated Chemical BioPhysics Research, Faculty of Science, Universiti Putra Malaysia, 43400 UPM Serdang, Selangor, Malaysia, e-mail: mostafa.yousefzadeh@gmail.com
Mohd Basyaruddin Abdul Rahman, Integrated Chemical BioPhysics Research, Faculty of Science, Universiti Putra Malaysia, 43400 UPM Serdang, Selangor, Malaysia; and Foundry of Reticular Materials for Sustainability (FORMS), Institute of Nanoscience and Nanotechnology, Universiti Putra Malaysia, 43400 UPM Serdang, Selangor, Malaysia, e-mail: basya@upm.edu.my

https://doi.org/10.1515/9781501524721-012

Favorable MOF structures offer great opportunities for gas storage, catalysis, separation, biological imaging and sensing and drug delivery, among others [7]. MOFs gain benefits from their unique properties, which make them ideal support matrices for encapsulating various molecules and biomolecules. Specifically, structural features such as tunable pore size and shape, biodegradability, high loading capacity, low toxicity and versatile functionalities provide an edge for MOFs to be used in biomedical applications.

In this chapter, host-drug interactions in MOFs related to biomedical applications will be delineated, specifically from a computational point-of-view. Two categories of host-guest interactions are highlighted: (i) interactions of MOFs with drug biomolecules and (ii) MOF interactions with drug small molecules. Several computational methods that can be used to analyze drug-MOFs interactions will be presented, including quantum mechanics and molecular mechanics approaches. Lastly, the main factors affecting drug-MOFs interactions elucidated via computational studies will be presented.

12.2 Drug-metal-organic framework (MOF) interactions

12.2.1 Biomolecules and MOFs

Biological molecules, also known as biomolecules, are natural substances produced by living organisms. Proteins, enzymes, peptides, carbohydrates, DNA, nucleic acids and lipids are examples of biomolecules. Enzymes function as biological catalysts in living systems. Due to technological advancement, enzymes may now be isolated from biological materials, purified and used in a range of applications in industrial settings. The challenge though for their practical use is denaturation, which reduces their catalytic activity. To date, several enzymes have been developed, modified and immobilized for various industrial applications.

In short, there are two experimental approaches for encapsulating biomolecules using MOFs either after the MOF synthesis (post-synthetic modification), in which the entrapment of biomolecules is carried out within preformed MOFs or during the synthetic formation of MOFs (*de novo* encapsulation), in which MOFs assemble around active species [8, 9]. Each method has its own advantages with different effects on the framework formation and the encapsulated biomolecules.

Various methods for immobilizing biomolecules, such as enzymes and peptides, into the outer and internal surface structures of MOFs have been explored through pore encapsulation or entrapment, surface attachment or conjugation, covalent linkage using post-synthesis modification, and coprecipitation synthesis using *de novo* encapsulation [10, 11]. The most common is a *de novo* pore encapsulation approach

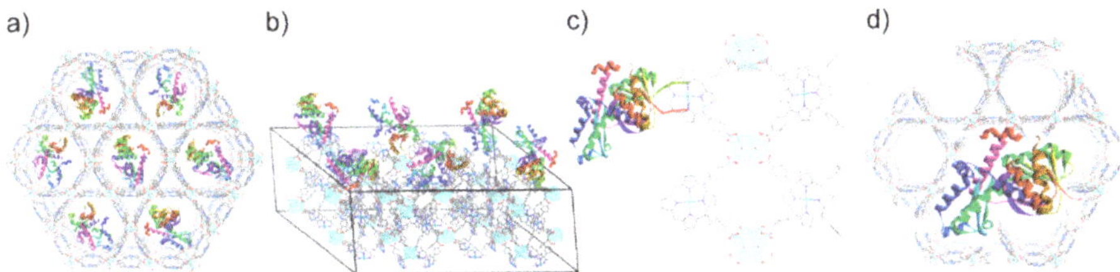

Figure 12.1: Common immobilization methods of guest molecules within or on MOFs. (a) pore encapsulation or entrapment, (b) surface attachment or conjugation, (c) covalent linkage and (d) coprecipitation or in situ synthesis.

with entrapment of enzymes occurring by diffusion into the MOFs pores or ZIFs cages (Figure 12.1a). This approach also has been widely used to entrap small organic molecules drugs and even inorganic complexes within the pores of MOFs. A major advantage is that biomolecules are protected from deactivation in conditions that cause denaturing due to their isolation within the MOF pores. In such cases, enzymes can maintain their structural dynamics and substrate accessibility. However, in certain applications, the MOFs must be structurally degraded to release the biomolecules for them to be active.

Another common technique is a simple and low-cost surface attachment or conjugation of biomolecules on the surface of MOFs (Figure 12.1b). Enzymes immobilized on MOF surfaces are mainly through weak interactions such as hydrogen bonding, electrostatic interaction, van der Waals forces and $\pi-\pi$ interactions. Surface attachment can be performed by conjugation or interaction with the metal clusters or functional groups of the organic linkers of the MOFs. Weak interactions between drug molecules and MOFs causes poor stability and detachment of such molecules from MOF surfaces. The covalent linkage method is recommended to ensure better stability between host and guest interaction (Figure 12.1c). Usually, the abundance of amino groups on an enzyme's surface can create peptide bonds with carboxyl functionalized MOFs. MOFs can be modified with specific functional groups, especially amino, carboxyl and hydroxyl groups, to serve as immobilization sites.

A more challenging, yet innovative approach of immobilizing different sized guests inside MOF pores is coprecipitation, also known as *de novo* encapsulation. Systematic exploration of reaction conditions must be pursued, especially the mixing ratio of molecules and MOF precursors (metal ions and organic linkers), for this to be successful. This approach allows for the nucleation and growth of MOFs simultaneously while embedding a guest enzyme inside it during the assembly of the MOF (Figure 12.1d). MOFs can retain the enzyme inside the pores via strong hydrophobic and hydrophilic interactions, which avoid leakage of the enzymes from the support whereas the lack of such specific interactions between enzymes and mesoporous silica nanoparticles typically lead to leaching of enzymes from the support [12].

Recent research in nanotechnology has resulted in a variety of different MOF-nano scaffolds that have the potential to preserve biomolecules with a high molecular entrapment loading [13]. Liang et al. described a variety of biocomposites involving ZIF-8 and different biomolecules, such as bovine serum albumin (BSA), human serum albumin (HSA), lysozyme, glucose dehydrogenase (PQQ-GDH), urease, lipase, insulin and oligonucleotides [14]. Several biomolecules induced the ZIF-8 biocomposites to move from a rhombic dodecahedron shape into different morphologies, such as truncated cubic, nanoleaf, nanoflower and nanostar shapes. The efficiency of biomolecule encapsulation was found to be between 80 and 100 %.

Li et al. successfully encapsulated a nerve detoxifying agent, organophosphorus acid anhydrolase (OPAA), into a water-stable zirconium MOF via an adsorption method [15]. Using ultraviolet-visible (UV-Vis) spectroscopy, the uptake of OPAA by PCN-128y was measured and a maximum loading was achieved after 24 h. Various enzymes can be encapsulated by matching the size of the enzyme to that of the big channel (4.4 nm). The effective encapsulation of OPAA resulted in a loading capacity of 12 wt%. Therefore, it can be concluded that the development and selection of protective materials such as MOFs are essential to achieving high enzyme loading and improving biomolecule stability.

12.2.2 Targeted drug delivery

Targeted drug delivery systems selectively direct a drug to specific cells and tissue, especially for cancer treatment [16]. MOFs effectively enable researchers to overcome treatment-related issues by preventing premature drug release in the body, prolonging the drug circulation time in the bloodstream, enhancing the treatment efficacy of the anticancer drug by optimizing the drug dose at the tumor site while sparing the healthy tissues, and improving drug uptake and intracellular delivery. Thus, reticular material-based drug delivery systems offer a replacement for conventional chemotherapy in improving patients' quality of life [17].

Many studies have explored MOFs in cancer applications including imaging, radiotherapy, phototherapy and drug delivery [18–20]. For example, in optical applications, specific luminescent groups are used to activate the specific targeting properties in MOFs [21]. Indeed, biomolecules such as proteins, peptides, DNA and antibodies can be functionalized at the surface of MOFs to enhance the therapeutic value [22, 23]. Significant advantages over other matrices are as follows:

(i) MOF frameworks can simply be altered via the specific combination of numerous metal ions and organic linkers to produce MOFs with different morphologies, sizes, compositions and physicochemical characteristics.
(ii) Even with weak coordination bonds, MOFs can be stable while biodegradable MOFs making them suitable for bioavailability.

(iii) High surface areas and tailorable porosities increase the potential to carry all sizes of drugs; and
(iv) High potential for surface functionalization.

The appropriate functionalization approach on the outer surface of MOFs is very important as it strongly defines the overall efficacy of the desired application. In general, the outer surfaces of MOFs can be functionalized using three approaches [24]: (i) presynthesis, (ii) intersynthesis and (iii) post-synthesis. In presynthesis functionalization, the organic linker acts as the molecular topography to allow desired functional groups to be attached. Likewise, in post-synthetic functionalization, the MOFs are first synthesized as a reactive site backbone allowing functionalization in a controlled manner. In some cases, the intersynthetic functionalization occurs in parallel with the framework development allowing the attachment of functional groups to the pores. For all functionalization approaches, either a covalent bond, intermolecular interaction, or mechanical docking is produced (Figure 12.2).

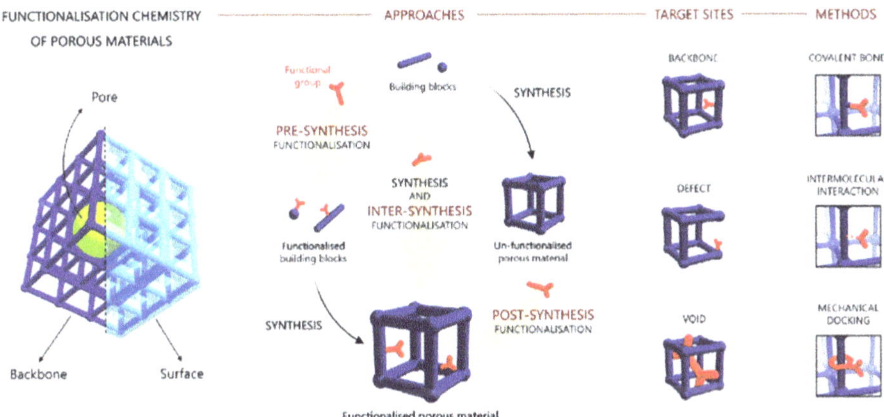

Figure 12.2: Three main approaches of the synthetic pathway for the functionalization chemistry of porous materials [24]. Copyright 2020 by John Wiley and Sons. Figure adapted with permission.

Histidine, which can be easily integrated into proteins or peptides, is an example of a targeting ligand for functionalization of MOFs. Läechelt and coworkers present a versatile functionalization approach by coordinative interaction between the outer surfaces of MOFs and histidine residues [25]. Functionalization of HKUST-1 ($[Cu_3(BTC)_2]_2$ where BTC = 1,3,5-benzenetricarboxylate), MIL-88A(Fe) ($[Fe_3O(fum)_6(H_2O)_2(OH)]_2$, where fum = fumarate) and MOF-801 ($[Zr_6O_4(OH)_4(fum)_6]_2$) exhibit binding strength dependent on the length of the histidine residues. In this mild surface functionalization approach, the MOFs were simply immersed with histidine residues in HEPES-buffered glucose (pH = 7.4). All histidine functionalized MOFs show pH-controlled release, allowing intracellular delivery and enhanced cytotoxicity toward HeLa cancer cells [25].

Arginylglycylaspartic acid (RGD) homing peptide can be integrated on the outer surface of MOFs for a targeted purpose. Wang et al. demonstrated a high-efficiency functionalization by surface covalent between MIL-101(Fe) and RGD peptide. This functionalized MIL-101(Fe) displays a strong covalent bond with the modified RGD (K(ad)RGDS-PEG1900). In this facile method, the as-prepared wet β-CD-MIL-101(Fe) was reimmersed with modified (K(ad)RGDS-PEG1900) in Tris-HCl buffer (10 mM, pH = 8.0) solution. This multifunctional MIL-101(Fe) was used as a premature drug release controller, significantly enhancing tumor uptake, and decreasing cytotoxicity induction in normal cells [26]. Another work by Kamal et al. showed that functionalizing RGD on the surface of nanoZIF-8 loaded with GEM (GEM⊂nZIF-8) exhibited efficient uptake within cancerous human adenocarcinoma alveolar epithelial cells (A549). The GEM⊂RGD@nZIF-8 system induced high cytotoxicity (75 % at a concentration of 10 $\mu g\, mL^{-1}$) and apoptosis (62 %) after 48 h treatment, compared with nonfunctionalized GEM⊂nZIF-8 [27]. RGD surface functionalization also increased the surface roughness of the nanoparticles, which can be correlated with a higher penetration to cell membrane for influencing inhibition of cancer cells. It was more selective toward lung cancer cells (A549) than normal human lung fibroblast cells (MRC-5) with a selectivity index (SI) of 3.98.

Galactose (G) derivative possesses a specific binding due to its high affinity toward a class of protein called galactin that overexpresses hepatoma cancer cells [28]. Yang et al. synthesized DOX-loaded ZIF-8 (ZIF-8@DOX) in a one-pot technique, and then modified it with the water-soluble carboxylated pillar-6 arene (WP6). The coordination of the carboxyl groups on WP6 with the metal nodes on ZIF-8@DOX resulted in ZIF-8@DOX@WP6, which improved the water dispersion of ZIF-8@DOX. The G and ZIF-8@DOX@WP6 were combined to produce ZIF-8@DOX@WP6@G via WP6 and G's host-guest interaction. The targeting function of G can improve the drug delivery carrier's targeting ability to HepG2, because the ZIF-8@DOX@WP6@G has good selectivity and high efficiency to kill hepatoma cells when compared to free DOX [29].

Nucleic acids are a form of biomolecule that comprises both deoxyribonucleic acid (DNA) and ribonucleic acid (RNA), essential for the storage and expression of genetic information. DNA is tremendously important and has a wide application in the targeted binding of drug delivery applications. Chen et al. utilized the "click chemistry" approach to covalently connect DNA to the MOFs [30]. This technique successfully hybridized the UiO-68 ($[Zr_6O_4(OH)_4(TPDC)_6]_2$, where TPDC = p-terphenyl-4,4"-dicarboxylate) with DNAzymes and worked as functional units for multiplexed ion-sensing and logic-gate systems. In this functionalization approach, the UiO-68 loaded doxorubicin (DOX) was incubated with DNA in PBS buffer having pH = 7.4. This DNAzyme functionalized UiO-68 displayed stimuli-responsive properties with regard to particular analytes. Also, the DOX was selectively released from the DNA-coated

MOFs and toxically induced into MDA-MB-231 breast cancer cells in response to increased concentrations of both H^+ and Mg^{2+} adenosine triphosphate [30].

Folic acid is a simple and robust active targeting ligand that can be used to target the folate receptor, which is expressed in most tumor cells. Chowdhuri et al. synthesized $NAYF_4$ up-conversion nanoparticles (UCNPs) by integrating YCl_3, $YbCl_3$ and $ErCl_3$ with a dual solution of sodium hydroxide and ammonium fluoride. By utilizing the one-pot method, the UCNP@ZIF-8 was surface-functionalized with folic acid, followed by encapsulation of the model drug 5-fluorouracil (5-FU). The UCNP@ZIF-8/FA showed pH-responsive characteristics when the 5-FU released was higher in the acidic compared to a neutral environment for both 12 and 24 h. The cellular internalization was increased after treatment with functionalized UCNPs@ZIF-8/FA compared to UCNPs@ZIF-8, thus inducing higher cytotoxicity in cancer cells [31].

Hyaluronic acid is a hydrophilic natural polysaccharide with negative charge and actively targets tumor cells and adheres to the overexpressed CD44 receptor on the tumor surface. Li and coworkers synthesized the CCM@ZIF-8/HA by coating ZIF-8 with hyaluronic acid, followed by mixing in methanol solution to encapsulate the curcumin (CCM). The CCM@ZIF-8/HA demonstrated pH-stimuli when the CCM released was marked higher in acidic condition (pH = 5.5) within a few days in comparison with the neutral condition. Based on the fluorescent data, it was found that hyaluronic acid promotes the uptake of CCM by Hela cells, resulting in higher growth inhibition against Hela cells [32].

Taking all these points into consideration, it is a viable approach to functionalize MOFs and ZIFs with targeting ligands in order to improve cancer-targeting efficacy. These systems possess unique features such as a huge surface area, exceptionally high porosity (up to 90 % of the free volume), and simple functionalization methods. Therefore, the systems have the capability to resolve some issues in single-carrier applications (organic-based or inorganic-based systems) associated with low loading efficacy, poor biocompatibility, or nonbiodegradability [33, 34]. However, to fulfill the need of biological function such as targeted and localized delivery systems, many of these MOFs and ZIFs needs to be surface functionalized as they do not inherently possess their own the targeting properties [35].

12.3 Computational modeling of metal-organic frameworks

Computational chemistry has been developed for large systems and materials [36]. Implementing computational methods for materials requires powerful computing resources. Computer simulations have become the primary tools to study the physical properties, dynamics of molecular conformations, structure-function relationships and intermolecular interactions. They are extensively used to provide information on the fluctuations and conformational changes at the atomic level. Since these methods allow for the investigation of the structure, dynamics and thermodynamics of molecules, it has been an excellent strategy to explore the behavior of molecular systems, proteins and complexes under certain conditions that are otherwise impossible to achieve in the laboratory.

In general, computational methods used for MOFs materials are reliant on two main theories: (i) quantum mechanics (QM) [37] and (ii) molecular mechanics (MM) [38]. As shown in Figure 12.3, these theories have been expanded into further algorithmic solving methods such as molecular dynamics (MD), Monte Carlo (MC), *ab initio* and semiempirical calculations. In addition, the idea from mixed functions of quantum mechanics and molecular mechanics has provided a valuable method called quantum mechanics/molecular mechanics (QM/MM). Table 12.1 lists examples of different approaches used to investigate host-guest interactions involving MOFs [39–46].

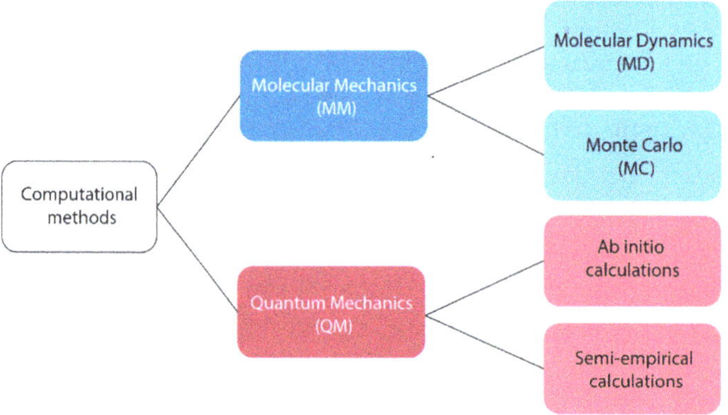

Figure 12.3: Horizontal hierarchical diagram for general computational methods.

12.3.1 Quantum mechanics approach

Quantum mechanics (QM) is known as a first principles method, in which the electronic properties of atoms and molecules are obtained by solving the Schrödinger equation (equation (12.1)) [47]:

$$\hat{H}\Psi = E\Psi \tag{12.1}$$

where $\hat{H}$, Ψ and E are Hamiltonian operators, wavefunctions and energy value (or eigenfunction), respectively. QM methods attempt to accurately describe the nuclei, the surrounding electrons and their energy, which generates an ultimate electron density. Hence, the produced electron density is responsible for estimating polarization, rotational energy states, vibrational states, precise molecular structure, among others

Table 12.1: Some computational approaches for studying MOFs systems.

MOFs	Substrates	Methods	Conditions	Applications	Ref.
IRMOFs[a]	Trp-cage (polypeptide)	MD[d]	T = 280–400 K P = 1 Bar	Catalysis	[39]
	Peptides	MD[d]	T = 298 K P = 1 Bar		[40]
	Proteins	MD[d]	T = 300–500 K P = 1 Bar		[41]
	Gemcitabine	QM[e]	T = 298 K P = 1 Bar	Drug delivery	[42]
	Hydrogen	GCMC[f]	T = 298 K P = 1 Bar	Gas storage	[43]
	SO_2		T = 298 K P = 1 Bar		[44]
	Natural Gases		T = 298 K P = 1 Bar		[45]
CD-MOFs[b]	Ibuprofen	GCMC[f]	T = 300 K P = 1 Bar	Drug delivery	[46]
	Caffeine		T = 300 K P = 1 Bar	Gas storage	
	Urea		T = 300 K P = 1 Bar		
bioMOFs[c]	Ibuprofen	GCMC[f]	T = 300 K P = 1 Bar	Drug delivery	
	Caffeine		T = 300 K P = 1 Bar	Gas storage	
	Urea		T = 300 K P = 1 Bar		

[a]isoreticular MOFs, [b]cyclodextrin-based MOFs, [c]biocompatible MOFs, [d]Molecular Dynamics, [e]Quantum Mechanics, [f]grand canonical Monte Carlo.

[48]. For MOFs, the *ab initio* and semiempirical calculations are the two QM methods that have been most employed.

Ab initio calculations allow for investigating the structural parameters and properties of framework materials. Density functional theory (DFT) and Hartree-Fock (HF) theory are two *ab initio*-based methods that calculate the energy levels through different molecular orbital approximations. The energy of the system taken by the DFT method is calculated from equation (12.2) [49]:

$$E_{\text{DFT}} = E_{\text{NN}} + E_T + E_V + E_{\text{coul}} + E_{\text{exch}} + E_{\text{corr}} \tag{12.2}$$

According to the above equation, E_{NN}, E_T, E_V, E_{coul}, E_{exch} and E_{corr} are repulsion energy between nuclei, the kinetic energy of the electrons, attraction energy between nucleus and electron, classical electron-electron Coulomb repulsion energies, nonclassical electron-electron exchange energies and correlated movement of electrons of different spin, respectively. Although the terms of E_{NN}, E_V and E_{coul} described in DFT are the same as in the HF method, E_T and $E_{exa\ ch}$ terms have different interpretation in DFT and HF methods. Moreover, the correlated movement of electrons of different spin (E_{corr}) is only explained in DFT calculations [50]. Consequently, DFT has been presented as a promising *ab initio* method to compute the electronic structure of MOF materials [51]. In the 1960s, DFT was first introduced by Hohenberg, Kohn and Sham to find an approximate solution for the many-electron Schrödinger equation [52]. The key feature of the DFT calculation is a precise prediction of ground state energy (E_0) as a function of the atomic positions. Thus, further properties such as molecular structures, vibrational frequencies, electric and magnetic properties, atomization energies, ionization energies and reaction paths can be accurately extracted [53].

The semiempirical calculation is another useful method in QM, which is formalized from HF theory and available empirical parameters [54]. Semiempirical calculations are highly demanding for larger systems. Semiempirical methods can be categorized into several groups. Zero differential overlap (ZDO) is defined as reducing the two-electron integral [55]. Modified neglect of differential overlap (MNDO) has produced many other novel methods such as parametric method number 3 (PM3) and Austin Model 1 (AM1) [56–58]. More atomic types and specific parameters are assigned in the MNDO method for generating further properties such as dipole moments, heats of formation, first ionization energies and geometrical variables. The parameters are taken separately from similar compounds for instance hydrocarbons, CHN systems, and CHO systems, among others.

12.3.2 Classical mechanics approach

Molecular mechanics (MM) is referred to as a nonquantum method for predicting conformational properties by calculating potential energy [59]. The lowest potential energy is obtained by computing the best approximation of bonded terms (bond lengths, bond angles and dihedral angles) and nonbonded terms (electrostatic and van der Waals interactions). Electrostatic and van der Waals interactions arise from Coulomb and Lennard–Jones potentials, respectively. Coulomb and Lennard–Jones potentials are obtained by solving equations (12.3) and (12.4), respectively [60, 61]:

$$u^{LJ}(r) = 4\varepsilon\left[\left(\frac{\sigma}{r}\right)^{12} - \left(\frac{\sigma}{r}\right)^{6}\right] \tag{12.3}$$

$$u^{Coulomb}(r) = \frac{Q_1Q_2}{4\pi\varepsilon_0 r} \tag{12.4}$$

According to equation (12.3), σ is the van der Waals radius and ε is well depth. Also, based on equation (12.4), ε_0 is the permittivity of free space and Q is the charge between particles. Minimizing the potential energy complies with empirical correction schemes that only consider the atoms' positions. Thus, it can be exploited for molecular systems with thousands of atoms [62]. Also, MM has been employed for the statistical distribution of MOF crystal structures at different temperatures and pressures. MD and MC simulations are the most used methods to describe the interaction between the MOF framework and its guest molecules.

12.3.2.1 Molecular dynamics

MD simulations are carried out by generating atomic and molecular time-dependent motions (trajectories) via numerical integration of Newton's laws of motion [63]. These trajectories are required to be captured sequentially from simulated atomic coordinates at specific periods. The total time duration for MD simulation should be long enough to determine the natural processes being studied. MD simulation can be implemented based on NVE (constant number of particles, volume and energy), NVT (constant number of particles, volume and temperature) and NPT (constant number of particles, pressure and temperature) ensembles [64]. NVE provides energetic relaxation for the system, and it is suitable for adiabatic processes. The thermostat applied for NVT can exchange the energy of endothermic and exothermic processes. The NPT ensemble set the desired pressure for the system by regulating the volume of the simulation box. MD simulations have been used to detect interactions of guest molecules with MOF frameworks [65], diffusion of drug molecules inside MOF pores [46] and the flexibility of MOF structures [66]. Many software programs have been developed to conduct MD simulations depending on the structures, simulation time and corresponding uses of the molecule types. To obtain the potential energy of the system studied, a potential function or "force field" is needed. Some common potential functions applied for MD simulations are GROMOS, UFF, CHARMM, AMBER and OPLS-AA [67–69]. These force fields have been tested with experimental data and are suitable for simulating the behaviour of most chemical systems.

12.3.2.2 Grand canonical Monte Carlo simulations

MC simulations are a mathematical technique that employ the laws of classical mechanics to compute the equilibrium properties of the many-body system [70]. MC simulations generate randomized displacement in the variables of a system. Given a certain criterion, it subsequently decides whether the accepted or rejected changes lead to a new state of the system. The most common ensembles involved in MC calculations are micro-canonical, canonical, isothermal-isobaric and grand canonical ensembles [71]. In the grand canonical ensemble, chemical potential (μ), volume (V) and temperature (T) are constant, except for the number of particles (N). GCMC, an MC method based

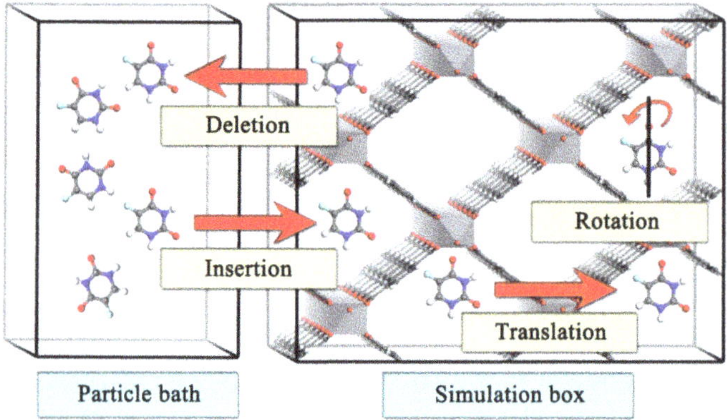

Figure 12.4: The movements of guest molecules during MC cycles.

on the grand canonical ensemble, has been firmly established for exploring drug loading and adsorption inside MOF pores [72–75]. In each MC cycle, drug molecules move four ways through MOF pores during the simulation: (i) deletion, (ii) insertion, (iii) rotation and (iv) translation (Figure 12.4) [76, 77]. Particles are allowed to move either to the particle bath or to the simulation box. Deletion and insertion continue if the term 'μ (particle in the bath) = μ (confinement)' condition is satisfied. The Metropolis algorithm is then used to assess the probability of the particles that are changing the configuration in the simulation box [78]. The accuracy of the GCMC simulation is determined by the ratio of the accepted and rejected moves, which must be around 1, and it is a requirement for a suitable selection of the rotation angle and displacement rate. After the new arranged system passes the evaluation, it is employed for the calculation of the average thermodynamic properties of the system being modelled.

12.3.3 QM/MM hybrid approach

A combination of QM and MM principles has produced state-of-the-art computational techniques called the quantum mechanics/molecular mechanics (QM/MM) method [48]. QM/MM is an appropriate way for investigating chemical reactions and electronic properties in MOF systems. The advantage of combining both principles is that a higher accuracy can be achieved by QM calculations while also utilizing the speed of MM calculations where the interactions are not important. In this method, the active site in MOFs (e. g., where the chemical reaction takes place or the molecule whose properties are going to be calculated) is treated by the QM level, whereas the rest of the surrounding framework and the explicit solvent molecules are treated by the MM level [79]. QM and QM/MM methods has been used to explore different MOF-catalyzed reactions, gas adsorption, gas separation and storage [80, 81]. However, the mechanism of drug adsorption on MOFs internal surfaces as studied by QM/MM calculation is scarcely reported. Different functionalized MOF structures provide different mechanisms of chemical reactions with drug molecules, which can be elucidated by QM/MM calculation. As experimental evidence is insufficient for strongly concluding the mechanisms of drug adsorption and drug release, computational tools such as QM/MM can be utilized to provide atomic-level insight into various complex drug@MOF systems [82]. This opportunity helps explore the thermodynamic behaviour of MOFs, electronic properties of MOF crystal structure, drug loading in MOFs and framework-drug interactions.

12.3.4 Molecular docking

The conformational and binding affinity of small biomolecules on the active site of macromolecules (e. g., proteins and lipids) are frequently explored by employing molecular docking simulations [83]. This method is similarly implemented for complex biomolecule@MOF systems based on two main components described by MM parameters: (i) scoring function that calculates the free energy of binding between the biomolecules and the frameworks and (ii) search algorithms to explore the configurational and conformational degrees of freedom for biomolecules within MOF systems [84]. Molecular docking has been used to describe how a guest molecule can be placed on the surface of $[Zn(BDC)(H_2O)_2]_2$, for example, by the best conformation and binding affinity [85]. Gemcitabine (Gem), ibuprofen (Ibu), methylene blue (Mtb) and amoxicillin (Amx) were selected for study because of their familiarity in medical research. V-shaped clefts in $[Zn(BDC)(H_2O)_2]_2$ were found to be the favorable sites for the biomolecules. The lowest binding affinity estimated for each biomolecule@MOF system illustrated the best conformation of the guest molecules within the frameworks. Specifically, Ibu and Mtb fit entirely inside the clefts and induced strong interactions with coordinated zinc (Zn) metal and weak π–π interactions with the phenyl rings comprising the backbones of the frameworks. Alternatively, Gem and Amx did not fit into the V-shaped clefts due to their larger molecular size.

Docking is then considered a fast approach to identifying the interactions involved in MOF systems. The results from the docking simulation agreed with experimental observations that imatinib (Imt) achieved better adsorption in MIL-101(Cr) compared to MIL-100(Fe) [86]. The binding energy for Imt@MIL-101(Cr) and Imt@MIL-100(Fe) systems was obtained as –9.90 and –9.17 kcal mol^{-1}, respectively. The Imt molecule had a strong interaction with the coordinatively unsaturated metal sites (CUSs), Cr(III) and Fe(III), through its amide group. This phenomenon occurred since MIL-101(Cr) had a larger surface area than MIL-100(Fe). Therefore, more interactions and adsorption on the MIL-101(Cr) surface was expected. In a different report, doxorubicin (Dox) was introduced into ZIF-8 (Figure 12.5a) and the best conformation of the guest molecule was generated by using a docking simulation [87]. Phenolic oxygen atoms created the strongest interaction with Zn(II) in ZIF-8. Four tautomeric

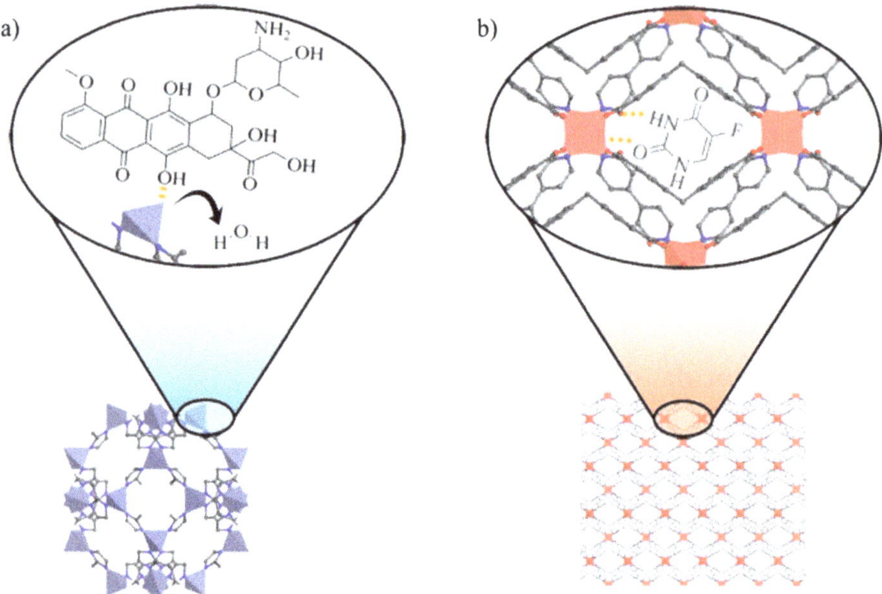

Figure 12.5: Schematic representation for the interaction of a drug molecule (a) onto the metal site and (b) within the pore.

forms of 5-FU were also docked on the MOF's pore, who had a chemical formulae of $[Cu(L)(4,4'\text{-bipy})(H_2O)]_2 \cdot 1.5nCH_3CN$ (H_2L = diphenylmethane-4,4'-dicarboxylic acid) [88] (Figure 12.5b). The best confirmation for the 5-FU molecules presented strong electrostatics and HB interactions with Cu(II) and atoms surrounding Cu(II), respectively. A recent molecular docking and DFT study on 5-FU with MIL-101(Mg) showed that the size of the pore played a major role on the electronic stability of the drug, with π–π stacking being the strongest interactions [89].

CD-MOFs (cyclodextrin-based MOFs) embody two different pockets including γ-CD pairs (0.8 nm) and spherical cages (1.7 nm). Encapsulating azilsartan (Azl) into the CD-MOFs revealed several critical findings [90]. All molecules initially preferred to be docked in the smaller pockets as a result of a stronger framework-guest interaction (HB interaction between the hydroxyl groups of γ-CDs and Azl molecules). However, three Azl molecules were packed as an assembly in the spherical cages at higher loadings, which was governed by guest-guest interactions. Adsorption of biomolecules into the MOF is attributed to framework-guest interactions and the docking technique is useful for exploring atomic-level insight of these interactions. According to the MOF's cage and the pockets available in the MOF's internal environment, sufficient sized biomolecules can be docked via their best conformation within the pockets. Also, biomolecules are further stabilized by different framework-guest interactions. CUSs are the most important sites for making strong interactions with biomolecules.

12.4 Factors affecting host-guest interactions in MOFs

In MOF chemistry, a variety of framework-guest interactions have been elucidated by computational tools, which are useful to support experimental results. Metal sites in MOFs are proven suitable for coordination bond interactions whereas organic linkers can provide intermolecular interactions. Functionalizing organic linkers with polar groups, such as hydroxyl (–OH) and amine ($–NH_2$) groups, can yield more hydrogen-bonding (HB) interactions. Designing MOFs for better interactions with the guest molecule is therefore a crucial step in the development of these materials for applications concerning biomolecule encapsulation. Improving framework-guest interactions not only increases the loading of molecules into MOF pores but can also reduce the release rate of the molecules if necessary [91]. Although framework-guest interactions are governed by many factors, MOF cage size and shape, the presence of different functional groups and CUSs should have the highest consideration.

12.4.1 The Effect of Cage and Pore Size

Appropriate MOF cages (e. g., size, shape) is known to be crucial for achieving better framework-guest interactions, subsequent higher drug adsorption and slower drug release. The largest average potential energy of adsorption realized for a framework-guest interaction was that of ibuprofen (Ibu) on the MIL-53 surface ($-140\,kJ\,mol^{-1}$) [92]. Smaller values of –130 and $-85\,kJ\,mol^{-1}$ were calculated for MIL-100 and MIL-101, respectively. This result showed that MIL-53 has more framework-guest interactions and a higher potential to preserve Ibu molecules. This can be attributed to a smaller and narrower cage in MIL-53 compared to MIL-100 and MIL-101. Three MOF, $[(CH_3)_2NH_2][(Sm_3(L1)_2(HCOO)_2(DMF)_2(H_2O)]\cdot 2DMF\cdot 18H_2O$ (1), $[(Cu_2(L2)(H_2O)_2]\cdot$ 2.22DMA (2) and $[Zn_2(L1)(DMA)]\cdot 1.75DMA$ (3) (H_4L1=2,6-di(3',5'-dicarboxylphenyl)-pyridine and H_4L2 = 2,5-di(3',5'-dicarboxylphenyl)pyridine) were chosen for study due to their different window sizes [93]. The window sizes are 4.4 × 4.1 and 5.5 × 5.0 Å, 4.2 × 4.6 and 5.8 × 7.3 Å and 4.3 × 4.5 and 6.4 × 7.8 Å for MOFs 1, 2 and 3, respectively. 5-FU was loaded into MOF 2, which showed a release rate of 96 % after 120 h in phosphate saline buffer solution at pH = 7.4 and 37 °C. When 5-FU was loaded in MOFs 1 and 3, release rates of 72 and 79 % were achieved, respectively, after 96 h under the same conditions. GCMC simulations for 5-FU loading into MOFs 1, 2 and 3 elucidated two key points. First, the 5-FU molecules had strong electrostatic interactions with Zn(II) when they were accommodated in the MOFs' smaller cage, which led to slower drug release from MOF 1 compared to MOF 3. Second, 5-FU molecules were preferably aggregated by van der Waals (guest-guest) interactions when they filled up the larger pores in MOF 3. The release mechanism of 5-FU from MOF 3 was described as initially fast with the aggregated drugs in the large pore being released first, followed by a slower release from the smaller pore. The same cannot be said for MOF 1 as the rate of drug release was inhibited by the organic linkers.

Another report on the 5-FU study showed that the drug initially preferred to load in the smaller tetrahedral cage of UiO-66 than filling the larger octahedral cage, which was occupied at higher loadings [94]. The drug molecule was well stabilized by HB interactions with the MOF's metal-oxide ($C{=}O_{5\text{-}FU}\cdots H\text{-}O_{UiO\text{-}66}$) SBU and π–π interactions with the organic linkers. It is important to note that a strong binding between the drug molecules and MOF cages can be a good indicator of framework-guest interaction. In another report, a comparison of three different MOFs, including MIL-100(Fe), UiO-66(Zr) and MIL-127(Fe), were considered for loading of aspirin (Asp) and ibuprofen (Ibu) inside their cage [56]. Each MOF structure has two different cage sizes: (i) MIL-100(Fe) has the largest cage sizes of 25 and 29 Å; UiO-66(Zr) has cage sizes of 8 and 11 Å and MIL-127(Fe) has the smallest cage sizes of 6 and 10 Å. The pore volume values of these MOFs are 1.42, 0.64 and 0.47 $cm^3\,g^{-1}$ for MIL-100(Fe), UiO-66(Zr) and MIL-127(Fe), respectively. It was observed that Asp@MIL-127(Fe) and Ibu@MIL-127(Fe) released the least amount of loaded drugs (e. g., 37 and 48 %, respectively, after 24 h). This behavior is related to cage size as the drug molecules formed framework-guest interactions at the same level as drug aggregation (guest-guest interactions) occurred due to the insufficiently small cage sizes. In contrast, MIL-100(Fe) released the greatest amount of Asp and Ibu by 99 and 84 %, respectively, after 24 h. Failure to correlate the drug molecule size and MOF pore size may lead to inefficient encapsulation and poor framework-guest interactions. HKUST-1, which is composed of a small cage (10 Å) and a large cage (14 Å) was used for encapsulating caffeine (Caf) [95]. A DFT geometry optimization revealed that the Caf molecule could not be accommodated within the small cage due to its larger molecular size, thus it preferred to be housed within the large cage. Accordingly, Caf molecules were observed to be strongly adsorbed on Cu(II) CUSs with a calculated interaction energy of $-84\,kJ\,mol^{-1}$.

12.4.2 Functionalization of organic linkers

The organic linker is another potential site for enhancing framework-guest interaction. When 5-FU was loaded into the mixed-linker MOF, $[Zn(BTC)(HME)]\cdot(DMAc)(H_2O)$ (H_3BTC = 1,3,5-benzene tricarboxylic acid, HME = protonated melamine, DMAc =N,N-dimethylacetamide) [96], the organic linkers containing N-donor sites were found to enhance the framework-guest interactions. GCMC simulations demonstrated that 5-Fu molecules created strong HB interactions with the N-donor sites on the backbone of the frameworks ($N\text{-}H_{5\text{-}FU}\cdots N_{MOF}$) at zero coverage (pressure $\rightarrow$ 0). Furthermore, 5-Fu molecules formed HB and π–π interactions at near-saturation levels of loading. The authors succeed in reducing the time of 5-FU release in the mixed-linker MOF compared to the parent MOF [97]. In another example, Gem was found to prefer interactions with the organic linker of HO-IRMOF-74-III rather than available CUSs [42]. Geometry optimization illustrated that the Gem molecule was preferential to interactions with O atoms surrounding Mg(II) in the parent MOF (IRMOF-74-III). Gem@HO-IRMOF-74-III recorded a lower binding energy of -33.8 kcal mol^{-1} compared to Gem@IRMOF-74-III with an energy binding of -30.9 kcal mol^{-1}. The impact of functionalizing IRMOF-16 using hydroxyl groups (–OH) was further investigated utilizing several computational techniques [91]. According to the DFT-based geometry optimization, Gem molecules tended to be optimized over metal-oxide clusters supported by framework-guest interactions. Also, Gem molecules were stabilized by three primary HB interactions with hydroxyl functional groups (–OH) grafted on the organic linker: (i) $N\text{-}H_{Gem}\cdots O_{IRMOF\text{-}16}$ (2.0 Å), (ii) $N_{Gem}\cdots H\text{-}O_{IRMOF\text{-}16}$ (2.2 Å) and $O\text{-}H_{Gem}\cdots O_{IRMOF\text{-}16}$ (2.5 Å). These interactions enhanced the interaction energy of the Gem by 40 % compared to the parent IRMOF-16. Additionally, GCMC simulations estimated that the release of the drug molecule from the Gem@HO-IRMOF-16 system was as slow as Gem@IRMOF-16. Although functionalizing organic linkers increases interaction sites for biomolecules, selecting desirable functional groups for better framework-guest interactions and release must be considered.

To understand the influence of various functional groups (e. g., –H, $-CF_3$, $-CH_3$, –OH, –F, $-NH_2$, $-NO_2$ and –Br) were grafted on MIL-88B(Fe) for trapping caffeine (Caf). A quantitative structure-activity relationship (QSAR) analysis was then performed [98]. This study suggested that protic functional groups such as amine ($-NH_2$) and hydroxyl (–OH) can enhance Caf adsorption remarkably by serving as anchoring points. Also, Caf adsorption was improved by enrichment of methyl substitutes (mono, di, tri and tetra) via van der Waals interactions. This theoretical consequence was reflected later in laboratory experiments. Oridonin (Ori) was used for encapsulation within various mono-substituted functionalized MOF-5 structures [99]. The drug release profile demonstrated cumulative Ori release percentage in order of MOF-5 > Br-MOF-5 > H_3C-MOF-5 > H_2N-MOF-5 > CH_2CH-MOF-5 > HO-MOF-5 at pH = 5.5 and 7.4.

The rotational dynamics behavior of functionalized organic linkers was studied by Maurin et al. [100]. They found that perturbation of the ligand flip in H_2N-UiO-66(Zr) was as much as in $(HO)_2$-UiO-66(Zr), Br-UiO-66(Zr) and UiO-66(Zr) owing to the nature of the functional group. The highest rotational dynamics for H_2N-UiO-66(Zr) enhanced the transition rate of caffeine (Caf) from one cage to the next cage. In addition, DFT optimization demonstrated that Caf molecules adopted different orientations inside the H2N-UiO-66(Zr) cage compared to other MOF cages. This was to achieve better optimization and framework-guest interactions. A long-functionalized linker can be favourable for further stabilization of biomolecules on the framework's surface as suggested by Froudakis and colleagues [79]. The N atom in Tamoxifen (Tam) formed major HB interactions with a hydroxyl group in HO-IRMOF-14 ($N_{Tam}\cdots H\text{-}O$, 2.8 Å) and HO-IRMOF-16 ($N_{Tam}\cdots H\text{-}O$, 1.94 Å). Although the strongest HB interaction was found for Tam@IRMOF-16, being a longer and more flexible linker in OH-IRMOF-16 allowed for freely accessible rotating phenyl rings and a greater formation of π–π interactions with the Tam molecules. This condition promoted the overall interaction energy of the system.

12.4.3 Coordinatively unsaturated metal sites

MOFs having coordinatively unsaturated metal sites are quite active for the adsorption of biomolecules [101]. MOFs containing CUSs take advantage of strong coordination interactions with biomolecules and result in a slow-release rate. This property has enabled MOFs to be considered in medical nanofabrication. For example, DFT calculations demonstrated that heparin (Hep) can replace the weakly bound H_2O molecules on the CUSs of MIL-101(Fe) [102] (Figure 12.6a). The slow release of Hep and low toxicity of the Hep@MIL-101(Fe) system allowed it to be combined with thrombolytic alumina-based sol-gel coating for further biomedical applications. DFT calculations are shown to be capable of properly explaining the CUS-guest interaction. Although this computational technique has been frequently used to evaluate the impact of various types of CUSs (e. g., Fe, Ni, Cr, Co, Cu, Zn, etc.) on gas adsorption, it has not been well described for biomolecular adsorption [103, 104].

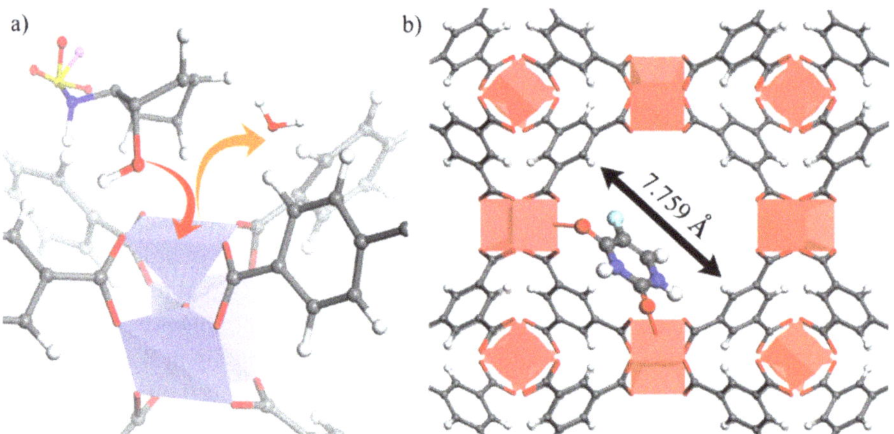

Figure 12.6: (a) Interaction mechanism of (a) Hep in MIL-101(Fe) [102] and (b) 5-FU in HKUST-1 [107].

DFT geometry optimization of ibuprofen (Ibu) encapsulated within MIL-101(Cr) and UMCM-1(Zn) made for a better comparison of framework-guest interactions [105]. The carboxylic groups of Ibu molecules were found to situate near to the metal-oxide SBU in MIL-101(Cr) and UMCM-1(Zn) by 2.141 and 7.267 Å, respectively. The Ibu molecule was able to form a coordination bond with the Cr(II) CUSs in MIL-101(Cr) by the lowest binding energy (–73.17 kJ mol^{-1}). In contrast, the Ibu molecule did not create any chemical bond on the saturated metal-oxide SBU in UMCM-1(Zn), thus, generating a higher binding energy of $-34.97\ \text{kJ mol}^{-1}$. Observation of the highest occupied molecular orbital (HOMO) gave further evidence for the possibility of creating a bond between the carboxylic group of Ibu and MIL-101(Cr). This bonding mechanism explained the delayed release of Ibu from MIL-101(Cr) as reported by experimental results [106].

Encapsulating 5-FU into HKUST-1 showed a noticeable effect on the window size of the frameworks, which was elucidated by geometry optimization using the DFT level of computation [107]. Two negatively charged O atoms in 5-FU molecules were able to create strong dative bonding with two positively charged Cu(II) CUSs. The 5-FU@HKUST-1 system was further stabilized by HB interactions between amine in 5-FU and O atoms surrounding Cu(II) ($\text{N-H}_{5\text{-FU}}\cdots\text{O}_{\text{HKUST-1}}$). Analysis of the optimized 5-FU@HKUST-1 system indicated that the 5-FU molecule built a bridge between two paddlewheel SBUs and reduced their distance to each other (Figure 12.6b). A wide series of nonsteroidal antiinflammatory drugs (NSAIDs) were chosen for adsorption on three Zr(IV)-based MOFs: (i) UiO-66, (ii) MOF-808 and (iii) MOF-802 [108]. According to the report, poor adsorption of the drugs into MOF-802 arose from the low porosity of the framework. NSAID drugs had high adsorption on UiO-66 and MOF-808. DFT calculations found that Zr(IV) CUSs in UiO-66 and MOF-808 had the highest affinity for the adsorption of NSAIDs. Furthermore, π–π interactions between the phenyl rings of the guest NSAID molecules and the phenyl rings of the organic linkers were the second most important interaction for achieving reasonable adsorption. Therefore, understanding the framework geometry, the chemistry of the organic linkers and metal sites, as well as engineering the organic linker with proper functional groups is critical for promoting framework-guest interactions.

12.4.4 Diffusivity

An important factor for determining whether a MOF can be a promising drug carrier is having a slow, consistent drug release. Hence, diffusion inside the MOF pores is a property that must be evaluated for candidate drug molecules. This property should be compared with the freely moving drugs in the same media. Although diffusion of drugs obtained from computational analysis can be extended to their experimental release, this section is focused on the mobility of drug molecules inside of MOFs. Here, MD simulation is commonly used to analyze the mobility of drug molecules. MD trajectories allow for calculations of mean square displacement (MSD) and self-diffusion coefficient (D) for drug molecules.

Che and colleagues explored 28 different MOFs in order to determine the textural properties of the MOF structures and their relationship with adsorption and diffusion of the drug molecule amlodipine (Aml) [109]. Among the MOFs studied, the MOF-74 series produced noteworthy results for self-diffusion of Aml (Table 12.2). It was found that the mobility of the drug molecule was the fastest inside IRMOF-74-II due to lower chemical affinity and heat of adsorption (Q_{st}) (including framework-drug and

Table 12.2: Pore volume (V_p) and surface area (SA) for MOF-74 series, and diffusion coefficient (D) and heat of adsorption (Q_{st}) for Aml in the corresponding MOFs.

MOFs	V_p ($cm^3\ g^{-1}$)	SA ($m^2\ g^{-1}$)	D ($10^{-11}\ m^2\ s^{-1}$)	Q_{st} ($kJ\ mol^{-1}$)
Zn-MOF-74	0.57	1256.53	1.08	154.88
Mg-MOF-74	0.70	1572.19	1.62	161.73
IRMOF-74-II	1.23	2215.05	33.53	9.03
IRMOF-74-III	1.42	2620.76	3.13	75.85
IRMOF-74-IV	1.82	2759.10	0.26	116.04
IRMOF-74-V-hex	2.11	2919.84	0.22	102.39
IRMOF-74-V	2.24	3017.80	1.64	61.00
IRMOF-74-VII-oeg	2.54	2766.01	0.11	128.45
IRMOF-74-VI	2.55	2934.08	0.94	95.28
IRMOF-74-VII	2.96	3342.32	0.15	101.97
IRMOF-74-IX	3.74	3011.88	0.06	122.19

drug-drug interactions). MOF-74 had a more desirable pore size than IRMOF-74-II for Aml since the drug molecules fit well inside the pores leading to a higher heat of adsorption. In the largest IRMOF-74 pores, IRMOF-74-IX, the diffusion coefficient of Aml reached the lowest value ($0.06\times10^{-11}\ m^2\ s^{-1}$) with a high Q_{st} of 122.19 kJ mol^{-1}. This was attributed to the larger surface area that leads to greater frameworks-drug interactions as well as tighter drug packing for more drug-drug interactions. As drug molecules initially leave the MOF pores by loosening the drug-drug interactions at an early stage of release, IRMOF-74-IX is presumed to release a huge amount of drug molecules during this time. Therefore, it is important to note that a lower diffusion coefficient value is not an indicator of a slow and controlled drug release. Although this issue may seem challenging, carefully selecting a MOF having a suitable pore size according to the drug's molecular size can provide the lowest diffusion and the slowest release.

The strength of drug-drug interactions in the drug packing inside MOF pores should impact the mobility of drug molecules. In general, hydrophilic drug molecules form stronger interactions and drug aggregations, whereas hydrophobic drug molecules make more loosely-held aggregations through weak interactions. For example, four types of drugs, including valproate (Val), gabapentin (Gab), levetiracetam (Lev) and phenytoin (Phe), were stored in ZIF-8, ZIF-67 and ZIF-90 [110]. As the authors reported, Val, Gab, Lev and Phe molecules were loaded into ZIF-8 at 25, 25, 24 and 21 mol uc^{-1} loadings, respectively. The values of diffusion coefficient for Val, Gab, Lev and Phe into ZIF-8 were examined to be 1.04, 1.07, 0.91 and 1.03 ($10^{-4}\ cm^2\ s^{-1}$), respectively. Although ZIF-8 was filled by Val and Gab as much as Lev, the mobility of Lev molecules was slowest inside the pore. Figure 12.7 shows the 3D representation of the molecular structures for the drugs. The most polar chemical groups, such as carbonyl and amine, can be seen in the structures of Lev and Phe. In comparison, Lev molecule is more flexible than Phe, which is caused by rotation of the chiral center and a methyl group, whereas the rigidity in Phe molecule is due to the presence of

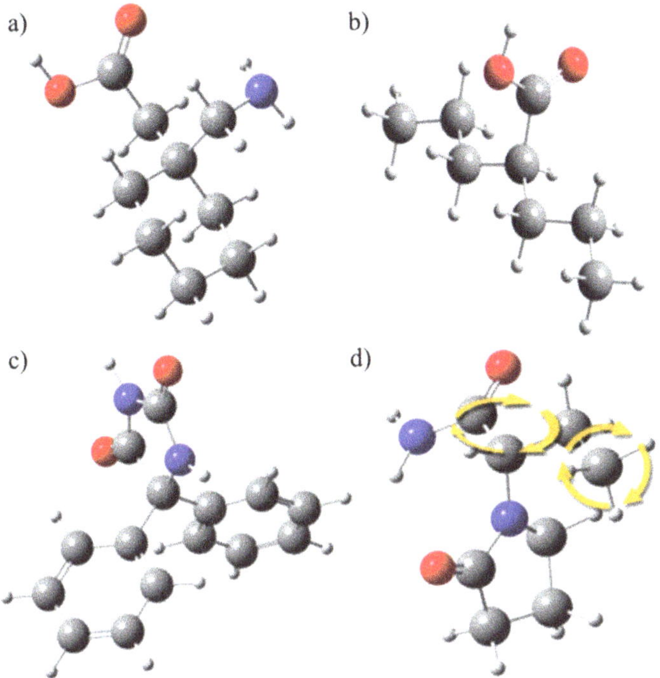

Figure 12.7: 3D illustration of (a) gabapentin, (b) valproate, (c) phenytoin and (d) levetiracetam.

two phenyl rings connected to imidazolidine-2,4-dione. Therefore, slow movement of Lev molecules can be clearly explained by having more polar chemical groups and flexibility that assist to build stronger and tighter hydrophilic drug packing inside the ZIF-8 pore.

The suspension media can also affect the mobility and diffusivity of drugs. In absence of solvent, MD simulation recorded an 8.6 wt% and 7.353×10^{-6} $cm^2\,s^{-1}$ for the loading and the release processes, respectively, for doxepin (Dox) inside ZIF-8 [111]. However, Dox molecules were not able to properly accommodate interaction sites in the presence of ethanol. As a result, the self-diffusion coefficient of Dox molecules was reduced to 1.004×10^{-4} $cm^2\,s^{-1}$, meaning that the drug molecules moved faster inside ZIF-8 pores. A MD simulation had also revealed the impact of drug concentrations and drug-drug interactions on drug mobility. As reported by Dekamin et al. [112], the mobility of drug molecules (including temozolomide (Tmz), alendronate (Ald) and 5-fluorouracil (5-FU)) inside the UiO-66 pore at low concentrations were getting faster nearing the end of MD simulation runs. In contrast, the drug molecules at high concentrations always indicated consistent slow mobility. More negative drug-drug interaction energy in high drug concentration systems was reasoned for its slower mobility. The drug-drug interaction energy for Tmz, Ald and 5-FU were examined as −0.48, −0.97 and −1.68 kJ mol^{-1} in low drug concentration, respectively, and as −15.69, −6.67

and −8.88 kJ mol^{-1} in high drug concentration, respectively. In high concentration, 5-FU had the slowest mobility due to stronger drug packing. 5-FU is the smallest drug molecule compared to Tmz and Ald. Drugs having smaller molecular sizes are able to strongly aggregate inside MOF pores, which can influence their mobility at high concentrations.

Nevertheless, mobility of drug molecules inside MOF pores at low concentration may be governed by framework-drug interactions as well as rotational linker dynamics. For instance, diffusion of low concentration (32 wt%) of gemcitabine (Gem) into IRMOF-74-III and its −OH functionalized linker were estimated by using MD simulation [42]. The values for the diffusion coefficient of Gem in IRMOF-74-III and OH-IRMOF-74-III were recorded as 6.6×10^{-12} and $12.5 \times 10^{-12}\ m^2\ s^{-1}$, respectively, meaning that Gem molecules had faster mobility in OH-IRMOF-74-III. Although many computational works [91, 99] have proposed that polar functional groups (such as −OH and $-NH_2$) incorporated on organic linkers made for stronger HB interactions with drug molecules, IRMOF-74-III suggested that having bulkier functional groups was more efficient to slow down the drug diffusion via wider contact surface. Devautour-Vinot and colleagues implemented DFT calculations and dielectric relaxation spectroscopy in order to find UiO-66(Zr)-caffeine interactions and the impact of functional groups on rotational linker dynamics [101]. It was realized that a bulkier functional group used in the study, −Br, slowed down the ligand flip, which significantly reduced the mobility of caffeine molecules inside the UiO-66(Zr) pore. Functional groups such as $-NH_2$ enhanced the flipping of phenyl ring resulting in accelerated drug mobility from one cage to another. Rotational linker dynamics are mainly divided into four types: (i) complete rotation, (ii) partial rotation, (iii) rotation of side groups and (iv) mechanically interlocked molecule rotation [113]. In MOF chemistry, however, understanding of these four types of rotational linker dynamics toward adsorption and delivery of drugs remains limited.

12.4.5 Loading capacity

In order to achieve a sustainable duration of therapeutic activity and a reasonable dose for drug delivery, carrier systems should have slow-release kinetics and high loading capacity. So far, it has been clear that MOFs are capable of controlling drug release by providing various framework-drug interactions (including ligand-drug and metal node-drug). In the interest of higher drug loading inside MOF pores, more factors must be considered. Framework-drug interactions are stronger than drug-drug interactions, which can be determinative for a consistent rate of drug release. However, drug-drug interactions may become important after the sudden loading of drug molecules and the saturation of MOF surfaces. As an example, after CD-MOF-1, MOF-74 and BioMOF-100 surfaces were almost saturated by ibuprofen (Ibu) (framework-drug interactions), aggregation of Ibu molecules began in the center of MOF pores [92]. Subsequently, Ibu molecules made a tighter packing as the MOF pore was filled by the drug molecules. Being closer to Ibu molecules assisted them to build strong HB interactions via carboxylic groups (O-H···O=C).

It is well established that the size of the drug molecule is a major factor affecting the loading within MOFs pores. Simulated maximum loading for 5-FU reached to 477 mg g^{-1} in ZIF-8, which was as high as Caf in ZIF-8 (422 mg g^{-1}) [114]. The same trend was experimentally observed for maximum loading of Hydroxyurea (Hxu), 5-FU and Ibu in HKUST-1and H_2N-MIL-53(Al) [115]. For each drug@MOFs complex system, the values of loading were obtained in the order of Hxu > 5-FU > Ibu, which corresponded to the size of drug molecules. In another study, the loading of 5-FU, Hxu and Mercaptopurine (Mpt) were compared inside ZIF-7, ZIF-8 and ZIF-9 cages [116]. It should be noted that the molecular size of 5-FU and Hxu is relatively smaller than Mpt. Computational adsorption isotherms estimated loading values of 0.65 and 0.61 g g^{-1} for 5-FU and Hxu in all ZIFs, respectively. The loading values had a good agreement with experimental reports [117]. Interestingly, 5-FU and Hxu molecules did not show a significant difference in loading among the ZIFs, even though ZIF-8 (0.485 cm^3 g^{-1}) had a significantly larger pore volume than ZIF-7 (0.078 cm^3 g^{-1}) and ZIF-9 (0.076 cm^3 g^{-1}). In contrast, Mpt loaded into ZIF-7, ZIF-8 and ZIF-9 pores by 0.36, 0.55 and 0.44 g g^{-1}, respectively. Smaller drug molecules produced a higher loading, which was less dependent on MOF pore volume, whereas bigger drug molecules were more affected by MOF pore volume size. Furthermore, as mentioned previously, it is expected that smaller drug molecules will build stronger drug-drug interactions and tighter drug packing.

Although drug-drug interactions and drug packing can be similarly found for conventional porous materials such as mesoporous silica, carbonaceous materials and polymeric matrixes [118], MOFs standout as having large pore size and pore volume. Researchers have demonstrated that MOFs having larger pore size and pore volume are able to load more drugs to induce larger drug packing, but drug-drug interactions have not been well presented. For instance, in a computational investigation on drug loading inside a series of IRMOF-74s [119], IRMOF-74-XI was identified as having the highest loading of methotrexate (Mtx) (2.78 g g^{-1}) and 5-FU (4.24 g g^{-1}) as a result of the largest pore size. Following the trend, MOF-74-I demonstrated the lowest loading Mtx (0.31 g g^{-1}) and 5-FU (0.71 g g^{-1}) owing to the smallest pore size (Figure 12.8). All IRMOF-74 family members built strong linear correlations for values of Mtx (R^2 = 0.98) and 5-FU (R^2 = 0.99) loading. It was clearly shown that increasing the pore size of MOFs led to higher loading of drug molecules. From the same point of view, calculations for the loading of Ibu for the IRMOF-74 family showed the highest loading (2.55 g g^{-1}) for the IRMOF-74-XI [46]. Other notable findings are the loading of Ibu in MIL family, which was obtained in the order of MIL-53(Fe) (0.22 g g^{-1}) > MIL-100(Fe) (0.57 g g^{-1}) > MIL-101(Cr) (1.03 g g^{-1}); a trend that was in line with their pore size. The maximum loading of Ibu in MIL-101(Cr) was in agreement with another computational prediction (1.11 g g^{-1}) [106] and experimental observation (1.37 g g^{-1}) [107].

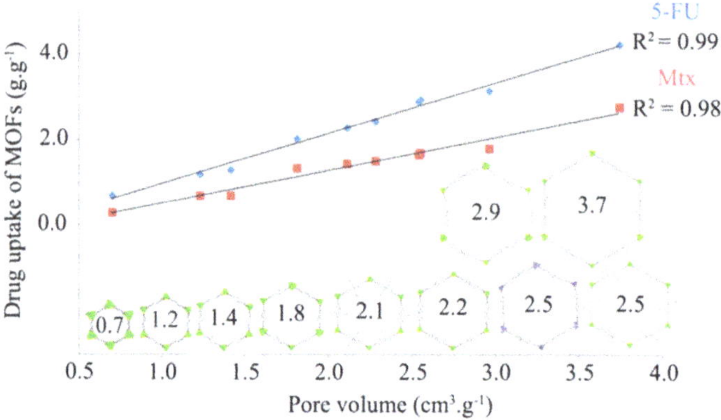

Figure 12.8: Correlation between drug loaded into IRMOF-74 family and their pore volume [122].

Although a MOF's pore volume and pore size are always taken into account when studying loading capacity for drugs, developing the surface area using functional groups may also enhance the drug loading capacity. To do so, functionalization of the frameworks using a post-synthetic modification is an appropriate strategy. It is known that functionalization of MOFs within their pores reduces the porosity and drug loading capacity, whereas surface coating of MOF crystals leads to increase in surface area and drug loading capacity. UiO-66 and its carboxylic acid-functionalized pore (UiO-66-COOH) were employed for the adsorption of Caf molecules [120]. Nitrogen adsorption isotherms and Brunauer–Emmett-teller (BET) surface area analysis proved that functionalizing UiO-66 with carboxylic acid groups reduces the pore volume (from 0.44 to 0.32 $cm^3 g^{-1}$) and surface area (from 1044 to 627 $m^2 g^{-1}$), respectively. Consequentially, loading of Caf molecules into UiO-66 and UiO-66-COOH were determined to be 21.4 and 17.5 wt%, respectively. In another example, GCMC simulations estimated the maximum loading of Gem molecules into OH-functionalized-IRMOF-16 and its unmodified framework as 4343 and 4749 $mg g^{-1}$, respectively [91]. In another study, MOF-5 provided the highest loading capacity for oridonin (Ori) compared to all functionalized-MOF-5 structures [99]. The results for maximum loading capacity of Ori were observed in the order of MOF-5 > HO-MOF-5 > H_3C-MOF-5 = Br-MOF-5 > H_2N-MOF-5 > CH_2=CH-MOF-5. It seems that porosity is a dominant factor over framework-drug interactions for the drug loading capacity of MOFs.

That aside, surface coating of MOFs can be an alternative method to enhance the surface area and drug capacity. One-pot synthesis of ZIF-8 capped with poly(ethylene glycol) (PEG) indicated an improved drug loading capacity for Dox molecules (Dox@ZIF-8/PEG, 10 wt%) in acidic solution (pH = 3) [121]. another study, ZIF-8 pores were able to receive only 4.9 wt% Dox molecules in an aqueous solution [87]. In the case of ZIF-8, coating its surface using large functional groups is recommended for greater drug loading and controlled release [122]. Moreover, hydroxyapatite (HAp) was used to coat Fe_3O_4@Fe-MOF's surface with the purpose of loading anticancer drug,

Dox, and controlling the drug release in different pH solutions [123]. The new composite microspheres, Fe_3O_4@Fe-MOF/HAp, successfully improved the loading capacity of Dox compared to core-shell nanospheres, Fe_3O_4@Fe-MOF. In pH = 6.0, 6.5, 7 and 7.4, the loading capacity of Dox molecules on Fe_3O_4@Fe-MOF surface was given approximately 25, 34, 43 and 53 mg g^{-1}, respectively, while they were increased on Fe_3O_4@Fe-MOF/HAp surface as 33, 43, 62 and 75 mg g^{-1}, respectively. Yang et al. presented that engineering of the MOF's outer surfaces using different functional groups with the aim of increasing loading capacity and controlling anticancer drug release is frequently reported [124]. However, almost all of these reports are experimental works and understanding the mechanism of drug adsorption onto MOF's outer surface at the atomic scale is very limited. To do so, the most common computational tools in this research path such as MD and GCMC simulations could be pursued.

Incorporating any functional group that involves MOF pores, even small and polar groups such as hydroxyl (–OH), decreases the loading capacity of drugs (Figure 12.9a). This restriction is attributed to the reduced porosity of MOF. In contrast, engineering the MOF outer surface using functional groups that have a greater van der Waals surface promotes the loading capacity of drugs due to expanding the surface area (Figure 12.9b). This approach is especially recommended for MOFs having small pore volume (e. g., ZIF-8) and for drugs having large molecular size (e. g., Dox).

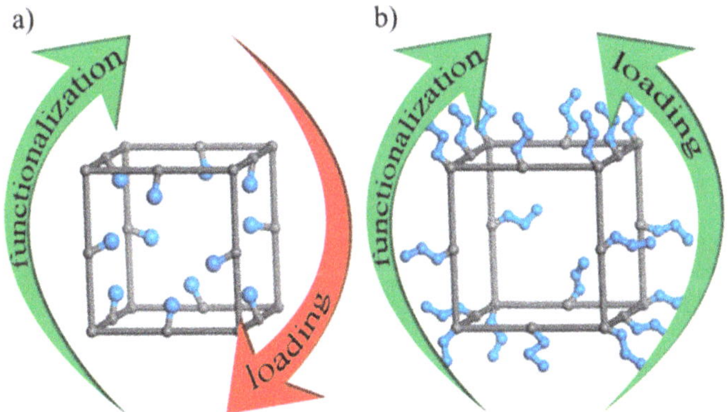

Figure 12.9: Schematic representation for (a) functionalization involved in MOF's pore and (b) functionalization of MOF's outer surface.

12.5 Future directions

Recent studies have shown MOFs, as a versatile and tuneable mesoporous/nanoporous materials, can potentially act as the vehicle for targeted drug delivery. However, the actual applications may depend on the delivery mechanism and the toxicity of the MOFs themselves. For a particular MOF to be applied for drug delivery, many other factors such as the physical state of the drug@MOF complex, temperature and pH stability, and bioavailability should be considered. There should be an immediate need for a full study on the applicability of a MOF-based drug delivery system to be designed and standardized.

Computational methods have played a major role in the advancement of medical and pharmaceutical industries. By leveraging the time- and cost-efficiency of theoretical and computational predictions, faster development of drugs and drug delivery systems has been accomplished. Due to the rapid advancement of computer hardware and artificial intelligence (AI) technology, machine-learning methods are currently "the way forward" as they are being continuously developed by many scientists, including in MOF research. Although it may take some time to optimize the algorithms and the processes, these are exciting times for computational research to truly make an impact on the global scale. Hopefully soon, an optimized drug carrier can be modeled in just a few clicks.

Bibliography

[1] Song Y, Wang F, Liu G, Luo S, Guo R. Promotion effect of carbon nanotubes-doped SDS on methane hydrate formation. Energy Fuels. 2017;31:1850–7.

[2] Wu Z, Zhao D. Ordered mesoporous materials as adsorbents. Chem Commun. 2011;47:3332–8.

[3] Reeve PJ, Fallowfield HJ. Natural and surfactant modified zeolites: a review of their applications for water remediation with a focus on surfactant desorption and toxicity towards microorganisms. J Environ Manag. 2018;205:253–61.

[4] Yaghi OM, Eddaoudi M, O'Keeffe M, Li H. Design and synthesis of an exceptionally stable and highly porous metal–organic framework. Nature. 1999;402:276–9.

[5] Yaghi OM. Reticular chemistry – construction, properties, and precision reactions of frameworks. J Am Chem Soc. 2016;138(48):15507–9.

[6] Abtab SMT, Alezi D, Bhatt PM et al. Reticular chemistry in action: a hydrolytically stable MOF capturing twice its weight in adsorbed water. Chem. 2018;4:94–105.

[7] Yaghi OM. Reticular chemistry in all dimensions. ACS Cent Sci. 2019;5(8):1295–300.

[8] Juan-Alcañiz J, Gascon J, Kapteijn F. Metal–organic frameworks as scaffolds for the encapsulation of active species: state of the art and future perspectives. J Mater Chem. 2012;22:10102–18.

[9] Majewski MB, Howarth AJ, Li P, Wasielewski MR, Hupp JT, Farha OK. Enzyme encapsulation in metal–organic frameworks for applications in catalysis. CrystEngComm. 2017;19:4082–91.

[10] Xia H, Li N, Zhong X, Jiang Y. Metal–organic frameworks: a potential platform for enzyme immobilization and related applications. Front Bioeng Biotechnol. 2020;8:695.

[11] Lian X, Fang Y, Joseph E et al. Enzyme-MOF (metal–organic framework) composites. Chem Soc Rev. 2017;46:3386–401.

[12] Chen Y, Han S, Li X, Zhang Z, Ma S. Why does enzyme not Leach from metal–organic frameworks (MOFs)? Unveiling the interactions between an enzyme molecule and a MOF. Inorg Chem. 2014;53(19):10006–8.

[13] Al-Sharabati M, Sabouni R, Husseini GA. Biomedical applications of metal–organic frameworks for disease diagnosis and drug delivery: a review. Nanomaterials. 2022;12(2):277.

[14] Liang K, Ricco R, Doherty C et al. Biomimetic mineralization of metal–organic frameworks as protective coatings for biomacromolecules. Nat Commun. 2015;6:7240.
[15] Li P, Moon SY, Guelta MA, Harvey SP, Hupp JT, Farha OK. Encapsulation of a nerve agent detoxifying enzyme by a mesoporous zirconium metal–organic framework engenders thermal and long-term stability. J Am Chem Soc. 2016;138(26):8052–5.
[16] Danhier F, Le Breton A, Préat V. RGD-based strategies to target alpha(v) beta(3) integrin in cancer therapy and diagnosis. Mol Pharm. 2012;9(11):2961–73.
[17] Pisano C, Cecere SC, Di Napoli M et al. Clinical trials with pegylated liposomal Doxorubicin in the treatment of ovarian cancer. J Drug Deliv. 2013;2013:898146.
[18] Beg S, Rahman M, Jain A et al. Nanoporous metal organic frameworks as hybrid polymer-metal composites for drug delivery and biomedical applications. Drug Discov Today. 2017;22(4):625–37.
[19] Peller M, Böll K, Zimpel A, Wuttke S. Metal–organic framework nanoparticles for magnetic resonance imaging. Inorg Chem Front. 2018;5(8):1760–79.
[20] Della Rocca J, Lin W. Nanoscale metal–organic frameworks: magnetic resonance imaging contrast agents and beyond. Eur J Inorg Chem. 2010;24:3725–34.
[21] Cui Y, Yue Y, Qian G, Chen B. Luminescent functional metal–organic frameworks. Chem Rev. 2012;112(2):1126–62.
[22] Chowdhury MA. Metal–organic-frameworks for biomedical applications in drug delivery, and as MRI contrast agents. J Biomed Mater Res, Part A. 2017;105(4):1184–94.
[23] Xing Q, Pan Y, Hu Y, Wang L. Review of the biomolecular modification of the metal–organ-framework. Front Chem. 2020;8:642.
[24] Canossa S, Wuttke S. Special issue: functionalization chemistry of porous materials. Adv Funct Mater. 2020;30:2070271.
[25] Röder R, Preiß T, Hirschle P et al. Multifunctional nanoparticles by coordinative self-assembly of his-tagged units with metal–organic frameworks. J Am Chem Soc. 2017;139(6):2359–68.
[26] Wang XG, Dong ZY, Cheng H et al. A multifunctional metal–organic framework based tumor targeting drug delivery system for cancer therapy. Nanoscale. 2015;7(38):16061–70.
[27] Kamal NAMA, Abdulmalek E, Fakurazi S, Cordova KE, Abdul Rahman MB. Surface peptide functionalization of zeolitic imidazolate framework-8 for autonomous homing and enhanced delivery of chemotherapeutic agent to lung tumor cells. Dalton Trans. 2021;50(7):2375–86.
[28] Dings RPM, Miller MC, Griffin RJ, Mayo KH. Galectins as molecular targets for therapeutic intervention. Int J Mol Sci. 2018;19(3):905.
[29] Yang K, Yang K, Chao S, Wen J, Pei Y, Pei Z. A supramolecular hybrid material constructed from pillar[6]arene-based host-guest complexation and ZIF-8 for targeted drug delivery. Chem Commun. 2018;54(70):9817–20.
[30] Chen WH, Yu X, Cecconello A, Sohn YS, Nechushtai R, Willner I. Stimuli-responsive nucleic acid-functionalized metal–organic framework nanoparticles using pH- and metal-ion-dependent DNAzymes as locks. Chem Sci. 2017;8(8):5769–80.
[31] Chowdhuri AR, Laha D, Pal S, Karmakar P, Sahu SK. One-pot synthesis of folic acid encapsulated upconversion nanoscale metal organic frameworks for targeting, imaging and pH responsive drug release. Dalton Trans. 2016;45(45):18120–32.
[32] Li Y, Zheng Y, Lai X, Chu Y, Chen Y. Biocompatible surface modification of nano-scale zeolitic imidazolate frameworks for enhanced drug delivery. RSC Adv. 2018;8(42):23623–8.
[33] Li S, Wang K, Shi Y, Cui Y et al. Novel biological functions of ZIF-NP as a delivery vehicle: high pulmonary accumulation, favorable biocompatibility, and improved therapeutic outcome. Adv Funct Mater. 2016;26(16):2715–27.
[34] Wang S, Chen Y, Wang S, Li P, Mirkin CA, Farha OK. DNA-functionalized metal–organic framework nanoparticles for intracellular delivery of proteins. J Am Chem Soc. 2019;141(6):2215–9.

[35] Cai W, Chu CC, Liu G, Wáng Y-XJ. Metal–organic framework-based nanomedicine platforms for drug delivery and molecular imaging. Small. 2015;11(37):4806–22.
[36] Jensen F. Introduction to computational chemistry. 3rd ed. John Wiley & Sons, Inc.; 2017.
[37] Becker TM, Lin LC, Dubbeldam D, Vlugt TJH. Polarizable force field for CO_2 in M-MOF-74 derived from quantum mechanics. J Phys Chem C. 2018;122:24488–98.
[38] Sikora BJ, Winnegar R, Proserpio DM, Snurr RQ. Textural properties of a large collection of computationally constructed MOFs and zeolites. Microporous Mesoporous Mater. 2014;186:207–13.
[39] Zhang H, Lv Y, Tan T, van der Spoel D. Atomistic simulation of protein encapsulation in metal–organic frameworks. J Phys Chem B. 2016;120(3):477–84.
[40] Hu Z, Jiang J. A helical peptide confined in metal–organic frameworks: microscopic insight from molecular simulation. Microporous Mesoporous Mater. 2016;232:138–42.
[41] Tuan Kob TNA, Ismail MF, Abdul Rahman MB, Cordova KE, Mohammad Latif MA. Unravelling the structural dynamics of an enzyme encapsulated within a metal–organic framework. J Phys Chem B. 2020;124(18):3678–85.
[42] Kotzabasaki M, Galdadas I, Tylianakis E, Klontzas E, Cournia Z, Froudakis GE. Multiscale simulations reveal IRMOF-74-III as a potent drug carrier for gemcitabine delivery. J Mater Chem B. 2017;5:3277–82.
[43] Bueno-Pérez R, García-Pérez E, Gutiérrez-Sevillano JJ, Merkling PJ, Calero S. A simulation study of hydrogen in metal–organic frameworks. Adsorp Sci Technol. 2010;28(8–9):823–35.
[44] Song XD, Wang S, Hao C, Qiu JS. Investigation of SO_2 gas adsorption in metal–organic frameworks by molecular simulation. Inorg Chem Commun. 2014;46:277–81.
[45] Wei M, Dahuan L, Qingyuan Y, Chongli Z. Computational study of the effect of organic linkers on natural gas upgrading in metal–organic frameworks. Microporous Mesoporous Mater. 2010;130(1–3):76–82.
[46] Erucar I, Keskin S. Efficient storage of drug and cosmetic molecules in biocompatible metal organic frameworks: a molecular simulation study. Ind Eng Chem Res. 2016;55(7):1929–39.
[47] Gori P, Pulci O, de Lieto Vollaro R, Guattari C. Thermophysical properties of the novel 2D materials graphene and silicene: insights from ab-initio calculations. Energy Proc. 2014;45:512–7.
[48] De Gosson MA. Principles of Newtonian and quantum mechanics: the need for Planck's constant, H (second edition). World Scientific; 2016.
[49] Kwong M, Abdelrasoul A, Doan H. Controlling polysulfone (PSF) fiber diameter and membrane morphology for an enhanced ultrafiltration performance using heat treatment. Results Mater. 2019;2:100021.
[50] Piquemal J, Marquez A, Parisel O, Giessner–Prettre C. A CSOV study of the difference between HF and DFT intermolecular interaction energy values: the importance of the charge transfer contribution. J Comput Chem. 2005;26:1052–62.
[51] Yu J, Xie LH, Li JR, Ma Y, Seminario JM, Balbuena PB. CO_2 capture and separations using MOFs: computational and experimental studies. Chem Rev. 2017;117(14):9674–754.
[52] Kohn W, Sham LJ. Self-consistent equations including exchange and correlation effects. Phys Rev. 1965;140:A1133.
[53] Grimme S, Bannwarth C, Shushkov P. A robust and accurate tight-binding quantum chemical method for structures, vibrational frequencies, and noncovalent interactions of large molecular systems parametrized for all spd-block elements (Z = 1–86). J Chem Theory Comput. 2017;13(5):1989–2009.
[54] Kim SG, Horstemeyer MF, Baskes MI et al. Semi-empirical potential methods for atomistic simulations of metals and their construction procedures. J Eng Mater Technol. 2009;131:41210.

[55] Zerner MC. Semiempirical molecular orbital methods. In: Lipkowitz KB, Boyd DB, editors. Reviews in computational chemistry. John Wiley & Sons, Inc.; 1991. p. 313–65.
[56] Dewar MJ, Thiel W. Ground states of molecules. 38. The MNDO method. Approximations and parameters. J Am Chem Soc. 1977;99:4899–907.
[57] Dewar MJ, Zoebisch EG, Healy EF, Stewart JJP. Development and use of quantum mechanical molecular models. 76. AM1: a new general purpose quantum mechanical molecular model. J Am Chem Soc. 1985;107:3902–9.
[58] Stewart JJP. Optimization of parameters for semiempirical methods II. Applications. J Comput Chem. 1989;10:221–64.
[59] Kostal J. Computational chemistry in predictive toxicology: status quo et quo vadis? In: Fishbein JC, Heilman JM, editors. Advances in molecular toxicology. vol. 10. Elsevier; 2016. p. 139–86.
[60] Migliorati V, Serva A, Terenzio FM, D'Angelo P. Development of Lennard-Jones and buckingham potentials for lanthanoid ions in water. Inorg Chem. 2017;56(11):6214–24.
[61] Hou Q, Zhang L. Biomimetic design of peptide neutralizer of Ebola virus with molecular simulation. Langmuir. 2020;36(7):1813–21.
[62] Vanommeslaeghe K, Guvench O, MacKerell AD Jr. Molecular mechanics. Curr Pharm Des. 2014;20(20):328–9.
[63] Vlachakis D, Bencurova E, Papangelopoulos N, Kossida S. Current state-of-the-art molecular dynamics methods and applications. Adv Protein Chem Struct Biol. 2014;94:269–313.
[64] Momenzadeh L, Belova IV, Murch GE. Prediction of the lattice thermal conductivity of zircon and the cubic and monoclinic phases of zirconia by molecular dynamics simulation. Comput Mater Sci. 2020;76:109522.
[65] Nalaparaju A, Khurana M, Farooq S, Karimi IA, Jiang JW. CO_2 capture in cation-exchanged metal–organic frameworks: holistic modeling from molecular simulation to process optimization. Chem Eng Sci. 2015;124:70–8.
[66] Haldoupis E, Watanabe T, Nair S, Sholl DS. Quantifying large effects of framework flexibility on diffusion in MOFs: CH_4 and CO_2 in ZIF-8. ChemPhysChem. 2012;13:3449–52.
[67] Siu SWI, Pluhackova K, Bockmann RA. Optimization of the OPLS-AA force field for long hydrocarbons. J Chem Theory Comput. 2012;8(4):1459–70.
[68] Perez A, Marchan I, Svozil D et al. Refinement of the AMBER force field for nucleic acids: improving the description of α/γ conformers. Biophys J. 2007;92(11):3817–29.
[69] Vanommeslaeghe K, Hatcher E, Acharya C et al. CHARMM general force field: a force field for drug-like molecules compatible with the CHARMM all-atom additive biological force fields. J Comput Chem. 2009;31:671–90.
[70] Al-Janabi N, Fan X, Siperstein FR. Assessment of MOF's quality: quantifying defect content in crystalline porous materials. J Phys Chem Lett. 2016;7:1490–4.
[71] Burgot JL. Statistical thermodynamics in brief. In: Burgot JL, editor. Notion of activity in chemistry. Springer; 2017. p. 251–3.
[72] Orellana-Tavra C, Marshall RJ, Baxter EF et al. Drug delivery and controlled release from biocompatible metal–organic frameworks using mechanical amorphization. J Mater Chem B. 2016;4:7697–707.
[73] Liu F, Xu B. A three-dimensional DyIII-based metal–organic framework: smart drug carrier and anti-lung cancer activity. Z Anorg Allg Chem. 2018;644:821–6.
[74] Lu K, Aung T, Guo N, Weichselbaum R, Lin W. Nanoscale metal–organic frameworks for therapeutic, imaging, and sensing applications. Adv Mater. 2018;30:1707634.
[75] André V, Quaresma S. Bio-inspired metal–organic frameworks in the pharmaceutical world: a brief review. In: Zafar F, Sharmin E, editors. Metal–organic frameworks. London: IntechOpen; 2016.

[76] Ross GA, Bodnarchuk MS, Essex JW. Water sites, networks, and free energies with grand canonical Monte Carlo. J Am Chem Soc. 2015;137:14930–43.
[77] Wu Y, Chen H, Liu D, Qian Y, Xi H. Adsorption and separation of ethane/ethylene on ZIFs with various topologies: combining GCMC simulation with the ideal adsorbed solution theory (IAST). Chem Eng Sci. 2015;124:144–53.
[78] Chib S, Greenberg E. Understanding the Metropolis-Hastings algorithm. Am Stat. 1995;49:327–35.
[79] Cui K, Schmidt JR. Enabling efficient and accurate computational studies of MOF reactivity via QM/MM and QM/QM methods. J Phys Chem C. 2020;124:10550–60.
[80] Borzehandani MY, Abdulmalek E, Abdul Rahman MB, Mohammad Latif MA. First-principles investigation of dimethyl-functionalized MIL-53 (Al) metal–organic framework for adsorption and separation of xylene isomers. J Porous Mater. 2021;28(2):579–91.
[81] Xu K, Moeljad AMP, Mai BK, Hirao H. How does CO_2 react with styrene oxide in Co-MOF-74 and Mg-MOF-74? Catalytic mechanisms proposed by QM/MM calculations. J Phys Chem C. 2017;122:503–14.
[82] Vasconcelos IB, Wanderley KA, Rodrigues NM, da Costa NB Jr, Freire RO, Junior SA. Host-guest interaction of ZnBDC-MOF+ doxorubicin: a theoretical and experimental study. J Mol Struct. 2017;1131:36–42.
[83] Chen R, Li L, Weng Z. ZDOCK: an initial-stage protein-docking algorithm. Prot Struct Funct Bioinform. 2003;52:80–7.
[84] Pagadala NS, Syed K, Tuszynski J. Software for molecular docking: a review. Biophys Rev. 2017;9:91–102.
[85] Rodrigues MO, de Paula MV, Wanderley KA, Vasconcelos IB, Alves S Jr, Soares TA. Metal organic frameworks for drug delivery and environmental remediation: a molecular docking approach. Int J Quant Chem. 2012;112:3346–55.
[86] Qi C, Cai Q, Zhao P et al. The metal–organic framework MIL-101 (Cr) as efficient adsorbent in a vortex-assisted dispersive solid-phase extraction of imatinib mesylate in rat plasma coupled with ultra-performance liquid chromatography/mass spectrometry: application to a pharmacokine. J Chromatogr A. 2016;1449:30–8.
[87] Vasconcelos IB, da Silva TG, Militão GCG et al. Cytotoxicity and slow release of the anti-cancer drug doxorubicin from ZIF-8. RSC Adv. 2012;2:9437–42.
[88] Liu JQ, Wu J, Jia ZB et al. Two isoreticular metal–organic frameworks with $CdSO_4$-like topology: selective gas sorption and drug delivery. Dalton Trans. 2014;43:17265–73.
[89] Borzehandani MY, Sathar MHA, Abdulmalek E, Abdul Rahman MB, Mohammad Latif MA. In Silico identification of the mechanism of fluorouracil adsorption inside MIL-101 (Mg) metal–organic framework. AIP Conf Proc. 2022;2506:060003.
[90] He Y, Zhang W, Guo T et al. Drug nanoclusters formed in confined nano-cages of CD-MOF: dramatic enhancement of solubility and bioavailability of azilsartan. Acta Pharmacol Sin B. 2019;9:97–106.
[91] Kotzabasaki M, Tylianakis E, Klontzas E, Froudakis GE. OH-functionalization strategy in metal–organic frameworks for drug delivery. Chem Phys Lett. 2017;685:114–8.
[92] Bernini MC, Fairen-Jimenez D, Pasinetti M, Ramirez-Pastor AJ, Snurr RQ. Screening of bio-compatible metal–organic frameworks as potential drug carriers using Monte Carlo simulations. J Mater Chem B. 2014;2:766–74.
[93] Liu JQ, Li XF, Gu CY et al. A combined experimental and computational study of novel nanocage-based metal–organic frameworks for drug delivery. Dalton Trans. 2015;44:19370–82.
[94] Nazari M, Rubio-Martinez M, Tobias G et al. Metal–organic-framework-coated optical fibers as light-triggered drug delivery vehicles. Adv Funct Mater. 2016;26:3244–9.

[95] Supronowicz B, Mavrandonakis A, Heine T. Interaction of biologically important organic molecules with the unsaturated copper centers of the HKUST-1 metal–organic framework: an ab-initio study. J Phys Chem C. 2015;119:3024–32.
[96] Xin XT, Cheng JZ. A mixed-ligand approach for building a N-rich porous metal–organic framework for drug release and anticancer activity against oral squamous cell carcinoma. J Coord Chem. 2018;71:3565–74.
[97] Wang J, Ma D, Liao W et al. A hydrostable anionic zinc-organic framework carrier with a bcu topology for drug delivery. CrystEngComm. 2017;19:5244–50.
[98] Gaudin C, Cunha D, Ivanoff E et al. A quantitative structure activity relationship approach to probe the influence of the functionalization on the drug encapsulation of porous metal–organic frameworks. Microporous Mesoporous Mater. 2012;157:124–30.
[99] Cai M, Qin L, You L et al. Functionalization of MOF-5 with mono-substituents: effects on drug delivery behavior. RSC Adv. 2020;10(60):36862–72.
[100] Devautour-Vinot S, Martineau C, Diaby S et al. Caffeine confinement into a series of functionalized porous zirconium MOFs: a joint experimental/modeling exploration. J Phys Chem C. 2013;117:11694–704.
[101] Tao Y, Fan Y, Xu Z, Feng X, Krishna R, Luo F. Boosting selective adsorption of xe over kr by double-accessible open-metal site in metal–organic framework: experimental and theoretical research. Inorg Chem. 2020;59:11793–800.
[102] Vinogradov VV, Drozdov AS, Mingabudinova LR et al. Composites based on heparin and MIL-101 (Fe): the drug releasing depot for anticoagulant therapy and advanced medical nanofabrication. J Mater Chem B. 2018;6:2450–9.
[103] Zong S, Zhang Y, Lu N, Ma P, Wang J, Shi XR. A DFT screening of M-HKUST-1 MOFs for nitrogen-containing compounds adsorption. Nanomaterials. 2018;8:958.
[104] Vlaisavljevich B, Huck J, Hulvey Z et al. Performance of van der Waals corrected functionals for guest adsorption in the M2 (dobdc) metal–organic frameworks. J Phys Chem A. 2017;121:4139–51.
[105] Babarao R, Jiang J. Unraveling the energetics and dynamics of ibuprofen in mesoporous metal–organic frameworks. J Phys Chem C. 2009;113:18287–91.
[106] Horcajada P, Serre C, Vallet-Regí M, Sebban M, Taulelle F, Férey G. Metal–organic frameworks as efficient materials for drug delivery. Angew Chem. 2006;118:6120–4.
[107] Souza BE, Donà L, Titov K et al. Elucidating the drug release from metal–organic framework nanocomposites via in situ synchrotron microspectroscopy and theoretical modeling. ACS Appl Mater Interfaces. 2020;12:5147–56.
[108] Lin S, Zhao Y, Yun YS. Highly effective removal of nonsteroidal anti-inflammatory pharmaceuticals from water by Zr(IV)-based metal–organic framework: adsorption performance and mechanisms. ACS Appl Mater Interfaces. 2018;10:28076–85.
[109] Liu J, Yang Z, Che Y, Zhang Y, Zhang Z, Zhao CX. Computational investigation of metal organic frameworks as potential drug carriers for antihypertensive amlodipine. Alchem J. 2022;68(1):e17474.
[110] Xiaodong S, Keywanlu M, Tayebee R. Experimental and molecular dynamics simulation study on the delivery of some common drugs by ZIF-67, ZIF-90, and ZIF-8 zeolitic imidazolate frameworks. Appl Organomet Chem. 2021;35(11):e6377.
[111] Dou X, Keywanlu M, Tayebee MB. Simulation of adsorption and release of doxepin onto ZIF-8 including in vitro cellular toxicity and viability. J Mol Liq. 2021;329:115557.
[112] Boroushaki T, Dekamin MG, Hashemianzadeh SM, Naimi-Jamal MR, Koli MG. A molecular dynamic simulation study of anticancer agents and UiO-66 as a carrier in drug delivery systems. J Mol Graph Model. 2022;113:108147.
[113] Gonzalez-Nelson A, Coudert FX, van der Veen MA. Rotational dynamics of linkers in metal–organic frameworks. Nanomaterials. 2019;9(3):330.

[114] Proenza YG, Longo RL. Simulation of the adsorption and release of large drugs by ZIF-8. J Chem Inf Model. 2020;60(2):644–52.
[115] Sose AT, Cornell HD, Gibbons BJ, Burris AA, Morris AJ, Deshmukh SA. Modelling drug adsorption in metal–organic frameworks: the role of solvent. RSC Adv. 2021;11(28):17064–71.
[116] Gomar M, Yeganegi S. Adsorption of 5-fluorouracil, hydroxyurea and mercaptopurine drugs on zeolitic imidazolate frameworks (ZIF-7, ZIF-8 and ZIF-9). Microporous Mesoporous Mater. 2017;252:167–72.
[117] Chun-Yi S, Chao Q, Xin-Long W et al. Zeolitic imidazolate framework-8 as efficient pH-sensitive drug delivery vehicle. Dalton Trans. 2012;41(23):6906–9.
[118] Jiang J. Molecular simulations in metal–organic frameworks for diverse potential applications. Mol Simul. 2014;40:516–36.
[119] Erucar I, Keskin S. Computational investigation of metal organic frameworks for storage and delivery of anticancer drugs. J Mater Chem B. 2017;5(35):7342–51.
[120] Sarker M, Jhung SH. Zr-MOF with free carboxylic acid for storage and controlled release of caffeine. J Mol Liq. 2019;296:112060.
[121] Wang H, Li T, Li J, Tong W, Gao C. One-pot synthesis of poly(ethylene glycol) modified zeolitic imidazolate framework-8 nanoparticles: size control, surface modification and drug encapsulation. Colloids Surf A, Physicochem Eng Asp. 2019;568:224–30.
[122] Wang Q, Sun Y, Li S, Zhang P, Yao Q. Synthesis and modification of ZIF-8 and its application in drug delivery and tumor therapy. RSC Adv. 2020;10(62):37600–20.
[123] Yang Y, Xia F, Yang Y et al. Litchi-like Fe3O4@fe-MOF capped with HAp gatekeepers for pH-triggered drug release and anticancer effect. J Mater Chem B. 2017;5(43):8600–6.
[124] Wu MX, Yang YW. Metal–organic framework (MOF)-based drug/cargo delivery and cancer therapy. Adv Mater. 2017;29:1606134.

Youssef Belmabkhout

13 Future perspectives: are metal-organic frameworks deployable beyond the laboratory?

13.1 Introduction

The chapters presented in this book have clearly demonstrated how the field of MOFs has grown to become one of the most explored fields in solid-state chemistry. Very recently, reticular chemistry has matured to more and more closely combine with various research fields within the engineering discipline as the power of MOFs has been clearly shown in addressing problems in many applications related to energy efficiency and environmental sustainability. Such diversity in applications where MOF materials can be deployed has been an obvious outcome of the high degree of modularity of this solid-state chemistry, vested by the reign of reticular chemistry. The fundamental concepts underpinning reticular chemistry has driven the discovery of tens of thousands of structures both hypothetically and experimentally. Regardless, most reported studies involving reticular chemistry are not truly made by design as claimed, do not target specific technology-deployment objectives, and remain limited to low technology readiness levels (TRLs). This chapter discusses (i) the growing trend of MOF research, (ii) the power of combining trial-and-error synthetic approaches and machine learning to practice reticular chemistry by design and (iii) a general road map for scale-up and extrapolation of technology-driven processes built around MOFs.

13.2 Blossoming of MOF research

In its first stage of development, an extraordinarily large number of chemists from around the world have directed their research to new MOF synthesis [1–12]. Although few of the discovered new structures were discovered by rational topological analysis [5–7, 9], the fact remains that the rich structural and topological possibilities of MOFs have enabled chemists to discover and populate the Cambridge Crystallographic Data Centre with structures realized from experimental trial-and-error. With structures and

Youssef Belmabkhout, Applied Chemistry and Engineering Research Centre of Excellence (ACER CoE), Mohammed VI Polytechnic University (UM6P), Lot 660 – Hay Moulay Rachid, 43150 Ben Guerir, Morocco; and Technology Development Cell (TechCell), Mohammed VI Polytechnic University (UM6P), Lot 660 – Hay Moulay Rachid, 43150 Ben Guerir, Morocco, e-mail: Youssef.belmabkhout@um6p.ma

https://doi.org/10.1515/9781501524721-013

preliminary properties studied and reported, physical chemists by chemical and mechanical engineers joined the field to complement ongoing work in terms of advanced characterization [5, 7, 11–15]. This led to the widespread appreciation for the intrinsic properties of MOFs that allowed these materials to be assessed in a large number of applications: adsorption, catalysis, separation, sensing, medical among many others [8, 12–14]. However, at first, applications were pursued for a limited number of well-known MOFs platforms, such as the Zn-based zeolitic imidazolate frameworks or Zr-based *fcu*-MOFs (e. g., UiO-66) [16, 17], primarily because of their relatively high thermal and chemical stability. In fact, most other MOFs, at that time, were found to not have the required stability to reach a sufficiently advanced stage for multidisciplinary study.

Because of the porous character of this solid state-materials platforms, adsorption studies (experiments and simulations) are often the starting research endeavor before directing the discovered MOFs to other adsorptive and nonadsorptive related research enterprises [18, 19].

13.3 The power of machine learning combined with trials learning

As the library of known MOF structures (and their associated physical properties) has expanded, molecular simulation (MS) work has been increasingly used to screen thousands of structures, in the last decade primarily for insights into the best performing materials in terms of adsorptive separation and carbon capture. However, most reported MS-based studies have lacked depth with respect to structural properties as a function of optimal adsorptive mechanisms [20]. In addition, the few developed force fields applied to the large number of MOFs are debatable. In fact, most of the MS work considers all molecules as free to diffuse into the pores without any resistance via a purely equilibrium adsorption mechanism, regardless of the structure, pore size, topology or pore environment, which make such studies non-exploitable for screening. Contrary to MS work, trials learning has been useful in delineating advanced concepts behind structural properties and topologies of interest for particular applications [13, 14, 20]. However, these identified concepts have yet reached full potential in being universally true for all MOFs with specific charge and structural properties. Accordingly, MS-related work needs to be more strongly linked to lessons and outputs gleaned from trials learning at the batch or dynamic scale in order to correctly and accurately direct the screening of MOFs for particular applications. Because only a few MOF platforms have been screened by considering stability factors, the results from trials learning need to be analyzed for consistency before use as inputs in MS-related work. This will generate a large number of additional experimental trials that can be carried out in the laboratory as proof-of-concept or proof-of-discovery.

13.4 MOF development: the next stage

Despite a large interest in multidisciplinary MOF-related research, there remain key disciplines that are underused or unused for valorizing the results of research around MOFs and to take them to the market stage. Among these fields are reactor engineering and process design, intensification and scale-up (down). If equipment research can generate ca. 20 % improvement for any process, the solid-state related chemistry research in MOFs associated with equipment research can multiply by at least one factor of magnitude the process improvement. Thus, the bottom-up strategy that associates the process metrics to specific structural, porosity and topology metrics is the only way to exploit the power of reticular chemistry to a deployment-level and to realize the benefit of reticular chemistry for technology-design. Nevertheless, as is true for any first stage, low TRL emerging technology, there remains a large canyon to cross to avoid the valley of technology death [21] that prohibits the advancement of many unindustrialized, low maturity technologies. Obstacles to market penetration are classified into either technical or economical as detailed below.

13.4.1 Form and scale of MOFs

Most low TRL lab scale work on MOFs are made using materials in powder form at very small scales. Regardless of the process scale, this makes deploying any MOF-related technology impractical because a process cannot operate in optimal condition when MOFs are in powder form. So far there is a limited number of works addressing the problem of shaping and forming where the amount of MOFs required is small, such as applying MOFs in sensing [22]. In such applications, the MOFs are shaped in the form of pure MOF or composite thin films. In this case, there is so far no studies that demonstrate any negative effect of mass transfer challenges on sensing signal performances (e. g., resistance, capacitance, sensitivity or selectivity). Although it is less critical for water harvesting applications [23], when MOFs are considered for catalytic or separation applications, there is so far few works that address the problem of MOF shaping/structuring [24], particularly at large scale. In fact, there is no single published conclusive or convincing endeavor showing the technical feasibility of MOF structuring and/or shaping at large scale. In the meantime, many announced projects at the government level aims to address the forming and shaping of MOFs for use in small-to-large scale applications. It is important to note that this important step of forming and shaping is a prerequisite for testing the emerging MOFs in real process conditions.

13.4.2 Process scale-up involving MOFs

Forming and shaping are not the only obstacles facing the deployment of MOFs. In fact, process extrapolation and scale-up studies using existing or emerging materials like MOFs is still heavily lacking. This is a major step in the overall maturation process of any new or advanced technology. History shows that a great majority of emerging technologies have been discontinued or put on stand-by at this stage. In fact, going straight from the lab to the industrial scale usually involves a lot of risk. Most often, one or more complementary and intermediate experiments are generally necessary [25]. The problem is to define precisely these works in such a way as to collect all the necessary information with a minimum cost and in as short a time as possible. It is at this level that the methodology of process development becomes decisive to ensure complete industrial success.

How can the results obtained in the laboratory be reproduced on a large scale? This is, in general, the great question of process design. To respond adequately, it is necessary, among other things, to build and operate a pilot unit. The analysis of techno-economic data extracted from the pilot stage are critical before moving to demonstration or deployment stage. The aim of a pilot unit allows the simultaneous study of physical and chemical phenomena where: (i) all industrial constraints are taken into account: raw material impurities, recycling flows, reliability of materials, operation for long period, (ii) the process must be broken down into subsets which can be studied separately with particular attention needing to be placed on systems for which extrapolation poses issues, (iii) the phenomena occurring in each of the subsystems must be analyzed and finally (iv) the size of the pilot unit whereby it must be as large as possible to allow easy extrapolation and as small as possible to reduce investment and operating costs [25].

13.4.3 Techno-economic feasibility

Based on the process extrapolation phase involving both process simulation and experimentation at the pilot phase, it will be possible to determine a more refined value of capital expenditure/operating expenditure (Capex/Opex) that will help in the decision-making process. Such expertise is generally not common and available in the academic world. This point creates a big gap with the industrial world and most of the developed emerging technologies remain not mature enough technically and economically particularly for processes where large amounts of MOF are needed. Because of the hybrid nature of MOFs, it is evident that MOF are still expensive materials and will represent most of the Opex. Nevertheless, there is an exception for high investment sectors, such as the example of air recycling in space shuttles that requires a 0.1–1 Kg scale of MOFs [26]. In such a case, it is clear that the use of MOFs for trace CO_2 removal using physical adsorption can be technically and economically feasible when replacing chemical absorption and hydrolytic adsorption systems.

Bibliography

[1] Chen Z et al. Reticular chemistry for highly porous metal–organic frameworks: the chemistry and applications. Acc Chem Res. 2022;55:579–91.

[2] Ghasempour H et al. Metal–organic frameworks based on multicarboxylate linkers. Coord Chem Rev. 2021;426:213542–88.

[3] Assen AH, Karim A, Kyle C, Belmabkhout Y. The chemistry of metal–organic frameworks with face-centered cubic topology. Coord Chem Rev. 2022;468:214644–99.

[4] Furukawa H, Cordova KE, O'Keeffe M, Yaghi OM. The chemistry and applications of metal–organic frameworks. Science. 2013;341:1230444.

[5] Xue DX et al. Topology meets MOF chemistry for pore-aperture fine tuning: ftw-MOF platform for energy-efficient separations via adsorption kinetics or molecular sieving. Chem Commun. 2018;54:6404–7.

[6] Chen Z et al. Applying the power of reticular chemistry to finding the missing alb-MOF platform based on the (6,12)-coordinated edge-transitive net. J Am Chem Soc. 2017;139:3265–74.

[7] Alezi D et al. Reticular chemistry at its best: directed assembly of hexagonal building units into the awaited metal–organic framework with the intricate polybenzene topology, pbz-MOF. J Am Chem Soc. 2016;138:12767–70.

[8] Cai G et al. Metal–organic framework-based hierarchically porous materials: synthesis and applications. Chem Rev. 2021;121(20):12278–326.

[9] Feng L et al. Strategies for pore engineering in zirconium metal–organic frameworks. Chem. 2020;6:2902–23.

[10] Krause S et al. A pressure-amplifying framework material with negative gas adsorption transitions. Nature. 2016;532:348–52.

[11] Kolobov N, Metal–Organic Goesten M G GJ. Frameworks: molecules or semiconductors in photocatalysis? Angew Chem, Int Ed. 2021;60:26038–52.

[12] Chen Z et al. Balancing volumetric and gravimetric uptake in highly porous materials for clean energy. Science. 2020;368:297–303.

[13] Cadiau A et al. Metal–organic framework–based splitter for separating propylene from propane. Science. 2016;353:137–40.

[14] Belmabkhout et al. Natural gas upgrading using a fluorinated MOF with tuned H_2S and CO_2 adsorption selectivity. Nat Energy. 2018;3:1059–66.

[15] Liu L et al. Imaging defects and their evolution in a metal–organic framework at sub-unit-cell resolution. Nat Chem. 2019;11:622–8.

[16] Park KS et al. Exceptional chemical and thermal stability of zeolitic imidazolate frameworks. Proc Natl Acad Sci. 2006;103:10186–91.

[17] Bai Y et al. Zr-based metal–organic frameworks: design, synthesis, structure, and applications. Chem Soc Rev. 2016;45:2327–67.

[18] Adil K et al. Gas/vapor separation using ultra-microporous metal–organic frameworks: insights into the structure/separation relationship. Chem Soc Rev. 2017;46:3402–30.

[19] Cuadrado-Collados C et al. Quest for an optimal methane hydrate formation in the pores of hydrolytically stable metal–organic frameworks. J Am Chem Soc. 2020;142:13391–7.

[20] Nugent P, Belmabkhout Y et al. Porous materials with optimal adsorption thermodynamics and kinetics for CO_2 separation. Nature. 2013;495:80–4.

[21] Gbadegeshin SA et al. Overcoming the valley of death: a new model for high technology startups. Sustain Futures. 2022;4:100077–92.

[22] Tchalala M, Concurrent R. Sensing of CO_2 and H_2O from air using ultramicroporous fluorinated metal–organic frameworks: effect of transduction mechanism on the sensing performance. ACS Appl Mater Interfaces. 2019;11(1):1706–12.

[23] Almassad HA et al. Environmentally adaptive MOF-based device enables continuous self-optimizing atmospheric water harvesting. Nat Commun. 2022;13:4873–83.
[24] Mallick A et al. Advances in shaping of metal–organic frameworks for CO_2 capture: understanding the effect of rubbery and glassy polymeric binders. Ind Eng Chem Res. 2018;57:16897–902.
[25] Chaouki J, Rahmat S-G. Scale-up processes: iterative methods for the chemical, mineral and biological industries. Berlin Germany: De Gruyter; 2021.
[26] Bhatt PM et al. A fine-tuned fluorinated MOF addresses the needs for trace CO_2 removal and air capture using physisorption. J Am Chem Soc. 2016;138:9301–7.

Index

https://doi.org/10.1515/9781501524721-014

www.ingramcontent.com/pod-product-compliance
Lightning Source LLC
Chambersburg PA
CBHW080703220326
41598CB00033B/5293

* 9 7 8 1 5 0 1 5 2 4 7 0 7 *